# SOIL ACIDITY AND LIMING

## Second Edition

# AGRONOMY

## *A Series of Monographs*

The American Society of Agronomy and Academic Press published the first six books in this series. The General Editor of Monographs 1 to 6 was A. G. Norman. They are available through Academic Press, Inc., 111 Fifth Avenue, New York, NY 10003.

1. C. EDMUND MARSHALL: The Colloid Chemical of the Silicate Minerals, 1949
2. BYRON T. SHAW, *Editor*: Soil Physical Conditions and Plant Growth, 1952
3. K. D. JACOB: Fertilizer Technology and Resources in the United States, 1953
4. W. H. PIERRE and A. G. NORMAN, *Editors*: Soil and Fertilizer Phosphate in Crop Nutrition, 1953
5. GEORGE F. SPRAGUE, *Editor*: Corn and Corn Improvement, 1955
6. J. LEVITT: The Hardiness of Plants, 1956

The Monographs published since 1957 are available from the American Society of Agronomy, 677 S. Segoe Road, Madison, WI 53711.

7. JAMES N. LUTHIN, *Editor*: Drainage of Agricultural Lands, 1957 *General Editor,* D. E. Gregg
8. FRANKLIN A. COFFMAN, *Editor*: Oats and Oat Improvement *Managing Editor,* H. L. Hamilton
9. C. A. BLACK, *Editor-in-Chief*, and D. D. EVANS, J. L. WHITE, L. E. ENSMINGER, and F. E. CLARK, *Associate Editors*: Methods of Soil Analysis, 1965
   Part 1—Physical and Mineralogical Properties, Including Statistics of Measurement and Sampling
   A. L. PAGE, *Editor*: Methods of Soil Analysis, 1982
   Part 2—Chemical and Microbiological Properties, Second Edition *Managing Editor,* R. C. Dinauer
10. W. V. BARTHOLOMEW and F. E. CLARK, *Editors*: Soil Nitrogen, 1965
    (Out of print; replaced by no. 22) *Managing Editor,* H. L. Hamilton
11. R. M. HAGAN, H. R. HAISE, and T. W. EDMINSTER, *Editors*: Irrigation of Agricultural Lands, 1967 *Managing Editor,* R. C. Dinauer
12. FRED ADAMS, *Editor*: Soil Acidity and Liming, Second Edition, 1984
    *Managing Editor,* R. C. Dinauer
13. K. S. QUISENBERRY and L. P. REITZ, *Editors*: Wheat and Wheat Improvement, 1967
    *Managing Editor,* H. L. Hamilton
14. A. A. HANSON and F. V. JUSKA, *Editors*: Turfgrass Science, 1969
    Managing Editor, H. L. Hamilton
15. CLARENCE H. HANSON, *Editor*: Alfalfa Science and Technology, 1972
    *Managing Editor,* H. L. Hamilton
16. B. E. CALDWELL, *Editor*: Soybeans: Improvement, Production, and Use, 1973
    *Managing Editor,* H. L. Hamilton
17. JAN VAN SCHILFGAARDE, *Editor*: Drainage for Agriculture, 1974
    *Managing Editor,* R. C. Dinauer
18. GEORGE F. SPRAGUE, *Editor*: Corn and Corn Improvement, 1977
    *Managing Editor,* D. A. Fuccillo
19. JACK F. CARTER, *Editor*: Sunflower Science and Technology, 1978
    *Managing Editor,* D. A. Fuccillo
20. ROBERT C. BUCKNER and L. P. BUSH, *Editors*: Tall Fescue, 1979
    *Managing Editor,* D. A. Fuccillo
21. M. T. BEATTY, G. W. PETERSEN, and L. D. SWINDALE, *Editors*: Planning the Uses and Management of Land, 1979 *Managing Editor,* R. C. Dinauer
22. F. J. STEVENSON, *Editor*: Nitrogen in Agricultural Soils, 1982
    *Managing Editor,* R. C. Dinauer
23. H. E. DREGNE and W. O. WILLIS, *Editors*: Dryland Agriculture, 1983
    *Managing Editor,* D. A. Fuccillo
24. R. J. KOHEL and C. F. LEWIS, *Editors*: Cotton, 1984
    *Managing Editor,* D. A. Fuccillo

# SOIL ACIDITY AND LIMING

## Second Edition

*Edited by*

**FRED ADAMS**

*Managing Editor:* RICHARD C. DINAUER

*Assistant Editor:* KRISTINE E. GATES

*Editor-in-Chief ASA Publications:* DWAYNE R. BUXTON

Number 12 in the series

**AGRONOMY**

**American Society of Agronomy, Inc.**
**Crop Science Society of America, Inc.**
**Soil Science Society of America, Inc.**
**Publisher**
**Madison, Wisconsin USA**

**1984**

American Society of Agronomy, Inc.
Crop Science Society of America, Inc.
Soil Science Society of America, Inc.
677 South Segoe Road, Madison, Wisconsin 53711 USA

Second Printing 1990

**Library of Congress Cataloging in Publication Data**

Soil acidity and liming.

(Agronomy; no. 12)
Includes bibliographies and index.
1. Soil acidity. 2. Liming of soils. I. Adams, Fred, 1921–
II. Series.
S593.5.S67 1984 631.8′21 84-11013
ISBN 0-89118-080-X

Printed in the United States of America

# CONTENTS

# FOREWORD

Our understanding of the nature, causes, and management of soil acidity continues on the "upward, spiralling merry-go-round" so elegantly described by Hans Jenny in 1961. It is good that it does. Only by properly managing our soils can we be assured of plentiful supplies of food, fiber and shelter, and a clean and healthy environment.

Excessive soil acidity is a continuing problem in many agricultural areas and is increasingly becoming a yield-limiting factor in other areas. The pH of soils of the dryland wheat–growing area of the Pacific Northwest has dropped an average of one unit in the last 20 years. Similar effects are reported from other areas and even more dramatic changes are observed in the surface layers of minimum-tilled fields. Greater understanding and appreciation of these long-term, agriculturally intensified reactions are essential to maintain optimal fertility of our soils.

The factors of acid soil fertility and crop responses to acid soils are complex. Not all acid soils need lime, nor should all lime-responsive soils be brought to the same pH. Soil tests for lime requirement must be selected and calibrated to provide for the differences among soils and among the crops to be grown on them. This monograph highlights knowledge in these and other areas and points out the many challenges that lie ahead. It, as its predecessor, will serve as a benchmark from which future progress can be established.

On behalf of the membership of ASA, CSSA, and SSSA, we express our appreciation to the editor and the authors for their important functions leading to this publication. We acknowledge the help of our society members who reviewed the manuscripts. We are also grateful to the Headquarters staff for editorial and production efforts that allowed this volume to be a reality.

April 1984

K. J. Frey, *president*
*American Society of Agronomy*

W. F. Keim, *president*
*Crop Science Society of America*

D. R. Nielsen, *president*
*Soil Science Society of America*

# PREFACE

The first edition of *Soil Acidity and Liming* (published in 1967) was developed and written in response to the need for a reference text on an important, worldwide, agricultural topic. It covered topics ranging from basic chemical concepts to soil acidity to the practical usage of agricultural liming materials. The text was intended to serve as a source of information for agronomists, horticulturists, and others who are interested in efficient crop production on acid soils.

Knowledge about soil acidity and its effects on plant growth and crop production has continued to expand. The basic tenets of acid soil chemistry have not changed during the last 2 decades, but additions and modifications of concepts involving the chemistry of solution aluminum make updating of this topic highly desirable. Significant advances have been made in identifying the role in plant nutrition of elements responsible for acid soil infertility, in particular the effects of too much aluminum and too little calcium.

The relative importance of different crops produced on the acid soils of the United States has changed markedly during the last 30 years. Concurrently, plant breeders have initiated programs to develop cultivars of increasing tolerance to acid soils. Cropping systems have also changed, including the adoption of no-till practices in some areas. These changes have affected liming practices and have created a need for continuing field research so that liming recommendations are updated for new practices.

The intent of the editor and authors of the revised edition of *Soil Acidity and Liming* was to maintain the concept of a reference text on soil acidity, one that is suitable for those involved in research, both soils and plants, and as a supplemental text for graduate students. The edition was expanded to include a chapter on tropical acid soils, an area of increasing concern and interest. A key feature of the text is an effort to explain soil chemical properties and physiological responses of plants. This text was never intended to be a mere catalog of reported data in the literature; authors, instead, have striven to offer interpretations wherever possible.

March 1984

FRED ADAMS, *editor*
Auburn University
Auburn, Alabama

# CONTRIBUTORS


| | |
|---|---|
| **Fred Adams** | Professor of Soil Chemistry, Department of Agronomy and Soils, Auburn University, Auburn, Alabama |
| **Stanley A. Barber** | Professor of Agronomy, Department of Agronomy, Purdue University, West Lafayette, Indiana |
| **J. R. Brown** | Professor of Agronomy, Department of Agronomy, University of Missouri, Columbia, Missouri |
| **Ralph B. Clark** | Plant Physiologist, Agricultural Research Service, U.S. Department of Agriculture, Department of Agronomy, University of Nebraska, Lincoln, Nebraska |
| **Charles D. Foy** | Research Soil Scientist, Plant Stress Laboratory, Agricultural Research Service, U.S. Department of Agriculture, Beltsville, Maryland |
| **William L. Hargrove** | Assistant Professor, Department of Agronomy, University of Georgia, Georgia Agricultural Experiment Station, Experiment, Georgia |
| **Thomas L. Jackson** | Professor of Soil Science, Department of Soil Science, Oregon State University, Corvallis, Oregon |
| **Eugene J. Kamprath** | Professor of Soil Science, Department of Soil Science, North Carolina State University, Raleigh, North Carolina |
| **Douglas J. Lathwell** | Professor of Soil Science, Department of Agronomy, Cornell University, Ithaca, New York |
| **E. O. McLean** | Professor of Soil Chemistry, Department of Agronomy, Ohio State University, Columbus, Ohio |
| **W. Shaw Reid** | Professor of Soil Science, Department of Agronomy, Cornell University, Ithaca, New York |
| **H. M. Reisenauer** | Professor of Soil Science, Department of Land, Air and Water Resources, University of California, Davis, California |
| **Grant W. Thomas** | Professor of Agronomy, Department of Agronomy, University of Kentucky, Lexington, Kentucky |

**Factors for Converting Non-SI Units to Acceptable SI Units**

| Non-SI Units | | SI Units |
|---|---|---|
| Multiply | By | To obtain |
| acre | $4.05 \times 10^3$ | square meter, $m^2$ |
| acre | 0.405 | hectare, ha ($10^4$ $m^2$) |
| acre | $4.05 \times 10^{-3}$ | square kilometer, $km^2$ ($10^6$ $m^2$) |
| Angstrom unit | 0.1 | nanometer, nm ($10^{-9}$ m) |
| atmosphere | 0.101 | megapascal, MPa ($10^6$ Pa) |
| bar | $10^{-1}$ | megapascal, MPa ($10^6$ Pa) |
| British thermal unit | $1.05 \times 10^3$ | joule, J |
| calorie | 4.19 | joule, J |
| calorie per square centimeter minute (irradiance) | 698 | watt per square meter, W $m^{-2}$ |
| calorie per square centimeter (langley) | $4.19 \times 10^4$ | joules per square meter, J $m^{-2}$ |
| centimeter per day (elongation rate) | 0.116 | micrometer per second, $\mu m$ $s^{-1}$ ($10^{-6}$ m $s^{-1}$) |
| cubic feet | 0.028 | cubic meter, $m^3$ |
| cubic feet | 28.3 | liter, L ($10^{-3}$ $m^3$) |
| cubic inch | $1.64 \times 10^{-5}$ | cubic meter, $m^3$ |
| cubic inch | 16.4 | cubic centimeter, $cm^3$ ($10^{-6}$ $m^3$) |
| curie | $3.7 \times 10^{10}$ | becquerel, Bq |
| degrees (angle) | $1.75 \times 10^{-2}$ | radian, rad |
| dyne | $10^{-5}$ | newton, N |
| erg | $10^{-7}$ | joule, J |
| foot | 0.305 | meter, m |
| foot-pound | 1.36 | joule, J |
| gallon | 3.78 | liter, L ($10^{-3}$ $m^3$) |
| gallon per acre | 9.35 | liter per hectare, L $ha^{-1}$ |
| gram per cubic centimeter | 1.00 | megagram per cubic meter, Mg $m^{-3}$ |
| gram per square decimeter hour (transpiration) | 27.8 | milligram per square meter second, mg $m^{-2}$ $s^{-1}$ ($10^{-3}$ g $m^{-2}$ $s^{-1}$) |
| inch | 25.4 | millimeter, mm ($10^{-3}$ m) |
| micromole ($H_2O$) per square centimeter second (transpiration) | 180 | milligram ($H_2O$) per square meter second, mg $m^{-2}$ $s^{-1}$ ($10^{-3}$ g $m^{-2}$ $s^{-1}$) |
| micron | 1.00 | micrometer, $\mu m$ ($10^{-6}$ m) |
| mile | 1.61 | kilometer, km ($10^3$ m) |
| miles per hour | 0.477 | meter per second, m $s^{-1}$ |
| milligram per square decimeter hour (apparent photosynthesis) | 0.0278 | milligram per square meter second, mg $m^{-2}$ $s^{-1}$ ($10^{-3}$ g $m^{-2}$ $s^{-1}$) |
| milligram per square centimeter second (transpiration) | 10 000 | milligram per square meter second, mg $m^{-2}$ $s^{-1}$ ($10^{-3}$ g $m^{-2}$ $s^{-1}$) |
| millimho per centimeter | 1.00 | decisiemen per meter, dS $m^{-1}$ |
| ounce | 28.4 | gram, g ($10^{-3}$ kg) |
| ounce (fluid) | $2.96 \times 10^{-2}$ | liter, L ($10^{-3}$ $m^3$) |
| pint (liquid) | 0.473 | liter, L ($10^{-3}$ $m^3$) |
| pound | 454 | gram, g ($10^{-3}$ kg) |
| pound per acre | 1.12 | kilogram per hectare, kg $ha^{-1}$ |
| pound per acre | $1.12 \times 10^{-3}$ | megagram per hectare, Mg $ha^{-1}$ |
| pound per bushel | 12.87 | kilogram per cubic meter, kg $m^{-3}$ |
| pound per cubic foot | 16.02 | kilogram per cubic meter, kg $m^{-3}$ |
| pound per cubic inch | $2.77 \times 10^4$ | kilogram per cubic meter, kg $m^{-3}$ |
| pound per square foot | 47.9 | pascal, Pa |
| pound per square inch | $6.90 \times 10^3$ | pascal, Pa |
| quart (liquid) | 0.946 | liter, L ($10^{-3}$ $m^3$) |
| quintal (metric) | $10^2$ | kilogram, kg |
| square centimeter per gram | 0.1 | square meter per kilogram, $m^2$ $kg^{-1}$ |
| square feet | $9.29 \times 10^{-2}$ | square meter, $m^2$ |
| square inch | 645 | square millimeter, $mm^2$ ($10^{-6}$ $m^2$) |
| square mile | 2.59 | square kilometer, $km^2$ |
| square millimeter per gram | $10^{-3}$ | square meter per kilogram, $m^2$ $kg^{-1}$ |
| temperature (°F − 32) | 0.556 | temperature, °C |
| temperature (°C + 273) | 1 | temperature, K |
| tonne (metric) | $10^3$ | kilogram, kg |
| ton (2000 lb) | 907 | kilogram, kg |
| ton (2000 lb) per acre | 2.24 | megagram per hectare, Mg $ha^{-1}$ |

# 1 The Chemistry of Soil Acidity

**GRANT W. THOMAS**

*University of Kentucky*
*Lexington, Kentucky*

**WILLIAM L. HARGROVE**

*University of Georgia*
*Experiment, Georgia*

Current concepts about the nature of acid soils and their ion exchange chemistry were developed and validated mostly during the 1950s and 1960s. The rediscovery of exchangeable $Al^{3+}$ (Coleman & Harward, 1953; Low, 1955) was a basic contribution to understanding the behavior of acid soils. However, it did not explain acidity that was titratable but not replaceable by a neutral, unbuffered salt. It soon became apparent that this type of acidity was not caused by exchangeable $Al^{3+}$ but was a result of pH-dependent hydrolytic reactions involving hydroxy-Al, organic matter, and hydrated oxides of Al and Fe.

The work of Rich and Obenshain (1955), Rich (1960), and Jackson (1963a, b) showed that much of the Al in many soils was present as solid-phase hydroxy-Al polymers of variable positive charge and size. These polymers neutralized negative charges on exchangers but could not be replaced by $K^+$ ions. Most of the early work on hydroxy-Al was done from a mineralogical viewpoint, but the papers of Thomas (1960b) and Coleman and Thomas (1964) clearly showed its role in soil acidity. Similar observations were made much earlier by Kelley and Brown (1926), but they did not recognize the significance of their work at that time.

The works of Martin and Reeve (1958) and Schnitzer and Skinner (1963a, b) showed that organic matter reactions with Al also were important in soil acidity. These early works helped to explain why topsoils tend to have lower KCl–exchangeable $Al^{3+}$ than subsoils do. Since that time, the effect of organic matter on Al availability has been the subject of considerable work (Evans & Kamprath, 1970; Thomas, 1975; Bloom et al., 1979; Hargrove & Thomas, 1981a). All of these authors showed that exchangeable $Al^{3+}$ was reduced in the presence of organic matter.

The Fe and Al oxides, which can be major contributors to soil acidity, were first investigated extensively by Mattson and Wiklander (1940a, b). They believed that soil clays were merely mixtures of $SiO_2$, $Fe_2O_3$, and


*Soil Acidity and Liming*—Agronomy Monograph no. 12 (2nd Edition).

$Al_2O_3$ and that their behavior was related to electrical properties of the resulting mixture. This idea was shown to be untenable after the clay mineralogical discoveries of the 1930s (Kelley, 1943), and interest in the Mattson-Wiklander concept declined until mineralogical and chemical work in the southeastern United States, Hawaii, Africa, Latin America, and Southeast Asia showed that, indeed, many soils acted very much like Mattson and Wiklander had predicted. Following the work of Parks and de Bruyn (1962), the doctrines exposued by Mattson and Wiklander, albeit deceptively clothed in a new vocabulary, have been widely accepted. The most recent chemical concepts of amorphous soils and soils in which crystalline clay minerals are heavily coated with Fe and Al oxides began with the work of van Raij and Peach (1972); these concepts continue to the present (Uehara & Gillman, 1980). Parker et al. (1979) have sown some refreshing doubts on the applicability of the concepts in acid soils.

Many years ago, Jenny (1961) aptly described the study of soil acidity as a "merry-go-round." Terminology changes, analytical methods become more precise, and our overall understanding of both chemistry and soils improves. Nevertheless, the observations made by Vietch in 1904 are still confirmed. He showed that the soil acidity that directly affected plant growth was extractable by a neutral salt and was primarily $Al^{3+}$. We have yet to improve on those findings in a practical way.

## I. MATERIALS AND REACTIONS RESPONSIBLE FOR SOIL ACIDITY

Soil acidity is determined largely by soil composition and the ion exchange and hydrolysis reactions associated with the various soil components. A brief description of the soil components important to soil acidity and a discussion of their roles follow. These are divided into inorganic and organic categories.

### A. Inorganic Components

The inorganic substances of importance to soil acidity include the layer silicates, the oxide minerals (including allophane), and soluble acids. The composition and structure of the layer silicates and oxide minerals have been described in detail in many articles and books, most recently by Dixon and Weed (1977). The discussion here focuses on cation exchange and buffer capacities.

#### 1. LAYER SILICATES

In brief, layer silicate minerals are made up of sheetlike arrangements of Si tetrahedral and Al octahedral units. Based on the number and arrangement of tetrahedral and octahedral layers, silicate clays are classified as 1:1, 2:1, and 2:2 minerals. For detailed descriptions of clay mineral structures, see Dixon and Weed (1977).

Table 1—Important layer silicate minerals, their CEC, and their distribution in soil orders.

| | Permanent-charge CEC | Soil orders |
|---|---|---|
| | $cmol(+)\ kg^{-1}$ | |
| Mica (illite) | 40 | Entisols, Inceptisols, Aridisols, Mollisols, Alfisols |
| Vermiculite (dioctahedral) | 150 | Aridisols, Mollisols, Alfisols, Ultisols |
| Hydrobiotite | 80 | Entisols, Inceptisols, Aridisols, Mollisols |
| Vermiculite (trioctahedral) | 150 | Entisols, Inceptisols, Aridisols, Mollisols |
| Montmorillonite | 100 | Entisols, Mollisols, Alfisols, Vertisols |
| Kaolinite | 5 | Alfisols, Ultisols |
| Halloysite | 5 | Inceptisols, Oxisols, Alfisols |

The important layer silicate minerals are listed in Table 1 along with the typical permanent charge cation exchange capacity (CEC) for each and the soil orders in which each is a dominant mineral. In the United States, kaolin–dioctahedral vermiculite (extensively coated and interlayered) is prominent in the older soils of the Southeast. In the Midwest, loessial soils are dominated by montmorillonite; glacial till soils are rich in illite. Soils of the Southwest and West either contain trioctahedral micaceous minerals and vermiculite or are almost monomineralic with respect to montmorillonite. Soils of the humid Northwest and Hawaii have rather high proportions of allophane, along with kaolinite and other clay minerals. In many acid soil areas, completely chloritized or sesquioxide-coated minerals occur more frequently in the surface soil, and vermiculite-like minerals with a larger apparent permanent charge CEC occur more frequently in subsoils.

Before the advent of x-ray diffraction in the1930s, soil clays were believed to be a complex mixture of $SiO_2$, $Al_2O_3$, and $Fe_2O_3$ (Thomas, 1977). However, the seminal works of Hendricks and Fry (1930) and Kelley et al. (1931) established that soil clays were primarily crystalline. Marshall (1935) subsequently showed that isomorphous substitution in the lattice structure of montmorillonite accounted for its CEC.

Mehlich (1942, 1943) first recognized the importance of clay mineralogy to soil acidity. He noted that the percent base saturation of North Carolina soils varied with soil type at given pH values. From plots of pH vs. base saturation for several pure clay minerals and soils, he deduced the apparent mineralogy of the soils and correctly concluded that the base saturation required for practical liming would vary from 20 to 80%.

The buffering capacity of inorganic soil components is related to both the permanent charge CEC and to the pH-dependent charge. The term *permanent charge CEC,* referring to the component of cation exchange capacity caused by lattice charge from isomorphous substitution, is due to Schofield (1949). He showed that the CEC of a clay subsoil from Rothamsted was relatively constant between pH 2.5 and 5 but increased between 5 and 7. He also found the CEC of montmorillonite to be invariant below pH 6 and to increase somewhat as pH was raised from 6 to 9. Schofield attributed the constant or permanent charge component to isomorphous substitution and the variable or high pH component to the ionization of $H^+$ from SiOH groups.

With layer silicates such as montmorillonite or even kaolinite, nearly all of the CEC measured at pH 7 or 8 is in the permanent charge category. The ionization of SiOH groups, once thought to be an important source of CEC (particularly in the kaolin minerals), now is discounted. In part, this is due to the realization that silicic acid is so weak that SiOH groups are not expected to ionize appreciably in the pH range of concern in soils (Goates & Anderson, 1956). In addition, no correlation was found between CECs of kaolinite samples and their surface areas.

Schofield's (1949) conclusion about the permanent charge was and still is perfectly valid as long as one deals with moderately pure layer silicates. However, its extension to soils—where organic matter, intergrade minerals, and perhaps allophane are present—is not. In soils, pH-dependent charge also contributes to the buffering capacity. Researchers now recognize that the major source for pH-dependent charge of soil (other than organic matter or allophane) is sesquioxides that occur as a layer silicate–sesquioxide complex (Coleman & Thomas, 1964; Schwertmann & Jackson, 1964; Volk & Jackson, 1964).

## 2. OXIDE MINERALS

Ferric and Al oxides, either hydrous or anhydrous, are present in nearly all soils. They occur as discrete crystals, coatings on layer silicates or other minerals, and as mixed gels that may also incorporate silica, $H_2PO_4^-$, and perhaps organic colloids. Iron oxide coatings on kaolinite appear to be amorphous until more than a limiting amount of the sesquioxide is present, under which circumstances hematite crystals form.

Apparently, coatings of Al or Fe hydrous oxides (or both) develop on the surfaces of layer silicates in soils. The coatings tend to be thin, corresponding more or less to a monolayer of hydrous oxide. They are believed to be positively charged and bound electrostatically to the aluminosilicate layers. These coatings, or interlayers, lead to the widely distributed, and important, intergrade minerals (Barnhisel, 1977). Their composition is intermediate between montmorillonite (or vermiculite) and chlorite. Montmorillonite–vermiculite–chlorite intergrades occur in vast variety. Many of the intergrade materials found in soils give 1.4 nm (14 Å) *c*-axis spacings that collapse only partly on $K^+$ saturation and heating, suggesting the presence of gibbsite-like interlayers. Amounts of interlayer material usually are smaller than would be required for a chlorite structure, leading to the conclusion that the gibbsite-like material is in islands rather than continuous sheets (Rich, 1960). The CECs of the intergrade minerals usually are smaller than would be expected for uncoated layer silicates (Rich & Obenshain, 1955). Removal of interlayer material (with NaOH or $F^-$ solutions) increases CEC and leads to the expansion-contraction properties of the parent layer silicate, which usually is a montmorillonite or dioctahedral vermiculite. Intergrade minerals from soils have been described by Dixon and Jackson (1962), Klages and White (1957), Rich and Obenshain (1955), Sawhney (1960), and Tamura (1957). Their occurrence is summarized by Rich (1968). The preparation of synthetic intergrade minerals and their properties are described by Barnhisel and Rich (1963), Rich (1960), Shen and Rich (1962), Slaughter and Milne (1960), Coleman and Thomas (1964), Hsu and Bates (1964b), and Veith (1977).

A pH-dependent component of CEC is characteristic of clays from many acid soils and of clay minerals, e.g., kaolinite or montmorillonite, that have been artificially coated (interlayered) with Al or Fe hydrous oxide (Thomas & Swoboda, 1963; Coleman et al., 1964). It appears that the hydrous oxide coatings are positively charged (Hsu & Rich, 1960; Jackson, 1963a, b; Hsu & Bates, 1964b), with the magnitude of the positive charge varying inversely with pH.

The characteristic cation exchange behavior of layer silicate–sesquioxide complexes results from the fact that the negative charge of the layer silicate is practically invariant with pH, but the opposite charge on the clay–sesquioxide complexes is not. The result is that the net charge varies from positive at low pH to negative at high pH. It is zero when the positive charge of the sesquioxide just equals the lattice charge of the layer silicate and reaches a net negative charge that essentially equals lattice charge of the layer silicate at pH near 8.

pH-dependent charge leads to large variation in effective CEC with pH and prevents the identification of CEC at pH 6 with lattice charge. The variation in CEC with pH arises because the number of $OH^-$ groups on the sesquioxides is increased as pH goes up, and the positive charge of the polycation is reduced. This liberates negative sites on which other cations can be retained. Other tightly held anions besides $OH^-$ have the same effect of increasing CEC. Thus, $F^-$ or $H_2PO_4^-$ ions, which are bound more tightly by Fe and Al oxides than is the clay, result in apparent increases in CEC. Sulfate and even $Cl^-$ or $NO_3^-$ salts also interact with clay mineral–sesquioxide complexes, with the sorption of the anion on the hydrous oxide component and the cation on the layer silicate (Thomas & Swoboda, 1963; Chang & Thomas, 1963).

Mixed gels containing primarily silica ($SiO_2$) and alumina ($Al_2O_3$) in mole ratios usually between 0.5 and 1.3 are termed *allophane* and occur extensively in volcanic ash soils. Allophane-like materials with appropriate silica/alumina ratios, low crystallinity, and high CEC are found in many soils.

Although isomorphous substitution accounts for most of the charge in crystalline clays, the source of charge in allophanes cannot be explained in this way. The work of Birrell (1958), Birrell and Gradwell (1956), and Wada and Ataka (1958) showed that allophanes have a combination of two kinds of apparent CEC. The first is directly related to salt adsorption. Allophanes appear to be poorly crystallized halloysites, and as such they take up salts rather indiscriminately. Once they become dry, their power to adsorb salts ceases, as in the case of halloysite (Thomas, 1960a). Imposed on the general salt uptake is a pH-dependent CEC of rather large proportions, which disappears entirely on acid treatment (Aomine & Jackson, 1959). Allophane-containing soils have very large capacities for dissolving limestone (Rixon & Sherman, 1962). The pH-dependent charge probably is similar to that which is found in synthetic catalysts used in the petroleum industry, the source of which is thought to be coordination shifts or selective $OH^-$ adsorption (Milliken et al., 1950; Tamele, 1950; Fripiat, 1964). Fripiat (1964) summarized the cation exchange behaviors of such substances in terms of the Al content (whether the Al was in 4 or 6 coordination) and the dependence of coordination number on pH.

In volcanic ash soils and probably in many others, amorphous aluminosilicates contribute significantly to the CEC and buffer characteristics (Birrell, 1958; 1961). The pronounced pH dependence of their CEC leads to quite different buffer characteristics than those obtained for layer silicates.

### 3. SOLUBLE ACIDS

Soluble acids can, under certain circumstances, strongly influence soil pH and other soil properties. Although the importance of soluble acids is often transient, their effect on soil acidity can be large. Some circumstances where soluble acids play a significant role follow.

**a. Large Applications of $NH_4^+$ Fertilizers, Especially on Slightly Buffered Soils**—High concentrations of $HNO_3$ or $H_2SO_4$ [from $(NH_4)_2SO_4$] can accumulate as a result of nitrification of $NH_4^+$ to $NO_2^-$:

$$NH_4^+ + \tfrac{3}{2} O_2 \rightarrow NO_2^- + 2H^+ + H_2O.$$

Resulting pH values can be $< 4$, and rapid loss of exchangeable $Ca^{2+}$ and $Mg^{2+}$ can occur. This effect can be especially dramatic under no-tillage management (Blevins et al., 1977).

**b. Applications of Monocalcium Phosphate Fertilizer**—Monocalcium phosphate hydrolyzes in water to yield dicalcium phosphate and phosphoric acid:

$$Ca(H_2PO_4)_2 \xrightarrow{H_2O} CaHPO_4 + H_3PO_4.$$

The effect is especially dramatic when the fertilizer is applied in a band.

The $H_3PO_4$ diffuses away from the site of fertilizer application and produces a zone around the fertilizer band where the pH can be as low as 1.5 (Lindsay & Stephenson, 1959). The extremely acid solution can dissolve Al, Fe, and Mn from soil minerals. The effects are transient, and within a few weeks pH rises again to a value near that prevailing before the fertilizer addition. However, the solvent action of concentrated $H_3PO_4$ can cause drastic alterations of minerals and the production of new solid phases.

**c. Oxidation of Pyrite in Mine Spoils and Drained Tideland Soils**—Very high concentrations of $H_2SO_4$ and very low pH values can result from the oxidation of pyrite (in soils that contain significant quantities of pyrite):

$$2FeS_2 + 7H_2O + 7\tfrac{1}{2}O_2 \rightleftharpoons 4SO_4^{2-} + 8H^+ + 2Fe(OH)_3.$$

Strip mine spoils often contain a large amount of pyrite. As a result, soil pH values can be as low as 2 in these spoils (McFee et al., 1981; Massey, 1972).

Acid-sulfate soils, or cat clays, also have very low pH values as a result of pyrite oxidation. These soils are common in marine floodplains in the temperate and tropical zones of the world. When these soils are drained and exposed to oxidation, pH values as low as 2 have been observed. Most cat clays are formed in deposits that are high enough in organic matter to be classified as peat or muck but that also contain appreciable amounts of mineral material, which dissolve in the $H_2SO_4$ to yield large amounts of soluble Al, Mn, and Fe. High quantities of lime are needed to neutralize the acidity of such soils. Amounts of free $H_2SO_4$ as large as 300 cmol $(H^+)$ $kg^{-1}$ of soil have been reported. The usual management scheme is to keep the soils flooded as much of the time as possible to prevent oxidation of S and to reduce $SO_4^{2-}$ that may have been formed under nonflooded conditions.

**d. Acid Precipitation**—On most agricultural soils where liming and heavy fertilization practices are common, the effects of acid precipitation (about pH 4) are likely to be insignificant. However, in soils that have a very low CEC, low organic matter content, and low clay content and are very shallow, acid precipitation may have a significant impact (Cowling, 1980; McFee & Kelley, 1977).

## B. Organic Matter

Soil organic matter is formed by the decomposition of plant and animal tissue in the soil. Much effort has been made to characterize soil organic matter and elucidate its structure using a variety of analytical tools such as elemental analyses, infrared spectroscopy, nuclear magnetic resonance spectroscopy, and differential thermal analyses. However, the variety of plant and animal tissue and the diversity of conditions under which soil organic matter forms result in materials that are heterogeneous with respect to elemental composition, kinds and amounts of functional groups, and chemical structure. For detailed reviews on the chemical composition and structure of organic matter, see Allison (1973), Flaig et al. (1975), Kononova (1966), and Schnitzer and Khan (1972). Only a brief description of soil organic matter is given here.

In general, soil organic matter may be described as a condensed polymeric, aromatic backbone to which are attached aliphatic side chains such as peptides, sugars, and carboxylic acids. It is both colloidal and amorphous. A model structure is shown in Figure 1. As shown in Figure 1, soil organic matter contains a high density of carboxyl groups and phenolic hydroxyls. These are important in exchange reactions and as sources of acidity and are discussed in more detail later.

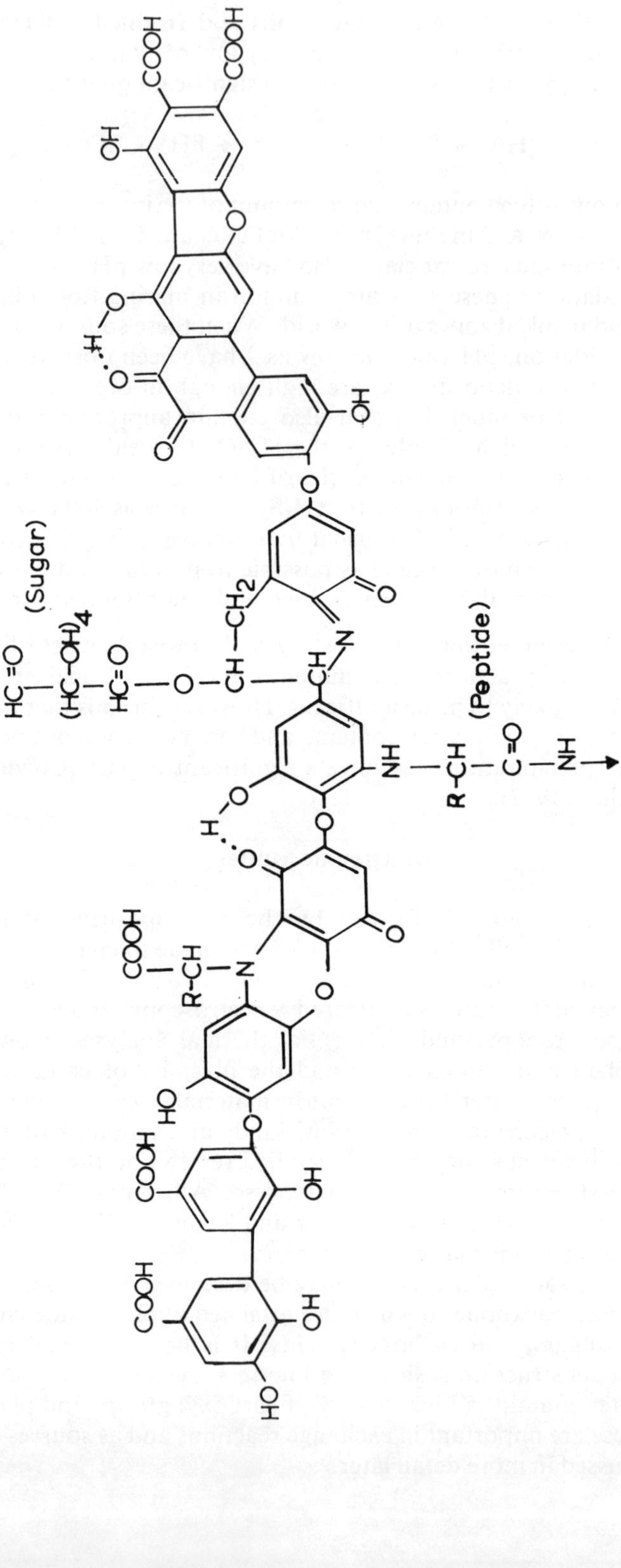


Fig. 1—Model-type structure for soil organic matter (Stevenson, 1972).

Based on solubility differences, soil organic matter is divided into three fractions:

1) The fraction that is soluble in both strong base and strong acid is fulvic acid. In general, fulvic acid is composed of molecules of lower molecular weight and contains more oxygen, and thus more reactive groups, than the other fractions.
2) The fraction that is soluble in strong base but precipitates at about pH 2 is humic acid. Humic acid is more aromatic in nature and is thus more stable and resistant to decomposition than fulvic acid. It also has a higher molecular weight, in general, than fulvic acid.
3) The third fraction, humin, is insoluble in either strong base or strong acid. It consists of macromolecules and organic compounds that are not extractable primarily because of their intimate association with the inorganic components of the soil.

These are rather arbitrary divisions based only on solubility. Therefore, each fraction is by no means homogeneous but is a complex mixture of organic compounds. Progress has been made, but the accumulation of knowledge of the structure of these fractions still does not approach the exactness of the knowledge of clay mineral structures.

The British chemist J. T. Way (1850), who first discovered cation exchange in soils, did not recognize that soil organic matter exhibited cation exchange behavior in his original experiments. However, it was found soon afterward that organic matter had an even larger so-called base exchange capacity than the mineral component of the soil (Johnson, 1859). In this century, McGeorge (1931) was one of the first soil scientists to study soil organic matter quantitatively in regard to cation exchange. He observed that, in general, divalent cations were held more tightly by humic acid than monovalent cations and thereby was first to propose a lyotropic series for organic matter. He also made another important observation: leaching humic acid with strong HCl as a pretreatment increased the effective CEC. Leaching with HCl probably removed much of the complexed metal (Al and Fe) that was otherwise nonexchangeable. McGeorge (1931) did not know exactly which functional groups were responsible for cation exchange and simply related CEC to the lignin content. Later, Gillam (1940) and Broadbent and Bradford (1952) showed that the main functional groups involved in cation exchange in organic matter were carboxyl groups with some minor contributions from phenolic, enolic, and alcoholic hydroxyls. Since these groups are weak acids, their dissociation, and thus the CEC, is variable and dependent on pH. Measured CECs at pH 7 for extracted soil humic and fulvic acids ranged from 570 to 1420 cmol (+) $kg^{-1}$ (Schnitzer & Khan, 1972). Estimates of CEC of organic matter in whole soils, however, are not as high, ranging from 60 to 280 cmol (+) $kg^{-1}$ of organic matter (Kamprath & Welch, 1962).

Many researchers have shown that humic and fulvic acids extracted from acid soils do not appear to have many free–COOH groups but are present largely as Fe and Al complexes (Martin & Reeve, 1958; Bhumbla & McLean, 1965; Schnitzer & Gupta, 1964; Griffith & Schnitzer, 1975). The

Al held by organic matter seems to be quite stable as an Al–organic matter complex since the Al is not exchangeable with KCl and apparently not easily titratable (Bloom et al., 1979; Hargrove & Thomas, 1981a, 1982; McLean et al., 1965). However, a large portion of the Al can be removed with strong (2*M*) HCl or 0.5*M* $CuCl_2$ solution (Hargrove & Thomas, 1981a). In addition, Juo and Kamprath (1979) have used $CuCl_2$ solutions as an extractant for estimating the potentially reactive Al pool in acid soils.

Since the bound Al is not very exchangeable, the effective CEC of organic matter is inversely proportional to the amount of Al present. This relationship can be seen best in the data of Posner (1966). Thus, in many acid soils in which a considerable amount of Al may be complexed with soil organic matter, organic matter may contribute very little to the effective CEC of the soil.

The presence of soluble organic acids dissolved from litter or produced from plant residues may also play an important role in soil acidity. They are important weathering agents and lead to the development of acid soils. This is especially important in Spodosol formation.

### C. Hydrolysis and Its Role in Soil Acidity

Hydrolysis is a molecular phenomenon that is very important to soil acidity. A large charge/size ratio for cations results in a large hydration energy. Hydrolysis occurs when the charge/size ratio (or acidity) of the cation is so great as to cause rupture of H–O bonds. The result is ionization of the hydrate to yield hydronium ions. This can be illustrated for $Al^{3+}$ as follows:

$$Al^{3+} + 6H_2O \rightarrow [Al(H_2O)_6]^{3+} \xrightarrow{H_2O} H_3O^+ + [Al(H_2O)_5OH]^{2+}.$$

The cations that hydrolyze are those that are highly charged, small, or both. Table 2 lists some of the important cations found in soils, their charge-size function ($Z^2/r$), and their hydrolysis constant ($pK_H$). The correlation of the charge-size function with the hydrolysis constant is good except that the

Table 2—Charge-size functions and hydrolysis constants for several cations found in soils (Huheey, 1978, p. 266).

| Cation | $Z^2/r$† | $pK_H$ |
|---|---|---|
| $Na^+$ | 2.2 | 14.48 |
| $Ca^{2+}$ | 10.3 | 12.70 |
| $Mn^{2+}$ | 12.5 | 10.70 |
| $Fe^{2+}$ | 13.3 | 10.10 |
| $Zn^{2+}$ | 13.7 | 9.60 |
| $Cu^{2+}$ | 14.1 | 7.53 |
| $Mg^{2+}$ | 14.3 | 11.42 |
| $Fe^{3+}$ | 35.5 | 2.19 |
| $Al^{3+}$ | 43.6 | 5.14 |

† Units are $C^2\ m^{-1} \cdot 10^{28}$.

transition metals have slightly lower $pK_H$ values than expected. This is due to their electronic structure.

The important soil cations that hydrolyze and contribute significantly to soil acidity include $Al^{3+}$ and $Fe^{3+}$. Monomeric $Al^{3+}$ ions displaced from clay minerals by cations hydrolyze in solution, and almost certainly the hydrolysis products are readsorbed, causing increased hydrolysis (Ragland & Coleman, 1960; Frink & Peech, 1963b; Jackson, 1963a, b). Further, $H^+$ ions produced by hydrolysis react with and dissolve or decompose soil minerals. The pH measured in such a system is a transient index and cannot really be interpreted. Similarly, salt addition to sesquioxide-coated, interlayered minerals appears to increase the hydrolysis of nonexchangeable Fe or Al, resulting in an increase in $H^+$ ion concentration of the solution and lower pH (Coleman et al., 1964). Kissel et al. (1971) found that extensive hydrolysis of nonexchangeable Al occurred during the extraction of soils with neutral unbuffered salts. The hydrolysis reaction continued as the extractant volume increased. Furthermore, they found that the proportion of hydroxy-Al that hydrolyzed was related to the stability of the hydroxy-Al interlayers. Veith (1977) also found that $H^+$ in the exchangeable acidity of various $Al^{3+}$-saturated exchangers was a result of hydrolysis of exchanged $Al^{3+}$ and of adsorbed nonexchangeable Al. This hydrolysis reaction can account for the anomalous reports of exchangeable $H^+$ in soils.

Hydrolysis is responsible for both the low pH of solutions containing $Al^{3+}$ ions and the neutralization or buffer capacity of such solutions. Schofield and Taylor (1954) and Frink and Peech (1963a) carefully examined the first-stage hydrolysis of $AlCl_3$ by measuring pHs of solutions of different concentrations. They describe the first-stage hydrolysis as follows:

$$Al^{3+} + H_2O \rightarrow AlOH^{2+} + H^+.$$

The hydrolysis constant is

$$K_h = [AlOH^{2+}][H^+]/[Al^{3+}]$$

Since equal amounts of $H^+$ and $AlOH^{2+}$ are produced, the right-hand side of the hydrolysis equation can be written as $[H^+]^2/Al^{3+}$. Taking negative logarithms yields $pH = ½\ pK + ½\ pAl$, with $pK$ around 5. The hydrolysis equation can be written in a different way, showing $H^+$ ions coming specifically from water molecules coordinated around Al (Jackson, 1963a). However, the $pK$ is the same.

In Al solutions with low OH/Al ratio, most of the Al is found as $Al^{3+}$, $Al(OH)^{2+}$ and $Al(OH)_2{}^+$ (Smith & Hem, 1972). As the number of hydroxyls per Al becomes larger (near OH/Al of 1.0), polymerization increases significantly, and poorly defined, large hydroxyl-Al ions form (Brosset et al., 1954; Raupach, 1957). These vary in numbers of OH per Al and in size, becoming larger as the OH content increases (Ruff & Tyree, 1958). At even higher OH/Al ratios, the major product is gibbsite, although it forms very slowly below a pH of about 4.8 and an OH/Al ratio of 2. The gibbsite is formed from the larger hydroxy-Al polymers that decline in quantity with time. The pH declines due to hydrolysis, and one is left with a mixture of $Al^{3+}$ and gibbsite according to the dismutation reaction:

$$[Al(OH)_2]_n^{n+} \rightarrow \frac{2n}{n} Al(OH)_3 + \frac{n}{3} Al^{3+}.$$

Dismutation of interlayer hydroxy-Al material to form gibbsite also can occur. Barnhisel and Rich (1963) found that gibbsite formed in synthetic Al-OH-montmorillonite when the amount of interlayer material was large. Although hydroxy-Al polymers are stabilized by interaction with clay, the amounts that can be held are limited. From the data of Hsu and Bates (1964b), the limit for montmorillonite seemed well below that for vermiculite. However, Veith (1977) has shown that montmorillonite adsorbed relatively large OH-Al polymers and therefore accumulated much more Al in the interlayers than vermiculite or a sulfonic resin, which excluded large OH-Al polymers because of their smaller *d*-spacings, high charge densities, or both. Furthermore, relatively large amounts of $Al(OH)_{3-x}^{x+}$ become exchangeable, whereas the Al exchanged from vermiculite or resin was always the $Al^{3+}$ ion.

Hsu and Bates (1964a, b) proposed that a definite hydrolysis product, $Al_6(OH)_{12}^{6+} \cdot 12H_2O$, forms when base is added to an $Al^{3+}$ salt solution and that the above cation, with one positive charge per Al, is the nonexchangeable Al adsorbed by vermiculite from partly neutralized $AlCl_3$. More recent work casts doubt on this simple picture, however.

Hydrolysis is also important in Al–organic matter. Schnitzer and Skinner (1963a, b) first showed that Al was hydroxylated in organic matter with an average composition of $Al(OH)_2^+$ rather than being present as the neutral salt-exchangeable $Al^{3+}$ ion as Chernov (1964) postulated in his treatise on soil acidity. We also have found that not only does the amount of Al adsorbed by organic matter increase with an increase in pH but also the degree of hydrolysis, or the OH/Al ratio, of the adsorbed Al increases with pH (Hargrove & Thomas, 1981b). The OH/Al ratio quickly approached 2 as the pH at which Al adsorption occurred increased from 3.0 to 4.0, and the Al was almost always hydrolyzed to some extent even though pH was generally low ($< 4$).

Apparently, Al is hydrolyzed to a greater extent at low pH in organic matter because the hydrolytic product reacts with carboxyl groups more extensively than with clay and because the organic matter serves as a sink for $H^+$ ions produced by hydrolysis according to the following reactions:

$$Al(OOC)_3R + H_2O \rightleftharpoons Al(OH)(OOC)_2R + RCOOH$$

$$Al(OH)(OOC)_2R + H_2O \rightleftharpoons Al(OH)_2(OOC)R + RCOOH.$$

With an increase in the OH/Al ratio, each mole of adsorbed Al counters less negative charge in organic matter, freeing additional functional groups for more Al adsorption. Schnitzer and Skinner (1965) also calculated the average OH/Al ratio for extracted soil organic matter with varying Al contents and found that for low amounts of complexed Al (< 5% by weight) the OH/Al ratio was one, but that for moderate to high amounts (10–15% by weight) it approached two.

The degree of hydrolysis was especially high for samples that were leached with $Al^{3+}$ solutions compared with a single equilibration (Hargrove, 1980). Removal of $H^+$ by leaching apparently allowed more extensive hydrolysis of the adsorbed $Al^{3+}$. Furthermore, the estimated OH/Al ratio for Al–organic matter that was formed by leaching Ca–organic matter with $AlCl_3$ solution was about 2. Since all of the $Ca^{2+}$ was replaced by an equivalent amount of $Al^{3+}$, the net reaction, including hydrolysis, must be

$$3Ca(OOC)_2R + 2AlCl_3 + 4H_2O \rightleftharpoons 2Al(OH)_2(OOC)R + 4HOOCR + 3CaCl_2.$$

This means that it is impossible for Al to occupy all of the reactive sites in organic matter since it will always hydrolyze to some extent, and the $H^+$ produced by hydrolysis will lower the pH, causing fewer carboxyl groups to be ionized. Hence, the failure of Al to occupy all the carboxylic groups occupied by Ca is due to a combination of Al hydrolysis and less dissociation of $H^+$ from carboxyls.

## II. BUFFER CURVES AND APPARENT ACID STRENGTHS

Titration curves have been used for many years in an attempt to characterize soil acidity, both as to amount and to acid strength. However, it is apparent from a historical perspective (Thomas, 1977) that their interpretation often lagged far behind the actual recording of data. In fact, Jenny (1961) commented with some astonishment on the different interpretations that had been made from the same data and wondered if present views would suffer the same drastic alterations accorded older ones.

In many cases, titration curves of clays, organic matter, and whole soils have led to erroneous ideas about the true strength of the soil acids. However, they are useful in determining buffer capacities and effective acid strengths as they exist in nature. Combined with other information, buffer curves can be very useful; alone or in the wrong context they can be quite misleading. Titration properties and apparent acid strengths can be divided into mineral and organic components.

### A. Minerals

Potentiometric titrations of clays suggested to Bradfield (1923, 1925) that beidellite-containing clays are weakly ionized acids with apparent $pK_a$ values around 6.5. Later work, of course, showed that the apparent weak

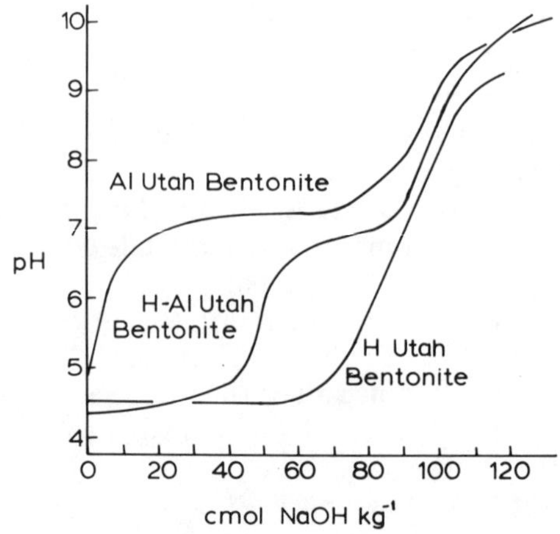


Fig. 2—Titration curve of H- and Al-montmorillonite (Coleman & Harward, 1953).

acid character of electrodialyzed clays is due to Al hydrolysis. Chernov (1964) and Coleman and Harward (1953), followed by many others, showed that when montmorillonite is truly $H^+$ saturated, the titration curve resembles that of a strong mineral acid, whereas the apparent weak acid character of clays is due to the presence of adsorbed $Al^{3+}$. Titration curves for $H^+$-bentonite and $Al^{3+}$-bentonite are shown in Figure 2. Coleman and Harward (1953), Harward and Coleman (1954), and Lin and Coleman (1960), from limited data obtained with montmorillonite, presumed that the sorbed $Al^{3+}$ was present as an exchangeable monomeric hydrated ion. Later work showed that in addition to such exchangeable $Al^{3+}$, polymeric nonexchangeable hydroxy-Al and hydroxy-Fe ions also contribute to buffer capacity in many cases and are quantitatively more important than exchangeable $Al^{3+}$ (Schwertmann & Jackson, 1964; Coleman et al., 1964; Coleman & Thomas, 1964; Volk & Jackson, 1964).

Work with electrodialyzed, acid-treated, or $AlCl_3$-leached montmorillonite led Coleman and his co-workers to suggest that the buffer properties of acid clays and soils, in the absence of organic matter, are largely due to the exchange and hydrolysis of $Al^{3+}$ ions. With acid montmorillonites and subsoil material from shale-sandstone derived soils of the North Carolina Piedmont, which have been found not to be extensively interlayered with Al and Fe hydrous oxide, they observed reasonably close correspondence between quantities of $Al^{3+}$ that could be exchanged by salt solutions and amounts of acidity titrated potentiometrically or conductimetrically with base (Lin & Coleman, 1960). It also was observed that H-saturated clays, prepared by mineral acid or $H^+$-exchange resin treatment, were not stable but changed spontaneously into $Al^{3+}$ or $Al^{3+}$-$Mg^{2+}$ clay (Barshad, 1960; Chernov, 1959; Chernov et al., 1956; Coleman & Craig, 1961). This explains why appreciable amounts of neutral salt exchangeable $H^+$ are not found in mineral soils.

Examples of buffer curves, with $Ca(OH)_2$, for $Al^{3+}$-saturated clay minerals are shown in Figures 3 and 4, which were constructed with data

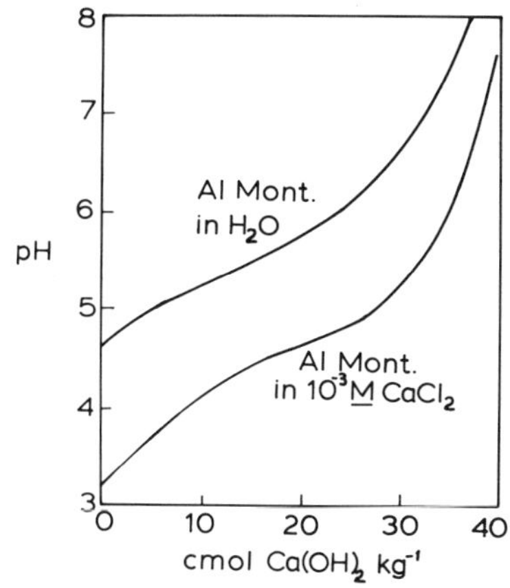


Fig. 3—Titration curve for Al-montmorillonite using $Ca(OH)_2$ (Turner & Nichol, 1962a).

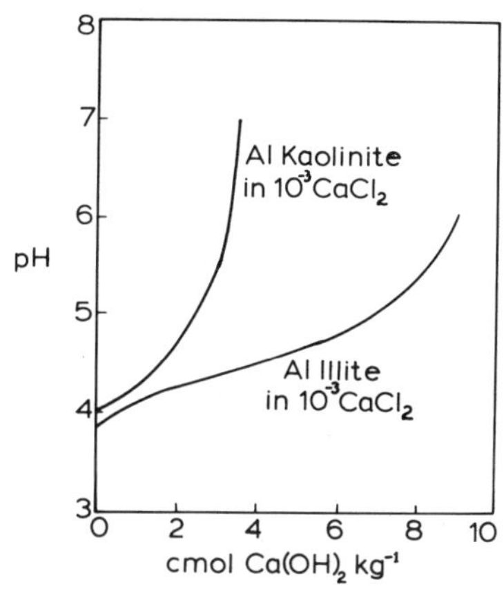


Fig. 4—Titration curve for Al-kaolinite and Al-illite with $Ca(OH)_2$ (Turner & Nichol, 1962a).

given by Turner and Nichol (1962a). Figure 3 is for $Al^{3+}$-montmorillonite in water or $10^{-3}M$ $CaCl_2$, whereas Figure 4 is for kaolinite and illite in $10^{-3}M$ $CaCl_2$. The buffer characteristics and the effect of electrolyte are representative for clays and soils relatively free of organic matter and dominated by monomeric exchangeable $Al^{3+}$. Apparently this situation exists in nature only in special cases (Jackson, 1964). The highly acid subsoil layers studied by Coleman et al. (1959b), with their large amounts of exchangeable $Al^{3+}$, are not typical.

Apparently the presence of hydroxy-Al or hydroxy-Fe interlayers or surface coatings that interact with layer silicates is nearly universal in acid soils and leads to weak acid character for titration curves. A considerable amount of data on the buffer characteristics of clay–hydrous sesquioxide complexes has been obtained. Coleman and Thomas (1964) observed pronounced buffering in the pH range 5 to 8 when Al and Fe hydrous oxide complexes of montmorillonite were titrated with NaOH in $1M$ NaCl.

Schwertmann and Jackson (1964) and Volk and Jackson (1964) also have presented pertinent data on the potentiometric titration of acid soils and clays. Their titrations were performed continuously, i.e., base was added at a constant rate to stirred suspensions in 0.5*M* KCl, and no attempt was made to obtain "equilibrium" pHs corresponding to given additions of base. Their curves were interpreted as showing four buffer ranges that correspond to postulates made earlier by Jackson (1963a). For titration in 0.5*M* KCl, the four buffer ranges and the clay components responsible are as follows:

| Range | pH | Reactant with added base |
|---|---|---|
| I | $<4$ | $H_3O^+$ |
| II | 4–5.6 | $Al(H_2O)_6^{3+}$ |
| III | 5.6–7.6 | "Strong" hydroxy-Al |
| IV | $>7.6$ | "Weak" hydroxy-Al |

Jackson and co-workers (Schwertmann & Jackson, 1964; Volk & Jackson, 1964) have shown that the proportions of the titratable acidity in clays and soils that fall into the first three ranges depend on pretreatment. Treatment of montmorillonite or vermiculite with $H^+$-resin yields material with most of its acidity in range I and II. Aging for some weeks causes a large increase in range III at the expense of ranges I and II. With naturally acid soils, Jackson's third buffer range, the pH-dependent acidity, ranged from 0.6 to 2.6 cmol (+) $kg^{-1}$, less than would be expected from results of more traditionally conducted titrations or exchange capacity determinations. The claim that rapid titration differentiates between proton sources by their rate of release is speculative. We have observed and Page et al. (1965) have documented the fact that rates of neutralization vary greatly, depending on whether one is neutralizing exchangeable $H^+$ or exchangeable monomeric $Al^{3+}$ (rapid) or interlayer hydroxy-Al (slow, requiring days to reach a constant pH). From this, it seems that Jackson's method (Schwertmann & Jackson, 1964; Volk & Jackson, 1964) would under-estimate the "pH-dependent" charge that comes from hydroxy-Al.

Much experimental work, of which that of Schwertmann and Jackson (1963, 1964), Coleman and Thomas (1964), and Coleman et al. (1964) is the most directly pertinent, illustrated that hydroxy-Al interlayers or surface coatings are important sources of pH-dependent charge in soils. Along with organic matter, they probably account for most of the pH-dependent charge (and buffer capacity) developed between pH 5.5 and 8. Some apparent pH-dependent charge also is created during the neutralization of soils or clays containing appreciable exchangeable monomeric $Al^{3+}$. That is, hydroxy-Al ions produced during neutralization may be adsorbed by the clay, or remain adsorbed if produced by surface hydrolysis, and be further hydrolyzed to produce $Al(OH)_3$ at a higher pH than would be required for completion of the reaction in solution (Thomas, 1960b). A large buffer capacity between pH 5 and 8 that cannot be attributed to organic matter suggests the presence of layer silicate–hydrous sesquioxide complexes.

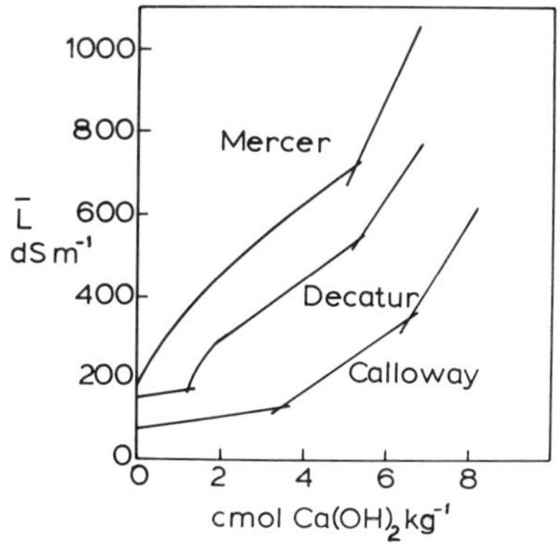


Fig. 5—Conductimetric titration curves [using $Ca(OH)_2$] for selected Kentucky subsoils (Kissel & Thomas, 1969).

Conductimetric titrations (Low, 1955; Harward & Coleman, 1954) with NaOH were found early to be useful in distinguishing between exchangeable $H^+$ and $Al^{3+}$. When $H^+$ clay is titrated with a base, conductivity decreases until all the $H^+$ ions are neutralized, because of the larger mobility of $H^+$ ions than of cations of the base. As excess $OH^-$ ions are added, the conductivity rises rapidly due to the high mobility of those ions. Aluminum clays, like weakly ionized acids, have little free $H^+$ ion present at any one time. Therefore, conductivity rises as base is added, due to the formation of salt. When the end point is reached, the conductivity rises sharply due to the mobility of excess $OH^-$.

Coleman and Thomas (1967) prematurely asserted that conductimetric titration curves of acid clays, which distinguish so well between exchangeable $H^+$ and $Al^{3+}$, did not distinguish at all between exchangeable $Al^{3+}$ and nonexchangeable hydroxy-Al. Soon after that, Kissel and Thomas (1969) showed that conductimetric titrations using $Ca(OH)_2$ did distinguish between exchangeable and nonexchangeable Al in 10 Kentucky subsoils. Examples of their titrations are shown in Figure 5. Furthermore, they found that titration of KCl extracts to a phenolphthalein end point gave estimates of exchangeable acidity that were too high due to hydrolysis of nonexchangeable Al and to overtitration to reach the phenolphthalein end point (Kissel & Thomas, 1969; Kissel et al., 1971). Rich (1970) and Dewan and Rich (1970) also used conductimetric titrations to distinguish between exchangeable and nonexchangeable acidity using $Ba(OH)_2$ in place of $Ca(OH)_2$.

## B. Organic Matter

Most of what has been learned about acidity in organic matter results from titrations of carboxylic resins, extracted humic and fulvic acids, and organic soils. Katchalsky and Spitnik (1947) titrated a weakly acidic exchange resin, polymethacrylic acid, and used the following equation to calculate the acid strength from the titration data:

$$pH = pK_a + n \log [\alpha/(1 - \alpha)]$$

where $n$ = constant (which = 1 for monoprotic acids such as acetic acid but is usually about 2 for polyprotic acids such as soil organic matter) (Martin & Reeve, 1958), and $\alpha$ = fractional dissociation of the organic acid. The above equation is known as the modified Henderson-Hasselbalch equation and has also been used to describe titrations of natural materials such as peat soils and humic and fulvic acids (Bloom, 1979; Martin & Reeve, 1958; Schnitzer & Skinner, 1963a, b; Gilmour & Coleman, 1971; Posner, 1964; Hargrove & Thomas, 1981b, 1982a).

Katchalsky and Spitnik (1947) developed the modified Henderson-Hasselbalch equation empirically from the generalized Henderson-Hasselbalch equation in which the constant $n$ does not appear. The generalized Henderson-Hasselbalch equation was developed for titrations of monoprotic acids. A plot of the pH as a function of log $[\alpha/(1 - \alpha)]$ for a monoprotic acid resulted in a straight line with a slope of 1. However, similar plots of polyprotic acids such as polymethacrylic acid resulted in straight lines with slopes > 1. Therefore, Katchalsky and Spitnik (1947) added the empirical constant $n$, which is the numerical value of the slope, to the Henderson-Hasselbalch equation and used it to describe titrations of polyprotic acids.

After plotting titration data for some H- and Al-mucks according to the Henderson-Hasselbalch equation [pH vs. log $\alpha/(1 - \alpha)$], we found that the value of $n$ (the slope of the line) did not vary much for $H^+$-saturated organic matter (2.0–2.3). However, $n$ was considerably reduced as the Al content of the organic matter increased (Hargrove & Thomas, 1982a). This reaction indicates that the behavior of organic matter is changing from a typical polyprotic acid to a weaker monoprotic acid. This, in turn, implies that in a heterogeneous material (e.g., soil organic matter) that contains carboxylic acids of widely varying strength, $Al^{3+}$ reacts with the strongest carboxyls first, leaving carboxyl groups that are weaker and more nearly of uniform acid strength (c.f. Fig. 7).

When the main source of acidity in organic matter was assumed to be carboxylic groups, early work with titration curves of soil organic matter showed an abnormal weakness compared with titration curves of pure carboxylic acids. Most carboxylic acids have acid dissociation constants ($K_a$s) of $10^{-4}$ to $10^{-5}$, whereas titration curves of organic matter gave values of about $10^{-6}$. Martin and Reeve (1958) showed that this abnormal weakness was due to Al and Fe on exchange sites, just as the abnormal weakness of soil clays was caused by exchangeable $Al^{3+}$ (Harward & Coleman, 1954) and even more so by hydroxylated Al (Coleman & Thomas, 1964). Removal of Fe and Al from organic matter gave an apparent $K_a$ of $10^{-4}$. The effect of Al on the titration of organic matter is shown in Figure 6. These results pointed to the importance of Al in cation exchange and acidity in organic matter as well as in clay minerals.

We have also conducted conductimetric titrations of samples of muck of varying Al content (Hargrove & Thomas, 1982a). Figure 7 shows con-

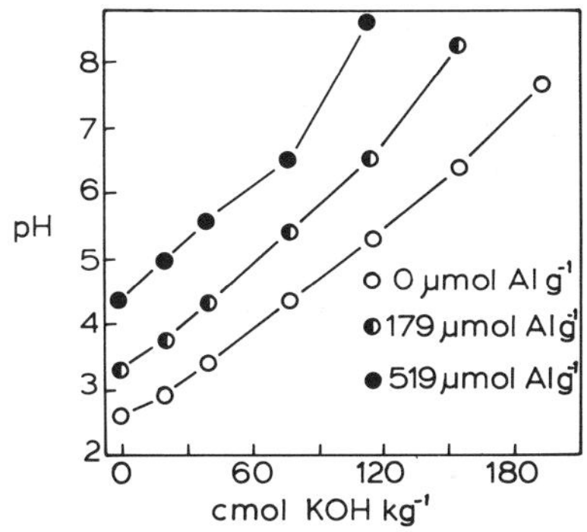


Fig. 6—Titration curves of organic matter of varying Al content (Hargrove & Thomas, 1982).

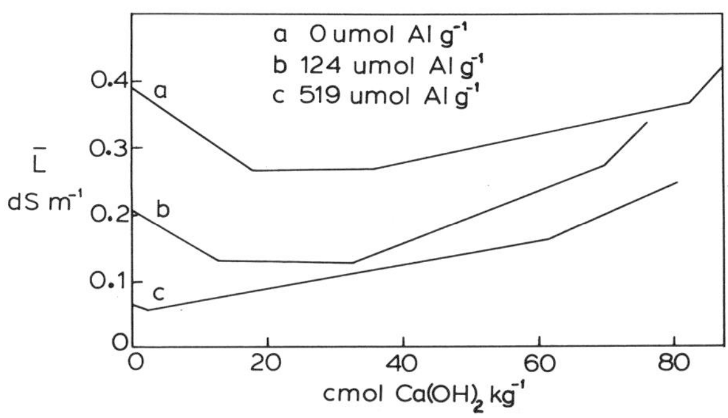


Fig. 7—Conductimetric titration curves of organic matter (Hargrove & Thomas, 1982).

ductimetric titration curves using $Ca(OH)_2$ for three samples ranging in Al content from 0 to 0.52 mmol $g^{-1}$. Just as conductimetric titrations of acid subsoils with $Ca(OH)_2$ distinguished between exchangeable and weaker nonexchangeable acidity, the titration curves in Figure 7 show three types of acidity that titrate differently. The first type, in which the curve has a negative slope, may be considered relatively "strong" acidity. The next segment of the curve is more or less horizontal and may be considered "weak" acidity. The third segment has a very small positive slope and may be considered "very weak" acidity. It is interesting to note that a conductimetric titration of a H-saturated weak exchange resin (IRC-50) showed only very weak acidity (Hargrove & Thomas, 1982a). It would, therefore, seem that a carboxylic resin is not a very good model for soil organic matter.

The effect of increasing Al content was to decrease the amounts of strong and weak acidity, whereas the amount of very weak acidity was increased or remained about the same. Apparently, Al reacts with the two stronger acid groups first and reacts very little, if any, with the very weak acid groups. Furthermore, the total amount of acidity titrated decreased as the Al content increased. These results show again that Al complexed by organic matter is not exchangeable and apparently not titratable at pH values below about 8.0. Also, since $Al^{3+}$ is apparently adsorbed at the expense of the more easily dissociated functional groups, in acid soils that contain significant amounts of Al–organic matter, the organic matter may contribute very little, if any, to the effective CEC of the soil, especially at low pH values.

## III. CATION EXCHANGE PROPERTIES OF ACID SOILS

### A. Components of CEC

There are two definitions for an acid soil. One states that the pH is < 7; the other states that basic metal cations make up < 100% of the CEC. Defining soil acidity in terms of base saturation would be simpler and more specific if there were a uniform convention for defining the CEC. Unfortunately, there is not.

The two most common methods of determining CEC are the adsorption of $NH_4^+$ at pH 7 and the sum of basic cations replaced by $NH_4OAc$ plus acidity neutralized by $BaCl_2$ + triethanolamine (TEA) solution (Thomas, 1982). The former method has the rationale of neutrality, which means very little in soils. The latter roughly corresponds to Bradfield and Allison's (1933) definition of a Ca-saturated soil. This has more practical utility in calcareous than in acid soils but at least represents a limit to soil acidity that can be neutralized by $CaCO_3$. Empirically, a pH of 8 to 9 appears to correspond to complete neutralization of adsorbed $Fe^{3+}$ or $Al^{3+}$ ions (Coleman & Thomas, 1964) and to nearly coincide with formation of $Al(OH)_3$ from $[Al(OH)_x]_n$ polymers (Keren et al., 1977).

Thus, CEC determination by salt solutions buffered at pH 8.2, such as the widely used $BaCl_2$–TEA method (Mehlich, 1948), may give a reasonably close estimate of the lattice charge of layer silicates plus contributions from any organic matter or allophane that may be present.

Table 3—Cation exchange capacities of Ohio surface soils (Bhumbla & McLean, 1965).

| Soil | Lime added | pH | Organic matter | CEC | | | CEC after organic matter removal | |
|---|---|---|---|---|---|---|---|---|
| | | | | $BaCl_2$–TEA | $NH_4OAc$ | KCl | $BaCl_2$–TEA | KCl |
| | cmol (+) kg$^{-1}$ | | % | cmol (+) kg$^{-1}$ | | | | |
| Clairemont (Typic Ustifluvents) | 4 | 5.7 | 1.4 | 10.2 | 7.2 | 6.4 | 7.3 | 4.9 |
| | 8 | 7.2 | ... | 10.1 | 7.4 | 8.2 | 8.4 | 5.5 |
| Bennington (Aeric Ochraqualfs) | 0 | 4.8 | 2.4 | 18.0 | 10.1 | 7.3 | 11.0 | 6.4 |
| | 8 | 6.3 | ... | 16.7 | 10.3 | 11.1 | 11.6 | 7.3 |
| Venango (Aeric Fragiaqualfs) | 3 | 4.8 | 3.6 | 21.7 | 11.8 | 6.0 | 16.8 | 6.1 |
| | 18 | 6.8 | ... | 19.9 | 12.7 | 11.6 | 16.9 | 7.4 |
| Mahoning (Aeric Ochraqualfs) | 4 | 4.7 | 3.9 | 21.2 | 14.2 | 9.1 | 14.5 | 8.2 |
| | 20 | 2.1 | ... | 18.7 | 14.8 | 15.4 | 16.4 | 9.3 |
| Unclassified | 8 | 4.7 | 7.5 | 41.3 | 23.4 | 12.7 | 20.3 | 8.8 |
| | 28 | 6.3 | ... | 36.7 | 25.6 | 23.0 | 17.5 | 8.3 |

A third method for determining CEC involves leaching with an unbuffered salt solution such as KCl or $CaCl_2$ (Coleman et al., 1959a, b; Pratt & Bair, 1962; Bhumbla & McLean, 1965; Grove et al., 1982). The determination is made at the pH of the soil, or at least the pH of a soil-salt solution mixture after exhaustive leaching, and has been called *effective CEC*.

Wide experience has shown that CECs measured for a given soil by the three methods are quite different. Exchange capacity measured at pH 8.2 is invariably larger than that measured at pH 7 (Coleman et al., 1959a; Mehlich, 1942, 1943; Pratt & Bair, 1962; Hanna & Reed, 1948), and the CEC measured with an unbuffered salt solution usually is found to be quite dependent on soil pH, i.e., to be larger in a limed than unlimed sample of the same soil (Bhumbla & McLean, 1965; Grove et al., 1982).

Data from a paper by Bhumbla and McLean (1965), showing the CECs measured for several limed and unlimed surface soils from Ohio, using $BaCl_2$–TEA at pH 8.2, $NH_4OAc$ at pH 7, and unbuffered KCl, are shown in Table 3.

The data show the following:

1) Cation exchange capacity at pH 8.2 is larger than CEC at pH 7, with the difference correlating directly with organic matter content.
2) Liming to change pH from around 5 to 6 to 7 did not appreciably affect CEC measured with $BaCl_2$–TEA or $NH_4OAc$.
3) Cation exchange capacity measured with unbuffered KCl was increased approximately twofold by liming.
4) Removal of organic matter reduced $BaCl_2$–TEA–CEC around 1.5 cmol (+) $kg^{-1}$ for each percentage organic matter originally present.
5) Removal of organic matter greatly reduced the pH effect on KCl–CEC, leading to values that were only slightly larger for limed than for unlimed soils and were close to those obtained with the original acid soils.

Hanna and Reed (1948) had previously shown large differences in $BaCl_2$–TEA–CEC and $NH_4OAc$–CEC for limed and unlimed surface soils from the North Carolina Coastal Plain, attributing the differences largely to organic matter.

Coleman et al. (1959b) measured effective CEC for Ap and B horizons of a number of North Carolina Piedmont soils and also determined the amounts of acidity not displaced by 1*M* KCl but neutralized on leaching with $BaCl_2$-TEA, pH 8.2.

Data for three Ultisols are shown in Table 4. For these soils, all with pHs between 5.3 to 6.1, effective CEC measured with KCl was from 30 to 50% of CEC at pH 8.2. Surface and subsoil samples were not very different, and organic matter cannot be solely responsible for the additional CEC that develops between pH 6 and 8 in such soils.

In the same work (Coleman et al., 1959b), CECs of six subsoil samples from the North Carolina Piedmont also were measured. Results for these subsoils, which contain montmorillonite-like minerals that are not extensively coated with sesquioxides are shown in Table 5. For the latter group of soils, the effective CEC averaged 70% of CEC at pH 8.2, as contrasted

Table 4—The pH, effective CEC, and CEC at pH 8.2 for several surface and subsoils from the North Carolina Piedmont (Coleman et al., 1959b).

| Soil | pH | Horizon | CEC with KCl | CEC with $BaCl_2$–TEA |
|---|---|---|---|---|
| | | | cmol (+) kg$^{-1}$ | |
| Cecil (Typic | 5.8 | Ap | 1.7 | 4.9 |
| Hapludults) | 5.6 | B | 2.7 | 5.7 |
| Georgeville (Typic | 6.1 | Ap | 4.9 | 10.9 |
| Hapludults) | 5.3 | B | 3.5 | 10.2 |
| Davidson (Rhodic | 6.1 | Ap | 5.5 | 12.0 |
| Paleudults) | 5.9 | B | 4.7 | 12.3 |

Table 5—The pH, effective CEC, and CEC at pH 8.2 for several subsoils from the North Carolina Piedmont (Coleman et al., 1959b).

| Soil series | pH | CEC with KCl | CEC with $BaCl_2$–TEA |
|---|---|---|---|
| | | cmol (+) kg$^{-1}$ | |
| Iredell (Typic Hapludults) | 6.3 | 16.1 | 23.7 |
| Granville (Typic Hapludults) | 4.8 | 13.4 | 19.1 |
| Mayodan (Typic Hapludults) | 4.9 | 11.7 | 16.0 |
| Wadesboro (inactive series) | 4.8 | 6.7 | 13.2 |
| White Store (Vertic Hapludalfs) | 4.6 | 22.4 | 29.7 |

with around 40% for the Cecil-Georgeville-Davidson group. McLean's results (Bhumbla & McLean, 1965) for acid Ohio soils showed KCl–CEC to be 30 to 40% of CEC at pH 8.2. Removal of organic matter increased the percentage to around 60, which is intermediate between the values obtained for the two groups of North Carolina soils.

The difference between the KCl–CEC and that at pH 8.2 varies tremendously with organic matter content. One of McLean's samples with 75 g kg$^{-1}$ organic matter gave a CEC difference of 28.6 cmol (+) kg$^{-1}$ (Bhumbla & McLean, 1965). This dropped to 11.5 cmol (+) kg$^{-1}$ when organic matter was removed. For McLean's 12 Ohio samples, the range was 2.4 to 11.5 cmol (+) kg$^{-1}$, with an average of 7.1 (Bhumbla & McLean, 1965). Subsoil samples for the Ultisols studied by Coleman et al. (1959b) gave an average of 5.7, with a range from 3.2 to 8.3. Bhumbla and McLean's (1965) acid soils derived from shale and sandstone gave an average of 6.4. Much larger differences are observed from some soils containing heavily interlayered vermiculite or allophane, but the values cited above appear typical for most situations where organic matter does not contribute appreciably to CEC.

The pH dependence of soil CEC is a function of the constituent minerals and organic matter. At very low pH values with clay minerals, $Al^{3+}$ is the predominant cation. As the pH is raised, the $Al^{3+}$ hydrolyzes and counters some sites, reducing the net CEC. Further additions of base (or lime) increase the hydrolysis of $Al^{3+}$, freeing the exchange sites for $Ca^{2+}$. This behavior was noted by Kelley and Brown (1926), Paver and Marshall (1934), and Thomas (1960b). Data from Kelley and Brown (1926) are given in Table 6, and data from three other sources are shown in Fig. 8.

Table 6—Effect of leaching a California soil with varying normalities of $AlCl_3$ on exchangeable $Ca^{2+}$, $Mg^{2+}$, and $Al^{3+}$ (Kelley & Brown, 1926).

| Normality of $AlCl_3$ | pH | $Ca^{2+}$ | $Mg^{2+}$ | $Al^{3+}$ | $\Sigma\ Ca^{2+} + Mg^{2+} + Al^{3+}$ |
|---|---|---|---|---|---|
| | | cmol (+) $kg^{-1}$ | | | |
| 0 | 6.95 | 21.8 | 6.1 | 0 | 27.9 |
| 0.001 | 5.96 | 18.0 | 5.4 | 0 | 23.4 |
| 0.002 | 5.65 | 16.9 | 3.8 | 0 | 20.7 |
| 0.004 | 5.05 | 15.0 | 4.0 | 0 | 19.0 |
| 0.01 | 4.94 | 10.1 | 2.8 | tr.† | 12.9 |
| 0.02 | 4.84 | 4.4 | 1.8 | 13.0 | 19.2 |
| 0.1 | 4.84 | 1.8 | 1.1 | 21.9 | 24.8 |

† tr. = trace.

The pH corresponding to minimum "effective" CEC varies considerably among the four soils (pH 4.2–5.2), but the experimental methods employed also vary in each case. In an experiment with Chambers montmorillonite, Keren et al. (1977) showed a minimum CEC of 49 cmol (+) $kg^{-1}$ at pH 7.7. This rose to 65 at pH 7.5 and to 70 at pH 8.5. Their experimental system was, of course, different still. In many soils, this minimum in effective CEC has not been observed, usually because the soil was at or near the minimum CEC as found in the field. Hence, unless the soil is artificially acidified, the increase in CEC with an increase in $Al^{3+}$ is not observed. Also, soils in which permanent negative charges on clays do not contribute significantly to CEC would not be expected to show behavior such as this.

The work of Parks and de Bruyn (1962) on hematite has led to a return of interest in the acidity produced by the adsorption of $OH^-$ ions by soils in which clay minerals with permanent charge are rather unimportant sources of acidity. The picture evoked is not very different from that of Mattson

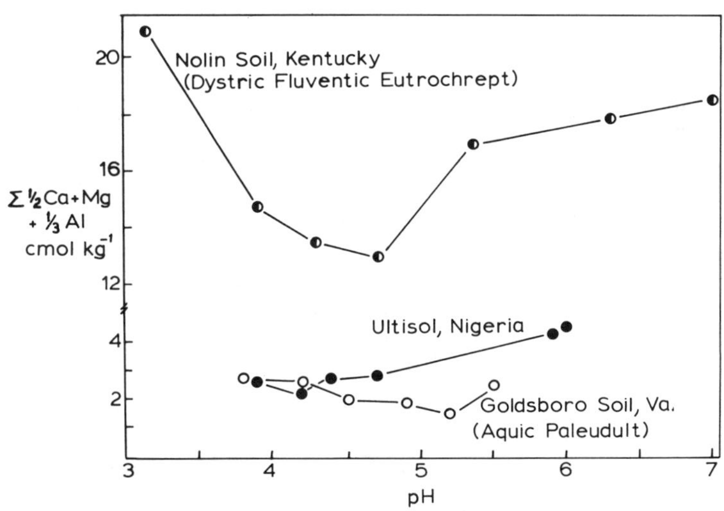


Fig. 8—Effect CEC vs. pH for three soils (Nolin soil, P. M. Bertsch, 1982, unpublished data, Univ. of Kentucky; Nigerian soil, Juo & Ballaux, 1977; Virginia soil, Bertsch & Alley, 1981).

and Wiklander (1940a, b) but the vocabulary is new. Certainly, there is much evidence to support the view that a major part of the acidity, particularly in tropical soils, comes from this source. The new interest, in part, has arisen from the increased chemical work done on tropical soils in the past 20 yr.

The basic feature of the pH-dependent acidity (Parks and de Bruyn, 1962) is that charge is developed on oxides by the adsorption of $H^+$ or $OH^-$ ions according to the scheme shown below.

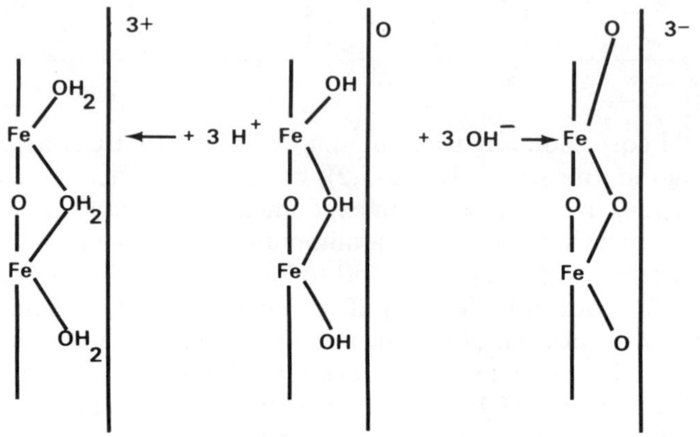


In this situation, both negative and positive charges arise only from the adsorption of $OH^-$ or $H^+$ ions, and there is no inherent charge on the oxide itself.

van Raij and Peech (1972) were among the first to apply this idea to soils in the post-Mattson era. Their results showed what Mattson had found out 30 yr earlier only too well. This approach works well only in soils in which there is essentially no source of permanent negative charge. In an Oxisol from Brazil, the negative and positive charges determined from titration coincided well with those determined by actual adsorption from an indifferent salt adjusted to various pH values. An Alfisol, however, did not show any agreement between titration and adsorption from a salt.

Perhaps the most enlightened view of what really is occurring is due to Parker et al. (1979). To understand their treatment, one must first know a basic definition. Figure 9 (Parks & de Bruyn, 1962) shows the ideal behavior of an oxide when titrated with acid and base in an indifferent electrolyte ($KNO_3$) of different ionic strengths. What is critical on this graph is the point at which all curves intersect (at pH 8.5 for $Fe^{3+}$ oxide). This is the zero point of charge (ZPC), i.e., the pH where there are equal quantities of cations and anions (or none of either). The ZPC corresponds to Mattson's equi-ionic point.

Parker et al. (1979) took four soils in a topographic and chronological sequence from southwest Virginia and titrated them according to van Raij and Peech (1972). The results for their most weathered soil are shown in Figure 10. The solid curves represent the titration curves with $H^+$ or $OH^-$ at three ionic strengths of $CaCl_2$, and the dashed line is the net charge de-

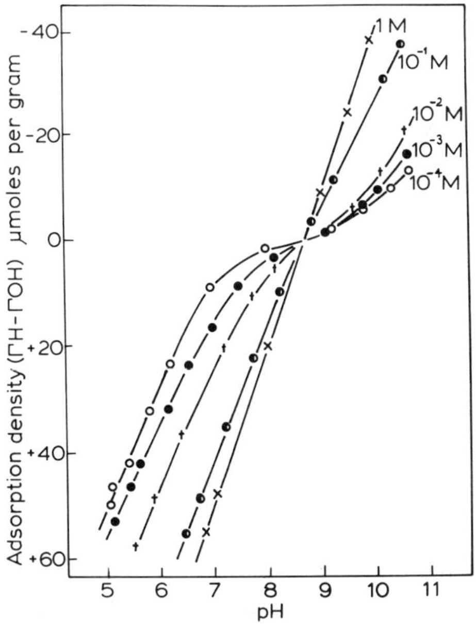


Fig. 9—Titration of $Fe_2O_3$ with $H^+$ and $OH^-$ in electrolyte of varying concentration (Parks & de Bruyn, 1962).

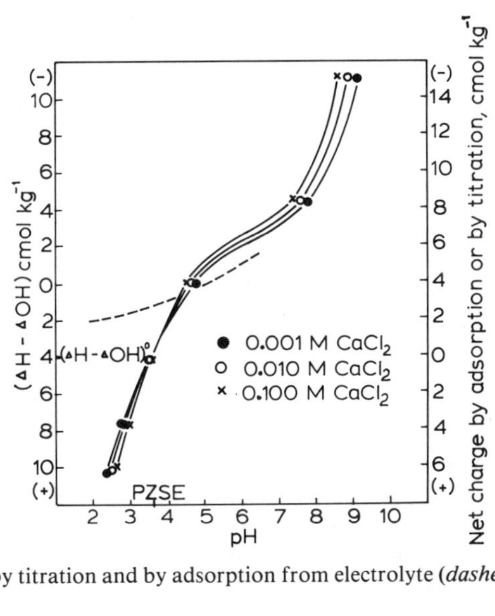


Fig. 10—Net charge by titration and by adsorption from electrolyte (*dashed line*) (Parker et al., 1979).

termined by adsorption of either $Ca^{2+}$ or $Cl^-$ from $0.1M$ $CaCl_2$. At the ZPC, the $CaCl_2$ equilibration shows an adsorption of about 3 cmol (+) $kg^{-1}$ more $Ca^{2+}$ than $Cl^-$. Accordingly, Parker et al. (1979) renamed the intersecting point from titration curves the point of zero salt effect, PZSE, since it clearly is not the ZPC. The less weathered soils were much farther off the mark than that illustrated.

The weaknesses in applying the Parks and de Bruyn (1962) titrations to soils are neatly summarized by Parker et al. (1979, p. 673) as follows:

> Errors in estimating . . . [ZPC] . . . from PZSE [point of zero salt effect] are likely to increase as the amount of permanent charge and exchangeable or hydrolyzable Al and Fe increase. The severity of the latter's effect will probably increase as the . . . [ZPC] . . . becomes further removed from neutrality. Even in cases where PZSE and . . . [ZPC] . . . coincide, predictions of net charge from titration curves should be made cautiously due to the questionable relationship between ($\Delta$H − $\Delta$OH) values and net charge.

Clearly, the behavior of oxidic soils cannot be explained by resorting to clay-mineral models. On the other hand, the behavior of most oxidic soils cannot be explained entirely by the approach of Parks and de Bruyn (1962) either. For a more optimistic view, however, see Sposito (1981).

Soil organic matter is a fairly weak acid with an experimental ionization constant ($K_a$) of around $10^{-4}$ (Martin & Reeve, 1958; Hargrove & Thomas, 1981b). However, as with all polyprotic acids, this $K_a$ is only an approximate average. Several researchers agree that the major reactive groups in soil organic matter are carboxyls attached to unsaturated rings that are arranged in polymers of variable and unknown sizes (Gillam, 1940; Broadbent & Bradford, 1952; Schnitzer & Gupta, 1964). It would be expected that an array of $K_a$ values would arise from different positions on the ring, different substitutions, and from polymerization. Although organic matter titrates practically linearly from pH 3 to 8 using potentiometric measurements, conductimetric titrations with $Ca^{2+}$ as the added base show a considerable portion of ionized $H^+$, followed by two progressively weaker acids (Schnitzer & Skinner, 1963a; Hargrove & Thomas, 1982).

What is significant about the unusually strong acidity in organic matter is that it is rarely seen in nature except in mucks and peats (White & Thomas, 1981). Organic matter in acid mineral soils reacts strongly with solution $Al^{3+}$. Results from Hargrove and Thomas (1982a) are shown below. As Al is adsorbed by the Michigan peat, the strongest acidity disappears almost completely, as does the second portion of the acidity.

| Al adsorbed | Strong acidity | Moderately weak acidity | Very weak acidity | Total | $\Delta$ |
|---|---|---|---|---|---|
| mmol g$^{-1}$ | cmol (+) kg$^{-1}$ | | | | |
| 0 | 35 | 37 | 93 | 165 | -- |
| 12.4 | 25 | 39 | 76 | 140 | 25 |
| 51.9 | 4 | 0 | 120 | 125 | 41 |

The increase in the very weak acidity at the highest $Al^{3+}$ adsorption suggests that some of the stronger acidity was transformed to weaker acidity by the adsorption of $Al^{3+}$. No exact mechanism can be postulated at present.

In any case, the net effect of Al on organic matter is the same as it is in clays, i.e., it weakens the acidity and thereby reduces the total charge available for CEC at a given pH.

The pH-dependence of soil CEC can be described qualitatively in terms of properties of the constituent minerals and organic matter. For the layer silicates, where most of the CEC results from lattice charge, neutral salt CEC, CEC at pH 7, and CEC at pH 8.2 are not very different. Following Schofield's (1949) terminology, most of the CEC of the layer silicates comes from permanent charge, and there is relatively little pH-dependent charge.

With the intergrade minerals or with $Fe^{3+}$ oxide-coated kaolinite (Sumner, 1963), neutral salt CEC is much smaller than CEC at pH 8.2 for samples that are initially acid (Coleman et al., 1964). The CEC is larger when the pH of the clay is adjusted upward before salt leaching to saturate for CEC determination, or when the salt solution is buffered at a pH higher than that of the soil.

In many southeastern soils with pHs near 5, neutral salt CEC is very small (c.f. preceding data for Ultisols). Removal of $Fe^{3+}$ oxide coatings or interlayer Al hydrous oxide (or both) increases neutral salt CEC and reduces the difference between it and CEC at pH 8.2. Synthetic intergrade minerals prepared by precipitating Al or Fe hydrous oxide between plates of montmorillonite or vermiculite have very small neutral salt CECs when acid but larger ones when neutralized (Coleman et al., 1964). The available data suggest that in such minerals, whether natural or synthetically prepared, interlayers or coatings of Fe or Al hydrous oxides lose $OH^-$ ions (or accept protons) at low pH, become positively charged, and neutralize part or all of the lattice charge of the layer silicate in the same way that charge in chlorite is internally compensated by excess cations in a gibbsite or brucite layer. As pH rises, the interlayer hydroxy-sesquioxide material gains $OH^-$ (or ionizes protons), loses positive charge, and frees exchange sites of the layer silicate, which then can adsorb and exchange cations. According to this, the effective CEC of such intergrade minerals will lie somewhere between zero and the lattice charge of the layer silicate, depending on the amount of interlayer material and the pH. Obviously, it is not correct to refer to the neutral salt CEC of an acid sample of such substance as permanent charge, as Coleman et al. (1959a) inferred some years ago.

Soil organic matter is rather weakly acidic by nature. Thus, effective CEC, as deduced from titration curves, is nearly a linear function of pH, being near zero at pH 3 to 4 and increasing regularly to pH 8. In most acid soils, organic matter behaves as if it has a $pK_a$ of around 6 due to the adsorption of Al, Fe, or both. Generally, the CEC of organic matter ($BaCl_2$-TEA at pH 8.2) is on the order of 200 cmol (+) $kg^{-1}$. It is about one half of that at pH 6 (Broadbent & Bradford, 1952; Kamprath & Welch, 1962; Mortensen & Himes, 1964).

Parenthetically, it is fortunate that the bulk of the CEC of most arable acid soils comes from organic matter, intergrade clays, and $Fe^{3+}$ oxide-coated kaolinite. Were all acid soils like the Mayodan–Granville–White Store complex referred to earlier (layer silicates without extensive interlayering, and thus with large amounts of salt-exchangeable $Al^{3+}$), agriculture could not have developed in the Southeast and other acid soil areas. Sharp lines of demarcation between productive and inherently unproductive soils are found where the shale-sandstone triassic basin borders igneous rock in North Carolina and Virginia and between limestone and acid shale soils in Kentucky, Virginia, and Tennessee. It is no accident that three of the first six U.S. Presidents came from the Davidson soil in Virginia.

## B. CEC and pH of Soils

Pratt and Bair (1962) measured CECs of 15 acid California surface soils at pHs between 3 and 8. Some of their data are reproduced in Figure 11. The soils varied in organic matter content (0.7–15%) and in clay mineralogy. Cation exchange capacity at pH 3 varied from a low of 2 to a high of 23 cmol (+) $kg^{-1}$. For all of the soils, CEC increased slowly with pH values to around pH 5 and then more rapidly to pH 8. Pratt and Bair (1962) concluded that the pH-dependent CEC was due to both organic matter and clay. Cation exchange capacity of organic matter at pH 8.0 varied from 150 to 380 (average 215) units. The pH-dependent CECs of $< 2$-$\mu$m clay were deduced as being between 11.2 and 24.2 cmol (+) $kg^{-1}$, with the median being 13 to 15 cmol (+) $kg^{-1}$.

For soils containing 30 to 50% clay and little organic matter, this would lead to values for ΔCEC (CEC at pH 8.2 minus neutral-salt CEC) 5 to 6, in agreement with the data for North Carolina and Ohio soils.

Helling et al. (1964) took an interesting approach to the study of CEC-pH relationship. They measured CECs of 60 Wisconsin soils at pH values between 2.5 and 8; by multiple correlation analysis they determined the contributions of organic matter and clay at each pH, and the variation in CEC of each as pH was changed. The average CEC of organic matter ranged from 36 cmol (+) $kg^{-1}$ at pH 2.5 to 213 cmol (+) $kg^{-1}$ at pH 8; that for clay was 38 cmol (+) $kg^{-1}$ at pH 2.5 and 64 cmol (+) $kg^{-1}$ at pH 8. Their data are summarized below and shown graphically in Fig. 12.

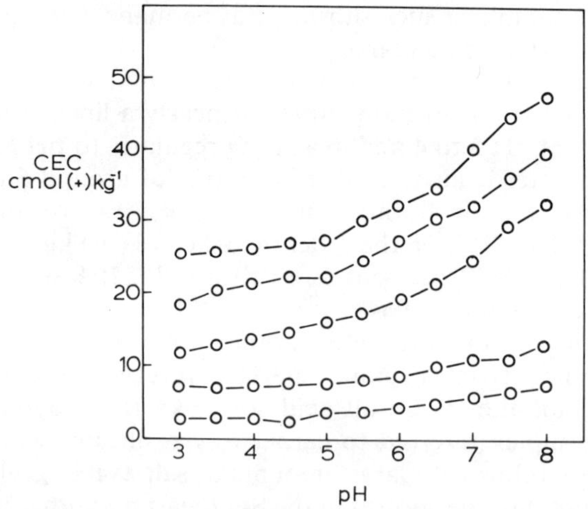


Fig. 11—Cation exchange capacity variations with pH for a group of acid California soils (Pratt & Bair, 1962).

| pH | CEC (means for 60 soils) | | |
|---|---|---|---|
| | Organic matter | Clay | Soil |
| | cmol (+) $kg^{-1}$ | | |
| 2.5 | 36 | 38 | 5.8 |
| 3.5 | 73 | 46 | 7.5 |
| 5.0 | 127 | 54 | 9.7 |
| 6.0 | 131 | 56 | 10.8 |
| 7.0 | 163 | 60 | 12.3 |
| 8.0 | 213 | 64 | 14.8 |

For this group of Wisconsin soils, organic matter was the major contributor to the pH-dependent CEC. In fact, the CEC of the clay was nearly constant in the pH range of 5 to 7, which is of most agricultural interest. Ross et al. (1964) obtained similar data for a group of Michigan soils.

As deduced from consideration of the properties of intergrade minerals and as shown earlier by Pratt and Bair (1962), the clay fraction of many soils has appreciable pH-dependent charge, even in the acid range. Apparently, the soils studied by Helling et al. (1964), and also those of Pratt and Bair (1962), were only slightly interlayered with hydrous sesquioxide and thus retained appreciable permanent charge at low pH. The sesquioxide–montmorillonite complexes prepared by Coleman and Thomas (1964) and many soils of the Southeast (Rich & Obenshain, 1955; Moschler et al., 1962; Coleman et al., 1959a, b) are more completely interlayered and may have effective CECs near zero at pH 4.

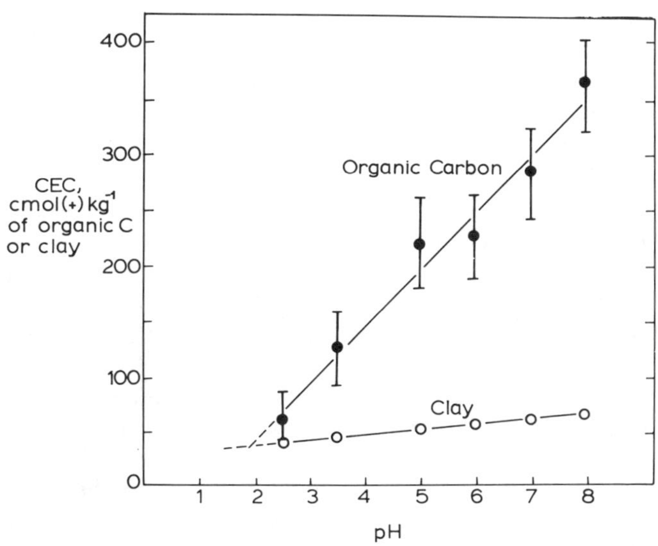


Fig. 12—Cation exchange capacities of organic matter and clay in Wisconsin soils (Helling et al., 1964).

### C. Soil pH and Percent Base Saturation

In the past, much attention has been given to establishing relationships between soil pH and the percent saturation of the CEC with $Ca^{2+}$ and $Mg^{2+}$ ions. This is termed percent base saturation, and considerable data have been accumulated to show that pH–percent base saturation relationships differ greatly from soil to soil. Much of the work on this topic is due to Mehlich (1942, 1943), whose ideas have found general acceptance.

Basing calculations of percent base saturation on CEC at pH 8.2, as determined with $BaCl_2$–TEA, Mehlich (1942, 1943) concluded that montmorillonite and illite (and soils containing those minerals) required approximately 80% base saturation to reach pH 7, whereas soils dominated by kaolinite or organic matter (or both) reached pH 7 when 30 to 50% base-saturated. Although Mehlich's findings are entirely correct, his ideas become more generally useful if interpreted in terms of concepts of effective CEC and pH-dependent charge that have developed since his early work. Now we can say with some assurance that for layer silicates, as those in Mehlich's montmorillonite, illite, and White Store soil, where most of the CEC results from lattice charge, the pH remains low as long as a significant amount of exchangeable acidity exists, i.e., base saturation is incomplete. That is because, with such materials, monomeric $Al^{3+}$ ions counter most of the permanent charge and account for most of the CEC.

With intergrade minerals and $Fe^{3+}$ oxide-coated kaolinite, on the other hand, lattice charge is neutralized mostly by basic metal cations or charged sesquioxides, with little monomeric $Al^{3+}$ being present. Due to the firm attachment of charged sesquioxides to the layer silicates, such complexes appear to have weak-acid character. For many such soils, percent base saturation, based on CEC at pH 8.2, is perhaps 35% at pH 6.

Organic matter, having weakly acidic groups, is rather like the intergrade minerals in its pH–percent base saturation (CEC at pH 8.2) relationships (Helling et al., 1964; Martin & Reeve, 1958; Turner & Nichol, 1962a). This is especially so when the functional groups are complexed with $Al^{3+}$ and $Fe^{3+}$. A pH of about 6 appears characteristic of half saturation of CEC at pH 8.2 with $Ca^{2+}$ and $Mg^{2+}$.

From the rather sketchy data available, it appears that many acid surface soils give pH–percent base saturation relationships characteristic of intergrade mineral–$Fe^{3+}$ oxide–coated kaolinite–organic matter materials. That is, the pH–percent saturation function has a shallow slope between pH 6 and 8, and there is relatively low percent base saturation of CEC at pH 8.2 (perhaps 50%) around neutrality.

Coleman et al. (1959a, b) attempted to modify concepts of pH–percent base saturation relationships by basing percent saturation on what they called effective CEC, or neutral-salt CEC, rather than CEC at pH 8.2. When propounding this, they were under the misapprehension that soil minerals have a permanent charge, analogous with lattice charge, which exists and is invariant below pH 6, and a pH-dependent charge that develops only above pH 6. They proposed that in the absence of soil organic matter, neutral salt CEC measured the permanent charge, and in the presence of organic matter it measured an "effective CEC," which was the sum of permanent charge and the amount of ionization of organic matter functional groups at the pH of the determination. It is now clear that their identification of neutral salt CEC as the permanent charge was wrong, in that it did not take into account interactions between layer silicates and sesquioxide coatings. Their accompanying suggestion, that lime rates might logically be based on the exchangeable $Al^{3+}$, perhaps has more merit. It is unfortunate that this has yet to be adopted outside of the state of Vermont.

### D. Anion Adsorption by Soils

The acid characteristics of soils containing $Fe^{3+}$ oxide–coated kaolinite and at least certain kinds of intergrade minerals are complicated by reactions involving certain anions. This is anion adsorption, which appears to involve the adsorption of anions such as $H_2PO_4^-$ or $SO_4^{2-}$ (or even $Cl^-$ or $NO_3^-$) by positively charged sesquioxide films on layer silicates (and perhaps also on edges of layer silicates themselves) with the concomitant freeing of cation exchange sites. Such reactions can affect apparent effective CEC, buffer curves, and other soil properties. Although incompletely understood, they are of great importance in many soils of the tropics and subtropics.

The retention of anions by soils was noted many years ago by Mattson (1931). He observed that $Cl^-$ and $SO_4^{2-}$ were held in quantity only by soils that had large amounts of $Al^{3+}$ and $Fe^{3+}$ oxides. Mattson (1931) and later Mattson and Wiklander (1940a, b) assumed that the anions were in competition with $OH^-$ ions found in the hydrous oxides that coated the soils. Therefore, at low pH, when the $OH^-$ concentration is low, the amount of $SO_4^{2-}$ or $Cl^-$ retained is relatively high. Anions such as $H_2PO_4^-$ or $F^-$ act much the same way in low concentration but also tend to cause dissolution and precipitation reactions that may be even more important.

Further work has substantiated Mattson's findings and some features of his interpretations of data. For example, Kamprath et al. (1956) and Ensminger (1954) showed that $SO_4^{2-}$ and $H_2PO_4^-$ compete for the same sites in soils when the $H_2PO_4^-$ concentration is small, and that more $SO_4^{2-}$ is adsorbed at lower pH. The results of Chao et al. (1964) show that $Fe^{3+}$ and $Al^{3+}$ compounds or coatings are responsible for $SO_4^{2-}$ adsorption. Several observations (Berg & Thomas, 1959; Chang & Thomas, 1963) suggest that $NO_3^-$ and $Cl^-$ are adsorbed rather weakly compared with $SO_4^{2-}$, and it appears that there is a great range in anion affinities for the hydrous oxides.

Apparently differences exist between Fe and Al hydrous oxides as to their preference for anions. The work done on this aspect of the problem is somewhat sketchy, but clearly hydrous Fe oxides adsorb anions such as $Cl^-$ and $SO_4^{2-}$ in larger quantities than an equivalent weight of $Al(OH)_3$ (Thomas & Swoboda, 1963). On the other hand, $Al(OH)_3$ seems to cover negative sites on clays more effectively than Fe oxide (Coleman & Thomas, 1964).

The exact mechanism by which anions are adsorbed by soils is probably impossible to prove. Mattson's idea of $OH^-$ displacement can be shown to be valid in most experiments. Yet, at least with $SO_4^{2-}$ and $H_2PO_4^-$, it is equally plausible to argue that a somewhat insoluble salt is formed that merely gives the appearance of anion adsorption. For example, Adams and Rawajfih (1977) and Wolt (1981) have shown that $SO_4^{2-}$ "adsorption" can be explained as the formation of basaluminite and alunite, or both. Regardless of the mechanism, the anions have a measurable effect on buffer capacity of soils.

In acid soils containing adsorbed $SO_4^{2-}$, liming decreases the amount of $SO_4^{2-}$ adsorbed, and appreciable amounts of $CaCO_3$ can be consumed in producing $CaSO_4$. The $OH^-$ ions consumed during the replacement of $SO_4^{2-}$ have not neutralized any $H^+$, but they have been removed from the system just as surely as if they had. This contribution to soil acidity (or more correctly, to lime requirement) has been pointed out by Mehlich (1964), who refers to it as acidity due to anions. Mehlich's data suggest that its magnitude in North Carolina Piedmont soils may be as great as 3 to 4 cmol (+) $kg^{-1}$.

In the case of $H_2PO_4^-$ adsorption, the effect is considerably greater. Recent work by Gillman and Fox (1980) shows that the CEC, hence potential acidity, is increased by $H_2PO_4^-$ additions. This is explained by the lowering of the ZPC in the presence of adsorbed P.

Whatever the mechanism, the effect is clear and is important in soils with low CEC. As with many aspects of soil acidity, this "modern" observation has a long history (Mattson and Wiklander, 1940a, b; Coleman & Mehlich, 1948).

## IV. EXCHANGEABLE IONS IN ACID SOILS

Researchers generally agree that the effective CECs of acid soils, corresponding to neutral salt CEC, are countered primarily by metal cations—largely $Ca^{2+}$, $Mg^{2+}$, and monomeric $Al^{3+}$ hexahydrate. There may be large amounts of nonexchangeable $H^+$, but if one defines the exchangeable ions as those displaced by a neutral salt solution, $Ca^{2+}$, $Mg^{2+}$, and exchangeable $Al^{3+}$ are by far the most abundant. Additional CEC develops as soil pH is raised through the addition of base. Thus, acid soils behave as though they have two components of acidity, that exchangeable with neutral salt and that which is titratable but not exchangeable.

### A. Exchangeable and Titratable Acidity

Exchangeable acidity is that portion of the soil acidity that can be replaced with a neutral, unbuffered salt such as KCl, $CaCl_2$, or NaCl. The acidity replaced by such salts depends only slightly on the method of extraction, except in the case of a micaceous mineral or where KCl is a poor extractor (Kaddah & Coleman, 1967). Titratable acidity is the amount of acid neutralized at a selected pH, which often is 8.2. According to these definitions, titratable acidity corresponds approximately to acidity measured with $BaCl_2$–TEA at pH 8.2. Pratt and Bair (1962) and Bhumbla and McLean (1965) have shown this to be true.

Exchangeable acidity in soils is due almost entirely to monomeric $Al^{3+}$ ions. This was first shown by Vietch (1904), who extracted soil with NaCl and titrated with base to a phenolphthalein end point. On checking the composition of the precipitate that formed, he found that it consisted primarily of Al, with some minor quantities of Fe, Mn, and Zn. Other early workers, e.g., Hartwell and Pember (1918), also found $Al^{3+}$ to be present in large quantities in acid soils where plants did not grow well. However, other researchers (Kelley and Brown, 1926) found acid soils with no exchangeable $Al^{3+}$. Their results showed that soils made extremely acid contained large quantities of exchangeable $Al^{3+}$, but moderately acid soils had no acidity that could be replaced (Fig. 13 & 14).

Paver and Marshall in 1934, and later Schofield (1949), Chernov (1959, 1964), Harward and Coleman (1954), and Low (1955), demonstrated the importance of $Al^{3+}$ in the exchangeable portion of soil acidity. Their work showed that the exchangeable acidity extracted from clays was almost entirely $Al^{3+}$. Further work showed that the $Al^{3+}$ that was exchanged by neutral salts was trivalent (Lin & Coleman, 1960; Thomas, 1960b) and that hydroxy-Al was not exchangeable by neutral, unbuffered salts (Rich, 1960; Chernov, 1964). An exception to this is the work of Veith (1977) who showed that even though only $Al^{3+}$ was replaced from Al-montmorillonite subjected to various levels of $Ca(OH)_2$, in Ca-montmorillonite to which an excess of $Al(OH)_x$ had been added (where $x$ varied from 0 to 2.2), the value of $x$ in the extracted Al varied from 0 to 2.6. The extractability of these polymers attributed to the ability of montmorillonite to expand to 2.0 nm (1.0 nm between each plate), and the low charge of the polymers per $Al^{3+}$ ion. However, whether this is much of a factor in natural montmorillonite systems found in soils is doubtful.

In soils where organic matter is an important contributor to CEC, it appears that a higher proportion of the exchangeable acidity is $H^+$ (Yuan, 1959, 1963), but this may be more apparent than real. Yuan's curves show that Al is extremely difficult to extract from organic matter. This is borne out by work of Martin and Reeve (1958), Schnitzer and Skinner (1963a), and Hargrove and Thomas (1981a).

It appears that much of the $H^+$ that apparently is exchangeable from organic soils arises from hydrolysis of $Al^{3+}$ that is difficult to replace with KCl. Paver and Marshall (1934) obtained similar results when they used $0.01M$ $BaCl_2$ to extract $Al^{3+}$ from clay.

In soils with practically no organic matter but containing large quantities of hydroxy-Al, similar results are obtained (Kissel et al., 1971). Furthermore, the exact amount of "exchangeable" $Al^{3+}$ is difficult to determine because it tends to increase as extraction volume increases. Some of their data (Kissel et al., 1971) for four Kentucky subsoils are shown in Table 7. It probably is true that no neutral salt extraction gives very exact results for $Al^{3+}$ or $H^+$ because of hydrolysis of $Al^{3+}$ and subsequent dissolution of

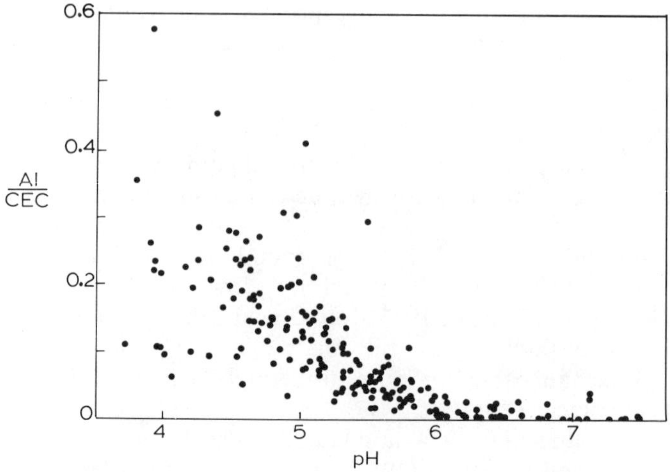


Fig. 13—KCl-exchangeable Al as a proportion of CEC at pH 8.2 for surface soils for Virginia (Coleman & Thomas, 1967).

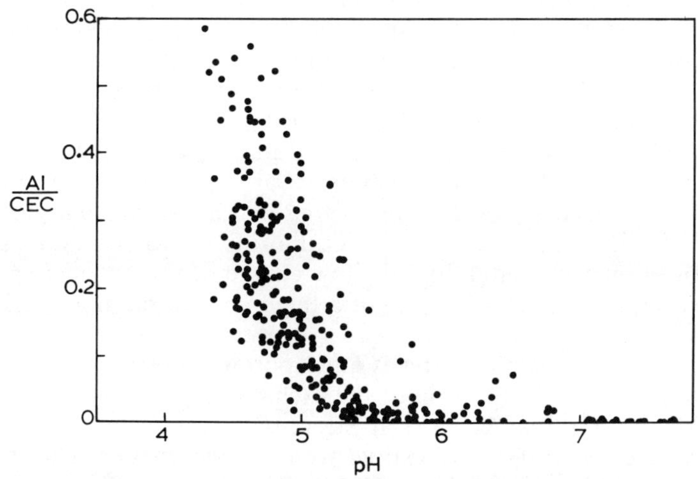


Fig. 14—KCl-exchangeable Al as a proportion of CEC at pH 8.2 for subsoils from Virginia (Coleman & Thomas, 1967).

Table 7—Amounts of exchangeable acidity as affected by extraction volume for four Kentucky subsoils (Kissel et al., 1971).

| Soil | 150 mL 1*M* KCl | | 400 mL 1*M* KCl | |
|---|---|---|---|---|
| | $H_3O^+$ | $Al^{3+}$ | $H_3O^+$ | $Al^{3+}$ |
| | cmol (+) kg$^{-1}$ | | | |
| Henry B3x (Typic Fragiaqualts) | 1.45 | 6.75 | 2.10 | 7.60 |
| Calloway Bx2 (Glossaquic Fragiudalfs) | 1.05 | 7.15 | 1.77 | 8.03 |
| Mercer Bx2 (Typic Fragiudalfs) | 0.20 | 1.00 | 0.64 | 1.46 |
| Shelbyville B22 (Mollic Hapludalfs) | 0.50 | 2.30 | 1.50 | 2.05 |

more Al as the extraction proceeds. Amedee and Peech (1976) have suggested the following two reactions to account for the inexact extraction of $Al^{3+}$ from soils:

$$ROH + KCl \rightarrow ROK + HCl$$

$$Al(OH)_3 + 3HCl \rightarrow AlCl_3 + 3H_2O.$$

These equations help explain both organic matter and hydroxy-Al soil behavior.

In trioctahedral vermiculites, Kaddah and Coleman (1967) also found very poor replaceability of $Al^{3+}$, especially in one from South Africa. After removal of $Al^{3+}$ + $H^+$ by NaCl, appreciable titratable acidity still remained on the clay. As in the case of organic matter, much of the hydrolysis occurred during the extraction itself, illustrating further the ambiguity associated with determining exchangeable $Al^{3+}$.

Despite the slow release of $H^+$ by neutral salt extractions, large quantities of $H^+$ are never found by extraction with unbuffered salt unless the clay is carefully $H^+$ saturated or unless a soluble acid is present (Dewan & Rich, 1970).

The exchangeable acidity as a proportion of the total acidity varies with the nature of the soil and the percent base saturation. Among the clay minerals, the ratio of exchange acidity/total acidity is highest for montmorillonite, intermediate for the dioctahedral vermiculites, and lowest for the kaolinite minerals. The ratio in some trioctahedral vermiculites appears to be quite low (Pratt & Bair, 1962), whereas in others it is comparable with the dioctahedral vermiculites. Where oxides of Al and Fe have accumulated as surface coatings on clays, the proportion of exchangeable $Al^{3+}$ to total acidity is reduced. Coleman et al. (1959a) found that montmorillonitic soils consistently had higher exchange acidity contents than did kaolinitic soils. Other data from Coleman et al. (1964), Thomas (1960b), Rich (1960), and an immense amount of unpublished work indicate that montmorillonites require a higher content of Fe or Al oxides (or both) than any other clay to reduce the proportion of exchangeable acidity by a comparable amount. The following tabulation illustrates the oxide effects on exchangeable acidity. The montmorillonite–sesquioxide complexes were leached with 1*M* $AlCl_3$ and washed with water prior to the acidity determinations.

Data from Coleman et al. (1964) are as follows:

| Exchanger | Percent surface $R_2O_3$ content | | Exchange acidity/ total acidity at pH 8.2 |
|---|---|---|---|
| | $Fe^{3+}$ | $Al^{3+}$ | |
| | | | cmol (+) kg$^{-1}$ |
| Wyoming bentonite | 0 | 0 | 95 |
| Wyoming bentonite | 10.7 | 0 | 55 |
| Wyoming bentonite | 21.4 | 0 | 43 |
| Wyoming bentonite | 0 | 6.8 | 48 |
| Wyoming bentonite | 0 | 13.6 | 12 |

The data show that both Fe and Al hydrous oxides reduce the exchange acidity, with Al oxide being more effective. Montmorillonite coated with hydrous Al oxide in amounts slightly less than would be required for a complete gibbsite interlayer had essentially no exchangeable acidity.

The exchangeable part of soil acidity corresponds approximately to the strong acid component (discussed in the section on buffer properties), whereas the nonexchangeable part corresponds to the weak acid component. Since the strong acid component is neutralized at a lower pH, partial neutralization of soil acidity reduces the exchangeable acidity/total acidity ratio. Ordinarily, soils or clays neutralized with $Ca(OH)_2$ to pH 5 to 5.5 have very little exchangeable acidity but may have large amounts of titratable acidity.

In soils containing appreciable organic matter (c.f. Fig. 13 & 14), the proportion of exchangeable acidity generally is low for two reasons. First, most of the acidity of organic matter is ionized only by raising the pH, i.e., it is truly a weakly ionized acid. Second, in acid soils much of the ionized CEC of organic matter is countered by Al, Fe, or both, which are not easy to exchange. These metals probably are present as hydroxy-metal-organic complexes that are difficult or impossible to replace (Hargrove & Thomas, 1981a, b; Schnitzer & Skinner, 1963a; Mortensen, 1963; White & Thomas, 1981). As with the clays, organic matter that is free of trivalent metal ions and their oxides has better-defined behavior than that which has $Fe^{3+}$ and $Al^{3+}$ compounds entrained in it. In acid soils, however, the latter situation prevails.

### B. Affinities of Cations for Acid Soils

The relative sorption affinities of ions in acid soils are determined, in part, by soil properties that have been discussed in other sections of this chapter. In montmorillonitic clay systems and exchange resins, the $H^+$ ion appears to be held with as little affinity as cations such as $Li^+$ and $Na^+$ (Coleman et al., 1959b; Foscolos, 1968). Hydrogen-saturated montmorillonite is dispersed and shows swelling characteristics similar to those of $Li^+$-montmorillonite. Hydrogen also appears to be held weakly by kaolinite, judging from titration curves of Coleman and Craig (1961).

In the micas, there is some disagreement about the place of $H^+$ in the lyotropic series. Rich (1964), for example, has shown that $H^+$ ions have the ability to scoop out $K^+$ ions in vermiculite, suggesting that their effective size must be about the same as $K^+$. Hydrogen monohydrate ($H_3O^+$) has this size. In the same vein, Jackson (1964) suggested that $H^+$ ions have replaced $K^+$ in weathered micas. If this is true, which is doubtful, $H^+$ ions must be held strongly, and the mineral must not decompose in their presence. In trioctahedral vermiculite, Foscolos (1968) found that $H^+$ is held more tightly than $Mg^{2+}$ or $Ca^{2+}$.

Many of the apparent anomalies concerning the position of $H^+$ ions in the lyotropic series have arisen largely because of failure to distinguish between $H^+$ and $Al^{3+}$. The half-life of $H^+$-clay (montmorillonite) is relatively short at room temperature—about 1000 h at 30°C (Coleman & Craig, 1961). It is probably much shorter for soil clays. The decomposition rate increases sharply at higher temperature, and it is doubtful that any experiments with clays have ever been carried out with pure $H^+$ systems. To compound the matter, most of the older work was done with clays that were treated in such a way that they were not fully $H^+$ saturated to begin with (Low, 1955). Thus, in many instances, apparent large adsorption affinities for $H^+$ have been due to decomposition of $H^+$-clay, with appearance of lattice cations, such as $Al^{3+}$, $Mg^{2+}$, and perhaps $Fe^{3+}$ (Thomas & Coleman, 1964). Of these, $Al^{3+}$ is most often predominant. Its triple charge makes it an extremely good competitor for cation adsorption sites on clays. Poorly crystalline soil clays that are coated with Al compounds probably are transformed from $H^+$ to $Al^{3+}$ saturation almost instantaneously, acting as $H^+$ ion "sinks."

At the same time, as shown by Kissel et al. (1971), the hydrolysis of $Al^{3+}$ causes the formation of $H^+$ ions during the extraction. As pointed out in Section IIIA, much of the so-called exchangeable $H^+$ is a result of such hydrolysis and really does not exist on the exchange sites of clays or organic matter.

Of course, there are sites in soils to which $H^+$ ions cling with great tenacity. Examples of these are the weaker carboxylic and all of the phenolic hydrogens on organic matter and truly weak-acid $H^+$ on clays and oxides. Jackson (1963a, 1964) suggests that water molecules in hydrous Al oxide interlayers can ionize $H^+$ and that $OH^-$ in the hydrous oxides can accept $H^+$.

Nye et al. (1961) and Clark and Turner (1965) have shown that monomeric $Al^{3+}$ ions sorbed on permanent charge sites of montmorillonite and other clay minerals participate normally in ion exchange reactions with other cations. Exchanges involving $Al^{3+}$ are described with fair precision by the equivalent fraction, ion exchange equations. The $Al^{3+}$ is quite strongly sorbed compared with $Ca^{2+}$ or $Na^+$, and the equilibrium is strongly dependent on total electrolyte concentration. At least a part of the dependence on electrolyte concentration may be the increased tendency of $Al^{3+}$ to hydrolyze at low ionic strength (McBride & Bloom, 1977).

In any soil or clay system, ion exchange reactions involving $Al^{3+}$ are complicated by the inevitable hydrolysis and readsorption reactions that occur (Ragland & Coleman, 1960; Jackson, 1963a). These may not seriously affect the results of short-term experiments, and ion exchange considerations lead to reasonable predictions of $Al^{3+}$ concentrations to be expected in soil solution of very acid soils. However, salt-exchangeable $Al^{3+}$ is scarce in surface soils, even acid ones, and too much reliance should not be put on an ion exchange treatment that may have limited applicability.

In montmorillonite that has been oversupplied with hydroxy-Al ions, Veith (1977) found that much of the hydroxy-Al was exchangeable by $Mg(NO_3)_2$. This was not observed with Al-saturated montmorillonite or vermiculite. Apparently, it occurs only when the montmorillonite spacing increases to about 2.0 nm, thus permitting freer exchange of the polymers.

The ion exchange behavior of $H^+$ in acid soils cannot be described in a meaningful way because of the instability of $H^+$-clay. However, one feature of $H^+$ ion adsorption is quite important. The $H^+$ ion is adsorbed very strongly by weakly acidic, functional groups whenever original soil pH is high. Hydrogen ions added to neutral or alkaline soil often are adsorbed quantitatively, with equivalent release of basic metal cations. This is essentially due to covalent bonding of the added $H^+$.

## V. SOIL pH

### A. General Aspects of Soil pH

The pH is perhaps the most commonly measured soil characteristic. Certainly it is the most widely used criterion for judging whether a soil is acid and, if so, how acid (Seatz & Peterson, 1964). The pH range for acid soils ordinarily is from around 4 to 7. Values much below 4 are obtained only when free acids are present (e.g., $H_2SO_4$ in cat clays and mine spoils). Values above 7 would indicate alkalinity rather than acidity, but it is clear that appreciable and sometimes very large amounts of "soil acidity" exist above pH 7.

Although soil pH is a very useful index, it is not completely understood and perhaps should be regarded as more empirical than otherwise. The pH of soil usually is measured in a slurry of soil and water. In routine procedures, as those used in soil testing laboratories, one part of soil is mixed with 1 or 2 parts of water and the electrodes are immersed in the stirred suspension.

The experimental factors affecting soil pH in the acid region generally are thought to be the soil/water ratio and the salt content of the soil water slurry. The effect of variable soil/water ratio is to yield lower pH values with larger concentrations of soil in the mixture. Differences of several tenths of a pH unit are observed in going from 1 part of the soil in 10 parts of water to a 1:1 mixture. This has been thought due to more concentrated soil acids in concentrated suspensions or perhaps to the development of a larger diffusion potential in more concentrated slurries. Associated with the soil/water ratio, and with the same cause, is the suspension effect. This refers to the fact that when soil particles settle from an unstirred suspension, the pH measured by inserting electrodes into the layer of sediment is different from (and usually lower than) that measured in the supernatant liquid. The magnitude of the suspension effect depends on the nature of the soil and the salt concentration. High CEC, loosely bound exchangeable ions, and low electrolyte concentrations favor large $\Delta$pH.

The main effect of salt concentration on soil pH is not through its influence on the suspension effect but rather through cation exchange. Soil pH is lower in salt solutions than in water and goes down as salt concentration increases because of the displacement of $H^+$ and $Al^{3+}$ ions from exchange sites.

In extremely acid soils with appreciable amounts of exchangeable monomeric $Al^{3+}$, the displacement of that ion and its subsequent hydrolysis probably is responsible for pH lowering by salts. Often the pH of an acid soil–salt solution mixture is lower than can be accounted for by the $Al^{3+}$ concentration of the soil solution, if we assume $Al^{3+}$ hydrolysis according to Schofield and Taylor (1954), Frink and Peech (1963a), or Jackson (1963a). In some cases, at least, this is due to an increased hydrolysis of Al resulting from the sorption of hydroxy-Al by clay. Data bearing on this and obtained by Ragland and Coleman (1960) and Frink and Peech (1963b) are the same, although the two sets of authors did not agree on the semantics. Ragland and Coleman (1960) contended that Al-montmorillonite suspensions in dilute $AlCl_3$ solutions had pHs lower than $AlCl_3$ of corresponding concentration because increased hydrolysis of Al in the presence of clay had resulted from the sorption of hydroxy-Al. Frink and Peech (1963b) agreed on the lower pH for Al clay in $AlCl_3$ but argued that Al hydrolysis was less in the more acid solution because hydrolysis is reversed by increasing concentrations of $H^+$ ions.

In soils with salt-free pHs of 5 or above, the reduction of pH caused by the addition of salt often is not accompanied by the appearance of measurable quantities of $Al^{3+}$ in solution (Coleman et al., 1959b). The $H^+$ ions responsible for the lower pH may come from organic matter or from hydrolysis of hydroxy-Al or hydroxy-Fe in interlayers or on surfaces or edges of layer silicates. In some soils, allophane or other amorphous aluminosilicates may supply $H^+$ to salt solutions.

Some soils heavily coated or interlayered with sesquioxides may undergo only small reductions in pH on the addition of salts, and, in some cases, the pH may even increase (Mattson, 1931; Mattson & Wiklander, 1940a). This is especially so for sulfate and phosphate salts. This behavior is readily explicable, in a qualitative sense, according to the ideas developed earlier. Salts can exchange $Al^{3+}$ and increase $H^+$ through increased hydrolysis, whereas the anion exchanges $OH^-$ from sesquioxides. Aluminum or $H^+$ exchange by salts in soils of temperate regions ordinarily greatly exceeds $OH^-$ exchange, and pH falls. In many Oxisols and Ultisols, however, the reverse is true; an excess of $OH^-$ is released, and the pH rises.

### B. Suspension Effect

Perhaps the most uncertain aspect of pH measurement of soil has to do with suspension effect and thus with the definition of pH in terms of $H^+$ ion activity. Uncertainty along these lines relates more to abstract than practical questions, although interpretations of soil pH effects on solubilities of metal hydroxides and $H_2PO_4^-$ compounds often are difficult because one does not know what measured pH really means in terms of $H^+$ ion activity.

The suspension effect as an experimental fact is not questioned by anyone, but its cause and the implications have been given widely different interpretations. The first, which has been supported most strongly by Marshall (1964), is based on the premise that exchangeable $H^+$ ions of clay ionize to a certain extent and the glass electrode responds to the ionized $H^+$. According to this premise, pH should decrease as the concentration of soil or clay in a suspension rises, because the concentration of ionized $H^+$ also goes up. Among the many implications of this view, one of the most important is that for systems consisting of sedimented clay and supernatant solution, or clay suspension separated from an equilibrium dialysate by a membrane, there is an electrical potential difference across the boundary separating clay from the clear solution. Marshall (1964) regards this as a Donnan potential, similar to that purported to occur when mixtures of diffusible and nondiffusible electrolytes are dialyzed against water.

A second explanation that has been advanced to explain the suspension effect (and the Donnan potential) supposes that a paste or suspension of clay or a layer of sedimented soil acts as a "leaky" permselective membrane and imposes different mobilities, and thus different transport numbers, on cations and anions (Coleman et al., 1951). It is an experimental fact that when dense, compact membranes prepared from cation exchange resins separate single salt electrolyte solutions of different concentration, a potential difference develops across the membrane such that

$$E = -(RT/F) \ln (a_1/a_2)$$

where $a_1$ and $a_2$ are cation (or anion) activities in the two solutions. If the membrane matrix is negatively charged, e.g., a cation exchanger, the more concentrated solution is negative. If the membrane is made from an anion exchanger, the more concentrated solution is negative. If the membrane is made from an anion exchanger, the more concentrated solution is positive. The potential difference across the membrane is thought to develop because only ions of charge opposite to that of the membrane can pass through, which they do under the concentration gradient until the electrical potential difference balances the tendency toward diffusion.

Many charged membranes are not perfectly permselective. They allow the passage of both cations and anions in response to a concentration gradient but restrict one more than the other. Such a leaky membrane might be visualized as a negatively charged matrix, with positively charged counter ions and moderately large pores containing both cations and anions. When such a membrane separates salt solutions of different concentrations, cations and anions can diffuse from the concentrated to the dilute side through the pores. Cations also can diffuse along the negatively charged framework and through pores so small that they exclude anions. Two pathways for cations, as opposed to one for anions, result in a larger transference number for the positive ion and a membrane potential given by

$$E = -(t_+ - t_-)(RT/F) \ln (a_1/a_2)$$

where $t_+$ and $t_-$ are transference numbers for cations and anions, respectively.

A paste or suspension of clay might be expected to behave as a leaky membrane, and the junction between the KCl salt bridge of the calomel electrode and the dilute interparticle solution certainly is a region of concentration gradient. In the case of negatively charged soil or clay, any junction potential that results from its leaky membrane character is of such a sign as to make a measured pH lower than would correspond to the $H^+$ ion activity in the interparticle solution. Large junction potentials are expected for (i) concentrated suspensions or pastes, (ii) high exchange capacity materials, (iii) monovalent exchangeable ions, and (iv) dilute interparticle solutions.

Marshall (1964) has pointed out that one cannot decide from direct evidence whether the suspension effect is due to a junction potential or to Donnan potential. However, Olsen and Robbins (1971) showed that in a Ag–AgCl–calomel–resin–Cl system, both $Cl^-$ and $H^+$ activities differed by $10^4$ between resin slurry and solution, whereas both freezing point and osmotic measurements showed negligible differences in activity.

Van Olphen (1963) discusses the interpretation of pH measurement in colloidal systems, and Overbeek (1953) has developed an extensive theory attributing the Donnan potential to transport phenomena, i.e., differences between mobilities of cations and anions in systems containing large, charged, particles such as clays. Coleman et al. (1951) and Bower (1961) have obtained data that support Overbeek's theoretical conclusions.

A matter related to the suspension effect, but less accessible to direct measurement, concerns the pH at the surfaces of soil particles. Frink and Peech (1963b) and Swoboda and Kunze (1968) point out that the $H^+$ ion concentration at the particle surface must be greater than the concentration in bulk solution, and thus surface pH is lower than bulk pH. Frink and Peech (1963b) used the Boltzmann equation, with an assumed double-layer potential, to calculate surface $H^+$ ion concentrations. According to such considerations, surface $H^+$ ion concentrations may be several orders of magnitude higher than those in bulk solution when the surface charge density is large and the electrolyte concentration is small. The work of Swoboda and Kunze (1968) was based on ionization of weak organic bases by montmorillonite. They found Otay montmorillonite to have stronger acidity on the surface than Wyoming montmorillonite, which, in turn was stronger than hectorite. No such measurements have been made on soils.

### C. Lime Potential: pH − ½pCa

To avoid possible uncertainties due to the suspension effect and to obtain a pH index that was independent of electrolyte concentration, Schofield and Taylor (1955) proposed that pH be measured in the clear supernatant solution after soil mixed with $0.01M$ $CaCl_2$ had settled. They suggested that results be expressed not as pH but as pH − ½pCa where pCa is the negative logarithm of $Ca^{2+}$ ion molar activity and the entire quantity in brackets is called the *lime potential* because it corresponds to the mean activity of $Ca(OH)_2$. It was claimed that the lime potential should be independent of salt concentration, at least within the limits usually found in soil.

Chao and Harward (1962) and Turner and co-workers (Turner & Nichol, 1962a, b; Turner et al., 1963; Turner, 1965) have taken a close look at the lime potential in partly neutralized Al-clays. They have determined that pH $-$ ½pCa does vary with electrolyte concentration but not widely, and that it appears to be a reasonably good index of $Ca^{2+}$ saturation of the permanent charge.

Turner and co-workers (Turner & Nichol, 1962a, b; Turner et al., 1963; Turner, 1965) supposed that two reactions occur when $Ca(OH)_2$ is added to Al-clay ($AlX_3$):

1) $2AlX_3 + 3Ca^{2+} \rightarrow 3CaX_2 + 2Al^{3+}$.
2) $Al^{3+} + 3OH^- \rightarrow Al(OH)_3$.

Combining the ion exchange equation with the solubility product expression and using measured values for the equilibrium constants led to an equation that expressed the lime potential, pH $-$ ½pCa, as a function of the $Ca^{2+}$ saturation. The Turner-Nichol approach (1962a, b) appears quite satisfactory for generalizing the neutralization behavior of Al-montmorillonite. Unfortunately, their method is not directly applicable to those acid soils where neutralization with $Ca(OH)_2$ involves more than the displacement of $Al^{3+}$ by $Ca^{2+}$ and its precipitation as $Al(OH)_3$.

Soil pH also is sometimes measured in concentrated salt solutions as 1*M* KCl. This removes any uncertainty regarding junction potential and, in addition, gives a pH that indicates something about the exchange acidity.

## VI. NEUTRALIZATION OF SOIL ACIDITY

### A. Mechanisms and Products of Neutralization

The mechanisms through which $CaCO_3$ reacts with acid soils are complex. The rates of neutralization and final reaction products are not known with certainty, although the influence of factors such as the amount of soil acidity and the particle size and reactivity of limestone have been studied extensively (Coleman et al., 1959a).

In water, $CaCO_3$ dissolves and hydrolyzes to form $OH^-$ ions as follows:

$$CaCO_3 + H_2O \rightarrow Ca^{2+} + HCO_3^- + OH^-.$$

The $OH^-$ ions formed subsequently react with $H^+$ ions formed from hydrolysis of $Al^{3+}$ and with exchangeable $Al^{3+}$. The generalized overall reaction of lime with an acid soil may be written as

$$2\ \text{Al-soil} + 3CaCO_3 + 3H_2O \rightarrow 3\ \text{Ca-soil} + 2Al(OH)_3 + 3CO_2.$$

The products of a completed liming reaction are exchangeable $Ca^{2+}$ (and $Mg^{2+}$) and $Al(OH)_3$ [and $Fe(OH)_3$]. In such an idealized case, the soil pH is about 8.3, and complete base saturation is achieved. Under most conditions, however, this situation is not achieved, since soils commonly are limed to pHs no higher than 6.5. When acid soil has reacted with $CaCO_3$ to reach pH 6.5, much of the Ca or Mg (or both) from the limestone has become exchangeable and soil acidity has been neutralized, but large amounts of titratable acidity and often unreacted limestone remain. Mehlich (1942, 1943) found that North Carolina soils were from 40 to 90% base saturated at a pH of 6.5. Similar work with Virginia soils suggests values in the neighborhood of 45 to 85% base saturation at pH 6.5. Moschler et al. (1962) found an increase of 20 to 30% in base saturation for each 1.0 unit of pH increase in field experiments. None of the soils that they studied could be expected to be over 65% base saturated at pH 6.5.

The compounds that make up the titratable acidity in partly neutralized soil are the hydroxy compounds of Al and Fe and the unionized groups on organic matter and clay. In mineral soils, the most important of these products are the hydroxy-Al and hydroxy-Fe compounds of the general type $[R(OH)_x]_n$, where $x$ is $< 3$.

## B. Rate of Reaction of Liming Materials with Soil

The rate of reaction of liming materials with soil varies considerably. A discussion of some of the factors that influence rate of neutralization follows.

### 1. SOURCE OF LIMING MATERIAL

Although the bulk of agricultural lime comes from ground limestone, many other materials are being used to neutralize soil acidity, including ground marl and chalk, slag from iron and steel making ("basic slag"), flue dust from cement plants, and refuse from sugar beet factories, paper mills, and Ca-carbide plants (Adams, 1981). The effectiveness of these materials as liming agents depends largely on their $CaCO_3$ equivalence or the amount of pure $CaCO_3$ required to equal the reactiveness of 100 g of the material.

### 2. SOLUBILITY OF LIMING MATERIAL

The rate of neutralization is strongly dependent on the rate of dissolution and hydrolysis to form $OH^-$ ions. In general, calcitic limestone is more soluble than dolomitic limestone.

### 3. RATE OF REMOVAL OF $OH^-$ IONS

The rate of hydrolysis of $CaCO_3$ is directly related to the rate at which the $OH^-$ ions are removed from solution. As long as sufficient $H^+$ ions are in solution, the number of $Ca^{2+}$ and $HCO_3^-$ ions will increase. When the $H^+$ ion concentration is lowered, the solubility of $Ca^{2+}$ and $HCO_3^-$ is reduced.

### 4. RATE OF HYDROLYSIS OF $Al^{3+}$ AND $Fe^{3+}$

In acid soils, the concentration of $H^+$ ions in solution is a function of the hydrolysis rate of $Al^{3+}$ or hydroxy-Al or hydroxy-Fe, whether these ions be adsorbed by clay or organic matter. Thus, rates of reaction with lime are influenced by factors that determine the rate and extent of polyvalent metal ion hydrolysis.

Hydrolysis of $Al^{3+}$ and hydroxy-Al in soil apparently is increased by salts (Coleman et al., 1964; Kissel et al., 1971), and rates of dissolution of $CaCO_3$ should be increased in the same way. Conversely, the strong adsorption of intermediate reaction compounds, e.g., $[Al(OH)_2^+]_n$, reduces the hydrolysis, and the neutralization reaction requires much longer to complete. Data of Moschler et al. (1962) for a number of Virginia soils that adsorb hydroxy-Al in large amounts suggest that the pH of soils limed with 18 t of dolomitic limestone per hectare was still increasing 5 yr after lime was applied. In no case did the pH rise much above 7.0, and appreciable titratable acidity remained.

## 5. PARTIAL PRESSURE OF $CO_2$

The partial pressure of $CO_2$ influences the solubility of $CaCO_3$. Although the partial pressure of $CO_2$ is relatively constant in the atmosphere, wide fluctuations can occur in the soil atmosphere due to microbial activity.

## 6. PARTICLE SIZE

Because liming materials are generally only sparingly soluble, the rate and extent of dissolution in the soil are dependent on the amount of surface area in contact with the soil solution. The minimum fineness for optimum dissolution of ground limestone appears to be the 0.25- to 0.30-mm size (50- to 60-mesh) (Meyer & Volk, 1952).

In this connection, Bertsch (1980) found that a 0.25- to 0.425-mm size (40- to 60-mesh) lime reacted more slowly with exchangeable $Al^{3+}$ than did a 0.045- to 0.075-mm size (200- to 325-mesh) limestone. With hydroxy-Al, however, the rates of reaction (on soils, resin, and clay) were exactly the same. These results suggest that the rate-limiting factor was further hydrolysis of hydroxy-Al rather than the fineness of the ground limestone below a 0.25- to 0.425-mm size (40- to 60-mesh).

## 7. SOIL MIXING

Due to a combination of the factors already discussed, neutralization rates generally increase with incorporation and mixing with soil in the plow layer of field soils.

## 8. ENVIRONMENTAL FACTORS

Temperature and moisture, of course, influence the rate of $CaCO_3$ dissolution and thereby the rate of neutralization of soil acidity. Large amounts of water are also needed to move $Ca^{2+}$ and $HCO_3^-$ down into acid subsoils.

## VII. MEASUREMENT OF SOIL ACIDITY AND LIME REQUIREMENT

Amounts of soil acidity are most often measured in one of the following ways:

1) Titration with base or equilibration with lime.
2) Leaching with a buffered solution, followed by titration of the leachate.
3) Subtracting the sum of exchangeable bases from CEC.
4) Equilibration with a buffered solution and estimation of acidity by pH change.
5) Extraction with an unbuffered neutral salt.

The best known titration procedure is the one attributed to Bradfield and Allison (1933), where increments of $Ca(OH)_2$ are added to soil suspensions and then aerated to precipitate excess $Ca(OH)_2$ as $CaCO_3$. The titration curve levels off at pH 8.2, with the intersection of the ascending and horizontal curves corresponding to $Ca^{2+}$ saturation, and the amount of $Ca(OH)_2$ added to reach that condition being equilvalent to the soil acidity. Potentiometric titrations of soil with NaOH, $Ca(OH)_2$, or $Ba(OH)_2$ in the absence or presence of salt solutions have been used to determine soil acidity, but due to the slowness of the neutralization reaction near neutrality and the possibility of side reactions, such procedures have had limited practical use. However, titration with alkaline earth cations usually gives amounts of soil acidity that correspond closely to those measured by other methods, as does titration with NaOH as long as the soil is suspended in a salt solution (Kamprath & Welch, 1962; Turner & Nichol, 1962a, b; Lin & Coleman, 1960). Conductimetric titrations also can be used to determine the amount of soil acidity (Coleman & Thomas, 1964; Kissel & Thomas, 1969; Dewan & Rich, 1970).

Leaching methods that employ buffer solutions are widely used to measure what still is called exchangeable $H^+$ but that really includes the many component parts of soil acidity. Perhaps the most widely used is the $BaCl_2$–TEA method of Mehlich (1948) or one of its modifications. In this method, 0.1 to 0.3*M* $BaCl_2$, containing half-neutralized 0.2*M* TEA (pH 8.2) is leached through appropriate amounts of soil. The amount of HCl required to titrate the leachate to a methyl red–bromcresol green end point is subtracted from that needed to similarly neutralize an aliquot of the $BaCl_2$–TEA solution. A modification of the procedure, in which soil is leached first with a neutral salt solution to remove exchangeable $Al^{3+}$ (which can be measured as salt-exchangeable acidity) and then with $BaCl_2$–TEA, has been used to separate soil acidity into exchangeable and weak-acid components (Coleman et al., 1959a; Bhumbla & McLean, 1965).

Calcium acetate has been used to distinguish between carboxyl and weaker acidic groups of soil organic matter. This application is based on the fact that HOAc will be liberated quantitatively from $Ca(OAc)_2$ through reaction with acidic groups stronger than HOAc and will not be liberated at all by weakly acidic groups with $pK_a > 7$. Schnitzer and Gupta (1965) have successfully applied the method to determining the carboxyl content of some of their humic and fulvic acid samples.

Procedures for exchange acidity that involve titration to pH 8 or leaching with pH 8 buffer solution measure values that correspond to the Bradfield (1942) Ca-clay-$CaCO_3$-$CO_2$ equilibrium at full $Ca^{2+}$ saturation. Exchangeable monomeric $Al^{3+}$, salt-exchangeable $H^+$, weak-acid groups of organic matter, and hydrolyzable sesquioxides associated with clay mineral or organic matter are all measured together. Neutral salt, leaching solutions displace exchangeable, monomeric $Al^{3+}$ and perhaps some $H^+$ from organic matter (Lin & Coleman, 1960; Bhumbla & McLean, 1965), although it appears likely that the apparent $H^+$ found in some instances comes from hydrolysis of $Al^{3+}$.

Subtracting exchangeable bases from CEC gives precisely the same values for exchangeable acidity as do corresponding leaching methods. Pratt and Bair (1962) have shown conclusively that exchangeable acidity measured with $BaCl_2$–TEA equals CEC with $BaCl_2$–TEA minus basic metal cations displaced by $BaCl_2$–TEA. Of course, for this correspondence to occur, one must use a CEC method that is comparable with the one used to determine exchangeable acidity. For example, exchangeable acidity measured with $BaCl_2$–TEA is not equal to the difference between CEC with $NH_4OAc$ and the bases exchanged.

A number of methods for assessing soil acidity in trying to arrive at a lime requirement are based on equilibrating soil with the buffer solution. Seatz and Peterson (1964) summarized work on this topic, including the early contributions of Schofield and Woodruff. Schofield (1933) proposed a lime requirement determination that involved equilibrating a soil sample overnight in *p*-nitrophenol neutralized to pH 7 with $Ca(OH)_2$ and titrating to determine the amount of $OH^-$ consumed by reaction with the acid clay. A number of adaptations of the method have used pH measurement rather than titration to measure reaction between soil and buffer solution. The Woodruff method (Woodruff, 1948) employes a mixture of CaOAc-*p*-nitrophenol–MgO that gives a linear buffer curve between pH 6 and 7. The mixture is adjusted at pH 7, and each 0.2 mmol of $H^+$ added causes the pH to drop 0.1 unit in that range. Appropriate samples of soils are added, and the drop in pH of the buffer-soil mixture below 7 indicates the amount of soil acidity that has reacted with the buffer. Shoemaker et al. (1961) proposed a similar method but with a different composition of buffer, which is claimed to work better in soils containing much clay-sorbed hydroxy-Al. Keeney and Corey (1963) find the Shoemaker et al. (1961) buffer useful for Wisconsin soils. It is now used very widely in the United States. Adams and Evans (1962) devised a buffer method that appears to work well for Ultisols.

All buffer methods suffer from the obvious disadvantage that the pH at which soil acidity is determined is not constant but varies between pH 6 and 7, depending on the quantity of soil and the amount of acidity. Whenever weak-acid character exists in the pH range of 6 to 7, and it usually does, results for different samples are not comparable. Of course, in a practical way, the recommendations for lime cannot be made that closely, anyway, so that inaccuracies in the method are not likely to cause major problems.

Although Vermont is the only state that uses the method, it is widely accepted that $Al^{3+}$ and $H^+$ extracted by 1*M* KCl gives a reasonably accurate estimate of the soil acidity that must be neutralized to obtain normal plant growth (Kamprath, 1970; Moschler et al., 1960; Reeve & Sumner, 1970; Farina et al., 1980). The pH achieved by liming to rid the soil of exchangeable $Al^{3+}$ varies considerably between soils, being much lower where organic matter is high (Evans & Kamprath, 1970; Thomas, 1975) and usually around pH 5.5 for predominantly mineral soils (c.f. Fig. 13 for topsoils and Fig. 14 for subsoils).

Kamprath's proposal that removal of $Al^{3+}$ from exchange sites gives maximum yields was strongly challenged by McLean (1970) on the grounds that it ignores other benefits of liming in addition to removal of toxic $Al^{3+}$. Although toxicity caused by $Al^{3+}$ is usually the most important problem with acid soils, the presence of excess $Mn^{2+}$ and a deficiency of Mo can also be powerful factors in yield reduction with some soils and crops.

This is not really a puzzle. Soil testing laboratories must have rapid methods suitable for handling large numbers of samples. Extracting and titrating is much too demanding. Recommended pHs should correspond to nontoxic $Al^{3+}$ saturation levels.

Most lime requirement methods emphasize attaining a certain pH level rather than taking care of a particular problem. Perhaps the rationale for this is the feeling that a "good" soil has a pH of between 6 and 7. It is true that in this pH range problems do not exist because of soil acidity, but in many cases, particularly in Oxisols and Ultisols, it appears to be an unnecessary luxury and can result in lower crop yields (Farina et al., 1980).

In many ways, the uncertainties attendant to measuring lime requirement reflect the lack of accurate information on the nature of the acid soil system. Knowledge about acid soil has grown considerably in recent years. This growth of knowledge should lead to an increased capability to identify problems in the field and to design remedial measures.

## REFERENCES

Adams, F. 1981. Alleviating chemical toxicities: Liming acid soils. p. 269–301. *In* G. F. Arkin and H. M. Taylor (ed.) Modifying the root environment to reduce crop stress. ASAE Monogr. no. 4, Am. Soc. of Agric. Engr., St. Joseph, Mich.

Adams, F., and C. E. Evans. 1962. A rapid method for measuring lime requirement of Red-Yellow Podzolic soils. Soil Sci. Soc. Am. Proc. 26:355–357.

Adams, F., and Z. Rawajfih. 1977. Basaluminite and alunite: A possible cause of sulfate retention by acid soils. Soil Sci. Soc. Am. J. 41:686–692.

Allison, F. E. 1973. Soil organic matter and its role in crop production. Elsevier North-Holland, Inc., New York.

Amedee, G., and M. Peech. 1976. The significance of KCl-extractable Al(III) as an index to lime requirement of soils of the humid tropics. Soil Sci. 121:227–233.

Aomine, S., and M. L. Jackson. 1959. Allophane determination in Ando soils by cation-exchange delta value. Soil Sci. Soc. Am. Proc. 23:210–214.

Barnhisel, R. I. 1977. Chlorites and hydroxy interlayered vermiculite and smectite. p. 331–356. *In* J. B. Dixon and S. B. Weed (ed.) Minerals in soil environments. Soil Sci. Soc. of Am., Madison, Wis.

Barnhisel, R. I., and C. I. Rich. 1963. Gibbsite formation from aluminum-interlayers in montmorillonite. Soil Sci. Soc. Am. Proc. 27:632–635.

Barshad, I. 1960. The effect of the total chemical composition and crystal structure of soil minerals on the nature of the exchangeable cations in acidified clays and in naturally occurring acid soils. Int. Congr. Soil Sci., Trans. 7th (Madison, Wis.) 2:435–444.

Berg, W. A., and G. W. Thomas. 1959. Anion elution patterns from soils and clays. Soil Sci. Soc. Am. Proc. 23:348–350.

Bertsch, P. M. 1980. Lime-aluminum interactions and phosphorus relations with selected Coastal Plain soils. M.S. Thesis. Virginia Polytechnic Inst. and State Univ., Blacksburg.

Bertsch, P. M., and M. M. Alley. 1981. Conventional and suspension limestone influence on soil chemical properties and corn and soybean yields. Agron. J. 73:1075–1078.

Bhumbla, D. R., and E. O. McLean. 1965. Aluminum in soils: VI. Changes in pH-dependent acidity, cation-exchange capacity, and extractable aluminum with additions of lime to acid surface soils. Soil Sci. Soc. Am. Proc. 29:370–374.

Birrell, K. S. 1958. Reactions of amorphous soil colloids with ions in solution. N.Z. Soc. Soil Sci. Proc. 3:11–13.

Birrell, K. S. 1961. The adsorption of cations from solution by allophane in relation to their effective size. J. Soil Sci. 12:305–316.

Birrell, K. S., and M. Gradwell. 1956. Ion-exchange phenomena in some soils containing amorphous mineral constituents. J. Soil Sci. 7:130–147.

Blevins, R. L., G. W. Thomas, and P. L. Cornelius. 1977. Influence of no-tillage and nitrogen fertilization on certain soil properties after 5 years of continuous corn. Agron. J. 69: 383–386.

Bloom, P. R. 1979. Titration behavior of aluminum organic matter. Soil Sci. Soc. Am. J. 43: 815–817.

Bloom, P. R., M. B. McBride, and R. M. Weaver. 1979. Aluminum organic matter in soils: Salt-extractable aluminum. Soil Sci. Soc. Am. J. 43:813–815.

Bower, C. A. 1961. Studies on the suspension effect with a sodium electrode. Soil Sci. Soc. Am. Proc. 25:18–20.

Bradfield, R. 1923. The nature of the acidity of collodial clay of acid soils. J. Am. Chem. Soc. 45:2669–2678.

Bradfield, R. 1925. The chemical nature of collodial clay. J. Am. Soc. Agron. 17:253–270.

Bradfield, R. 1942. Calcium in the soil: I. Physico-chemical relations. Soil Sci. Soc. Am. Proc. 6:8–15.

Bradfield, R., and W. B. Allison. 1933. Criteria of base saturation in soils. Int. Soc. Soil Sci., Trans. Comm. II (Copenhagen) A:63–79.

Broadbent, F. E., and G. R. Bradford. 1952. Cation-exchange groupings in the soil organic fraction. Soil Sci. 74:447–457.

Brosset, C., G. Biedermann, and L. G. Sillen. 1954. Studies on the hydrolysis of metal ions: XI. The aluminum ion, $Al^{3+}$. Acta Chem. Scand. 8:1917–1926.

Chang, M. L., and G. W. Thomas. 1963. A suggested mechanism for sulfate adsorption by soils. Soil Sci. Soc. Am. Proc. 27:281–283.

Chao, T. T., and M. E. Harward. 1962. Nature of acid clays and relationships to ion activities and ion ratios in equilibrium solutions. Soil Sci. 93:246–253.

Chao, T. T., M. E. Harward, and S. C. Fang. 1964. Iron or aluminum coatings in relation to sulfate adsorption characteristics of soils. Soil Sci. Soc. Am. Proc. 28:632–635.

Chernov, V. A. 1959. On the genesis of the exchangeable aluminum in soils. Soviet Soil Sci. 113:25–33.

Chernov, V. A. 1964. The nature of soil acidity (English translation furnished by Dr. H. Jenny, Univ. of California, Berkeley; additional editing by D. T. Kubach). Soil Sci. Soc. of Am., Madison, Wis.

Chernov, V. A., N. F. Beliaeva, and V. C. Maximova. 1956. Rate of replacement of hydrogen ion by aluminum ion in clays. Sov. Soil Sci. (Engl. Transl.) 110:849–851.

Clark, J. S., and R. C. Turner. 1965. Extraction of exchangeable cations and distribution constants for ion exchange. Soil Sci. Soc. Am. Proc. 29:271–274.

Coleman, N. T., and D. Craig. 1961. The spontaneous alteration of hydrogen clay. Soil Sci. 91:14–18.

Coleman, N. T., and M. E. Harward. 1953. The heats of neutralization of acid clays and cation exchange resins. J. Am. Chem. Soc. 75:6045–6046.

Coleman, N. T., E. J. Kamprath, and S. B. Weed. 1959a. Liming. Adv. Agron. 10:475–522.

Coleman, N. T., and A. Mehlich. 1948. Some chemical properties of soils as related to their cation exchange-anion exchange ratios. Soil Sci. Soc. Am. Proc. 13:175–178.

Coleman, N. T., and G. W. Thomas. 1964. Buffer curves of acid clays as affected by the presence of ferric iron and aluminum. Soil Sci. Soc. Am. Proc. 28:187–190.

Coleman, N. T., and G. W. Thomas. 1967. The basic chemistry of soil acidity. *In* R. W. Pearson and F. Adams (ed.) Soil acidity and liming. Agronomy 12:1–41. Am. Soc. of Agron., Madison, Wis.

Coleman, N. T., G. W. Thomas, F. H. LeRoux, and G. Bredell. 1964. Salt-exchangeable and titratable acidity in bentonite–sesquioxide mixtures. Soil Sci. Soc. Am. Proc. 28:35–37.

Coleman, N. T., S. B. Weed, and R. J. McCracken. 1959b. Cation-exchange capacity and exchangeable cations in Piedmont soils of North Carolina. Soil Sci. Soc. Am. Proc. 23: 146–149.

Coleman, N. T., D. E. Williams, T. R. Nielsen, and H. Jenney. 1951. On the validity of interpretations of potentiometrically measured soil pH. Soil Sci. Soc. Am. Proc. 15:106–110.

Cowling, E. B. 1980. Acid precipitation and its effects on terrestrial and aquatic ecosystems. *In* T. J. Kneip and P. J. Lioy (ed.) Aerosols: Anthropogenic and natural, sources and transport. Ann. New York Acad. Sci. 338:540–555.

Dewan, H. C., and C. I. Rich. 1970. Titration of acid soils. Soil Sci. Soc. Am. Proc. 34:38–44.

Dixon, J. B. and M. L. Jackson. 1962. Properties of intergradient chlorite-expansible layer silicates of soils. Soil Sci. Soc. Am. Proc. 26:358–362.

Dixon, J. B., and S. B. Weed (ed.) 1977. Minerals in soil environments. Soil Sci. Soc. of Am., Madison, Wis.

Ensminger, L. E. 1954. Some factors affecting the adsorption of sulfate by Alabama soils. Soil Sci. Soc. Am. Proc. 18:259–264.

Evans, C. E., and E. J. Kamprath. 1970. Lime response as related to percent Al saturation, solution Al and organic matter response. Soil Sci. Soc. Am. Proc. 34:893–896.

Farina, M. P. W., M. E. Sumner, C. O. Plank, and W. S. Letzsch. 1980. Exchangeable aluminum and pH as indicators of lime requirement for corn. Soil Sci. Soc. Am. J. 44: 1036–1041.

Flaig, W., H. Beutalspacher, and E. Rietz. 1975. Chemical composition and physical properties of humic substances. *In* J. E. Gieseking (ed.) Soil components, Vol. 1. Organic components. Springer-Verlag New York, Inc., New York.

Foscolos, A. E. 1968. Cation-exchange equilibrium constants of aluminum-saturated montmorillonite and vermiculite clays. Soil Sci. Soc. Am. Proc. 32:350–354.

Frink, C. R., and M. Peech. 1963a. Hydrolysis of the aluminum ion in dilute aqueous solutions. Inorg. Chem. 2:473–478.

Frink, C. R., and M. Peech. 1963b. Hydrolysis and exchange reactions of the aluminum ion in hectorite and montmorillonite suspensions. Soil Sci. Soc. Am. Proc. 27:527–530.

Fripiat, J. J. 1964. Surface properties of alumino-silicates. Clays Clay Miner. 12:327–358.

Gillam, W. S. 1940. Study on the chemical nature of humic acid. Soil Sci. 49:433–453.

Gillman, G. P., and R. L. Fox. 1980. Increases in the cation exchange capacity of variable charge soils following superphosphate applications. Soil Sci. Soc. Am. J. 44:934–938.

Gilmour, J. T., and N. T. Coleman. 1971. Buffer curves and acid strength of humic acid and two resins. Soil Sci. Soc. Am. Proc. 35:710–714.

Goates, J. R., and K. Anderson. 1956. Acidic properties of quartz. Soil Sci. 81:277–282.

Griffith, S. M., and M. Schnitzer. 1975. The isolation and characterization of stable metal-organic complexes from tropical volcanic soils. Soil Sci. 120:126–131.

Grove, J. H., C. S. Fowler, and M. E. Sumner. 1982. Determination of the charge character of selected acid soils. Soil Sci. Soc. Am. J. 46:32–38.

Hanna, W. J., and J. F. Reed. 1948. A comparison of ammonium acetate and buffered barium chloride methods for determining cation-exchange capacity of limed soils. Soil Sci. 66: 447–458.

Hargrove, W. L. 1980. Reactions of aluminum with soil organic matter. Ph.D. Diss. Univ. of Kentucky, Lexington.

Hargrove, W. L., and G. W. Thomas. 1981a. Extraction of aluminum from aluminum-organic matter complexes. Soil Sci. Soc. Am. J. 45:151–153.

Hargrove, W. L., and G. W. Thomas. 1981b. Effect of organic matter on exchangeable aluminum and plant growth in acid soils. p. 151–166. *In* R. H. Dowdy (ed.) Chemistry in the soil environment. ASA Spec. Pub. 40, Am. Soc. of Agron., Madison, Wis.

Hargrove, W. L., and G. W. Thomas. 1982a. Titration properties of Al-organic matter. Soil Sci. 134:216–225.

Hargrove, W. L., and G. W. Thomas. 1982b. Conditional formation constants for aluminum-organic matter complexes. Can. J. Soil Sci. 62:571–575.

Hartwell, R. L., and F. R. Pember. 1918. The presence of aluminum as a reason for the difference of so-called acid soil on barley and rye. Soil Sci. 6:259–279.

Harward, M. E., and N. T. Coleman. 1954. Some properties of H- and Al-clays and exchange resins. Soil Sci. 79:181–188.

Helling, C. S., G. Chesters, and R. B. Corey. 1964. Contributions of organic matter and clay to soil cation-exchange capacity as affected by the pH of the saturating solution. Soil Sci. Soc. Am. Proc. 28:517–520.

Hendricks, S. B., and W. H. Fry. 1930. The results of x-ray and mineralogical examination of soil colloids. Soil Sci. 29:457–476.

Hsu, P. H., and T. F. Bates. 1964a. Formation of x-ray amorphous and crystalline aluminum "hydroxides." Mineral. Mag. 33:749–768.

Hsu, P. H., and T. F. Bates. 1964b. Fixation of hydroxy-aluminum polymers by vermiculite. Soil Sci. Soc. Am. Proc. 28:763–769.

Hsu, P. H., and C. I. Rich. 1960. Aluminum fixation in a synthetic cation exchanger. Soil Sci. Soc. Am. Proc. 24:21–25.

Huheey, J. E. 1978. Inorganic chemistry: Principles of structure and reactivity. Harper & Row, Publishers, Inc., New York.

Jackson, M. L. 1963a. Aluminum bonding in soils: A unifying principle in soil science. Soil Sci. Soc. Am. Proc. 37:1–10.

Jackson, M. L. 1963b. Interlayering of expansible layer silicates in soils by chemical weathering. Clays Clay Miner. 11:29–46.

Jackson, M. L. 1964. Chemical composition of soils. p. 71–141. *In* F. E. Bear (ed.) Chemistry of the soil. 2nd ed. Van Nostrand Reinhold Co., New York.

Jenny, H. 1961. Reflections on the soil acidity merry-go-round. Soil Sci. Soc. Am. Proc. 25: 428–432.

Johnson, S. W. 1859. On some points of agricultural science. Am. J. Sci. Arts 28:71–85.

Juo, A. S. R., and J. C. Ballaux. 1977. Retention and leaching of nutrients in a limed Ultisol under cropping. Soil Sci. Soc. Am. J. 41:757–761.

Juo, A. S. R., and E. J. Kamprath. 1979. Copper chloride as an extractant for estimating the potentially reactive aluminum pool in acid soils. Soil Sci. Soc. Am. J. 43:35–38.

Kaddah, M. T., and N. T. Coleman. 1967. Salt displacement and titration of $AlCl_3$-treated tri-octahedral vermiculites. Soil Sci. Soc. Am. Proc. 31:328–332.

Kamprath, E. J. 1970. Exchangeable aluminum as a criterion for liming leached mineral soils. Soil Sci. Soc. Am. Proc. 34:252–254.

Kamprath, E. J., W. L. Nelson, and J. W. Fitts. 1956. The effect of pH, sulfate and phosphate concentrations on the adsorption of sulfate by soils. Soil Sci. Soc. Am. Proc. 20:463–466.

Kamprath, E. J., and C. D. Welch. 1962. Retention and cation-exchange properties of organic matter in Coastal Plain soils. Soil Sci. Soc. Am. Proc. 26:263–268.

Katchalsky, A., and P. Spitnik. 1947. Potentiometric titrations of polymethacrylic acid. J. Polym. Sci. 2:432–446.

Keeney, D. R., and R. B. Corey. 1963. Factors affecting the lime requirement of Wisconsin soils. Soil Sci. Soc. Am. Proc. 27:277–280.

Kelley, W. P. 1943. Mattson's papers on "The laws of soil colloidal behavior." Review and comments. Soil Sci. 56:443–456.

Kelley, W. P., and S. M. Brown. 1926. Ion exchange in relation to soil acidity. Soil Sci. 21: 289–302.

Kelley, W. P., W. H. Dore, and S. M. Brown. 1931. The nature of the base exchange material of bentonite, soils, and zeolites, as revealed by chemical investigation and x-ray analysis. Soil Sci. 31:25–55.

Keren, R., R. G. Gast, and R. I. Barnhisel. 1977. Ion exchange reactions in nondried chambers montmorillonite hydroxy-aluminum complexes. Soil Sci. Soc. Am. J. 41:34–39.

Kissel, D. E., E. P. Gentzsch, and G. W. Thomas. 1971. Hydrolysis of nonexchangeable acidity in soils during salt extractions of exchangeable acidity. Soil Sci. 111:293–297.

Kissel, D. E., and G. W. Thomas. 1969. Conductimetric titrations with $Ca(OH)_2$ to estimate the neutral salt replaceable and total soil acidity. Soil Sci. 108:117–179.

Klages, M. G., and J. L. White. 1957. A chlorite-like mineral in Indiana soils. Soil Sci. Soc. Am. Proc. 21:16–20.

Kononova, M. M. 1966. Soil organic matter. Pergamon Press, Inc., Elmsford, N.Y.

Lin, C., and N. T. Coleman. 1960. The measurement of exchangeable aluminum in soils and clays. Soil Sci. Soc. Am. Proc. 24:444–446.

Lindsay, W. L., and H. F. Stephenson. 1959. Nature of the reactions of monocalcium phosphate monohydrate in soils: I. The solution that reacts with the soil. Soil Sci. Soc. Am. Proc. 23:12–18.

Low, P. F. 1955. The role of aluminum in the titration of bentonite. Soil Sci. Soc. Am. Proc. 19:135–139.

Marshall, C. E. 1935. Layer lattices and the base exchange clays. Z. Kristallogr. Mineral. 91: 433–449.

Marshall, C. E. 1964. The physical chemistry and mineralogy of soils, Vol. 1. John Wiley & Sons, Inc., New York.

Martin, A. E., and R. Reeve. 1958. Chemical studies of podzolic illuvial horizons: III. Titration curves of organic matter suspensions. J. Soil Sci. 9:89–100.

Massey, H. F. 1972. pH and soluble Cu, Ni, and Zn in Eastern Kentucky coal mine spoil minerals. Soil Sci. 114:217–221.

Mattson, S. 1931. The laws of soil colloidal behavior: VI. Amphoteric behavior. Soil Sci. 32: 343–365.

Mattson, S., and L. Wiklander. 1940a. The laws of soil colloidal behavior: XXI A. The amphoteric points, the pH and the Donnan equilibrium. Soil Sci. 49:109–134.

Mattson, S., and L. Wiklander. 1940b. The laws of soil colloidal behavior: XXI B. Experimental. Soil Sci. 49:135–151.

McBride, M. B., and P. R. Bloom. 1977. Adsorption of aluminum by a smectite: II. An $Al^{3+}$-$Ca^{2+}$ exchange model. Soil Sci. Soc. Am. J. 41:1073–1077.

McFee, W. W., W. R. Byrnes, and J. G. Stockton. 1981. Characteristics of coal mine overburden important to plant growth. J. Environ. Qual. 10:300–308.

McFee, W. W., and J. M. Kelly. 1977. Air pollution: Impact on soils. Proc. Annu. Meet. Soil Conserv. Soc. Am. 32:203–207.

McGeorge, W. T. 1931. Organic compounds associated with base exchange reactions in soils. Arizona Agric. Exp. Stn. Tech. Bull. 31.

McLean, E. O. 1970. Lime requirements of soils—inactive substances or favorable pH range? Soil Sci. Soc. Am. Proc. 34:363–364.

McLean, E. O., D. C. Reicosky, and C. Lak-Shmanan. 1965. Aluminum in soils: VII. Interrelationship of organic matter, liming, and extractable aluminum with "permanent charge" (KCl) and pH-dependent cation-exchange capacity of surface soils. Soil Sci. Soc. Am. Proc. 29:374–378.

Mehlich, A. 1942. Base unsaturation and pH in relation to soil type. Soil Sci. Soc. Am. Proc. (1941) 6:150–156.

Mehlich, A. 1943. The significance of percentage base saturation and pH in relation to soil differences. Soil Sci. Soc. Am. Proc. 7:167–174.

Mehlich, A. 1948. Determination of cation and anion exchange properties of soils. Soil Sci. 65:429–445.

Mehlich, A. 1964. Influence of sorbed hydroxyl and sulfate on neutralization of soil acidity. Soil Sci. Soc. Am. Proc. 28:492–496.

Meyer, T. A., and G. W. Volk. 1952. Effect of particle size of limestone on soil reaction, exchangeable cations, and plant growth. Soil Sci. 73:37–52.

Milliken, T. H., G. A. Mills, and A. G. Oblad. 1950. Chemical characteristics and structure of cracking catalysts. Discuss. Faraday Soc. 8:279–290.

Mortensen, J. L. 1963. Complexing of metals by soil organic matter. Soil Sci. Soc. Am. Proc. 27:179–186.

Mortensen, J. L., and F. L. Himes. 1964. Soil organic matter. p. 206–241. *In* F. E. Bear (ed.) Chemistry of the soil. 2nd ed. Van Nostrand Reinhold Co., New York.

Moschler, W. W., G. D. Jones, and G. W. Thomas. 1960. Lime and soil acidity effects on alfalfa growth in a red-yellow podzolic soil. Soil Sci. Soc. Am. Proc. 24:507–509.

Moschler, W. W., R. K. Stivers, D. L. Hallock, R. D. Sears, H. M. Camper, M. J. Rogers, G. D. Jones, M. T. Carter, and F. S. McClaugherty. 1962. Lime effects on soil reaction and base content of eleven soil types in Virginia. Virginia Agric. Exp. Stn. Tech. Bull. 159.

Nye, P. H., D. Craig, N. T. Coleman, and J. L. Ragland. 1961. Ion-exchange equilibria involving aluminum. Soil Sci. Soc. Am. Proc. 25:14–17.

Olsen, R. A., and J. E. Robbins. 1971. The cause of the suspension effect in resin-water systems. Soil Sci. Soc. Am. Proc. 35:260–265.

Overbeek, J. Th. G. 1953. Donnan-E.M.F. and suspension effect. J. Colloid Sci. 8:593–605.

Page, A. L., F. F. Peterson, and J. P. Martin. 1965. Acidic properties of soils as influenced by their mineralogical characteristics. Soil Sci. Soc. Am. Proc. 29:246–250.

Parker, J. C., L. W. Zelazny, S. Sampath, and W. G. Harris. 1979. A critical evaluation of zero point of charge (ZPC) theory to soil systems. Soil Sci. Soc. Am. J. 43:668–674.

Parks, G. A., and P. D. de Bruyn. 1962. The zero point of charge of oxides. J. Phys. Chem. 66:967–973.

Paver, H., and C. E. Marshall. 1934. The role of aluminum in the reactions of the clays. J. Soc. Chem. Ind. London Trans. Commun. 53:750–760.

Posner, A. M. 1964. Titration curves of humic acid. Int. Congr. Soil Sci., Trans. 8th (Bucharest) 3:161–174.

Posner, A. M. 1966. The humic acid extracted by various reagents from a soil, Part I. Yield, inorganic components, and titration curves. J. Soil Sci. 17:65–78.

Pratt, P. F., and F. L. Bair. 1962. Cation-exchange properties of some acid soils of California. Hilgardia 33:689–706.

Ragland, J. L., and N. T. Coleman. 1960. The hydrolysis of aluminum salts in clay and soil systems. Soil Sci. Soc. Am. Proc. 24:457–460.

Raupach, M. 1957. Investigations into the nature of soil pH. CSIRO Soil Pub. no. 9, Melbourne, Australia.

Reeve, N. T., and M. E. Sumner. 1970. Lime requirements of natal oxisols based on exchangeable aluminum. Soil Sci. Soc. Am. Proc. 34:395–398.

Rich, C. I. 1960. Aluminum in interlayers of vermiculite. Soil Sci. Soc. Am. Proc. 24:26–32.

Rich, C. I. 1964. Effect of cation size and pH on potassium exchange in Nason soil. Soil Sci. 98:100–106.

Rich, C. I. 1968. Hydroxy interlayers in expansible layer silicates. Clays Clay Miner. 16:15–30.

Rich, C. I. 1970. Conductimetric and potentiometric titration of exchangeable aluminum. Soil Sci. Soc. Am. Proc. 34:31–38.

Rich, C. I., and S. S. Obenshain. 1955. Chemical and clay mineral properties of a red-yellow podzolic soil derived from mica schist. Soil Sci. Soc. Am. Proc. 19:334–339.

Rixon, A. J., and G. D. Sherman. 1962. Effects of heavy lime applications to volcanic ash soils of the humid tropics. Soil Sci. 94:19–27.

Ross, C. J., K. Lawton, and B. G. Ellis. 1964. Lime requirement related to physical and chemical properties of nine Michigan soils. Soil Sci. Soc. Am. Proc. 28:209–212.

Ruff, J. K., and S. Y. Tyree. 1958. Light-scattering studies on aqueous aluminum nitrate solutions. J. Am. Chem. Soc. 80:1523–1526.

Samson, H. R. 1954. The deflocculation of kaolinite suspensions. Ph.D. Diss. London Univ., London.

Sawhney, B. L. 1960. Weathering and aluminum interlayers in a soil catena: Hollis-Charlton-Sutton-Leicester. Soil Sci. Soc. Am. Proc. 24:221–226.

Schnitzer, M., and U. C. Gupta. 1964. Chemical characteristics of the organic matter extracted from the O and $B_2$ horizons of a gray wooded soil. Soil Sci. Soc. Am. Proc. 28:274–277.

Schnitzer, M., and U. C. Gupta. 1965. Determination of acidity in soil organic matter. Soil Sci. Soc. Am. Proc. 29:274–277.

Schnitzer, M., and S. U. Khan. 1972. Humic substances in the environment. Marcel Dekker, Inc., New York.

Schnitzer, M., and S. I. M. Skinner. 1963a. Organo-metallic interactions in soils: 1. Reactions between a number of metal ions and the organic matter of a podzol $B_h$ horizon. Soil Sci. 96:86–93.

Schnitzer, M., and S. I. M. Skinner. 1963b. Organo-metallic interactions in soils: 2. Reactions between different forms of iron and aluminum and the organic matter of a podzol $B_h$ horizon. Soil Sci. 96:181–186.

Schnitzer, M., and S. I. M. Skinner. 1965. Organo-metallic interactions in soils: 4. Carboxyl and hydroxyl groups in organic matter and metal retention. Soil Sci. 99:278–284.

Schofield, R. K. 1933. Rapid method of examining soils: II. J. Agric. Sci. 23:252–254.

Schofield, R. K. 1949. Effect of pH on electric charges carried by clay particles. J. Soil Sci. 1: 1–8.

Schofield, R. K., and A. W. Taylor. 1954. The hydrolysis of aluminum salt solutions. J. Chem. Soc. 18:4445–4448.

Schofield, R. K., and A. W. Taylor. 1955. The measurement of soil pH. Soil Sci. Soc. Am. Proc. 19:164–167.

Schwertmann, U., and M. L. Jackson. 1963. Hydrogen-aluminum clays: A third buffer range appearing in potentiometric titrations. Science (Washington, D.C.) 139:1052–1054.

Schwertmann, U., and M. L. Jackson. 1964. Influence of hydroxy-aluminum ions on pH titration curves of hydronium-aluminum clays. Soil Sci. Soc. Am. Proc. 28:179–182.

Seatz, L. F., and H. B. Peterson. 1964. Acid, alkaline, saline and sodic-soils. p. 292–319. *In* F. E. Bear (ed.) Chemistry of the soil. 2nd ed. Van Nostrand Reinhold Co., New York.

Shen, M. J., and C. I. Rich. 1962. Aluminum fixation in montmorillonite. Soil Sci. Soc. Am. Proc. 26:33–36.

Shoemaker, H. E., E. O. McLean, P. F. Pratt. 1961. Buffer methods for determining lime requirement of soils with appreciable amounts of extractable aluminum. Soil Sci. Soc. Am. Proc. 25:274–277.

Slaughter, M., and I. J. Milne. 1960. The formation of chlorite-like structures from montmorillonite. Clays Clay Miner. Proc. Conf. 7:114–124.

Smith, R. W., and J. D. Hem. 1972. Effect of aging on aluminum hydroxide complexes in dilute aqueous solutions. Geol. Surv. Water-Supply Paper 1827D, Dep. of Interior, U.S. Government Printing Office, Washington, D.C.

Sposito, G. 1981. The operational definition of the zero point of charge. Soil Sci. Soc. Am. J. 45:292–297.

Stevenson, F. J. 1972. Role and function of humus in soil with emphasis on adsorption of herbicides and chelation of micronutrients. BioScience 22:643–650.

Sumner, M. E. 1963. Effect of iron oxides on positive and negative charges in clays and soils. Clay Miner. Bull. 5:218–226.

Swoboda, A. R., and G. W. Kunze. 1968. Reactivity of montmorillonite surfaces with weak organic bases. Soil Sci. Soc. Am. Proc. 32:806–811.

Tamele, M. W. 1950. Chemistry of the surface and the activity of alumina-silica cracking catalysts. Discuss. Faraday Soc. 8:270–279.

Tamura, T. 1957. Identification of the 14Å clay mineral component. Am. Mineral. 42:107–110.

Thomas, G. W. 1960a. Factors affecting the removal of salts from halloysite. Soil Sci. 90:344–347.

Thomas, G. W. 1960b. Forms of aluminum in cation exchangers. Int. Congr. Soil Sci., Trans. 7th (Madison, Wis.) 2:364–369.

Thomas, G. W. 1975. The relationship between organic matter content and exchangeable aluminum in acid soil. Soil Sci. Soc. Am. Proc. 39:591.

Thomas, G. W. 1977. Historical developments in soil chemistry: Ion exchange. Soil Sci. Soc. Am. J. 41:230–238.

Thomas, G. W. 1982. Exchangeable cations. *In* A. L. Page et al. (ed.) Methods of soil analysis, Part 2, 2nd ed. Agronomy 9:159–165. Am. Soc. of Agron., Madison, Wis.

Thomas, G. W., and N. T. Coleman. 1964. The fate of exchangeable iron in acid clay systems. Soil Sci. 97:229–232.

Thomas, G. W., and A. R. Swoboda. 1963. Cation exchange in kaolinite-iron oxide systems. Clays Clay Miner. 11:321–326.

Turner, R. C. 1965. A study of the lime potential: 4. The lime potential during titration of Wyoming bentonite originally saturated with ferric ions. Soil Sci. 99:88–92.

Turner, R. C., and W. E. Nichol. 1962a. A study of the lime potential: 2. Relation between the lime potential and percent base saturation of negatively charged clays in aqueous salt suspensions. Soil Sci. 94:58–63.

Turner, R. C., and W. E. Nichol. 1962b. A study of the lime potential: 1. Conditions for the lime potential to be independent of salt concentration in aqueous suspensions of negatively charged clays. Soil Sci. 93:374–382.

Turner, R. C., W. E. Nichol, and J. E. Brydon. 1963. A study of the lime potential: 3. Concerning reactions responsible for the magnitude of the lime potential. Soil Sci. 95:186–191.

Uehara, G., and G. P. Gillman. 1980. Charge characteristics of soils with variable and permanent charge minerals: I. Theory. Soil Sci. Soc. Am. J. 44:250–252.

Van Olphen, H. 1963. Clay colloid chemistry. Interscience, New York.

van Raij, B., and M. Peech. 1972. Electrochemical properties of some Oxisols and Alfisols of the tropics. Soil Sci. Soc. Am. J. 36:587–593.

Veitch, F. P. 1904. Comparison of methods for the estimation of soil acidity. J. Am. Chem. Soc. 26:637–662.

Veith, J. A. 1977. Basicity of exchangeable aluminum, formation of gibbsite and the composition of the exchange acidity in the presence of exchangers. Soil Sci. Soc. Am. J. 41: 865–870.

Volk, V. V., and M. L. Jackson. 1964. Inorganic pH dependent cation exchange charge of soils. Clays Clay Miner. 12:281–285.

Wada, K., and H. Ataka. 1958. The ion-uptake mechanism of allophane. Soil Plant Food (Tokyo) 4:12–18.

Way, J. T. 1850. On the power of soils to absorb manure. J. R. Agric. Soc. Engl. 11:313–379.

White, R. E., and G. W. Thomas. 1981. Hydrolysis of aluminum on weakly acid organic exchangers. Implications for phosphorus adsorption. Fert. Inst. 2:159–167.

Wolt, J. D. 1981. Sulfate retention by acid sulfate-polluted soils in the Copper Basin area of Tennessee. Soil Sci. Soc. Am. J. 45:283–287.

Woodruff, C. M. 1948. Testing soils for lime requirement by means of a buffer solution and the glass electrode. Soil Sci. 66:53–63.

Yuan, T. L. 1959. Determination of exchangeable hydrogen in soil by a titration method. Soil Sci. 88:164–167.

Yuan, T. L. 1963. Some relationships among hydrogen, aluminum, and pH in solution and soil systems. Soil Sci. 95:155–163.

# 2 Physiological Effects of Hydrogen, Aluminum, and Manganese Toxicities in Acid Soil[1]

**CHARLES D. FOY**

*Plant Stress Laboratory*
*Agricultural Research Service, USDA*
*Beltsville, Maryland*

## I. ACID SOIL TOXICITY

Soil acidity is a major growth-limiting factor for plants in many parts of the world (McLean, 1976; Kamprath, 1978; Adams, 1978, 1981; Clark, 1982). Soils are acid because their parent materials were acid and initially low in the basic cations ($Ca^{2+}$, $Mg^{2+}$, $K^+$, and $Na^+$) or because these elements have been removed from the soil profile by normal rainfall leaching or the harvesting of crops (Kamprath & Foy, 1972). Even the normal growth of clover pasture for 50 yr decreased the pH of an Australian soil from 6.0 to 5.0 at a depth of 30 cm (Williams, 1980). Soil acidification is intensified by the use of acid-forming nitrogenous fertilizers (Pierre et al., 1971; Mason, 1980) and by acid deposition from polluted air (Ulrich et al., 1980).

Acid soil injury is an insidious problem; it may be mistaken for an ordinary nutrient deficiency, drought effect, herbicide injury, low-temperature damage, or even a plant disease. Extreme acidity in subsoils in particularly harmful because it can cause shallow rooting, drought susceptibility, and poor use of subsoil nutrients (Kauffman, 1977).

### A. Components

Acid soil toxicity is not a single factor but a complex of factors that may affect the growth of different plants through different physiological and biochemical pathways, which probably are controlled by different genes (Foy et al., 1978). Furthermore, acid soils having similar pH values may cause different mineral stress problems in a given plant genotype. The specific causes of poor plant growth on acid soils may vary with soil pH, clay mineral types and amounts, organic matter contents and kinds, levels

[1] Invitational contribution of the Plant Stress Laboratory, Plant Physiology Institute, USDA-ARS, Beltsville, MD 20705.


*Soil Acidity and Liming*—Agronomy Monograph no. 12 (2nd Edition).

of salts, and particularly, with plant species or genotype (Clark, 1982). Growth-limiting factors that have been associated with the acid soil infertility complex include toxicities of $Al^{3+}$, $Mn^{2+}$, and other metal ions, low pH ($H^+$ toxicity), and deficiencies or unavailabilities of certain essential elements, particularly Ca, Mg, P, and Mo (Jackson, 1967; Kamprath & Foy, 1972; Foy et al., 1978). Even Fe deficiency has been reported in upland rice on acid soils of pH 5.1 to 5.8 (IRRI, 1974). Aluminum-induced Fe deficiency has also been reported in sorghum (Clark et al., 1981; Furlani & Clark, 1981). In acid coal mine spoils and soils polluted with industrial wastes, Zn, Cu, Cd, Ni, or Pb may be present in toxic amounts for some plants (Bradshaw, 1970; Massey & Barnhisel, 1972; Brown & Jones, 1975; Chaney & Giordano, 1977). Acid soil factors may act somewhat independently, or more often together, to affect the growth of higher plants (Foy et al., 1978). They may also promote or inhibit the survival and function of rhizobia, mycorrhizae, and other microorganisms.

## II. HYDROGEN ION TOXICITY

### A. Effects on Plants

#### 1. GENERAL EFFECTS

The direct effects of the $H^+$ ion on plant growth are difficult to determine in acid soils, because at soil pH levels where it is considered harmful, Al, Mn, and other mineral elements may also be soluble in toxic concentrations, and the availabilities of essential elements, particularly Ca, Mg, P, and Mo, may be suboptimal. In most acid soils (pH $> 4.0$), $Al^{3+}$ and $Mn^{2+}$ toxicities are probably more important than $H^+$ ion toxicity in limiting the growth of higher plants, particularly the nonlegumes. However, $H^+$ ion toxicity may restrict the survival and activity of rhizobia or other soil microorganisms (Kamprath & Foy, 1984; Moore, 1974).

Because the effects of the $H^+$ ion are confounded with other factors in acid soils, investigators have used solution and sand culture techniques to characterize the responses of plants to low pH. In general, the roots of plants damaged by low pH ($< 4.0$) are short, thickened, fewer in number, and discolored brown or dull gray (Arnon et al., 1942; Islam et al., 1980). Lateral root growth may be severely inhibited to the point that root tips die. Affected plants may wilt in nutrient cultures during warm, sunny weather. Ekdahl (1957) found that root hair length was increased by 40% when the pH was increased from 5.5 to 7.2. Kerridge (1969) reported that the lengths of wheat roots were slightly greater at pH 5 than at 4, and this difference was greater in 14-day-old plants than in older plants. Howard and Adams (1965) showed that a solution pH $< 4.2$ greatly reduced the growth rate of primary cotton roots.

Excess $H^+$ ions have marked effects on root membrane permeability. They compete with other cations for absorption sites, interfere with ion transport, and cause root membranes to become leaky. Roots may lose previously absorbed cations as well as organic substances, and prolonged exposure to low pH may reduce their capacities for subsequent absorption of nutrients. Christiansen et al. (1970) found that a solution pH $< 4$ markedly enhanced the loss of organic substances from cotton radicles and that this effect could be reversed by raising the pH. Hussain et al. (1954) reported that pretreatment of barley roots with HCl at a pH $< 5$ caused a pronounced loss of K. This loss was largely prevented by the presence of the polyvalent cations $Ca^{2+}$, $Al^{3+}$, $La^{3+}$, and $Ce^{3+}$. Acid pretreatments at pH values $< 6$ also markedly reduced the capacities of roots for subsequent uptake of K. Exposure of barley roots to pH 3 caused them to lose the majority of their K, Ca, P, and soluble N contents and eliminated their capacities for subsequent K absorption.

Because of its effect on nutrient uptake and retention by plant roots, the $H^+$ ion (excess) can increase plant requirements for Ca, and perhaps other nutrients, in a growth medium. For example, Lund (1970) found that soybean taproots growing in the nutrient solution portion of a split medium required higher Ca levels at pH 4.5 than at 5.6. Maas (1969) reported that $H^+$ ion within the pH range of 3 to 5 markedly reduced Ca uptake by excised maize roots. Excess $H^+$ ions can also decrease the uptake of Mg (Arnon et al., 1942; Moore et al., 1961; Islam et al., 1980; Blamey et al., 1982), Mn (Robson & Loneragan, 1970), Zn (Rashid et al., 1976), and Cu (Bowen, 1969).

## 2. DIFFERENTIAL EFFECTS ON PLANT SPECIES AND VARIETIES

Plant species and genotypes within species differ widely in tolerance to excess $H^+$ ions in solution. Arnon and Johnson (1942) found that at pH 4 to 5, the growth of lettuce and tomato was reduced but the growth of bermudagrass was not. Lettuce and tomato absorbed less Ca at pH 4 to 5 than at higher pH, and additional Ca in the solution alleviated $H^+$ ion toxicity. At pH 6, plants grew well with either low or high Ca. Hence, the $H^+$ ion–induced growth reductions in these two species were attributed to reduced Ca uptake. Bermudagrass, which was tolerant to pH 4, was not responsive to additional Ca in solution. Nutrient solutions at pH 3 completely inhibited the growth of lettuce, tomato, and bermudagrass. In the absence of Al, rice roots were not affected by $H^+$ ion concentrations up to pH 3.5 (Thaworuwong & Van Diest, 1974), and yellow poplar grew well at solution pH levels of 4.0, 4.5, 5.0, and 5.5 (Patel & Mugwira, 1974).

Islam et al. (1980) grew six species in flowing solution cultures rigidly controlled at different pH levels between 3.3 and 8.5. Within the pH range of 5.5 to 6.5, all species produced maximal or near-maximal growth; however, outside this range, their responses were quite different. At pH 3.3, the roots of all species showed toxicity, and some were injured at pH 4. At pH 3.3, roots were short, thickened, fewer in number, and discolored brown or dull gray. Lateral root growth was inhibited in all species. At pH 4, the roots of ginger, cassava, and maize were healthy but those of french bean were gray, with short laterals and necrotic tips, and those of wheat were brown. In contrast, at pH 4.8 or above, roots were light colored and finely branched. In general, raising the solution pH was more beneficial to root growth than to top growth. Low pH caused suboptimal uptake of Mg in all six species, N in tomato and cassava, and Mn in maize. The order of species tolerance to $H^+$ ion toxicity (pH 3.3 and 4.0) was ginger > cassava > tomato > french bean > wheat > maize. This order agreed with that obtained under acid soil field conditions (CIAT, 1977), namely, cassava > bean > maize.

Blamey et al. (1982) found that a solution pH of 3.5 was lethal to four cultivars of sunflower, but cultivars differed in acid tolerance above pH 4.0. The critical pH levels above which yields were increased were 4.0 for 'Hysun 11', 4.5 for 'Hysun 32', and 5.0 for 'Hysun 21' and 'Hysun 31'. Root yields of cultivars also differed in response to pH increases above 3.5. Hysun 21 showed the greatest and Hysun 11 the least response to increased pH. Plant concentrations of P, Ca, and Mn were lower at low pH, but cultivars did not differ in P, Ca, Mg, Mn, N, K, or Fe concentrations.

## 3. EFFECTS ON RHIZOBIA-HOST PLANT RELATIONSHIPS

Bryan (1923a, b) reported that alfalfa bacteria were killed at a soil pH of 5.0, those of red clover at pH 4.5 to 4.7, and those of soybean at pH 3.5 to 3.9. He also found that these critical pH values, obtained in soils, were about the same as those in sand cultures. The critical pH for nodulation and growth of red clover was lower than that for alfalfa. Nodules formed at any pH where plants grew, but their numbers were greatest at a pH near 7.0. Sewell and Gainey (1930) concluded that within the pH range of 4.5 to 7.0, soil pH was a minor factor in alfalfa growth if all necessary elements were supplied. Doolas (1930) found that soybean rhizobia remained viable in soils at pH 4.8 to 8.3. Soybean roots showed depressed nodulation within the pH range of 3.8 to 4.6 but were healthy at pH 5.6. Spurway (1941) concluded that the optimal soil pH was 6.2 for alfalfa, 6.5 for red clover and sweetclover, 6.0 for soybean, 5.0 for cowpea, and 4.5 for lespedeza.

According to Andrew (1978) the $H^+$ ion is particularly important in legumes grown without fertilizer N. It affects rhizobial survival and multiplication in soils, root infection and nodule initiation, legume rhizobial efficiency, and nutrition of the host plant. The root infection process is particularly sensitive to acidity, requiring a higher pH than that for rhizobial survival (Munns, 1978). Low pH is also believed to influence the functioning of established nodules to varying degrees, depending on the nature of the host plant. Legumes are apparently much more tolerant to low pH when fertilizer N is supplied than when grown with only symbiotic N (Andrew, 1978; Munns, 1978). Franco and Munns (1982) found that decreasing the pH of nutrient solutions from 5.5 to 5.0 decreased the number of nodules formed by bean, but variable pH between 4.5 and 5.5 did not affect nodule growth and nitrogenase activity. Results suggested that low pH could be important in early nodulation of bean.

Andrew (1978) used sand cultures to determine the effects of low pH on nodulation and growth of several legumes. Alfalfa was the most sensitive species tested. It produced no nodules at pH 4.0 or 4.5, some nodules at pH 5.0, and full nodulation at pH 6.0. Glycine was also pH sensitive and showed a steady increase in nodulation between pH 4 and 6. Phasey bean and Miles lotononis were nodulated between pH 4 and 6. Silverleaf uncinatum, Townsville stylo, Kenya white clover, and *Trifolium rueppellianum* Fresen. had reduced nodulation at pH 4 but produced full nodulation at pH 4.5 and above. When inorganic N was supplied, the detrimental effects of low pH on growth were much smaller, but the order of tolerance to excess $H^+$ ion was similar to that for symbiotic plants.

Rhizobial strains within species also differ in their abilities to nodulate a given host plant at low pH (Munns, 1978). Keyser and Munns (1979a, b) showed that strains of cowpea rhizobia varied widely in tolerance to a soil pH of 4.6. These investigators concluded that low pH and Al toxicity were more important than Mn toxicity and Ca deficiency in regulating nodulation on cowpea and soybean roots.

Munns (1978) noted that wide variability in $H^+$ ion tolerance within both host plant and rhizobial species provide an opportunity to develop more effective rhizobia–host plant combinations but that better screening guidelines are needed for evaluating the effectiveness of such combinations under field conditions.

### B. Effects on Mineralization of Organic Matter

#### 1. MICROBIAL ACTIVITY

Soil pH is an important factor in determining the kinds, numbers, and activities of microorganisms involved in organic matter transformations (Sommers & Biederbeck, 1973; Alexander, 1980). By regulating microbial activity, pH affects the mineralization of organic matter and the subsequent availabilities of N, P, S, and micronutrients to higher plants. In general, organic matter decomposes more rapidly in neutral soils than in acid soils. At soil pH values below about 5.5 the fungi are most active in this decomposition, primarily because of a lack of competition from other organisms more sensitive to acid soil factors. Campbell and Lees (1967) suggested that fungi may carry out nitrification in strongly acid forest soils. At pH values $> 6.0$, actinomycetes and bacteria are more prominently involved in organic matter decomposition. Different bacteria and actinomycetes appear to have different optimum pH ranges for reproduction. For example, Corke and Chase (1964) noted differential pH requirements of certain actinomycetes, and other evidence indicates that acid tolerant strains of nitrifying bacteria have developed in certain soils (Jackson, 1967; Brar & Giddens, 1968).

#### 2. RELEASE OF N, P, AND S

Two important microbially induced reactions affecting N availability in soils are (i) ammonification, the production of $NH_4^+$ from the decomposition of organic matter; and (ii) nitrification, the oxidation of $NH_4^+$ to $NO_2^-$ and $NO_3^-$, brought about primarily by bacteria of the *Nitrosomonas* and *Nitrobacter* genera, respectively (Sommers & Biederbeck, 1973; Alexander, 1980). Ammonification can apparently occur over a wide range of soil pH, but nitrification is markedly reduced at pH values $< 6$ and $> 8$ (Chase et al., 1968; Alexander, 1980). According to Alexander (1980), nitrification rate falls rapidly with decreasing pH and is undetectable at pH values below 4.5. In studies involving a wide range of soils, Morrill and Dawson (1967) found that pH (with its related factors) was the best indicator of nitrification in soils.

In most mineral soils, one half to two thirds of the total P is in organic forms (Cosgrove, 1967). Extreme values reported range from 4% for a podzol to 90% for an alpine humus. A large part of this organic P is of microbial origin and must be mineralized to become available to plants. In acid soils the very insoluble Al and Fe phytates are believed to be the most abundant organic P compounds. Oliveira and Drozdowicz (1980) demonstrated the presence of bacteria and actinomycetes that can mineralize P from DNA, lecithin, and Ca phytate in the Cerrado soils of Brazil. Increasing the soil pH generally causes mineralization of phytate P and thus increases its availability to plants (Cosgrove, 1967), Thompson et al. (1954) found that the mineralization of P was positively correlated with soil pH. The amounts of P, C, and N mineralized in soils are generally roughly correlated with their proportions in the organic matter being decomposed, but the release of organic P is sometimes quite different from that of C and N. For example, Thompson et al. (1954) found that in 50 unlimed soils the mineralization of organic P increased markedly with increase in pH but the mineralization of organic C did not. On the basis of such evidence, Cosgrove (1967) suggested that alkaline soils should have higher C/P and N/P ratios than acid soils.

The plant availability of S contained in organic matter appears to be limited by the rate of organic matter breakdown rather than by the oxidation of the S compounds released (Burns, 1967). This oxidation is brought about largely through biological processes and can take place between pH 2.0 and 9.0. Nelson (1964) found that liming soil from pH 4.0 to 5.0 increased $SO_4^{2-}$ production threefold, but increasing the pH from 5.0 to 7.0 had very little additional effect. Freney (1967) stated that liming soils increased $SO_4^{2-}$ levels but noted that this could result from $SO_4^{2-}$ added in the lime, $SO_4^{2-}$ released from organic matter by hydrolysis at higher pH, or $SO_4^{2-}$ released from soil organic matter of bacteria growing more profusely at a favorable pH. Adding S to the soil may also increase the release of S from soil organic matter by increasing the growth of microorganisms. For additional information regarding S transformations in soils, see McLachlan (1975).

### 3. RELEASE OF MICRONUTRIENTS

Soil organisms may influence the availabilities of nutrients indirectly by changing the pH of the medium or by changing the oxidation state of an element. These reactions are important in determining mineral element solubilities and, hence, toxicities and deficiencies. Considerable evidence suggests that soil microorganisms alter the solubility or oxidation states of Mn, Zn, Cu, Al, and Mo (Alexander, 1961).

The availabilities of Mn, Cu, Zn, and Fe are closely related to the transformations of organic matter in soils (Alexander, 1961). For example, Zn deficiency of fruit trees can be corrected by sterilizing the soil, which releases Zn from microbial bodies and makes it available to plants. Zinc deficiency in flax has also been associated with microbial activities.

Copper availability is generally decreased by microbial metabolism during the decomposition of certain crop residues (Alexander, 1961); however, the decay of the more readily decomposable carbonaceous materials, e.g., glucose or asparagine, may increase the level of water-soluble and available Cu.

The availability of Fe may be increased by microbial action through solubilization of organic products that may chelate Fe or by the creation of a more reducing environment (Alexander, 1961). Ferrous Fe is likely to be more abundant at pH values $< 5.0$ and ferric Fe is favored $> 6.0$. Iron can also be precipitated by Fe bacteria in soils and made less available to plants.

Manganese deficiencies and toxicities are closely related to the activities of soil microorganisms in altering the pH and bringing about oxidation-reduction changes that determine the solubility of Mn (Foy, 1973).

Jones and Woltz (1960) reported an interesting microbial-micronutrient relationship. Liming a Leon soil (Aeric Haplaquods) from 4.8 to 8.0 reduced the incidence and severity of Fusarium wilt of tomatoes, caused by *Fusarium oxysporium* f., sp. *lycopersicon,* race 2. Soil additions of Mn plus Zn, or Fe plus Zn lignosulfates, which are plant available at high pH, reversed the disease control given by lime and increased the severity of the disease. The authors suggested that liming the soil reduced the severity of tomato wilt by creating an imbalance of micronutrients in the pathogen.

## III. ALUMINUM TOXICITY

Aluminum toxicity is probably the most important growth-limiting factor for plants in most strongly acid soils and mine spoils (Foy, 1974; McLean, 1976; Silva, 1976; Blue & Dantzman, 1977; Sanchez et al., 1979; Alam & Adams, 1979; Carvalho et al., 1980; Friesen et al., 1980). The problem is particularly severe below pH 5.0, but it may occur as high as pH 5.5 in kaolinitic soils. Much of the poor root development (and drought susceptibility) seen in soils with acid ($<$ pH 5.0) subsoil layers is probably due primarily to Al toxicity, which limits both rooting depth and degree of root branching.

Considerable evidence suggests that Al toxicity limits the microbial breakdown of organic matter in some strongly acid soils (Alexander, 1980). For examples, Mutatker and Pritchett (1966) found the $CO_2$ production was greater in a Ca muck than in a Ca-Al muck when both were maintained at pH 4.0. In a Latosol and in mixtures of Ca- and Al-saturated mucks, $CO_2$ production was significantly decreased with increasing levels of Al during incubation at pH 4.0 or below. Brar and Giddens (1968) concluded that acidity and the resulting low populations of nitrifiers were the causes of low nitrification rates in a Bladen soil (Typic Albaquults) at pH 4.6. They also suggested that Al was a likely factor in limiting the nitrifier population.

The critical soil pH at which Al becomes soluble or exchangeable in toxic concentrations depends on many soil factors, including the predominant clay minerals, organic matter levels, concentrations of other cations, amions and total salts, and particularly, the plant species or cultivar (Foy, 1974; Kamprath & Foy, 1972). Such complexity makes it difficult to devise a soil Al test that will accurately predict toxicity under all conditions.

### A. Plant Symptoms of Aluminum Toxicity

The symptoms of Al toxicity are not easily identified (Foy et al., 1978; Alam & Adams, 1979). In some plants the foliar symptoms resemble those of P deficiency (overall stunting; small, dark green leaves and late maturity; purpling of stems, leaves, and leaf veins; and yellowing and death of leaf tips). In others, Al toxicity appears as an induced Ca deficiency or reduced Ca transport problem within the plant (curling or rolling of young leaves and a collapse of growing points or petioles). Aluminum also induces Fe deficiency symptoms in rice and sorghum (IRRI, 1974; Clark et al., 1981; Furlani & Clark, 1981). Aluminum-injured roots are characteristically stubby and brittle. Root tips and lateral roots become thickened and may turn brown. The root system as a whole appears coralloid, with many stubby lateral roots but no fine branching. Such roots are inefficient in absorbing nutrients and water. Cotton plants affected by severe Al toxicity sometimes wilt in nutrient solutions (C. D. Foy, 1967, unpublished data). Young seedlings are generally more susceptible to Al toxicity than older plants (Thaworuwong & Van Diest, 1974).

### B. Physiological and Biochemical Effects of Aluminum

For plants in general, excess Al as been reported to interfere with cell division in root tips and lateral roots, increase cell wall rigidity by cross linking pectins, reduce DNA replication by increasing the rigidity of the DNA double helix, fix P in less available forms in soils and on root surfaces, decrease root respiration, interfere with enzymes governing sugar phosphorylation and the deposition of cell wall polysaccharides, and interfere with the uptake, transport, and use of several essential nutrient elements, including Ca, Mg, K, P, and Fe. For details on the older literature, see Foy (1974), Foy and Fleming (1978), and Foy et al. (1978).

Metal ions, e.g., $Al^{3+}$, are known to precipitate nucleic acid by forming strong complexes; in fact, they are used to complex and isolate polynucleotides from leaves (Trim, 1959). Matsumoto et al. (1979) found that Al was bound to the P in DNA and not to the histone proteins associated with the DNA double helix. Excess Al had two effects on susceptible wheat cultivars:

a rapid inhibition of root elongation, followed by inhibition of DNA synthesis (Ulmer, 1979). Naidoo (1977) found the highest Al concentrations in the nuclei of Al-injured snap bean cells and suggested that Al is bound to esteric P in nucleic acids and membrane lipids; he believed such binding inhibited cell division by interfering with nucleic acid replication. McLean (1980) found that Al caused an abnormal distribution of ribosomes on the endoplasmic reticulum of barley root cells; interference in protein synthesis was postulated. Schmandke et al. (1979) observed that $AlCl_3$ increased the firmness and decreased the solubility of protein/casein fibers in broadbean. Trivalent Al was believed to form coordination complexes with carboxyl and sulfhydryl groups of the protein, resulting in a cross linkage.

Many of the biochemical effects of Al on plants are probably associated with the alteration of root membrane structure and function (Hecht-Buchholz & Foy, 1981). Plant membranes are visualized as semifluid arrangements of proteins and lipids. Aluminum can bind to either proteins or lipids, depending on pH and other conditions. Vierstra and Haug (1978) found that Al decreased lipid fluidity in membranes of *Thermoplasma acidophilium*. In their study, 270 mg $L^{-1}$ of Al in solution produced dramatic changes, but detectable effects occurred with only 0.27 mg $L^{-1}$ of Al at pH 4.0. As little as 0.0027 mg $L^{-1}$ of $Al^{3+}$ severely reduced the transmembrane potential previously generated by ATP applications in barley roots (Siegal & Haug, 1982, p. 75). Gomez-Lepe et al. (1979) reported that Al binds to cell membrane proteins on the inner epidermal cells of onion.

Aluminum may also interfere with water use by plants. For example, Horton and Edwards (1976) found that increasing concentrations of Al in nutrient solutions increased the diffusive resistance to water in peach seedlings. Kauffman and Gardner (1978) concluded that the water potential of wheat plants increased when root growth was reduced by soil acidity (probably Al toxicity).

Excess Al may reduce the uptake of certain essential elements and increase that of others (Ben et al., 1976; Lau, 1979; Gurrier, 1979; Alam & Adams, 1979, 1980; Duncan et al., 1980; Alam, 1981). However, such data are difficult to interpret in terms of Al toxicity mechanisms; no one pattern of elemental accumulation can apply to all cases of Al injury. Duncan et al. (1980) found that decreased yields of sorghum in a Georgia soil (probably from Al toxicity) at pH 4.4 (compared with pH 5.5) were associated with decreased concentrations of Cu, Zn, Mg, and Ca and increased concentrations of Al, Fe, Mn, K, and P in plant tops. Aluminum toxicity in wheat was prevented by increasing the concentrations of Ca, Mg, K, or Na in nutrient solutions, either individually or collectively (Ali, 1973). The beneficial effects of these elements were probably due to a competitive reduction in Al-root contact or to a decrease in Al activity rather than to supplying nutrients.

Aluminum accumulates on or in the roots of Al-injured plants, often in association with P (McCormick & Borden, 1972), but it does not generally accumulate in the tops of Al-sensitive plants (Foy, 1974). Even when Al injury in a plant is correlated with increased Al concentrations in plant tops

(James et al., 1978; Alam & Adams, 1979; Duncan et al., 1980), the Al in the tops may not be directly harmful; it may be simply the result of passive accumulation caused by prior injury to the roots. In fact, the entire array of elements in the tops of Al-injured plants probably represents the accumulated, systemic effects of the initial root injury by Al. Such effects are generally too far removed from the initial root injury to reveal Al toxicity mechanisms. Thaworuwong and Van Diest (1974) concluded that Al concentrations of plant tops and roots were not useful indices of Al tolerance in rice. Roots washed with water or acid were not significantly different in Al concentrations.

Aluminum toxicity continues to be associated with P nutrition of plants. For example, James et al. (1978) concluded that Al-induced P deficiency reduced the growth of Sitka spruce in Scotland. Negative correlations were found between growth, beaded roots (Al injured), and foliar Al concentrations. Santana and Braga (1977) found that P concentrations in rice tops decreased with increasing Al saturation of the soil. Helyar (1978) concluded that Al toxicity effects were largely associated with Al interference with P metabolism and with Al binding to pectins in root cell walls, stopping root elongation. However, Ulmer (1979) found that increasing the P concentration in the pretreatment nutrient solutions for wheat did not affect the degree of injury in subsequent exposures to Al. Neither did the addition of indole-3-acetic acid (IAA) help Al-injured plants to recover. Trapp (1980) concluded that in yeast cells Al acts as a dead-end inhibitor of the Mg-dependent enzyme hexokinase, which is responsible for sugar phosphorylation. The formation of glucose-6-phosphate by the action of hexokinase on ATP apparently requires that the ATP first be combined with Mg, but the affinity of Al for ATP is 40 times that of Mg. By forming the highly stable Al–ATP complex, Al is believed to prevent the transfer of the terminal phosphoryl group to glucose by hexokinase.

Aluminum × Fe interactions are frequently mentioned in the literature. For example, Alam (1981) found that Al-induced Fe deficiency in oat and postulated that Al interfered with the reduction of $Fe^{3+}$ to $Fe^{2+}$ within the plant, a process essential for normal Fe metabolism. Clark et al. (1981) and Furlani and Clark (1981) found that Fe-deficiency chlorosis was a common symptom of Al toxicity in sorghum.

Aluminum × Ca interactions are important in acid soils. In fact, below a pH of 5.5, Al × Ca antagonism is probably the most important factor affecting Ca uptake by plants. Lance and Pearson (1969) showed that reduced Ca uptake was the first externally observable symptom of Al damage on cotton seedling roots. This effect was noted within 1 h after root exposure to a solution containing only 0.3 mg $L^{-1}$ of Al. The inhibition of Ca uptake was avoided by increasing the Ca concentrations of the nutrient solution to 600 mg $L^{-1}$. Aluminum was believed to interfere with the function of Ca in the plasmalemma. Lund (1970) found that Ca reduced the detrimental effects of Al in nutrient solutions. Aluminum reduced cotton root growth when the ratio of Al activity to Ca activity was $> 0.02$. Franco and Munns (1982) reported that increasing Ca ($CaCl_2$) concentrations from 8 to 80 mg

$L^{-1}$ suppressed Al toxicity in bean. Aluminum toxicity in peach seedlings was related to reduced Ca uptake but not to reduced transport (Edwards & Horton, 1977). Simpson et al. (1977) attributed poor root growth of alfalfa to Al × Ca interaction. Similar relationships were found in kikuyugrass grown in soil at pH 4.4 (Awad et al., 1976). These investigators concluded that the yield-reducing effects of Al toxicity and Al-induced Ca deficiency were confounded.

Aluminum toxicity has recently been implicated in grass tetany in cattle. For example, Allen and Robinson (1980) found very high concentrations of Al (2000–8000 mg $kg^{-1}$) in forage samples from annual ryegrass pastures on which grass tetany was prevalent. Rumen content samples of dead tetany animals contained an average of 2400 mg $kg^{-1}$ of Al compared with 400 mg $kg^{-1}$ from normal animals. The addition of Al to rumen fluid–ryegrass–buffer solution decreased Mg solubility by 56% and Ca solubility by 74% within 48 h. They concluded that Al levels in forages and in rumen contents are actively involved in the etiology of grass tetany in ruminants. However, subsequent work at the same location (Cherney et al., 1983) indicated that the very high Al levels reported previously in forage and rumen fluid samples collected at grass tetany sites were due to soil contamination of the forage (and ingestion by grazing animals) rather than high Al uptake by the plants. Recent work (D. L. Robinson, 1983, personal communication) has also shown that the administration of either $Al_2(SO_4)_3$ or $Na_2SO_4$ to cows depressed serum Mg levels. Hence, the decrease in Mg solubility induced by ($Al_2(SO_4)_3$, and previously attributed to Al, appears to be associated instead with the $SO_4^{2-}$ anion. Grass tetany is still consistently associated with high Al contents of rumen fluid, but the role of soil ingestion in the disease has not been determined. However, the $Al^+$ ion itself does not appear to be directly involved in the etiology of grass tetany as previously suggested.

Senile dementia (Alzheimer's disease) and renal dialysis encephalopathy in human beings have been associated with Al accumulation in brain cells (Karlik et al., 1980). Karlik et al. (1980) obtained evidence that Al formed three types of complexes with calf thymus DNA. Complex I formed at pH values > 6.0 and stabilized a portion of the DNA double helix; $Al(OH)^{2+}$ was believed to bind to phosphate oxygens. Complex II was observed at pH values < 6.0; it destabilized a fraction of the DNA double helix and produced intrastrand cross links; $Al^{3+}$ was presumably bound to hetocyclic N atoms and exocyclic carbonyl groups of purine and pyrimidine bases. Complex III formed at all pH values and had properties intermediate between I and II; structural evidence suggested that both $Al(OH)^{2+}$ and $Al^{3+}$ bound DNA simultaneously.

### C. Aluminum and Rhizobia

The rhizobia of some legume species appear to be more sensitive to Al than their host plants. For example, Pieri (1974) reported that the nodulation of peanut was reduced when Al saturation of the soil cation exchange capacity (CEC) reached 30% in sandy soils of Senegal. Higher levels of Al saturation were required for toxicity of the host plant. Carvalho et al. (1981) concluded that Al toxicity decreased the growth of *Stylosanthes* sp. more severely when plants were dependent on symbiotic $N_2$ fixation than when combined N was applied in fertilizer. The Al tolerances of six *Stylosanthes* sp. appeared to depend on the ability to nodulate and develop an efficient symbiosis in the presence of Al and on the inherent sensitivity of the host plant to Al.

The symbiotic $N_2$ fixation process itself is apparently less sensitive to Al than is the process of nodule formation. For example, Carvalho et al. (1982) found that a 10- to 20-d exposure to Al concentrations up to 2.7 mg $L^{-1}$ in nutrient solutions did not affect the $N_2$ fixation of well-nodulated plants of *S. hamata, S. humilis,* and *S. scabra.*

Mengel and Kamprath (1978) found that the shoot growth of soybean on eight organic North Carolina soils was significantly related to soil pH between 4 and 5. As pH increased, nodule numbers and weights and N concentration and total N uptake by plants increased markedly. The critical pH for shoot, root, and nodule growth was generally in the range of 4.6 to 4.8. The growth response to lime was attributed to decreased exchangeable and water-soluble Al, increased water-soluble Ca, and a more favorable pH for rhizobia.

The rhizobial colonization-infection process in bean roots grown in nutrient solution was decreased by Al concentrations above 0.89 mg $L^{-1}$ but was less sensitive to Al than root or shoot length growth of the host plant (Franco & Munns, 1982). Aluminum concentrations $> 0.51$ mg $L^{-1}$ significantly reduced shoot growth in 'Carioca' but not in 'Venezuela 350'. Aluminum at 0.51 mg $L^{-1}$ increased taproot elongation but decreased total root length. Higher Al concentrations produced stunting in taproots. Aluminum concentrations as high as 2.24 mg $L^{-1}$ at pH 4.5 did not affect root dry weight, nodule growth, and nitrogenase activity of the two cultivars studied.

Aluminum toxicity and low pH appeared more important than Mn toxicity and Ca deficiency in limiting the activities of rhizobia on cowpea and soybean roots (Keyser & Munns, 1979a). In that study, Al concentrations of 0.68 to 1.35 mg $L^{-1}$ produced more severe stress than did high Mn (10.8 mg $L^{-1}$) or low Ca (2 mg $L^{-1}$). Furthermore, Ca at 2 to 40 mg $L^{-1}$ did not protect rhizobia against Al toxicity. In general, cowpea rhizobia tended to be more tolerant to Al than *Rhizobium japonicum.* In another study, Keyser and Munns (1979b) found that 1.34 mg $L^{-1}$ of Al was more harmful to rhizobia than low pH (4.5) or low P (0.3 mg $L^{-1}$). Keyser et al. (1979) concluded that cowpea rhizobia contained a large, and perhaps continuous, variation in symbiotic tolerance to soil acidity at pH 4.6. Sixty-five percent of the acid soil sensitive strains were identified in nutrient solutions containing 1.35 mg $L^{-1}$ of Al. Strains that were highly tolerant to acid soils were not sensitive to Al in solution.

The extent to which N deficiency limits legume growth on acid soils is not easily determined. Munns et al. (1981) suggested that unlike other legumes, soybean grown on acidic soils may be limited by factors other than nodulation failure. In their studies with two soybean cultivars and 13 single-strain inoculants, liming the soil from pH 4.4 to 6.0 (aqueous paste) doubled plant growth regardless of N source, cultivar, rhizobial strain, or numbers. The inoculated plants were nodulated, green, and high in N even when growth was severely reduced by the acid soil. Plant symptoms indicated that soybean grown on acid soils were limited by Al toxicity to the host plant. This conclusion was supported by nutrient solution studies in which the pH, Al, and Ca were controlled at levels similar to those in soil paste extracts. In the solution experiments, soybean growth was unaffected by low Ca (8 mg $L^{-1}$) or low pH (4.5) but was depressed by 0.8 to 1.6 mg $L^{-1}$ of Al. From soil and nutrient culture studies, Munns et al. (1981) concluded that efforts to increase the acid soil tolerance of soybean should center on the plant and not on the rhizobia. This conclusion agrees with the observation that the relative acid soil (Al) tolerances of soybean cv. Perry and Chief were the same with symbiotic or fertilizer N (C. D. Foy & T. E. Devine, 1980, unpublished data). For additional information on rhizobial growth in relation to acid soil factors (excess Al and low availabilities of Ca, Mo, and P), see Andrew (1978) and Munns and Mosse (1980).

### D. Beneficial Effects of Aluminum

Aluminum is not regarded as an essential nutrient, but low concentrations can sometimes increase plant growth or produce other desirable effects (Foy et al., 1978; Foy & Fleming, 1978). Plants that have shown positive growth responses to Al include rice (Howeler & Cadavid, 1976), tropical legumes (Andrew et al., 1973), eucalyptus (Mullette, 1975), tea (Matsumoto et al., 1979), peach (Edwards et al., 1976), sugar beet (Kesar et al., 1975), maize inbreds (Clark, 1977), and wheat (Foy & Fleming, 1978). When beneficial effects of added Al were observed in rice cultivars, the growth stimulus was greater in Al-tolerant cultivars than in Al-sensitive cultivars (Howeler & Cadavid, 1976). An Al concentration (3 mg $L^{-1}$ at pH 4.5) that was highly detrimental to the Sonora 63 wheat cultivar from Mexico was markedly beneficial to the BH 1146 wheat cultivar from Brazil (Foy & Fleming, 1978). Kumar (1979) reported that 4 mg $L^{-1}$ Al in nutrient solution was beneficial to betal palm.

The mechanisms by which small quantities of Al benefit plant growth and other qualities are debatable and may be different for different plant genotypes and growth media. Possible explanations include (i) increasing the solubility and availability of Fe in calcareous soils (through hydrolysis of Al and lowering of pH), (ii) correcting or preventing Fe deficiency (by displacing Fe from bound, metabolically inactive sites within plants as suggested by Grime and Hodgson [1969] in *Scabiosa columbaria*), (iii) blocking negatively charged sites on cell walls and thereby promoting P uptake (Mullette, 1975), (iv) correcting or preventing P toxicity (Clark, 1977), (v) delaying root deterioration in low Ca solutions by slowing growth and preventing absolute depletion of Ca from the medium by sunflower (C. D.

Foy, 1979, personal observation), (vi) altering the distribution of growth regulator in roots of peach seedlings (Edwards et al., 1976), (vii) preventing toxicities of Cu (Liebig et al., 1942) and Mn (Rees & Sidrak, 1961), (viii) serving as a fungicide (Ko & Hora, 1972; Lewis, 1973; Muchovej et al., 1980), and (ix) reducing undesirable top growth in N-rich nursery stock (Borkenhagen & Iyer, 1972).

## E. Physiology of Differential Aluminum Tolerance

Plant species and cultivars within species differ widely in their tolerances to excess Al in the growth medium. For current reviews on differential tolerance, genetic control mechanisms, and the development of Al-tolerant plants, see Foy (1983a, b) and Kamprath and Foy (1984).

The exact physiological mechanisms of Al tolerance are still debated; tolerance may be controlled by different genes, acting through different biochemical pathways in different plants. Epstein (1969) suggested that an element present in excess can interfere with metabolism through competition for uptake, inactivation of enzymes, displacement of elements from functional sites, or alteration of the structure of water. Many of these effects probably involve modification of membrane structure and function. Obviously, Al tolerant plants must be able to prevent the absorption of excess Al or to detoxify the Al that is absorbed.

Aluminum tolerance has been associated with pH increases (solution cultures and soils), Al trapping in nonmetabolic sites within plants, greater P efficiency, greater Ca and Mg uptake and transport, lower root CEC, greater root phosphatase activity, higher internal concentrations of Si, higher $NH_4^+$ or $NO_3^-$ tolerance or preference, higher organic acid contents, greater Fe efficiency, and more resistance to drought. For citations from the earlier literature, see Foy (1974, 1983b), Foy and Fleming (1978), and Foy et al. (1978). This review emphasizes more recent developments in understanding Al tolerances in plants.

### 1. pH CHANGES IN ROOT ZONES

Some Al-tolerant cultivars of wheat, barley, rice, pea, and maize (inbreds) increase the pH of the nutrient solutions in which they grow and thus decrease the solubilgy and toxicity of Al by precipitation (Foy et al., 1978). In contrast, Al-sensitive cultivars of the same species decrease or have no effect on solution pH and are therefore exposed to Al for longer periods. In certain wheat genotypes, these pH differences have been demonstrated in thin layers of soil directly adjacent to plant roots and even in the bulk of soil in pot cultures. In wheat, evidence suggests that these differential pH changing abilities are associated with differential anion-cation uptake.

### 2. ALUMINUM UPTAKE AND DISTRIBUTION

Aluminum-tolerant plants may be grouped according to where Al accumulates within their tissues. In one group, Al concentrations in the tops are not consistently different from those of Al-sensitive plants, but the root Al concentrations are lower. Plants in this category include certain cultivars of wheat, barley, soybean, snapbean, triticale, and pea (Foy, 1974; Mugwira et al., 1981; Klimashevskii et al., 1976). In such cases, Al tolerance apparently involves an exclusion mechanism.

In a second group of plants, Al tolerance is associated with less Al in plant tops, entrapment of more Al in roots, or both. Examples are azalea; cranberry; rice; triticale; rye; alfalfa clones; certain cultivars of ryegrass, wheat, barley and potato (Foy et al., 1978); tomato genotypes (Baumgartner et al., 1976); populations of paper birch (Steiner et al., 1980); and kikuyugrass compared with Al-sensitive cabbage and lettuce (Huett & Menary, 1980).

In a third group, Al tolerance is directly associated with Al accumulation in tops; hence, such plants have high internal tolerance to Al. Examples of Al accumulators are tea, certain Hawaiian grasses, pine trees, and mangrove (Foy et al., 1978), flowering dogwood (Crum & Franzmeir, 1980), river birch (Bartuska & Unger, 1980), and mouse-ear cress (Tingey et al., 1982). Seven clones of Norway spruce were extremely tolerant to artificial acid rain in pots of soil; needles accumulated 1350 mg $kg^{-1}$ of Al without injury (Ogner & Teigen, 1980). Haridasan (1982) found that eight plant species from three families, native to the Cerrado region of central Brazil, accumulated 4310 to 14 120 mg $kg^{-1}$ of Al in their leaves when grown on a Red Latosol at pH 3.7 to 4.6. Aluminum accumulators were defined as those having more than 1000 mg $kg^{-1}$ of Al and nonaccumulators as those containing $< 400$. The lowest Al concentrations found were 150 to 200 mg $kg^{-1}$. Aluminum-accumulating species were no lower in foliar concentrations of Ca, Mg, K, Fe, Mn, Zn, and P than nonaccumulating species. Memon et al. (1981) concluded that in the Al-tolerant tea plant, Al is bound to cell walls of epidermal and mesophyll cells and thereby prevented from reaching critical metabolic sites within the cell.

Cases of differential internal tolerance to Al have been reported in rye genotypes (Aniol et al., 1980) and callus cell cultures of tomato lines (Meredith, 1978a, b). Mugwira (1980) reported that rye had a higher internal tolerance to Al than either wheat or triticale. Naidoo (1976) attributed the superior Al tolerance of 'Dade' snapbean (compared with 'Romano') to a greater ability to tolerate a given level of Al within the root cell nuclei where Al was presumably associated with esteric P in nucleic acids and membrane lipids.

Huett and Menary (1979) concluded that Al uptake by Al-sensitive cabbage and Al-tolerant kikuyugrass was nonmetabolic. In freeze-dried roots, Al was found uniformly distributed along the roots of Al-sensitive cabbage and lettuce and Al-tolerant kikuyugrass. Highest concentrations of Al were found in the root epidermis, followed by the cortex. Aluminum was found in the stele of all species and in the protoplasm of cortical cells (Huett & Menary, 1980). These investigators concluded that Al can enter the plant by moving into meristematic cells and the symplasm via the cortex, hence, bypassing the endodermal barrier. Unlike other investigators, Huett and

Menary (1980) found poor correlation between the distribution of Al and P in plant roots.

Some effort has been made to establish critical levels of Al for toxicity in plant tops. For example, Wallace and Romney (1977) reported that threshold concentrations of Al toxicity were 30 mg $kg^{-1}$ in soybean leaves and 20 mg $kg^{-1}$ in rice shoots. Aluminum concentrations of 70 mg $kg^{-1}$ in soybean leaves were reported to produce severe toxicity. However, it is doubtful that such concentrations are directly harmful to the tops. Normal alfalfa has been found to contain 300 mg $kg^{-1}$ of Al in its tops, and this species is much more sensitive to Al than either soybean or rice (Foy, 1964). Malavolta et al. (1979) stated that Al toxicity in sorghum was associated with 640 mg $kg^{-1}$ of Al in lower leaves and 1220 mg $kg^{-1}$ in upper leaves. Duncan (1982) found that sorghum genotypes that were tolerant to low soil pH (and probably Al) contained lower concentrations of Al, Fe, and Mn than did those that were more sensitive. An acid soil–sensitive cultivar had 5021 mg $kg^{-1}$ of Al and 1820 mg $kg^{-1}$ of Fe in its tops; a more tolerant cultivar contained 462 mg $kg^{-1}$ of Al and 403 mg $kg^{-1}$ of Fe. Duncan concluded that no single element comparison can be used to identify genotypes as tolerant or sensitive to acid soils. The assignment of critical Al levels in plant tops appears to be of questionable value. The accumulation of this element (and others, like Fe) under conditions of Al toxicity probably refects prior root damage, resulting in differential elemental transport. Aluminum concentrations directly harmful to tops have not been determined.

## 3. NITROGEN NUTRITION

In strongly acid soils, nitrification in inhibited, and $NH_4^+$ is an important source of N for plants. Many plants that are adapted to such soils, and hence, tolerant to Al, also tolerate $NH_4^+$ levels that are toxic to other plants. Some of these acid-tolerant plants seem to prefer $NH_4^+$ to $NO_3^-$. Examples are cranberry, sugarcane, European white birch, certain grasses, and blueberries. Both highbush and lowbush blueberries use $NH_4^+$ effectively but are often injured by $NO_3^-$. Nitrate toxicity in lowbush blueberry coincides with a lack of $NO_3^-$ reductase activity in its leaves and roots. Superior Al tolerance in certain wheat cultivars is characterized by the ability to use $NO_3^-$ efficiently in the presence of $NH_4^+$ and to increase the pH of the growth medium. The ratio between $NO_3^-$ and $NH_4^+$ in nutrient solution determines the rate and direction of plant-induced pH changes in the presence or absence of Al (Foy & Fleming, 1978).

Additional evidence suggesting an association between N metabolism and Al tolerance has been provided by Mesdag et al. (1970), who found that acid soil (Al) tolerance in wheat varieties was correlated with high protein in the grain. In contrast, Aniol and Kaczkowski (1979) found no such relationship in 55 spring wheat varieties. However, Foy and Fleming (1982) found that superior Al tolerance in the spring wheat variety UC-44-111 (compared with 'Anza') coincided with a higher $NO_3^-$ reductase (NR) activity in its tops, either in the presence or absence of Al, and with the ability to maintain a higher pH in nutrient solutions. Aluminum stress significantly reduced NR activity in the tops of Anza but had no effect on that in UC 44-111 tops.

## 4. CALCIUM NUTRITION

Aluminum tolerance in certain cultivars of wheat, barley, soybean, and snapbean has been associated with the ability to resist Al-induced Ca deficiency (Foy et al., 1978). Baumgartner et al. (1976) found that the Al-tolerant tomato cv. Santa Cruz Kada had a lower Ca requirement than the less Tolerant cv. Ronita. Aluminum-tolerant varieties of soybean accumulated more Ca, P, and Mg in their tops per unit of root length than did the more sensitive cultivars (Sartain, 1974). Huett and Menary (1980) found that 500 mg $kg^{-1}$ of Ca in nutrient solution completely overcame the detrimental effect of Al on root yields of cabbage and lettuce and on top yields of cabbage. Calcium did not affect Al concentrations in plant roots but decreased Al concentrations in plant tops of cabbage, lettuce, and kikuyugrass.

## 5. PHOSPHORUS NUTRITION

In many plants, Al tolerance appears to be closely related to P use efficiency. For example, in certain cultivars of wheat and tomato and inbred lines of maize, Al tolerance coincides with the ability to tolerate low P levels in nutrient solutions in either the presence or absence of Al (Foy et al., 1978). Salinas and Sanchez (1976) stated that Al toxicity and P deficiency frequently occur together on most of the Cerrado Oxisols of Brazil. They found that Al tolerance and low P tolerance coincided in certain cultivars of wheat and bean. Aluminum tolerance in cultivars of pea has been closely related to the ability to absorb and use P in the presence of Al (Klimashevskii et al., 1979). Baumgartner et al. (976) found that the Al tolerant tomato cv. Santa Cruz Kada had a lower P requirement than the less-tolerant 'Ronita'. Aluminum-tolerant soybean cultivars absorbed more P per unit of root length than did Al-sensitive cultivars (Sartain, 1974). According to Garcia and Leon (1978), differential Al tolerances of cocoa hybrids are probably due to differential abilities to extract P in the presence of Al. Fox (1979) noted that corn varieties developed for tropical America (probably Al tolerant) are more tolerant to low P soils than those developed for temperate climates and that varietal tolerance to Al (or Mn) toxicity appears closely related to resistance to P deficiency.

Cambrai and Calbo (1980) reported that pretreatment of sorghum roots with Al decreased P uptake by 42% in an Al-sensitive cultivar but did not affect that of the Al-tolerant cultivar. In both cultivars, Al reduced the ATPase activity of the plasma membrane. Aluminum in the root medium has been reported to greatly decrease concentrations of inorganic P, lipid-bound P, nucleic acid–bound P, protein-bound P, and total P in the tops of sorghum plants (Edgar Cunha, 1981, personal communication, Sete Lagoas, Brazil). These decreases were significantly lower in the Al-tolerant 'CMS 297' than in the Al-sensitive 'CMS 244'. Soybean selections differing in Al tolerance had similar levels of metabolic P-containing intermediates when grown with no Al; however, under Al stress, pyruvate and ATP levels were significantly increased in tolerant lines but not in sensitive lines (Hanson & Kamprath, 1979).

In contrast to the evidence cited above, Brauner (1979) found that Al tolerance in wheat cultivars was not related to concentrations of P, Ca, or Mg in plant tops.

Lowe and Bortner (1973) described an interesting interaction among Al, P, and cultivars of burley tobacco. The cv. L8 developed a "physiologic leaf spotting" when grown at low P levels in nutrient solutions or grown with low P and low pH in soils. Increasing either soil pH or soil P levels reduced the severity of spotting, and increasing both pH and P levels prevented the symptom. The cv. Burley 21 and Experimental 4 were resistant to this disorder. Reciprocal grafting studies with L8 and Burley 21 showed that the problem in L8 resided in the tops, not the roots. The tops of L8 contained normal P concentrations, but the spotting symptoms were significantly correlated with Al concentrations in the tops when plants were grown on an acid soil. The results indicated that the appearance of the spotting problem in L8 plants grown on acid soil was due to greater Al uptake or greater interference by Al in P metabolism. It was suggested that L8 has genetically reduced ability to metabolize P and that the problem is accentuated by Al.

## 6. INTERACTIONS OF Al WITH Fe, Mg, Si, AND K

Acid soil sensitivity in certain wheat and barley cultivars has been associated with Al-induced Fe-related chlorosis in nutrient cultures at pH 4.1 (Otsuka, 1969). Furlani and Clark (1981) observed that Al induced on Fe-related chlorosis in sorghum grown in nutrient solution. Aluminum sensitivity in a mutant strain of *Bacillus megaterium* was attributed to Al-induced Fe deficiency caused by a loss in the ability to produce a secondary hydroxamate (schizokinen) that chelates Fe and makes it available to the normal strain (Davis et al., 1971). Brown and Jones (1977b) found that certain sorghum lines grew poorly and developed chlorosis and purple pigmentation in Al-toxic Bladen soil at pH 4.3. Chlorotic lines contained significantly less Fe and P than those that remained green. Aluminum concentrations were higher in chlorotic plants than in green plants, but the highest concentration found was 127 mg $kg^{-1}$. Fleming and Foy (1982) found that the Al-tolerant barley cv. Dayton was more susceptible to Fe-related chlorosis than the Al-sensitive 'Kearney' when both were grown in nutrient cultures. Greater susceptibility to Fe stress in Dayton barley was characterized by leaf chlorosis, reduced top growth, and greater accumulations of Ca, Mn, and P in the tops of plants subjected to Fe treatments below 0.5 mg $L^{-1}$ in the nutrient solution.

Aluminum tolerance has been associated with greater uptake of K and Mg in potato cultivars and with greater Mg uptake in maize inbred lines. Certain Al-tolerant rice cultivars accumulate higher levels of Si in the epidermal cells of their leaves than do Al-sensitive cultivars. Silicon is known to reduce the toxicity of Mn in barley leaves and could play a similar role in detoxifying Al. For original references, see Foy et al. (1978).

## 7. ORGANIC Al COMPLEXES IN PLANTS

Jones (1961) suggested that naturally occurring organic acids in Al-tolerant species chelate Al and thereby reduce the Al-P precipitation expected at normal pH levels in plant sap. Small (1946) noted that acid-tolerant plants generally have strong organic acid buffer systems in their cells, but alkaline-tolerant plants have phosphate buffers dominating the system. The phosphate sysstems would be more susceptible to disruption by Al-P interaction than the organic acid systems. Aluminum tolerance in certain calcifuge species has been attributed to a chelating mechanism that also has an affinity for Fe (Grime & Hodgson, 1969). Lee (1977) noted that hydroxy-Al was toxic to maize seedlings but that Al citrate was not. The addition of humic acid also ameliorated Al toxicity at pH 4.7. Bartlett and Riego (1972) found that the toxicity of ionic $Al^{3+}$ for maize was prevented by complexing the Al with citrate, EDTA, or soil organic matter extract. High Al tolerance in the Al-accumulating tea plant may be due to chelation (and detoxification) of Al by organic acids and phenols (Sivasubramaniam & Talibudeen, 1972). Klimashevskii and Chernysheva (1980) found that the roots of Al-tolerant varieties of pea, maize, and barley contained substantially higher concentrations of citric acid than did those of Al-sensitive varieties of the same species.

## 8. ALUMINUM AND PLANT MEMBRANES

Hofler (1958) concluded that Al tolerance in *Spirogyra* depended on the structure of the plasmalemma and was genotype specific. Certain *Spirogyra* cells showed plasmolysis when treated for only a few seconds with a 0.1% $Al(NO_3)_3$ solution, but another strain of *Spirogyra* cells, found in a swamp, was essentially unaffected by the same treatment. Hecht-Buchholz and Foy (1981) found that the superior Al tolerance of Dayton barley (compared with Kearney) was associated with a longer resistance of the plasmalemma to Al stress. Aluminum-induced autolysis of root cells, beginning with the disorganization of the plasmalemma, required 2 to 4 d longer in Dayton than in Kearney. Horst et al. (1982) postulated that mucigel, produced on Al-treated cowpea roots, protected them against Al injury.

## 9. ALUMINUM-WATER RELATIONSHIPS

Excess Al can reduce water use by restricting plant root penetration and proliferation in acid subsoils when surface soils become dry. In addition, Al toxicities can damage roots to the extent that they cannot absorb adequate water, even in moist soils. In preliminary studies, the Al-sensitive Kearney barley (developed in the Kansas-Nebraska region) was more tolerant to ordinary soil water stress than was the Al-tolerant Dayton (developed in Ohio); however, Kearney was much more sensitive to Al-induced water stress than was Dayton (Krizek & Foy, 1981). When plants are grown in surface soils overlying acid, Al-toxic subsoils, both types of water stress tolerance are needed to minimize yield reductions from drought.

## IV. MANGANESE TOXICITY

Manganese toxicity is probably the second most important growth-limiting factor (after Al toxicity) in acid soils. The solubility and, hence, the potential toxicity of Mn to a given crop, depends on many soil properties, including total Mn content, pH, organic matter level, aeration, and microbial activity (Foy, 1973; Stahlberg et al., 1976). Manganese toxicity generally occurs in soils with pH values of 5.5 or below if the soil contains sufficient total Mn. However, it can also occur at higher soil pH values in poorly drained or compacted soils where reducing conditions favor the production of divalent Mn that plants absorb. Some soils do not contain sufficient Mn to produce toxicity, even at pH 5 or below. Soils of the Atlantic Coastal Plains of the United States are lower in total Mn than those of the Gulf Coastal Plains (Adams & Pearson, 1967). Hence, at a given soil pH, Mn toxicity is less likely in the former than in the latter. However, even within the Gulf Coastal Plains, soils vary widely in Mn contents (Sedberry et al., 1978).

Manganese toxicity is frequently induced or intensified by N fertilization, which lowers soil pH, fumigation with steam or methyl bromide, air-drying, or flooding (Nelson, 1977; Kluthcouski & Nelson, 1979). The addition of organic matter can reduce Mn toxicity, probably by chelating excess divalent Mn that plants absorb (Masui & Ishida, 1975). The availability of Mn in soils is closely related to the activities of microorganisms that can oxidize the soluble and toxic divalent Mn to the tetravalent, nontoxic form (Kamura & Nishitani, 1977; Bromfield, 1979). Sidorenko et al. (1979) reported that *Metallogenium* Perf. oxidized Fe and Mn on rice roots.

### A. Plant Symptoms of Manganese Toxicity

Unlike Al, excess Mn seems to affect plant tops more directly than roots (Foy, 1973). In addition, Mn produces more definitive symptoms in plant tops than does Al, and for a given plant, Mn accumulates somewhat in proportion to plant injury (Osawa & Ikeda, 1980; Pavan & Bingham, 1981). Plant symptoms of Mn toxicity include marginal chlorosis and necrosis of leaves (alfalfa, rape, kale, lettuce), leaf puckering (snapbean, soybean, cotton), chlorosis of young leaves (resembling Fe deficiency) and necrotic spots on leaves (barley, lettuce, soybean). Necrotic spots in barley and chlorotic leaf margins in lettuce contain much higher Mn concentrations than adjacent leaf tissue (Vlamis & Williams, 1973). In severe cases of Mn toxicity, plant roots turn brown, usually after the tops have been severely injured.

Specific plant physiological disorders associated with excess Mn are crinkle leaf of cotton (Foy, 1973), stem streak necrosis in potato (Lee &

McDonald, 1978), internal bark necrosis of apple trees (Miller & Schubert, 1977; Scibisz & Sadowski, 1979), growth retardation and leaf tip burn in carnation (Ishida & Masui, 1976) and fruit cracking at the blossom end of muskmelon (Masui & Ishida, 1975).

In several tropical pasture grasses, Mn toxicity has been associated with small, black or dark brown flecks on the lower leaves and sometimes with an induced Fe deficiency. Buffelgrass showed these symptoms, plus white bands 1 to 2 cm wide extending transversely across the leaves at regular intervals along the blades. No other species had this latter symptom (Smith, 1979). Similar bands were noted in certain maize genotypes grown on a high Mn Richland soil (R. B. Clark, 1981, personal communication). Manganese toxicity has been characterized by interveinal chlorosis on upper leaves in bean, eggplant, pepper, tomato, and spinach and by marginal leaf chlorosis on lower leaves of cabbage, lettuce, and celery. In Welsh onion, excess Mn induced chlorosis of the older leaves (Osawa & Ikeda, 1977, 1980).

Andrew and Pieters (1976) described Mn toxicity symptoms in glycine as follows: interveinal chlorosis of young expanding leaves similar to Fe deficiency; mottling and irregularly shaped brown spots in the vicinity of midrib and main veins of expanded leaves; puckering of leaf surfaces; and in severe cases, a browning of the midribs and main veins (particularly on the upper leaf surfaces) that sometimes extended to petioles and stems. Jones and Clay (1976) found that Mn injury in the tropical legume Townsville stylo was characterized by an overall yellowing, particularly in the youngest leaves and stems, with some interveinal chlorosis and leaf puckering. Under severe Mn toxicity, the tips of emerging leaves became necrotic before leaves were fully expanded, and the necrosis proceeded from the leaf tip downward. In Easter lily, high Mn levels in nutrient cultures produced shorter plants, leaf chlorosis, chlorotic mottling, and some leaf curling (Holmes & Coorts, 1980).

Manganese toxicity is common in burley tobacco because this crop is frequently grown on high Mn soils maintained at a low pH to control black root rot disease. In Virginia, this problem has been reported at soil pH values of 5.3 or below but not at higher pH. Manganese toxicity causes slow growth of tobacco seedlings, and toxicity symptoms appear 4 to 5 weeks after transplanting in the field (Link, 1975, 1979; Smiley, 1976). The symptoms include a light green or yellowish mottling of leaves between the large veins, dark green leaf veins, and a gradual development of necrotic spots, especially in the older leaves. Leaves may also have a hard, semiglossy appearance. Plants may recover but overall growth is reduced. Arnold (1977) described a disorder called "grey leaf" in tobacco grown on shallow, poorly drained soils at pH 4.7 to 5.1; the condition is characterized by high concentrations of both Mn and Fe in plant tops.

Plant symptoms of Mn toxicity are often detectable at stress levels that produce little or no reduction in vegetative growth. In contrast, Al toxicity can reduce yields greatly (by root damage) without producing clearly identifiable symptoms in plant tops. Hence, in acid soils that contain high levels of both Al and Mn in solution, the plant growth reductions observed may be largely or wholly attributed to Mn toxicity when Al toxicity is the more important of the two factors (Foy, 1976).

### B. Physiological and Biochemical Effects of Excess Manganese

Manganese toxicity has been associated with the following effects on plants: destruction of IAA by increasing the activity of IAA oxidase (cotton and Japanese morning glory); a possible amino acid imbalance (potato); increased replication of RNA in potato spindle tuber virus, increased activities of peroxidase and polyphenol oxidase, decreased activities of catalase, ascorbic acid oxidase, glutathione oxidase, and cytochrome C oxidase, lower ATP contents and lower respiration rates (cotton); reversal of growth inhibition of roots caused by gibberellic acid, which enhances auxin production (sorghum); and induction of erythromycin-resistant mitochondrial mutations (yeast). Manganese interacts with many other mineral elements in plant nutrition, and under certain conditions, the addition of these elements (Si, Fe, Ca, P) may alleviate Mn toxicity (Foy, 1973; Foy et al., 1978).

Recent literature provides additional evidence that excess Mn alters the activities of enzymes and hormones in plants. For example, Sirkar and Amin (1979) found that reduced respiration in young cotton seedlings, caused by Mn toxicity, could be reversed by adding IAA. (The addition of IAA did not affect the respiration of normal tissues.) They also reported that the respiration rates of Mn-injured tissue were increased by indole-3-butyric acid (IBA), indole-3-acetamide (IAC), and indole-3-propionic acid (IPA).

Horst and Marschner (1978a) proposed a Mn toxicity hypothesis that involves auxin and Ca. They observed that excess Mn inhibited Ca transport into the leaves of bean plants and induced typical Ca deficiency symptoms (crinkling in shoot apices and younger leaves). Manganese toxicity also reduced the CEC of leaf tissues and reduced Ca movement into the apparent free space of isolated leaf segments. From these observations, they concluded that Mn toxicity increased the activity of IAA oxidase, which decreased the auxin level (proposed earlier in cotton by Morgan et al., 1966, 1976). They reasoned further that the lowered auxin level then reduced cell wall expansion and that the subsequent reduction in formation of new, negative sites decreased Ca translocation into the tissues.

Helyar (1978) suggested that Mn toxicity problems are caused mainly when plants lose control of their Mn-activated enzyme systems. Heenan and Campbell (1981) found that Mn toxicity in soybean was associated with reduced nitrate reductase activity. Aoba and Sekiya (1977) reported that in Mn-injured satsuma trees the activities of leaf catalase, peroxidase and polyphenol oxidase, and $O_2$ uptake were higher, whereas the activity of ascorbic acid oxidase was lower than in healthy trees. In another study (Aoba et al., 1977), divalent Mn was detected in epidermal and xylem tissues of roots from Mn-injured satsuma trees. Treating the fine roots with heat or $NaN_3$ to inhibit their oxidizing powers greatly increased Mn absorption. By contrast, the untreated, healthy root oxidized divalent Mn to $MnO_2$, which remained on root surfaces. The site of greatest Mn oxidation was about 5 mm from the root apex and occurred in the epidermis, endodermis, and xylem. Masui et al. (1980) observed that Mn accumulation by muskmelon was much higher in fine roots than in coarse roots and especially high in root hairs. Manganese accumulation was greater in older leaves than in younger leaves and greater in the outer portion of the fruit (rind) than in inner portions. Memon et al. (1980) reported that in the Mn accumulator plant *Acanthopanax sciadophylloides* Franch. et Savat., Mn accumulated markedly in peripheral cells, leaf petiole, petiolule, and petiole tissues. High Mn concentrations were found in the epidermis, palisade, and spongy parenchyma cells. Low Mn was found in vascular bundle cells. In the petiolule and petiole, most of the Mn was concentrated in the cell walls of the epidermis, collenchyma, and bundle sheath cells. Such compartmentalization of Mn may act as a protective mechanism against toxicity by keeping excess Mn from key metabolic centers.

Satsuma tissues injured by excess Mn were stained dark brown and gave positive tests for pyrogall, catechol, and $Fe^{2+}$ (Aoba et al., 1977). The same investigators found that excess Mn in soil or nutrient solution increased plant levels of phenolic derivatives. The polyphenolic compounds identified in new leaves included five flavonoids and three phenolic acid derivatives. The latter were identified as 2,5-dihydroxybenzoylglucose, feruloylglucose, and feruloylrhamnoglucose. Rao and Gupta (1979) found that both Mn-deficient and Mn-intoxicated sugarcane accumulated more free amino acids than normal plants.

Manganese toxicity is often associated with a decrease in the Ca concentrations of plants (Smith, 1979; Horst & Marschner, 1978a; Takayanagi, 1976), and supplying additional Ca to the growth medium sometimes reduces Mn accumulation, decreases the severity of Mn-induced chlorosis and alleviates growth reductions (Osawa & Ikeda, 1977). Masui et al. (1976) reported that additional Ca decreased Mn concentrations in muskemelon plants but did not prevent yield decreases by excess Mn.

The addition of Si to nutrient solutions reduced Mn toxicity in barley (Williams & Vlamis, 1957) and bean (Horst & Marschner, 1978b, d). Apparently, this detoxification was not due to reduced Mn uptake but to the prevention of localized accumulations of Mn associated with necrotic spots on the leaves. Horst and Marschner (1978c) found that such spots on red kidney bean were precipitated Mn compounds that occurred primarily in the walls of cells adjacent to the vessels. These authors also reported that excess Mn induced secondary Fe and Ca deficiencies in red kidney bean. Kluthcouski and Nelson (1979) found that Si decreased the Mn concentrations of the youngest fully developed, soybean leaves. The effects of both Mn deficiency and Mn toxicity were slightly moderated by the addition of Si.

Manganese and Fe are closely related in plant nutrition, and some investigators have used the Fe/Mn ratio in plant tops as one indicator of Mn toxicity. For example, Hati et al. (1979) found that Mn toxicity occurred when Fe/Mn ratios were 1.36, 1.40 and 0.8 for cotton, wheat, and soybean, respectively, when each was grown on soil at pH 4.0. Iron and Mn contents were negatively correlated. Manganese toxicity occurred in snapbean, cv. Wonder Crop no. 2, when Fe/Mn ratios in leaves were $< 1.5$ (Kohno & Foy, 1983). Excess Mn often produces plant symptoms resembling Fe deficiency (Horst & Marschner, 1978c), and increasing the Fe supply may decrease Mn uptake (Masui et al., 1976). However, not all Mn toxicity is Fe related. Gupta and Rao (1977) concluded that Mn toxicity in sugarcane is not Mn-induced Fe deficiency. Van der Vorm and Van Diest (1979) stated that under acid, aerobic conditions, the absorption of Fe is little affected by large quantities of Mn. However, excess Mn uptake can interfere with normal Fe metabolism without reducing Fe uptake.

Temperature × Mn interactions affect the severity of Mn toxicity in a given plant. For example, Moraghan et al. (1980) reported that Mn toxicity in 'Bragg' soybean was more severe at a lower temperature. Ruffy et al. (1979) found that in tobacco, the lower the temperature, the lower the level of Mn in the leaf required for toxicity. Manganese concentrations in the leaf associated with toxicity at different day/night temperatures were as follows: 22/28°C, 700 to 1200 mg $kg^{-1}$; 26/22°C, 2000 to 3500 mg $kg^{-1}$; and 30/26°C, 5000 to 8000 mg $kg^{-1}$. Manganese toxicity was more severe in young leaves, and the severity decreased with age, even at the same internal Mn concentrations. They concluded that the increased tolerance of tobacco to high leaf Mn concentrations in warm temperatures was associated with a more rapid rate of leaf expansion, accompanied by increased vacuolar capacity for disposal of accumulated Mn. Thus, although Mn uptake may be increased with higher temperatures, as reported for oat (Cheng & Pesant, 1977), the increased internal tolerance to Mn at higher temperature may prevent toxicity. Such temperature × Mn interactions increase the difficulties of determining critical Mn concentrations in soils and plants.

Manganese toxicity can also be affected by the form of N ($NH_4^+$ or $NO_3^-$) supplied to plants (Osawa & Ikeda, 1980). Manganese toxicity symptoms on tomato, spinach, and cabbage were less severe when N was present as both $NH_4^+$ and $NO_3^-$ than when only $NO_3^-$ was present. In some cases the symptoms were less severe at pH 4 than at 6. The incidence of necrotic brown spots was reduced by increasing the proportion of $NH_4^+$, but all $NH_4^+$ reduced growth, even with normal Mn treatments. Osawa and Ikeda (1980) also found that $NH_4^+$ and $H^+$ at nontoxic levels greatly reduced Mn uptake and toxicity in vegetable crops. Another beneficial effect of $NH_4^+$ could have been a decreased pH (through root exchange for $H^+$), which would have protected plants against Fe deficiency whether Mn induced or not.

### C. Manganese and Rhizobia

Dobereiner (1966) suggested that the sensitivity of many legumes to acid soils could be explained by a specific Mn toxicity on $N_2$ fixation. She found that when bean plants were dependent on symbiotic $N_2$ fixation, 25 mg $L^{-1}$ of Mn in sand culture solutions decreased the total N contents of plants by 30 to 60%. However, when plants were adequately supplied with mineral N, the corresponding reduction was only 13%. In plants that were totally dependent on symbiotic $N_2$ fixation, the N contents of bean plants decreased linearly with the logarithm of the Mn concentration in plant tops; however, this was not observed in plants supplied with mineral N. The effects of Mn toxicity on $N_2$ fixation varied with inoculated strains of *Rhizobium phaseoli*. Souto and Dobereiner (1971) found that excess Mn markedly decreased nodulation in tropical legumes. Kliewer (1961) reported that 20 mg $L^{-1}$ of Mn in nutrient solutions reduced the number of nodulated alfalfa plants by 50% but did not affect the nodulation of birdsfoot trefoil. Dobereiner (1966) postulated that contradictions in the literature regarding the response of legumes to lime might be attributed to variable concentrations of soluble Mn in acid soils and differences in tolerance to Mn toxicity among both host plants and rhizobial strains.

Freire (1975) concluded that soil pH values of 4.0 to 4.5 were not harmful to the soybean nodulation process in the absence of excess Al and Mn. However, toxic Al and Mn, together with P and Ca deficiencies, were regarded as highly detrimental to nodule formation and $N_2$ fixation. In later work, Freire (1976) found that soybean cultivars with low tolerance to Mn toxicity showed greater responses to lime and had higher Ca uptake than Mn-sensitive varieties.

Recent studies by Keyser and Munns (1979a, b) with cowpea and soybean rhizobia indicate that in acid soils, Al toxicity and soil acidity are probably more important than Mn toxicity and Ca deficiency in restricting $N_2$ fixation. Aluminum at 0.68 to 1.35 mg $L^{-1}$ in solution produced a greater stress than high Mn (10.8 mg $L^{-1}$) or low Ca (2 mg $L^{-1}$). All rhizobial strains that were tolerant to Al were also tolerant to high Mn and low Ca. High Mn (10.8 mg $L^{-1}$) and low Ca (2 mg $L^{-1}$) solutions reduced the growth rates of some strains but did not stop the growth of any strain.

### D. Physiology of Differential Manganese Tolerance

Plant species and genotypes within species differ significantly in tolerance to excess Mn in the growth medium. For current reviews of differential Mn tolerance, genetic control, and the development of Mn tolerant plants, see Foy (1983a, b) and Kamprath and Foy (1984).

Manganese tolerance in plants has been associated with oxidizing powers of plant roots' (or associated microorganisms) Mn absorption and translocation rate, Mn entrapment in nonmetabolic centers, high internal tolerance to excess Mn, and the uptake and distribution of Si and Fe (Foy, 1973; Foy et al., 1978).

#### 1. MANGANESE UPTAKE AND TRANSPORT

Plant tissue Mn concentrations required to produce Mn toxicity symptoms vary with plant species, genotype within species, and environmental conditions, particularly temperature (Foy et al., 1978). Hence, "critical"

Mn concentrations in plants are useful only when applied within a limited set of circumstances. Internal Mn concentrations that have been associated with toxicity symptoms are as follows: apple leaves, 120 to 600 mg $kg^{-1}$ (Mask & Wilson, 1978); tomato, 450 to 500 mg $kg^{-1}$ in young leaves and 900 to 1000 mg $kg^{-1}$ in old leaves (Ward, 1977); sweet sorghum, 445 mg $kg^{-1}$ in upper leaves and 1440 mg $kg^{-1}$ in lower leaves (Malavolta et al., 1979); sunflower, 5300 mg $kg^{-1}$ (Edwards & Asher, 1982); flax genotypes, 500 mg $kg^{-1}$ (Moraghan & Freeman, 1978); cotton genotypes, 500 to 1959 mg $kg^{-1}$ (Foy et al., 1981a; Hati et al., 1979); Easter lily, 2000 mg $kg^{-1}$ (Holmes & Coorts, 1980); carnation cv. Coral, 2600 mg $kg^{-1}$ (Ishida et al., 1977); maize, 2500 to 6500 mg $kg^{-1}$ (Tanner, 1977), or 200 mg $kg^{-1}$ (Edwards & Asher, 1982); coffee, 1200 to 2600 mg $kg^{-1}$ (Pavan & Bingham, 1981); and snapbean cv. Wonder Crop no. 2, 120 mg $kg^{-1}$ (Kohno & Foy, 1983).

The difficulty of assigning critical internal Mn concentrations for toxicity are seen in the following example. Mask and Wilson (1978) reported that Bragg soybean had toxicity symptoms when leaves contained 171 to 191 mg $kg^{-1}$ of Mn; however, Jones and Nelson (1978) found that Bragg showed no toxicity symptoms when the leaves contained 320 mg $kg^{-1}$ of Mn. In a study of 20 lettuce cultivars in flowing solutions, Blatt and Van Diest (1981) concluded that Mn concentration in lettuce tops was not a useful index for Mn sensitivity.

Hutton et al. (1978) concluded that Mn tolerance in tropical legumes is associated with reduced Mn uptake in some species and greater internal tolerance in others. Bromfield (1978) noted that rape was more efficient than oat in absorbing Mn from $MnO_2$. Rape is also more sensitive to excess Mn than oat, and this may be related to greater Mn uptake. Flax is so efficient in absorbing Mn that it can accumulate sufficient Mn (500 mg $kg^{-1}$) in its tops to cause Mn toxicity, even on a calcareous soil at pH 8.1 (Moraghan & Freeman, 1978). Superior Mn tolerance in maize (compared with peanut) has been attributed to reduced transport of Mn from roots and stems to leaves; Mn-sensitive peanut accumulates high levels of Mn in its leaves (Benac, 1976). Cooke (1982) concluded that Mn toxicity in apple can result from a redistribution of Mn within the tree, as well as a Mn supply coming directly from the roots.

In some cases, differential Mn tolerances of plant genotypes within species have also been associated with differential Mn uptake. For example, Holmes and Coorts (1980) found that the cv. Ace of Easter lily was much more tolerant to excess Mn and absorbed less Mn than the cv. Nellie White. Gammon and McFadden (1979) reported that Mn accumulation by rose plants on *Rosa fortumana* rootstocks was five times that of rose plants on *R. odorata* rootstocks; however, Mn accumulation was not related to flower production.

### 2. INTERNAL TOLERANCE TO EXCESS Mn

Edwards and Asher (1982) concluded that differential Mn tolerances of several species grown in flowing nutrient solutions depended mainly on the abilities of plant tops to tolerate high Mn concentrations in tissue. However, in some species, low rates of intake, high retention of Mn (or both) in roots appeared important in tolerance. High Mn tolerance was associated with both high Mn uptake and high root retention in *Centrosema pubescens* Benth., with very high Mn uptake and internal tolerance and low root retention in sunflower, and high root retention of Mn in cassava. Low Mn tolerance in cowpea was associated with rapid uptake and transport of Mn to plant tops. Manganese-sensitive soybean absorbed Mn more slowly than cowpea and retained a higher proportion in its roots.

Temple-Smith and Koen (1982) found that plants vary in tolerance to excess Mn in nutrient solution. The degree of their tolerance was oat, sugar beet, and lupine > green pea > green bean > lucerne > barley > brussels sprout > poppy. Differences in tolerance were attributed largely to differences in internal tolerance to excess Mn. For example, Mn ''toxicity threshold'' concentrations in plant tops were 150, 280, 380, and 500 mg $kg^{-1}$ for poppy, brussels sprout, lucerne, and greenbean, respectively. High Mn sensitivities of poppy and brussels sprout were attributed to low shoot Mn toxicity thresholds as well as their tendencies to transport a high proportion of absorbed Mn to plant tops. French bean appeared to have a low internal tolerance to Mn.

Some plant genotypes within species also tolerate higher levels of Mn within their tops than do others. For example, Jones and Nelson (1978) concluded that the differential Mn tolerance of soybean cv. Bragg (sensitive) and Lee 68 (tolerant) was not reflected in Mn concentrations of plant tops. Heenan and Campbell (1981) found that Mn-tolerant Lee 68 and Mn-sensitive Bragg did not differ in Mn distribution among plant parts or in Mn concentrations in actively growing tissues. Brown and Devine (1980) reported that Mn-tolerant and Mn-sensitive soybean cultivars, and $F_2$ progeny from their reciprocial crosses, all contained about the same Mn concentrations in their tops (500 mg $kg^{-1}$). Similar findings have been reported for cultivars of wheat (Foy et al., 1978), cotton (Foy et al., 1981), and flax (Moraghan & Ralowicz, 1979). Horst (1982) concluded that Mn tolerance in cowpea genotypes depends on differential internal tolerances of leaves and not Mn exclusion. Tolerant genotypes were also characterized by a higher fraction of extractable Mn (80% methanol or 80% methanol plus 0.5*M* $CaCl_2$) and by a more homogenous distribution of Mn in leaves.

Smith (1979) concluded that different mechanisms were involved in explaining Mn tolerance in certain grasses. A *Setaria* sp. (foxtail and a *Paspalum* sp. (dallisgrass) were much more tolerant to Mn toxicity than rhodesgrass and buffelgrass. However, the tolerant species of *Paspalum* was the highest Mn accumulator, and the tolerant species of *Setaria* was the lowest. In sand culture the Mn concentrations in plant tops were 4880 and 2830 mg $kg^{-1}$ for *Paspalum* sp. and *Setaria* sp., respectively; for plants grown in soil the corresponding concentrations were 2837 and 1529 mg $kg^{-1}$.

Some plants accumulate high Mn concentrations in their tops without injury. Examples of Mn accumulators are sugar maple (Crum & Franzmeir, 1980); tropical plants of the genus *Rauwolfia* (Schchigel'ski et al., 1979); *Acanthopanax sciadophylloides* Franch. et Savat., peelberry poisonsumac, pogoda maple, redvein maple, and siebold maple (Memon et al., 1979); billberry (Bouwens & Longin, 1979); and white lupine (seeds) compared with blue lupine (seeds) (Oram et al., 1979). Jaffre (1979) found that the leaves of numerous New Caledonian species of the family Proteaceae contained abnormally high Mn concentrations (up to 5.5% of dry wt and 55.2% of ash wt in *Macadamia neurophylla* [Guillaumin] Virot.). Furthermore, such unusual Mn uptake occurred in soils that were not particularly high in Mn. Several of the Proteaceae also accumulated Al, but in general, the leaves of the Mn-accumulating species had relatively low Al contents. Species reported to be low in Mn-accumulating abilities include *Cryptomeria japonica* D. Don. and certain coniferous trees (Memon et al., 1979).

## 3. INTERACTIONS OF Mn WITH OTHER MINERAL ELEMENTS

Manganese has been reported to interact with Fe, Mo, P, Ca, and Si in affecting toxicity symptoms and growth. In some plants, excess Mn produces symptoms that resemble Fe deficiency, and these can be prevented by adding soluble Fe to the plant or soil. In others, Mn-induced symptoms are quite different and are not corrected by adding Fe. Brown and Jones (1977a) concluded that Mn tolerance was not related to Fe efficiency or inefficiency in soybean cultivars. The soybean cv. T203 is Mn tolerant and the soybean cv. Forrest is Mn sensitive, but both are equally susceptible to Fe-related chlorosis. Increased Ca in the growth medium may decrease Mn uptake and toxicity. Phosphorus additions can reduce the toxicity of Mn by rendering it inactive within the plant. Silicon can reduce Mn toxicity by preventing localized accumulations of Mn in leaves. From the foregoing, it appears that the toxicity of a given level of soluble Mn in the growth medium, or even within the plant, depends on interactions between Mn and several other mineral elements, particularly Fe and Si. Such interactions may account for the wide variety of plant symptoms and different growth reductions produced by excess Mn in different plant species and cultivars (Foy, 1973; Foy et al., 1978).

## 4. ORGANIC Mn COMPLEXES IN PLANTS

Little effort has been made to relate organic Mn complexes to Mn tolerance in plants, but this appears to be a promising area for research. Sutcliffe (1962) suggested that organic complexes can reduce the transport of Mn to plant tops. Tiffin (1967) found that Mn in tomato xylem exudate showed no binding to organic compounds, even when the exudate Mn content was enriched by 18-fold. However, Höfner (1970) found a close relationship between Mn and an amino acid and carbohydrate fraction of sunflower stem exudate. Gel filtration was used to fractionate the exudate into three associated components with molecular weights $< 1500$. Bremner and Knight (1970) and Bremner (1974) concluded that in ryegrass, Mn occurred as a single, cationic, and perhaps noncomplexed form. In phloem sap of castorbean, only limited organic binding of Mn was found (Van Goor & Wiersma, 1974). Gel filtration and electrophoresis showed that 60 to 70% of the Mn behaved as a simple cation; however, some organically bound Mn was associated with compounds having molecular weights between 1000 and 5000. White et al. (1981) concluded that Mn was bound primarily by citric and malic acids in soybean and tomato stem exudates.

Ebeid and Kutacek (1979) indicated that Mn did not occur as a free cation in xylem exudates of 20-day-old corn plants. Gel filtration showed that Mn was associated with three ultraviolet (254 nm) absorbing fractions. Two fractions contained amino acids, and the third (ca. 600 nm) appeared to be a carrier of a different type; it was practically free of amino acids but did contain 10.2% N. In this study, the addition of EDTA to xylem exudates favored Mn transport by fractions of relatively low molecular weight. Godo and Reisenauer (1980) reported that within a pH range of 4.5 to 6.5, the exudates from plant roots were important in controlling the uptake of soil Mn by plants. Exudate compounds, e.g., hydroxycarboxylase, increased soil Mn availability-solubility through reducing $MnO_2$ and complexing the divalent Mn released. Differential Mn uptake among plant genotypes might be explained by such a mechanism.

High-stability organic binding of Mn appears to be infrequent in plants but does occur. For example, Dieckert and Rozacky (1969) isolated a protein-Mn complex with a molecular weight of 56 300 and with one Mn atom from peanut seed; its function is unknown.

The overall evidence based on widely different plants indicates that Mn occurs as a free cation or is bound in soluble form with low molecular weight complexes in xylem fluid. However, the significance of this in relation to Mn tolerance has not been determined. It is worthwhile to note that the form of Mn occurring in xylem sap may be quite different from that in localized areas on plant leaves where toxicity occurs.

## V. CONCLUSIONS

Acid soil problems, e.g., $Al^{3+}$ and $Mn^{2+}$ toxicities, can be ameliorated by modifying the soil, the plant, or both (Foy, 1983a). The probability of success in any of these approaches would be greatly increased by a more complete understanding of the physiological and biochemical mechanisms by which plants adapt differentially to mineral stress. In the past, much of the research on mineral stress physiology has been confined to a single genotype of a particular plant species. A more promising approach for discovering stress tolerance mechanisms is to use pairs of stress-tolerant and sensitive genotypes (preferably near isogenic lines) of the same species. Knowledge concerning the physiology and biochemistry of such differential tolerances would be useful in modifying both plants and soils to achieve a desired result. Stress-tolerant and stress-sensitive genotypes are also valuable indicators of present and potential problems of mineral stress in soils, particularly when used in conjunction with conventional soil testing procedures (Long & Foy, 1970). Vose (1981) emphasized the potential value of plant mutants in studying physiological processes. He noted that 31 000 mutants of 130 species have been identified, but only a small percentage has been studied on a physiological basis.

Differential tolerances of plant genotypes to mineral stress undoubtedly involve differences in membrane structure, function, or both. For example, excess Al is known to inhibit DNA replication. But to gain access to the DNA, the Al must cross the outer cell membrane (plasmalemma) and also the nuclear membrane (tonoplast) or membranes surrounding other DNA-containing subcellular organelles that may be affected. In certain wheat genotypes the exclusion of Al appears to be an important feature of Al tolerance, but in other species, e.g., tea, which accumulates high internal levels of Al without injury, other mechanisms are obviously involved. Oxidation-reduction reactions may be implicated in differential Mn tolerances of some plant species (Mn exclusion by precipitation on roots), but the majority of evidence suggests that high internal tolerance is more important than exclusion in determining the Mn tolerances of genotypes within several species. To identify the mechanisms involved in such cases of differential internal tolerance, it will probably be necessary to determine the chemical forms as well as the physical locations and concentrations of toxic elements within the plants. Nuclear magnetic resonance and related techniques may be helpful in identifying such forms within intact plants or plant fractions.

To increase our understanding of differential mineral stress tolerance in plants, the following lines of study are suggested: characterization of cell membranes with respect to microstructure, electric potentials, enzyme activities, and permeabilities to various ions and molecules, with and without stress; chemical and physical compartmentalization of mineral elements in various plant fractions; levels and kinds of organic acids, amino acids and other ligands that may chelate metals and thereby regulate either mineral toxicity or deficiency; N metabolism ($NO_3^-$ reductase and protein accumulation); $NH_4^+$ vs. $NO_3^-$ tolerance or preference in relation to pH change and metal ion activities in root zones and plant fractions; the possible role of silicon in explaining differential Al and Mn tolerance; oxidation-reduction reactions in root zones as related to differential Mn tolerances; the nature of Mn-Fe interactions in producing differential Mn toxicity or Fe deficiency; the role of P and Ca nutrition in differential Al tolerance; interactions of mineral stress with water stress, temperature, and pathogens; and the role of rhizobia and mycorrhizae in differential plant adaptation to mineral stress.

## REFERENCES

Adams, F. 1978. Liming and fertilization of ultisols and oxisols. p. 377–394. *In* C. S. Andrew and E. J. Kamprath (ed.) Mineral nutrition of legumes in tropical and subtropical soils. CSIRO, East Melbourne, Australia.

Adams, F. 1981. Nutritional imbalances and constraints to plant growth in acid soils. J. Plant Nutr. 4:81–87.

Adams, F., and R. W. Pearson. 1967. Crop response to lime in the southern United States and Puerto Rico. *In* R. W. Pearson and F. Adams (ed.) Soil acidity and liming. Agronomy 12: 161–206. Am. Soc. of Agron., Madison, Wis.

Alam, S. M. 1981. Influence of aluminum on plant growth and mineral nutrition of barley. Commun. Soil Sci. Plant Anal. 12:121–138.

Alam, S. M., and W. A. Adams. 1979. Effects of aluminum on nutrient composition and yield of oats. J. Plant Nutr. 1:365–375.

Alam, S. M., and W. A. Adams. 1980. Effects of aluminum upon the growth and nutrient composition of oats. Pak. J. Sci. Ind. Res. 23:130–135.

Alexander, M. 1961. Introduction to soil microbiology. John Wiley & Sons, Inc., New York.

Alexander, M. 1980. Effects of acidity on microorganisms and microbial processes in soils. p. 363–364. *In* T. Hutchinson and M. Havas (ed.) Effects of acid precipitation of terrestrial ecosystems. Plenum Publishing Corp., New York.

Ali, S. M. E. 1973. Influence of cations on aluminum toxicity in wheat (*Triticum aestivum* Vill. Host). Ph.D. Diss. Oregon State Univ., Corvallis.

Allen, V. G., and D. L. Robinson. 1980. Occurrence of aluminum and manganese in grass tetany cases and their effects on the solubility of calcium and magnesium in vitro. Agron. J. 72:957–960.

American Society of Agronomy. 1980. Bragg soybeans develop abnormality. Crops Soils Mag. 32(5):18–19.

Andrew, C. S. 1978. Mineral characterization of tropical forage legumes. p. 93–112. *In* C. S. Andrew and E. J. Kamprath (ed.) Mineral nutrition of legumes in tropical and subtropical soils. CSIRO, East Melbourne, Australia.

Andrew, C. S., A. D. Johnson, and R. L. Sandland. 1973. Effect of aluminum on the growth and chemical composition of some tropical and temperate pasture legumes. Aust. J. Agric. Res. 24:325–339.

Andrew, C. S., and W. H. J. Pieters. 1976. Foliar symptoms of mineral disorders in *Glycine wightii.* CSIRO Australian Div. of Tropical Agron. Tech. Paper no. 18, p. 1–12.

Aniol, A., R. D. Hill, and E. N. Larter. 1980. Aluminum tolerance of spring rye inbred lines. Crop Sci. 20:205–208.

Aniol, A., and J. Kaczkowski. 1979. Wheat tolerance to low pH and aluminum: Comparative aspects. Cereal Res. Commun. 7:113–122.

Aoba, K., and K. Sekiya. 1977. Physiological studies on metallic trace elements in orchards: VI. Changes in inorganic nutrient contents and enzyme activities in leaves and fine roots of satsuma due to the worsening of soil chemical properties. (In Japanese with English summary.) Kaju Shikenjo Hokuku A 4:37–44.

Aoba, K., K. Sekiya, and R. Aime. 1977. Studies on metallic trace elements in orchards: III. Physiological studies on excess absorption of manganese by satsumas. (In Japanese with English summary.) Kaju Shikenjo Hokuku A 4:1–16.

Arnold, N. 1977. Grey tobacco. Lighter 47:27–28.

Arnon, D. I., W. E. Fratzke, and C. M. Johnson. 1942. Hydrogen ion concentration in relation to absorption of inorganic nutrients by higher plants. Plant Physiol. 17:515–524.

Arnon, D. I., and C. M. Johnson. 1942. Influence of hydrogen ion concentration on the growth of higher plants under controlled conditions. Plant Physiol. 17:525–539.

Awad, A. S., D. G. Edwards, and P. J. Mulhain. 1976. Effect of pH and phosphate on soluble soil aluminum and on growth and composition of kikuyu grass. Plant Soil 45:531–532.

Bartlett, R. J., and D. C. Riego. 1972. Effect of chelation on the toxicity of aluminum. Plant Soil 37:419–423.

Bartuska, A. M., and I. A. Unger. 1980. Elemental concentrations in plant tissues as influenced by low pH soils. Plant Soil 55:155–161.

Baumgartner, J. G., H. P. Haag, G.D. Oliveira, and D. Perecin. 1976. Tolerance of tomato (*Lycopersicon esculentum* Mill.) cultivars to aluminum and manganese. An. Esc. Super. Agric. Luiz de Queiroz Univ. Sao Paulo 33:513–541.

Ben, J. R., M. Morelli, and V. Estefanel. 1976. Effects of liming on the toxicity of aluminum to barley plants. Rev. Cent. Cienc. Rurais (Univ. Fed. St. Maria) 6:177–189.

Benac, R. 1976. Response of a sensitive (*Arachis hypogaea*) and a tolerant (*Zea mays*) species to different concentrations of manganese in the environment. (In French with English summary.) Cah. ORSTOM Ser. Biol. 11:43–51.

Blamey, F. P. C., D. G. Edwards, C. J. Asher, and M. K. Kim. 1982. Response of sunflower to low pH. p. 66–71. *In* A. Scaife (ed.) Proc. Ninth Int. Plant Nutrition Colloq., Vol. 1, Warwick, England. 22–27 August. Commonwealth Agric. Bureaux, Farnham House, Farnum Royal, Slough, U.K.

Blatt, C. R., and A. Van Diest. 1981. Evaluation of a screening technique for manganese toxicity in relation to leaf manganese distribution and interaction with silicon. Neth. J. Agric. Sci. 29:297–304.

Blue, W. G., and C. L. Dantzman. 1977. Soil chemistry and root development in acid soils. Proc. Soil Crop Sci. Soc. Fla. 36:9–15.

Borkenhagen, J. E., and J. G. Iyer. 1972. Aluminum sulfate as a stabilizer of nursery stock development. J. For. 70:33–34.

Bouwens, L., and J. Longin. 1979. Manganese foliar content in shrubs related to the ecophysiology of calcicole and calcifugal plants. Bull. Soc. R. Bot. Belg. 112:243–251.

Bowen, J. E. 1969. Absorption of copper, zinc, and manganese by sugarcane leaf tissue. Plant Physiol. 44:255–261.

Bradshaw, A. D. 1970. Pollution and plant evolution. New Sci. 48:497–500.

Brar, S. S., and J. Giddens. 1968. Inhibition of nitrification in Bladen grassland soil. Soil Sci. Soc. Am. Proc. 32:821–823.

Brauner, J. L. 1979. Tolerancia de cultivares de trigo (*Triticum aestivum* L.) ao alumínio e ao manganes: Suh determinacão, influência na concentracao de nutrientes e absorcão de cálcio e fósforo. Ph.D. Diss. Univ. of Sao Paulo, Sao Paulo, Brazil.

Bremner, I. 1974. Heavy metal toxicities. Q. Rev. Biophys. 7:75–124.

Bremner, I., and A H. Knight. 1970. The complexes of zinc, copper, and manganese present in rye grass. Br. J. Nutr. 23:279–289.

Bromfield, S. M. 1978. The effect of manganese oxidizing bacteria and pH on the availability of manganous ions and manganese oxide to oats in nutrient solutions. Plant Soil 49:23–39.

Bromfield, S. M. 1979. Manganous ion oxidation at pH values below 5.0 by cell free substances from *Streptomyces* sp. cultures. Soil Biol. Biochem. 11:115–118.

Brown, J. C., and T. E. Devine. 1980. Inheritance of tolerance or resistance to manganese toxicity in soybeans. Agron. J. 72:898–904.

Brown, J. C., and W. E. Jones. 1975. Heavy metal toxicity in plants: 1. A crisis in embryo. Commun. Soil Sci. Plant Anal. 6:421–438.

Brown, J. C., and W. E. Jones. 1977a. Manganese and iron toxicities dependent on soybean variety. Commun. Soil Sci. Plant Anal. 8:1–15.

Brown, J. C., and W. E. Jones. 1977b. Fitting plants nutritionally to soils: III. Sorghum Agron. J. 69:410–414.

Bryan, O. C. 1923a. Effect of reaction on growth, nodule formation and calcium content of alfalfa, alsike clover, and red clover. Soil Sci. 15:23–36.

Bryan, O. C. 1923b. Effect of acid soils on nodule forming bacteria. Soil Sci. 15:37–40.

Burns, G. R. 1967. Oxidation of sulphur in soils. Sulphur Inst. Tech. Bull. 13, Sulphur Inst., Washington, D.C.

Cambrai, J., and A. G. Calbo. 1980. Efeito do alumínio sobre a absorcão e sobre o transporte de fósforo em dois cultivares de sorgho (*Sorghum bicolor* L. Moench). Rev. Ceres 27:615–625.

Campbell, N. E. R., and H. Lees. 1967. The nitrogen cycle. p. 194–215. *In* A. D. McLaren and G. H. Peterson (ed.) Soil biochemistry. Marcel Dekker, Inc., New York.

Carvalho, M. M., C. S. Andrew, D. G. Edwards, and C. J. Asher. 1980. Comparative performances of six *Stylosanthes* species in three acid soils. Aust. J. Agric. Res. 31:61–76.

Carvalho, M. M., C. J. Asher, D. G. Edwards, and C. S. Andrew. 1982. Lack of effect of toxic aluminum concentrations on nitrogen fixation by nodulated Stylosanthes species. Plant Soil 66:225–231.

Carvalho, M. M., D. G. Edwards, C. S. Andrew, and C. J. Asher. 1981. Aluminum toxicity, nodulation and growth of *Stylosanthes* species. Agron. J. 73:261–265.

Chaney, R. L., and P. M. Giordano. 1977. Microelements as related to plant deficiencies and toxicities. p. 233–280. *In* L. F. Elliot and F. J. Stevenson (ed.) Soils for management and utilization of organic wastes and waste waters. Am. Soc. of Agron. and Soil Sci. Soc. of Am., Madison, Wis.

Chase, F. E., C. T. Corke, and J. B. Robinson. 1968. Nitrifying bacteria in soil. p. 593–611. *In* T. R. G. Gray and D. Parkinson (ed.) The ecology of soil bacteria. Univ. of Toronto Press, Toronto.

Cheng, B. T., and A. Pesant. 1977. Manganese and copper nutrients in oats as affected by natural color and texture of soil associated with temperature and aeration. Agrochimica 21:170–179.

Cherney, J. H., D. L. Robinson, L. C. Kappel, F. G. Hembry, and R. H. Ingraham. 1983. soil contamination and elemental concentrations of forages in relation to grass tetany. Agron. J. 75:447–451.

Christiansen, M. N., H. R. Carns, and D. J. Slyter. 1970. Stimulation of solute loss from radicles of *Gossypium hirsutum* L. by chillling, anaerobiosis, and low pH. Plant Physiol. 46:53–56.

CIAT. 1977. Annual report. CIAT (Int. Center for Tropical Agric.), Cali, Colombia.

Clark, R. B. 1977. Effect of aluminum on growth and mineral elements of Al-tolerant and Al-intolerant corn. Plant Soil 47:653-662.

Clark, R. B. 1982. Plant response to mineral element toxicity and deficiency. p. 71-142. *In* M. N. Christiansen and C. F. Lewis (ed.) Breeding plants for less favorable environments. John Wiley & Sons, Inc., New York.

Clark, R. B., P. A. Pier, D. Knudsen, and J. W. Maranville. 1981. Effect of trace element deficiencies and excesses on mineral nutrients in sorghum. J. Plant Nutr. 3:357-374.

Cooke, D. T. 1982. Differential susceptibility of apple cultivars to manganese toxicity. p. 123-128. *In* A. Scaife (ed.) Proc. Ninth Int. Plant Nutrition Colloq., Vol. 1, Warwick, England. 22-27 August. Commonwealth Agric. Bureaux, Farnham House, Farnum Royal, Slough, U.K.

Corke, C. T., and F. E. Chase. 1964. Comparative studies of actinomycete populations in acid podzolic and neutral mull forest soils. Soil Sci. Soc. Am. proc. 28:68-70.

Cosgrove, D. J. 1967. Metabolism of organic phosphates in soils. p. 216-226. *In* A. D. McLaren and G. H. Peterson (ed.) Soil biochemistry. Marcel Dekker, Inc., New Yoɪk.

Crum, J. R., and D. P. Franzmeir. 1980. Soil properties and chemical composition of tree leaves in southern Indiana. Soil Sci. Soc. Am. J. 44:1063-1069.

Davis, W. B., M. J. McCauley, and B. R. Byars. 1971. Iron requirements and aluminum sensitivity of hydroxamic acid requiring strain of *Bacillus megaterium.* J. Bacteriol. 105:589-594.

Dieckert, J. W., and E. Rozacky. 1969. Isolation and partial characterization of manganin, a new manganoprotein from peanut seeds. Arch. Biochem. Biophys. 134:473-477.

Dobereiner, J. 1966. Manganese toxicity effects on nodulation and N fixation in beans (*Phaseolus vulgaris* L.) in acid soils. Plant Soil 24:153-166.

Doolas, G. G. 1930. Soil acidity and soybean inoculation. Soil Sci. 30:275-288.

Duncan, R. R. 1981. Variability among sorghum genotypes for uptake of elements under acid soil field conditions. J. Plant Nutr. 4:21-32.

Duncan, R. R. 1982. Concentrations of critical nutrients in tolerant and susceptible sorghum lines under acid soil field conditions. p. 85-88. *In* M. R. Saric (ed.) Genetic specificity of mineral nutrition of plants. Serb. Scad. Sci. Art., Scientific Assemblies XII, Dep. Natl. Math. Sci. no. 3, Belgrade, Yugoslavia.

Duncan, R. R., J. W. Dobson, Jr., and C. D. Fisher. 1980. Leaf elemental concentrations and grain yield of sorghum grown on acid soil. Commun. Soil Sci. Plant Anal. 11:699-707.

Ebeid, M. M., and M. Kutacek. 1979. Effect of EDTA on transport forms of manganese in maize xylem exudate. Biol. Plant. 21:178-182.

Edwards, D. G., and C. J. Asher. 1982. Tolerance of crop and pasture species to manganese toxicity. p. 145-150. *In* A. Scaife (ed.) Proc. Ninth Int. Plant Nutrition Colloq., Vol. 1, Warwick, England. 22-27 August. Commonwealth Agricultural Bureaux, Farnham House, Farnum Royal, Slough, U.K.

Edwards, J. H., and B. D. Horton. 1977. Aluminum induced calcium deficiency in peach seedlings. J. Am. Soc. Hortic. Sci. 102:459-461.

Edwards, J. H., B. D. Horton, and H. C. Kirkpatrick. 1976. Aluminum toxicity symptoms in peach seedlings. J. Am. Soc. Hortic. Sci. 101:139-142.

Ekdahl, I. 1957. The growth of root hairs and roots in nutrient media and distilled water, and the effects of oxylate. Lantbrukshoegsk. Ann. 23:497-518.

Epstein, E. 1969. Mineral metabolism of halophytes. p. 381-397. *In* I. H. Rorison (ed.) Ecological aspects of the mineral nutrition of plants. Blackwell Scientific Pub. Ltd., Oxford.

Fisher, A. G., G. W. Eaton, and S. W. Porritt. 1977. Internal bark necrosis of Delicious apple in relation to soil pH and leaf manganese. Can. J. Plant Sci. 57:297-299.

Fleming, A. L., and C. D. Foy. 1982. Differential response of barley varieties to Fe stress. J. Plant Nutr. 5:457-468.

Fox, R. H. 1979. Comparative responses of field grown crops to phosphate concentrations in soil solutions. p. 81-106. *In* H. Mussel and R. C. Staples (ed.) Stress physiology in crop plants. Int. Conf. on Stress Physiology in Crop Plants. Boyce Thompson Inst. for Plant Res. and Rockefeller Foundation, John Wiley & Sons, Inc., New York.

Foy, C. D. 1964. Toxic factors in acid soils of the Southeastern United States as related to the response of alfalfa to lime. USDA Prod. Res. Rep. no. 80.

Foy, C. D. 1973. Manganese and plants. p. 51-76. *In* Manganese. Natl. Acad. of Sci.-Natl. Res. Coun., Washington, D.C.

Foy, C. D. 1974. Effects of aluminum on plant growth. p. 601–642. *In* E. W. Carson (ed.) The plant root and its environment. Univ. Press of Virginia, Charlottesville.

Foy, C. D. 1976. General principles involved in screening plants for aluminum and manganese tolerance. p. 255–267. *In* M. J. Wright and S. A. Ferrari (ed.) Plant adaptation to mineral stress in problem soils. Cornell Univ. Agric. Exp. Stn. Spec. Pub., Ithaca, N.Y.

Foy, C. D. 1983a. Plant adaptation to mineral stress in problem soils. Invitational lecture in a series of seminars on "plant breeding solutions for soil stress problems. Iowa J. Res. 57: 339–354.

Foy, C. D. 1983b. The physiology of plant adaptation to mineral stress. Invitational lecture in a series of seminars on "plant breeding for soil stress problems. Iowa J. Res. 57:355–391.

Foy, C. D., R. L. Chaney, and M. C. White. 1978. The physiology of metal toxicity in plants. Ann. Rev. Plant Physiol. 29:511–566.

Foy, C. D., and A. L. Fleming. 1978. The physiology of plant tolerance to excess available aluminum and manganese in acid soils. p. 301–328. *In* G. A. Jung (ed.) Crop tolerance to suboptimal land conditions. ASA Spec. Pub. no. 32, Am. Soc. of Agron., Madison, Wis.

Foy, C. D., and A. L. Fleming. 1982. Aluminum tolerance of two wheat cultivars related to nitrate reductase activities. J. Plant Nutr. 5:1313–1333.

Foy, C. D., A. L. Fleming, and J. W. Schwartz. 1973. Opposite aluminum and manganese tolerances in two wheat varieties. Agron. J. 65:123–126.

Foy, C. D., H. W. Webb, and J. E. Jones. 1981. Adaptation of cotton genotypes to an acid, manganese toxic soil. Agron. J. 73:107–111.

Franco, A. A., and D. N. Munns. 1982. Acidity and aluminum restraints on nodulation, nitrogen fixation and growth of *Phaseolus vulgaris* in nutrient solution. Soil Sci. Soc. Am. J. 46:296–301.

Freire, J. R. J. 1975. Compartamento de soja e do seu rizobio ao Al e Mn nos solos do Rio Grande do Sul. Cienc. Cult. (Sao Paulo) 28:169–170.

Freire, J. R. J. 1976. Inoculation of soybeans. p. 335–380. *In* J. M. Vincent, A. S. Whitney, and J. Bose (ed.) Exploiting the legume-*Rhizobium* symbiosis in tropical agriculture. College of Tropical Agric. Misc. Pub. no. 145, Univ. of Hawaii, Honolulu.

Freney, T. R. 1967. Sulfur containing organics. p. 229–253. *In* A. D. McLaren and G. H. Peterson (ed.) Soil biochemistry. Marcel Dekker, Inc., New York.

Friesen, D. K., A. S. R. Juo, and M. H. Miller. 1980. Liming and lime-phosphorus zinc interactions in two Nigerian ultisols: I. Interactions in the soil. Soil Sci. Soc. Am. J. 44:1221–1226.

Furlani, P. R., and R. B. Clark. 1981. Screening sorghum for aluminum tolerance in nutrient solutions. Agron. J. 73:587–594.

Gammon, N., Jr., and S. E. McFadden. 1979. Effect of rootstocks on greenhouse rose flower yield and leaf nutrient levels. Commun. Soil Sci. Plant Anal. 10:1171–1184.

Garcia, O. A., and S. A. Leon. 1978. Reaction of 5 cocoa hybrids (*Theobroma cacao*) to toxicity produced by aluminum in nutritive solution and on the plains of Colombia. Rev. Inst. Colomb. Agropecu. 13:219–228.

Godo, G. H., and H. M. Reisenauer. 1980. Plant effects on soil manganese availability. Soil Sci. Soc. Am. J. 44:993–995.

Gomez-Lepe, B. E., O. Y. Lee-Stadelman, J. A. Patta, and E. J. Stadelman. 1979. Effects of actylguanidine on cell permeability and other protoplasmic properties of *Allium cepa* epidermal cells. Plant Physiol. 64:131–138.

Grime, J. P., and J. S. Hodgson. 1969. An investigation of the significance of lime chlorosis by means of large scale comparative experiments. p. 381–397. *In* I. H. Rorison (ed.) Ecological aspects of the mineral nutrition of plants. Blackwell Scientific Pub. Ltd., Oxford.

Gupta, A. P., and G. S. G. Rao. 1977. Physiological studies on manganese nutrition. Indian Sugar 27:547–552.

Gurrier, G. 1979. Absorption of mineral elements in the presence of aluminum. Plant Soil 51: 275–278.

Hanson, W. D., and E. J. Kamprath. 1979. Selection for aluminum tolerances in soybeans based on seedling root growth. Agron. J. 71:581–586.

Haridasan, M. 1982. Aluminum accumulation by some Cerrado native species of central Brazil. Plant Soil 65:265–273.

Hati, N., T. R. Fisher, and W. J. Upchurch. 1979. Liming of soil: 1. Effect on plant available aluminum. J. Indian Soc. Soil Sci. 27:277–281.

Hecht-Buchholz, C., and C. D. Foy. 1981. Effect of aluminum toxicity on root morphology of barley. p. 343–345. *In* R. Brouwer et al. (ed.) Structure and function of plant roots. Martinus Nijhoff/Dr. W. Junk Publishers, The Hague.

Heenan, D. P., and L. C. Campbell. 1981. Soybean (*Glycine max*) nitrate reductase activity influenced by manganese nutrition. Plant Cell Physiol. 21:731–736.

Helyar, K. R. 1978. Effects of aluminum and manganese toxicity on legume growth. p. 207–231. *In* C. S. Andrew and E. J. Kamprath (ed.) Mineral nutrition of legumes in tropical and subtropical soils. CSIRO, Melbourne, Australia.

Hofler, K. 1958. Aluminumsalz Wirkung auf *Spirogonen* and *Zygnemen*. Protoplasma 49: 248–258.

Höfner, W. 1970. Elsen and manganhatlige verbindungen in blutungssaft von *Helianthus annuus*. Physiol. Plant. 23:673–677.

Holmes, R. L., and G. D. Coorts. 1980. Effects of excess manganese nutrition on the growth of Easter lily. Hortic. Sci. 15:497–498.

Horst, W. J. 1982. Factors responsible for genotypical manganese tolerance in cowpea (*Vigna unguiculata*). p. 99–104. *In* M. R. Saric (ed.) Genetic specificty of mineral nutrition of plants. Serb. Acad. Sci. Art. Scientific Assemblies XII, Dep. Natl. Math. Sci. no. 3, Belgrade, Yugoslavia.

Horst, W. J., and H. Marschner. 1978a. Effect of excessive manganese supply on uptake and translocation of calcium in bean plants. Z. Pflanzenphysiol. 87:137–148.

Horst, W. J., and H. Marschner. 1978b. Effect of silicon on manganese tolerance of bean plants (*Phaseolus vulgaris*). Plant Soil 50:287–304.

Horst, W. J., and H. Marschner. 1978c. Symptoms of manganese toxicity in *Phaseolus vulgaris*. Z. Pflanzenernaehr. Bokenkd. 141:129–142.

Horst, W. J., and H. Marschner. 1978d. Effect of silicon on the chemical state of manganese in bean leaf tissues. Z. Pflanzenernaehr. Bodenkd. 141:487–498.

Horst, W. J., A. Wagner, and H. Marschner. 1982. Mucilage protects root meristems from aluminum injury. p. 238–243. *In* A. Scaife (ed.) Proc. Ninth Int. Plant Nutrition Colloq., Vol. 1, Warwick, England. 22–27 August. Commonwealth Agric. Bureaux, Farnham House, Farnum Royal, Slough, U.K.

Horton, B. D., and J. H. Edwards. 1976. Diffusive resistance rates and stomatal aperatures of peach seedlings as affected by Al concentrations. Hort. Sci. 11:591–593.

Howard, D. D., and F. Adams. 1965. Calcium requirement for penetration of subsoils by primary cotton roots. Soil Sci. Soc. Am. Proc. 29:558–562.

Howeler, R. H., and L. F. Cadavid. 1976. Screening of rice cultivars for tolerance to Al toxicity in nutrient solutions as compared with a field screening method. Agron. J. 68: 551–555.

Huett, D. O., and R. C. Menary. 1979. Aluminum uptake by excised roots of cabbage, lettuce and kikuyu grass. Aust. J. Plant Physiol. 6:643–653.

Huett, D. O., and R. C. Menary. 1980. Effect of aluminum on growth and nutrient uptake of cabbage, lettuce and Kikuyu grass in nutrient solution. Aust. J. Agric. Res. 31:749–761.

Hussain, F., R. Overstreet, and L. Jacobson. 1954. The influence of hydrogen ion concentration on cation absorption by barley roots. Plant Physiol. 29:234–237.

Hutton, E. M., W. T. Williams, and C. S. Andrew. 1978. Differential tolerance to manganese in introduced and bred lines of *Macroptilium atropurpureum*. Aust. J. Agric. Res. 29:67–79.

International Rice Research Institute. 1974. International Rice Research Institute report, 1973. IRRI, Los Baños, Laguna, Philippines.

Ishida, A., and M. Masui. 1976. Studies on the manganese excess of carnation: II. Manganese excess in relation to certain soils, soil pH and two nitrogen forms. Engei Gakkai Zasshi 45: 283–288.

Ishida, A., M. Masui, T. Ogura, and A. Mikaya. 1977. Studies on the manganese excess of carnation: III. Effects of manganese concentration in nutrient solution on growth and flowering. Engei Gakkai Zasshi 45:383–388.

Islam, A. K. M. S., D. G. Edwards, and C. J. Asher. 1980. pH optima for crop growth: Results of a flowing solution culture experiment with six species. Plant Soil 54:339–357.

Jackson, W. A. 1967. Physiological effects of soil acidity. *In* R. W. Pearson and F. Adams (ed.) Soil acidity and liming. Agronomy 12:43–124. Am. Soc. of Agron., Madison, Wis.

Jaffre, T. 1979. The accumulation of manganese by the Proteaceae of New Caledonia. C. R. Hbd. Seances Acad. Sci. 289:425–428.

James, H., M. N. Court, D. A. MacLeod, and J. W. Parsons. 1978. Soil factors and mycorrhizal factors active on basaltic soils in Western Scotland. Forestry 51:105–119.

Jones, J. P., and S. S. Woltz. 1960. Fusarium wilt of tomato: Interaction of soil liming and micronutrient amendments on disease development. Phytopathology 60:812–813.

Jones, L. H. 1961. Aluminum uptake and toxicity in plants. Plant Soil 13:292–310.

Jones, R. K., and H. J. Clay. 1976. Foliar symptoms of nutrient disorders in Townsville stylo (*Stylosanthes humilis*). CSIRO Australian Div. of Tropical Agron. Tech. Paper no. 19, p. 1–11.

Jones, W. F., and L. E. Nelson. 1978. Response of field grown soybeans to lime. Soil Sci. Plant Anal. 9:607–614.

Kamprath, E. J. 1978. Lime in relation to aluminum toxicity in tropical soils. p. 233–245. *In* C. S. Andrew and E. J. Kamprath (ed.) Mineral nutrition of legumes in tropical and subtropical soils. CSIRO, East Melbourne, Australia.

Kamprath, E. J., and C. D. Foy. 1972. Lime-fertilizer-plant interactions in acid soils. p. 105–151. *In* R. W. Olsen et al. (ed.) Fertilizer technology and use. 2nd ed. Soil Sci. Soc. of Am., Madison, Wis.

Kamprath, E. J., and C. D. Foy. 1984. Lime-fertilizer-plant interactions in acid soils. *In* O. P. Engelstad (ed.) Fertilizer technology and use. 3rd ed. Soil Sci. Soc. of Am., Madison, Wis. (In press.)

Kamura, T., and K. Nishitani. 1977. Oxidation mechanisms of manganese in soils: II. Effects of soil reaction on manganese oxidation. (In Japanese.) Nippon Dojo Hiryogaku Zassi 48:103–106. (English abstract in Soil Sci. Plant Nutr. 23:391.)

Karlik, S. J., G. L. Eichhorn, and D. R. Crapper-McLachlan. 1980. Molecular interactions of aluminum with DNA. Neurotoxicology 1:83–88.

Kauffman, M. D. 1977. The effect of soil acidity and lime placement on root growth and yield of winter wheat and alfalfa. Ph.D. Diss. Oregon State Univ., Corvallis. Diss. Abstr. 37B: 7.3192–3193.

Kauffman, M. D., and E. H. Gardner. 1978. Segmental liming of soil and its effect on the growth of wheat. Agron. J. 70:331–336.

Kerridge, P. C. 1969. Aluminum toxicity in wheat (*Triticum aestivum* Vill. Host). Ph.D. Diss. Oregon State Univ. Univ. Microfilms 69-4086, Ann Arbor, Mich. (Diss. Abstr. 29B: 3159.)

Keser, M., F. Benedict, F. Newbauer, and F. E. Hutchinson. 1975. Influence of aluminum ions on developmental morphology of sugarbeet roots. Agron. J. 67:84–88.

Keyser, H. H., and D. N. Munns. 1979a. Effects of calcium, manganese and aluminum on growth of rhizobia in acid media. Soil Sci. Soc. Am. J. 43:500–503.

Keyser, H. H., and D. N. Munns. 1979b. Tolerance of rhizobia to acidity, aluminum and phosphate. Soil Sci. Soc. Am. J. 43:519–523.

Keyser, H. H., D. H. Munns, and J. S. Hohenberg. 1979. Acid tolerance of rhizobia in culture and in symbiosis with cowpea. Soil Sci. Soc. Am. J. 43:719–722.

Kliewer, W. M. 1961. The effects of varying combinations of molybdenum, aluminum, manganese, phosphorus, nitrogen, calcium, and hydrogen concentrations, lime and *Rhizobium* strain on growth, composition, and nodulation of several legumes. Ph.D. Diss. Cornell Univ., Ithaca, New York.

Klimashevskii, E. L., and N. F. Chernysheva. 1980. Content of organic acids and physiologically active compounds in plants differing in their susceptibility to the toxicity of $Al^{3+}$. Sov. Agric. Sci. 1980(2):5–8.

Klimashevskii, E. L., Yu A. Markova, and T. S. Lebedeva. 1979. Interaction of aluminum and phosphorus on root surfaces and cell walls. Sov. Agric. Sci. 1979(3):6–9.

Klimashevskii, E. L., Y. A. Markova, S. M. Zyabkina, G. K. Zirenka, T. E. Zolotuklhn, and S. E. Pavolova. 1976. Aluminum absorption and localization in root tissues of different pea varieties. (In Russian with English summary.) Fiziol. Biokhim. Kult. Rast. 8:396–401.

Kluthcouski, J., and L. E. Nelson. 1979. Variations in the manganese concentrations in soybean trifoliolates. Commun. Soil Sci. Plant Anal. 10:1299–1310.

Ko, W. H., and F. K. Hora. 1972. Identification of an Al ion as a soil fungotoxin. Soil Sci. 113:42–45.

Kohno, Y., and C. D. Foy. 1983. Manganese toxicity in bush bean as affected by concentration of manganese and iron in the nutrient solution. J. Plant Nutr. 6:363–386.

Krizek, D. T. 1982. Plant response to atmospheric stress caused by waterlogging. p. 293–334. *In* M. N. Christiansen and C. F. Lewis (ed.) Breeding plants for less favorable environments. John Wiley & Sons, Inc., New York.

Krizek, D. T., and C. D. Foy. 1981. Water stress: Role in differential aluminum tolerance of barley genotypes. Agron. Abstr., p. 181–182.

Kumar, P. H. 1979. Effect of aluminum on the growth of *Areca triandra* Roxb. in sand culture. Sci. Hortic. (Amsterdam) 10:183–185.

Lance, J. C., and R. W. Pearson. 1969. Effect of low concentrations of aluminum on growth and water and nutrient uptake by cotton roots. Soil Sci. Soc. Am. Proc. 33:95–98.

Lau, C. H . 1979. Effect of potassium and aluminum treatments on growth and nutrient uptake of rubber (*Hevea*) seedlings and on soils. J. Rubber Res. Inst. Malays. 27:92–103.

Lee, C. R., and M. L. McDonald. 1978. Influence of soil amendments on potato growth, mineral nutrition, and tuber yield and quality in very strongly acid soils. Soil Sci. Soc. Am. J. 41:573–577.

Lee, Y. S. 1977. Aluminum toxicity in corn seedlings. Hanguk Toyang Bilyo Hakhoe Chi 10: 75–78.

Lehova, E. 1979. Manganese excess induced by continuous application of high rates of ammonium nitrate to the apple tree. p. 188–191. *In* T. Kudrev et al. (ed.) Mineral nutrition of plants, Vol. 2. Proc. First Int. Symp. on Plant Nutrition, Varna, Bulgaria. 24–29 September. Bulgaria Acad. Sci., Inst. Plant Physiol. Sofia, Bulgaria.

Lewis, W. E. 1973. Effect of mineral salts on *Aphanomyces enteiches* and *Aphanomyces* root rot of peas. Phytopathology 63:989–993.

Liebig, G. F., A. P. Vanselow, and H. D. Chapman. 1974. Effects of Al on Cu toxicity as revealed by solution culture and spectrographic studies of citrus. Soil Sci. 53:341–357.

Link, L. 1975. Burley faces problems from manganese toxicity. Southeast Farm Press, Clarksdale, Miss., 25 June.

Link, L. 1979. Critical pH for the expression of manganese toxicity. Southeast Farm Press, Clarksdale, Miss., 25 June.

Long, F. L., and C. D. Foy. 1970. Plant varieties as indicators of aluminum toxicity in the $A_2$ horizon of a Norfolk soil. Agron. J. 62:679–681.

Lowe, R. H., and C. E. Bortner. 1973. Effect of phosphorus, nutrition, and soil pH on "physiologic spotting" of L8 burley tobacco. Agron. J. 65:263–265.

Lund, Z. F. 1970. The effect of calcium and its relation to some other cations on soybean root growth. Soil Sci. Soc. Am. Proc. 34:456–459.

Maas, E. V. 1969. Calcium uptake by excised maize roots and interactions with alkali cations. Plant Physiol. 44:985–989.

Malavolta, E., E. L. M. Coutinho, G. C. Vitti, N. V. Alejo, N. J. Novaes, and V. L. Furlani Netto. 1979. Studies on the mineral nutrition of sweet sorghum: 1. Deficiency of macro and micronutrients and toxicity of aluminum, chlorine and manganese. An. Esc. Super. Agric. Luiz de Queiroz Univ. Sao Paulo 36:173–202.

Mask, P. L., and D. O. Wilson. 1978. Effect of Mn on growth, nodulation and nitrogen fixation by soybeans grown in the greenhouse. Commun. Soil Sci. Plant Anal. 9:653–666.

Mason, M. G. 1980. An investigation of reduction in wheat yields after use of a high level of ammonium sulphate for a number of years. Aust. J. Exp. Agric. Anim. Husb. 20:210–219.

Massey, H., and R. I. Barnhisel. 1972. Copper, nickel and zinc released from acid, coal mine spoil materials of eastern Kentucky. Soil Sci. 116:207–212.

Masui, M., and A. Ishida. 1975. Studies on the manganese excess in muskmelon: IV. Manganese excess in relation to steam sterilization, soil pH and organic matter. Engei Gakkai Zasshi 44:41–46.

Masui, M., A. Nukaya, and A. Ishida. 1976. Studies on the manganese excess of muskmelon: V. The manganese, calcium and iron concentrations in nutrient solution. Engei Gakkai Zasshi 45:267–274.

Masui, M., A. Nukaya, A. Ishida, and T. Ogura. 1980. The manganese excess of muskmelons: 6. Manganese distribution in the plant parts. Engei Gakkai Zasshi 49:79–84.

Matsumoto, H., S. Morimura, and E. Hirasawa. 1979. Localization of absorbed aluminum in plant tissues and its toxicity studies in the inhibition of pea root elongation. p. 171–194. *In* K. Kudrev et al. (ed.) Mineral nutrition of palnts, Vol. 1. Proc. First Int. Symp. on Plant Nutrition, Varna, Bulgaria. 24–29 September. Bulgaria Acad. Sci., Inst. Plant Physiol., Sofia, Bulgaria.

McCormick, L. H., and F. Y. Borden. 1972. Phosphate fixation by aluminum in plant roots. Soil Sci. Soc. Am. Proc. 36:779–802.

McLachlan, K. D. 1975. Sulphur in Australian agriculture. Univ. of Sydney Library, Sydney, Australia.

McLean, E. O. 1976. Chemistry of soil aluminum. Commun. Soil Sci. Plant Anal. 7:619–636.

McLean, I. B. 1980. The toxic aluminum reaction in corn and barley roots: An ultrastructural and morphological study. Michigan State Univ., East Lansing. Univ. Microfilms Int. 1314642, Ann Arbor, Mich. (Masters Abstr. 18[4]:259.)

Memon, A. R., M. Chino, Y. Takeoka, K. Hora, and M. Yatazawa. 1980. Distribution of manganese in leaf tissues of a manganese accumulator: *Acanthopanax sciadophylloides* as revealed by electronprobe x-ray microanalyzer. J. Plant Nutr. 2:457–476.

Memon, A. R., M. Chino, and Y. Yatazawa. 1981. Microdistribution of aluminum and manganese in the tea leaf tissues as revealed by x-ray analyses. Commun. Soil Sci. Plant Anal. 12:441–452.

Memon, A. R., S. Ito, and M. Yatazawa. 1979. Absorption and accumulation of iron, manganese and copper in plants in the temperate forest of central Japan. Soil Sci. Plant Nutr. 25:611–620.

Mengel, D. B., and E. J. Kamprath. 1978. Effect of soil pH and liming on growth and nodulation of soybeans in Histosols. Agron. J. 70:959–963.

Meredith, C. P. 1978a. Response of cultural tomato cells to aluminum. Plant Sci. Lett. 12:17–24.

Meredith, C. P. 1978b. Selection and characterization of aluminum resistant variants from tomato cell cultures. Plant Sci. Lett. 12:25–34.

Mesdag, L., L. A. J. Slootmaker, and J. Post, Jr. 1970. Linkage between tolerance to high soil acidity and genetically high protein content in the kernel of wheat, *Triticum aestivum* L., and its possible use in breeding. Euphytica 19:163–174.

Miller, S. S., and O. E. Schubert. 1977. Plant manganese and soil pH associated with internal bark necrosis in apple. Proc. W.V. Acad. Sci. 49:97–102.

Moore, D. P. 1974. Physiological effects of pH on roots. p. 565–600. *In* E. W. Carson (ed.) The plant root and its environment. Univ. Press of Virginia, Charlottesville.

Moore, D. P., R. Overstreet, and L. Jacobson. 1961. Uptake of magnesium and its interaction with calcium in excised barley roots. Plant Physiol. 36:290–295.

Moraghan, J. T., and T. J. Freeman. 1978. Influence of FeEDDHA on growth and manganese accumulation in flax. Soil Sci. Soc. Am. J. 42:455–460.

Moraghan, J. T., and P. Ralowicz. 1979. Relative responses of four flax differential lines to FeEDDHA. Crop Sci. 19:9–11.

Morgan, P. W., H. E. Joham, and J. V. Amin. 1966. Effect of manganese toxicity on the indoleacetic acid-oxidase system of cotton. Plant Physiol. 41:718–724.

Morgan, P. W., D. M. Taylor, and H. E. Joham. 1976. Manipulations of IAA-oxidase activity and auxin deficiency symptoms in intact cotton plants with manganese nutrition. Physiol. Plant 37:149–156.

Morrill, L. G., and J. E. Dawson. 1967. Patterns observed for the oxidation of ammonium to nitrate by soil organisms. Soil Sci. Soc. Am. Proc. 31:757–760.

Muchovej, J. J., L. A. Maffia, and R. M. C. Muchovej. 1980. Effect of exchangeable soil aluminum and alkaline calcium salt on the pathogenicity and growth of *Phytophthora capsici* from green pepper—*Capsicum frutescens.* Phytopathology 70:1212–1214.

Mugwira, L. M. 1980. Growth and Ca, Mg, K and P uptake by triticale, wheat and rye at four Al levels. J. Plant Nutr. 2:591–606.

Mugwira, L. M., M. Floyd, and S. V. Patel. 1981. Tolerances of triticale lines to manganese in soil and nutrient solution. Agron. J. 73:319–322.

Mullette, K. J. 1975. Stimulations of growth in eucalyptus due to aluminum. Plant Soil 42: 495–599.

Munns, D. N. 1978. Legume-*Rhizobium* relations. p. 247–263. *In* C. S. Andrew and E. J. Kamprath (ed.) Mineral nutrition of legumes in tropical and subtropical soils. CSIRO, East Melbourne, Australia.

Munns, D. N., J. S. Hohenberg, T. L. Righetti, and D. T. Lauter. 1981. Soil acidity tolerance of symbiotic and nitrogen fertilized soybeans. Agron. J. 73:407–410.

Munns, D. N., and B. Mosse. 1980. Mineral nutrition of legume crops. p. 115–125. *In* R. J. Summerfield and A. H. Bunting (ed.) Advances in legume science. Kew Richmond Surrey, Royal Bot. Gardens, U.K.

Mutatker, V. K., and W. L. Pritchett. 1966. Effects of added aluminum on some microbial processes and on the growth of oats (*Avena sativa*) in Arredondo fine sand. Soil Sci. 103: 39–46.

Naidoo, G. 1977. Aluminum toxicity in two snapbean varieties. Ph.D. Diss. Univ. of Tennessee, Knoxville. Diss. Abstr. 27B:5478.

Nelson, L. E. 1964. The effect of pH on the acetate soluble sulfur content of a Mayhew soil in Mississippi before and after incubation. Soil Sci. Soc. Am. proc. 28:290–291.

Nelson, L. E. 1977. Changes in water soluble Mn due to soil sample preparation and storage. Commun. Soil Sci. Plant Anal. 8:479–487.

Ogner, G., and O. Tiegen. 1980. Effects of acid irrigation at different temperatures on seven clones of Norway spruce. Rep. of Norwegian Forest Res. Inst. 36.3, Box 61, 1432, As NLH Norway.

Oliveira, R. G. B., and A. Drozdowicz. 1980. Occurrence of microorganisms capable of decomposing organic phosphates in Cerrado soil in Brazil. Zentralbl. Bakteriol. Parasitenkd. Infektionskr. Hyg. Abt. 2 Naturwiss. Mikrobiol. Landwirtsch. Technol. Umweltschutzes 135:467–476.

Oram, R. N., D. J. David, A. G. Green, and B. J. Read. 1979. Selection in *Lupinus albus* for lower seed manganese concentration. Aust. J. Agric. Res. 30:467–476.

Osawa, T., and H. Ikeda. 1977. Heavy metal toxicities in vegetable crops: VI. The effect of potassium and calcium concentration in the nutrient solution on manganese toxicity in vegetable crops. Engei Gakkai Zasshi 46:181–188.

Osawa, T., and H. Ikeda. 1980. Heavy metal toxicities in vegetable crops: VIII. The effect of form of nitrogen supplied and pH level of the nutrient solution on manganese toxicities in vegetable crops. Engei Gakkai Zasshi 49:197–202.

Otsuka, K. 1969. Aluminum induced Fe chlorosis: IV. Aluminum and manganese toxicities for plants. Nippon Dojo Hiryogaku Zasshi 40:177–220. (English abstract in Soil Sci. Plant Nutr. 16:140, 1970.)

Patel, K. I., and L. M. Mugwira. 1974. Sudangrass and yellow poplar responses to lime, calcium, and magnesium applications in acid subsoil. Agron. Abstr., p. 177.

Pavan, M. A., and F. T. Bingham. 1981. Toxicity of metals in plants. I. Characterization of manganese toxicity in coffee (*Coffea arabica* L.). Pesq. Agropecu. Bras. 16:815–821.

Pieri, C. 1974. First experimental results on the susceptibility of ground nut to aluminum toxicity. Agron. Trop. (Paris) 29:685.

Pierre, W. H., J. R. Webb, and W. D. Schrader. 1971. Quantitative effects of nitrogen fertilizer on the development and downward movement of soil acidity in relation to level of fertilization and crop removal in a continuous corn cropping system. Agron. J. 63:291–297.

Rao, G. S. C., and A. P. Gupta. 1979. Influence of manganese on the accumulation of free amino acids in cane leaves. Int. Sugar J. 81:99–102.

Rashid, A., F. M. Chaudhry, and M. Sharif. 1976. Micronutrient availability to cereals from calcareous soils: III. Zinc absorption by rice and its inhibition by important ions of submerged soils. Plant Soil 45:613–623.

Rees, W. J., and G. M. Sidrak. 1961. Interrelationships of aluminum and manganese toward plants. Plant Soil 14:101–117.

Robson, A. D., and J. F. Loneragan. 1970. Sensitivity of annual *Medicago* species to manganese toxicity as affected by Ca and pH. Aust. J. Agric. Res. 21:223–232.

Ruffy, T., G. S. Miner, and C. D. Raper. 1979. Temperature effects on growth and manganese tolerance in tobacco. Agron. J. 71:638–644.

Salinas, J. G., and P. A. Sanchez. 1976. Soil plant relationships affecting varietal and species differences in tolerance to low available soil phosphorus. Cienc. Cult. (Sao Paulo) 28: 156–168.

Sanchez, S. L. F., P. N. Vivas, and B. R. Perez. 1979. The influence of lime, phosphorus and potassium on the yield of Napier grass *Pennisetum purpureum,* growth on the Oxisol in the "Llanos Orientales," Colombia. Rev. Inst. Colomb. Agropecu. 14:63–72.

Santana, M. B., and J. M. Braga. 1977. Aluminum-phosphorus interactions of acidic soils in southern Bahia. Rev. Ceres 24:200–211.

Sartain, J. B. 1974. Differential effects of aluminum on top and root growth, nutrient accumulation and nodulation of several soybean varieties. Ph.D. Diss. North Carolina State Univ., Raleigh,.

Schmandke, H., G. Muschiolik, M. Schultz, G. Schmidt, and H. D. Quade. 1979. The effect of aluminum ions on chemical and functional properties of spun protein fibers. Nahrun 23:229–236.

Sewell, M. C., and P. L. Gainey. 1930. Interrelation of nutrients and soil reaction on growth and inoculation of alfalfa. Soil Sci. 30:297–306.

Shchigel'ski, O. A., R. G. Butenko, A. G. Vollososovich, and V. K. Yashchenko. 1979. Aluminum content in cultures of isolated tissues and intact medicinal plants. Sov. Plant Physiol. (Engl. Transl.) 26:511–514.

Scibisz, K., and A. Sadowski. 1979. Internal bark necrosis in Stark Crimson Delicious apple trees in relation to mineral nutrition. Gartenbauwissenschaft 44:177–181.

Sedberry, J. E., Jr., P. G. Schilling, F. E. Wilson, and F. J. Peterson. 1978. Diagnosis and correction of zinc problems in rice production. Louisiana Agric. Exp. Stn. Bull. no. 708.

Sidorenko, O. D., V. V. Blinov, and V. T. Emtsev. 1979. Formation of iron-manganese concretions on rice roots. (In Russian with English summary.) Izv. Timiryazevski. Skh. Akad. 5:172–177.

Siegal, N., and A. Haug. 1982. Physiology and biochemistry of $Al^{3+}$ ion toxicity in barley roots. Annu. Rep. of 1981, MSU/DOE Plant Res. Lab., Michigan State Univ., East Lansing.

Silva, A. R. 1976. Application of the plant genetic approach to wheat culture in Brazil. p. 223–231. *In* M. J. Wright and S. A. Ferrari (ed.) Plant adaptation to mineral stress in problem soils. Cornell Univ. Agric. Exp. Stn. Spec. Pub., Ithaca, New York.

Simpson, J. R., A. Pinkerton, and J. Lazdovokis. 1977. Effects of subsoil calcium on the root growth of some lucerne genotypes (*medicago sativa* L.). Aust. J. Agric. Res. 29:629–638.

Sirkar, S., and J. V. Amin. 1979. Influence of auxins on respiration of manganese toxic cotton plants. Indian J. Exp. Biol. 17:618–619.

Sivasubramaniam, S., and O. Talibudeen. 1972. Effects of aluminum on the growth of tea (*Camellia sinensis*) and its uptake of potassium and phosphorus. Tea 43:4–13.

Small, J. 1946. pH and plants. D. Van Nostrand Co., New York.

Smiley, J. H. 1976. Manganese toxicity, major burley disease. Southeast Farm Press, Clarksdale, Miss., 7 July.

Smith, F. W. 1979. Tolerance of seven tropical pasture grasses to excess manganese. Commun. Soil Sci. Plant Anal. 10:853–867.

Sommers, L. E., and V. O. Biederbeck. 1973. Tillage management principles: Soil microorganisms in conservation tillage. p. 87–108. *In* Proc. Natl. Conserv. Tillage Conf., Des Moines, Iowa, 28–30 March. Soil Conserv. Soc. of Am., Ankeny, Iowa.

Souto, S. M., and J. Dobereiner. 1971. Manganese toxicity in tropical forage legumes. Pesqui. Agropecu. Bras. 4:128–138.

Spurway, C. H. 1941. Soil reaction (pH) preference of plants. Michigan Agric. Exp. Stn. Bull. no. 306.

Stahlberg, S., S. Sombatpanit, and J. Stahlberg. 1976. Manganese relationships in soil and plant: II. Studies on manganese fixation. Acta Agric. Scand. 26:65–81.

Steiner, K., D. H. McCormick, and D. S. Canavera. 1980. Differential response of paper birch (*Betula papyrifera*) provenances to aluminum in solution cultures. Can. J. For. Res. 10: 25–29.

Sutcliffe, J. F. 1962. Mineral salts absorption in plants. Pergamon Press, Inc., Elmsford, N.Y.

Takayanagi, H. 1976. Effects of EDTA on the absorption of manganese by tea plants and solubilization of manganese from tea soils. Chagyo Gijutsu Kenkyu 51:44–52.

Tanner, P. D. 1977. Toxic effects of manganese on maize as affected by calcium and molybdenum application in sand culture. Rhod. J. Agric. Res. 15:25–32.

Temple-Smith, M. G., and T. B. Koen. 1982. Comparative response of poppy (*Papaver somniferum* L.) and eight crop and vegetable species to manganese excess in solution culture. J. Plant Nutr. 5:1153–1169.

Thaworuwong, N., and A. Van Diest. 1975. Influence of high acidity and aluminum on the growth of lowland rice. Plant Soil 41:141–159.

Thompson, L. M., C. A. Black, and J. A. Zoellner. 1954. Occurrence and mineralization of organic phosphorus in soils, with particular reference to association of nitrogen, carbon and pH. Soil Sci. 77:185–196.

Tiffin, L. O. 1967. Translocation of manganese, iron, cobalt and zinc in tomato. Plant Physiol. 42:1427–1432.

Tingey, D. T., S. Raba, J. Wagner, and K. D. Rodecap. 1982. Vermiculite: A source of metals for *Arabidopsis thaliana.* J. Am. Soc. Hortic. Sci. 107:465–468.

Trapp, G. 1980. Studies of aluminum interaction with enzymes and proteins—the inhibition of hexokinase. Neurotoxicology 1:89–100.

Trim, R. R. 1959. Metal ions as precipitants for nucleic acids and their use in the isolation of polynucleotides from leaves. Biochem. J. 72:298–304.

Ulrich, B., R. Mayer, and P. K. Khanna. 1980. Chemical changes due to acid precipitation in a loess derived soil in Central Europe. Soil Sci. 130:193–199.

Ulmer, S. E. 1979. Aluminum toxicity and root DNA synthesis in wheat. Ph.D. Diss. Iowa State Univ., Ames.

Van der Vorm, P. D. J., and A. Van Diest. 1979. Aspects of the iron and manganese nutrition of rice plants: 1. Iron and manganese uptake by rice plants grown under aerobic and anaerobic conditions. Plant Soil 51:233-246.

Van Goor, B. J., and D. Wiersma. 1974. Redistribution of potassium, calcium, magnesium and manganese in the plant. Physiol. Plant. 31:163-168.

Vierstra, R., and A. Haug. 1978. The effect of $Al^{3+}$ on the physical properties of membrane lipids in *Thermoplasma acidophilium.* Biochem. Biophys. Res. Commun. 84:138-143.

Vlamis, J., and D. E. Williams. 1973. Manganese toxicity and marginal chlorosis of lettuce. Plant Soil. 39:245-251.

Vose, P. B. 1981. Induced mutants—a tool in plant research. p. 159-181. *In* Int. Symp. on Induced Mutations as a Tool for Crop Plant Improvement, Vienna. 9-13 March. IAEA and FAO, Vienna.

Wallace, A., and E. M. Rommey. 1977. Aluminum toxicity in plants grown in solution culture. Commun. Soil Sci. Plant Anal. 8:791-794.

Ward, G. M. 1977. Manganese deficiency and toxicity in greenhouse tomatoes. Canadian J. Plant Sci. 57:107-115.

White, M. C., F. D. Baker, R. L. Chaney, and A. M. Decker. 1981. Metal complexation in xylem fluid: II. Theoretical equilibrium model and computational computer program. Plant Physiol. 67:301-310.

Williams, C. H. 1980. Soil acidification under clover pasture. Aust. J. Exp. Agric. Anim. Husb. 20:561-567.

Williams, D. E., and J. Vlamis. 1957. The effect of silicon on yield and Mn uptake and distribution in the leaves of barley plants grown in culture solution. Plant Physiol. 32:404-409.

# 3 Physiological Aspects of Calcium, Magnesium, and Molybdenum Deficiencies in Plants[1]

**RALPH B. CLARK**

*Agricultural Research Service, USDA*
*University of Nebraska*
*Lincoln, Nebraska*

Plants grown on acid soils often exhibit many physiological disorders. These disorders are usually related to the changes and chemical forms of mineral elements that occur in acid soils. The higher rainfall common to acid soils often leaches essential mineral elements from the soil and enhances toxic conditions by increasing solubilization and activity of such elements as Al and Mn.

Physiological disorders that often appear in plants grown on acid soils are Al and Mn toxicities and Ca, Mg, Mo, and P deficiencies. This chapter addresses the physiological effects of Ca, Mg, and Mo deficiencies in plants.

Recent literature has classified mineral element functions in plants according to their biochemical and physiological behaviors rather than according to their concentrations (macronutrients or micronutrients) in plant tissues (Clarkson & Hanson, 1980; Hughes, 1972; Mengel & Kirkby, 1982). With the use of these approaches, elements required for plant growth have been classified into four major groups: (i) elements that are covalently bonded into constituents or organic matter (C, H, O, N, and S); (ii) elements that occur as oxyanions or are esterified with native alcohol groups (P, B, and Si); (iii) elements that have nonspecific, osmotic, and ionic balance functions plus specific functions in enzyme conformation and catalysis, i.e., these elements maintain an ionic identity as free or reversibly bound ions as metal-protein complexes (K, Na, Mg, Ca, Mn, and Cl); and (iv) elements that form metalloproteins, are present as structural chelates, and participate in redox reactions or valency changes (Fe, Cu, Zn, Mo, and Mn in its function in the photosystem II pathway). From the physiological point of view, this classification better describes the functions of Ca, Mg, and Mo in plant processes.

[1] Contribution of the U.S. Department of Agriculture, Agricultural Research Service, and the Department of Agronomy, University of Nebraska, Lincoln, as Paper no. 7159, Journal Series, Nebraska Agricultural Experiment Station.


*Soil Acidity and Liming*—Agronomy Monograph no. 12 (2nd Edition).

# I. CALCIUM

The chemical properties of Ca are relatively complex and $Ca^{2+}$ controls biochemical and physiological functions that are very different from those of other elements. Clarkson and Hanson (1980) and Williams (1974, 1976) have discussed many of the unique features of $Ca^{2+}$ and its chemistry in biological systems. The function of $Ca^{2+}$ depends on its binding with other compounds. Calcium has a relatively large ionic radius (0.099 nm) compared with other divalent ions and has a coordination number of six or higher (often seven or eight) to form octohedral complexes. Calcium has the ability to bind with proteins and other macromolecules with low geometrical demand. Calcium coordinates strongly with a variety of ligands and readily substitutes its water of hydration to produce complex cross-linking or bridging of macromolecules. Ligands of $Ca^{2+}$ complexes often consist of high amounts of carboxylates, little water, and usually with at least one protein chain carboxyl-O group. Ligand exchanges with $Ca^{2+}$ are relatively rapid, and the binding strength of $Ca^{2+}$ is not dependent on the number of carboxylate ligands. Calcium increases hydrophobicity of membranes, strengthens cell wall structures, increases membrane stability and integrity, reduces water permeability, wards off membrane damage and leakiness, delays senescence and abscission, and acts as a protein-pectin cement in the middle lamella. Clarkson and Hanson (1980) describe the function of $Ca^{2+}$, "Simply put, $Ca^{2+}$ can tie diverse macromolecules together."

## A. Calcium Uptake

### 1. MECHANISMS

The first step in $Ca^{2+}$ uptake, once it is at the root surface, is an attraction, or affinity, of the root cells for intake into the cell. Possible forces of attraction and intake are considered to be cation exchange or binding sites on cells exposed to the free space or charge differences between cellular fluids and external solutions (gradient concentrations), ion effluxes and influxes, cell or membrane specificity for ions, and ion carrier molecules. Whether any of these forces are responsible for uptake of $Ca^{2+}$ is debatable. Early research favored cation exchange or binding sites and physiochemical theories to the more recent theories of carrier molecules and ion pumps as the process of cation uptake (Kirkby, 1979; Marschner, 1974).

Absorption processes by roots are subject to conditions and variations in active root surfaces, permeability changes, and metabolism (Wiersum, 1979). Calcium moves across the cortical cells by diffusion or by displacement exchange in the free space (Bangerth, 1979). To enter the stele and xylem tissues, $Ca^{2+}$, water, and other elements must pass through the cortical cells to the endodermis. The suberization of endodermal cells and the casparian strip effectively restrict passage of materials in the free space (apoplast). Therefore, the elements must enter into cells and pass through the endodermis into the stele and xylem via the cytoplasm (symplast). Cytoplasms of these cells are interconnected by plasmodesmata. Because symplastic $Ca^{2+}$ movement is relatively limited (Bangerth, 1979; Russell & Clarkson, 1976), $Ca^{2+}$ transport into the stele and xylem occurs preferably near the root tip and at sites of branch root formation where cells are not as suberized and the casparian strip is not fully developed. The inability of $Ca^{2+}$ to move readily in the symplasm (Raven, 1977) is not well understood, but this may be because high proportions of Ca in the cytoplasm are in a bound form so that the activity of $Ca^{2+}$ is low (Russell & Clarkson, 1976).

Whether $Ca^{2+}$ is absorbed by active or passive processes is still uncertain. Active uptake has been defined as transport against an electrochemical gradient through biological membranes. Most electrochemical gradient data favor passive uptake of $Ca^{2+}$ into the protoplast (Higinbotham et al., 1967; Kirkby, 1979; Maas, 1969; Macklon, 1975). However, uptake of $Ca^{2+}$ at low concentrations (0.005–0.05 m*M*) is metabolically controlled (Maas, 1969), indicating active uptake. Measurement of $Ca^{2+}$ uptake is complicated by the relatively large fraction of $Ca^{2+}$ held on absorption sites in cell walls (Kirkby, 1979). Compared with $K^+$, the movement of $Ca^{2+}$ into the cells is less energy dependent. The uptake of $K^+$ ($K^+$ in the xylem sap) was dependent on the energy status of the roots, and the uptake of $Ca^{2+}$ was not. Calcium has also been found to move into and through root tissues down electrochemical potential gradients (Bowling, 1966, 1973; Bowling et al., 1966; Higinbotham et al., 1967).

The concept that absorption of mineral elements and water is restricted to the young, apical parts of the roots has been assumed for a long time. Other microscopic and physiological observations support this view. Early investigators reported thin-walled root hairs to be highly abundant near root tips, whereas endodermal cells away from root tips were coated with hydrophobic suberin and were highly thickened [see Russell and Clarkson (1976) for a review of these early findings]. These observations suggested that the endodermal cell layer regulated the movement of water and solutes to the steles and xylem. The rate of respiration in the apical root region was higher than in older portions. When relationships between metabolism and ion uptake were established, it was natural to conclude that ion uptake occurred near the root apex. Later studies have shown that these views were not accurate for all elements (Clarkson & Sanderson, 1970; Clarkson et al., 1968; Marschner & Richter, 1974; Russell & Clarkson, 1976). The concept that $Ca^{2+}$ (as well as $Mg^{2+}$) is absorbed near the root apical region is fairly accurate, but it does not apply to $H_2PO_4^-$ and $K^+$ (Russell & Clarkson, 1976).

Calcium uptake and movement to the xylem in both barley and garden marrow were restricted to the relatively young parts of the root (Russell & Clarkson, 1976). For barley, a close relationship was noted between the decrease in $Ca^{2+}$ movement in the roots and the deposition of suberin in the endodermis. It was concluded that $Ca^{2+}$ moved radially across the cortex to the xylem in the apoplast and not in the symplast. Autoradiographic measurements of Ca in nonsuberized root tissues showed that Ca increased more rapidly in the stele than in the cortex (Clarkson & Sanderson, 1971), but later differences in Ca accumulations in the inner cortex, endodermis, and xylem were small. These studies substantiate the concept that $Ca^{2+}$ uptake is a passive process.

## 2. FACTORS AFFECTING UPTAKE

**a. Calcium Concentration**—Calcium concentrations in soil solutions vary extensively. Adams (1974) reported Ca concentrations in soil solutions of 1.7 to 19.4 m*M*. Soil solution concentrations of 1 to 5 m*M* are generally required to protect plant roots against deleterious effects of low pH, toxic ions, and other ionic imbalances (Foy, 1974; Marschner, 1974; Moore, 1974; Raju et al., 1972). Calcium concentrations of 0.1 to 1 m*M* are adequate at the root surface to supply sufficient Ca to plants as long as other elements are balanced and interferences from other elements do not occur (Wiersum, 1979). Concentrations of Ca in acid soils vary, and many factors affect $Ca^{2+}$ availability to plants (Kamprath & Foy, 1984).

Some process must bring the ion from the external solution into the cellular solutions of the plant. Any factor that inhibits or enhances the uptake process will likewise affect the uptake of $Ca^{2+}$ or any other element ion into the root. Numerous review articles and symposia proceedings discuss the factors that affect ion uptake (Anderson, 1973; Baker & Hill, 1975; Clarkson, 1974; Frausto da Silva & Williams, 1976; Lüttge & Higinbotham, 1979; Lüttge & Pitman, 1976; Pitman, 1977; Wardlaw & Passioura, 1976).

**b. Other Ions**—Many reports on competitive effects of other cations on $Ca^{2+}$ uptake are available (De Wit et al., 1963; Foy, 1974; Hiatt & Leggett, 1974; Kirkby, 1979; Wiersum, 1979). Major cations depressing $Ca^{2+}$ uptake are $NH_4^+$, $K^+$, $Mg^{2+}$, and $Al^{3+}$. The physiological processes by which increased concentrations of one cation species depresses the uptake of other cations are not clearly understood. Some investigators explain this in terms of carrier site competition (Epstein, 1972), but Kirkby (1979) considers this explanation unlikely because $K^+$, $Ca^{2+}$, and $Mg^{2+}$ have different ionic sizes and structures. A more likely explanation is that cells produce anion equivalents (organic acids) that nonspecifically attract cations (Kirkby, 1979). The cation species absorbed fastest neutralizes the anion equivalents first and reduce electrostatic attractions for the other cation species. The absorption rate also depends on the concentration of the individual cation species in the root media as well as the uptake mechanism. For example, $K^+$ is usually taken up at a faster rate than other cation species, and because of this, $K^+$ competes strongly against both $Ca^{2+}$ and $Mg^{2+}$. Potassium, which is actively absorbed, competes effectively for the organic anions. This may explain why $K^+$ competes so strongly in cation uptake.

Ammonium is a monovalent cation similar in size to $K^+$ ($K^+ = 0.133$ nm and $NH_4^+ = 0.145$ nm). Ammonium inhibitory effects on $Ca^{2+}$ uptake may be explained in terms of general cation competition. However, $NH_4^+$-fed plants also have lower concentrations of inorganic cations, including $Ca^{2+}$ (Kirkby & Mengel, 1967) and reduced synthesis of organic acids (Hiatt & Leggett, 1974; Kirkby, 1979; Kirkby & Mengel, 1967), and higher concentrations of diffusable anions (Kirkby, 1968; Kirkby & Mengel, 1967;

Kirkby & Knight, 1977). Hence, $NH_4^+$ inhibition of $Ca^{2+}$ uptake may be an indirect effect of reduced cations for cation-anion balances. The rate of $Ca^{2+}$ uptake is also dependent on the counterion; it is highest with $NO_3^-$, lowest with $SO_4^{2-}$, and intermediate with $Cl^-$.

Strong associations between high N levels and Ca deficiencies have been reported (Kirkby, 1979). Besides the cation competition between $NH_4^+$ and $Ca^{2+}$ and the effect of $NO_3^-$ on organic acid synthesis, N stimulation of plant growth often induces Ca deficiencies in tissues. For example, Ca deficiencies are often reported for rapidly growing fruits (tomatoes and apples, in particular), heart tissues of many vegetables, and foliage, tubers, and seeds of some field crops. The rapid growth resulting from high N can create a Ca requirement that exceeds the rate of $Ca^{2+}$ uptake. These disorders may be aggravated by a simultaneous depression of root growth by too much N, which should impair $Ca^{2+}$ uptake. Nitrogen-stimulated shoot growth may divert $Ca^{2+}$ from fruits or developing seeds at a critical stage of development. In the field, Ca deficiencies have been more closely associated with $NH_4^+$ than with $NO_3^-$ fertilization (Shear, 1975; Wilcox et al., 1973), probably because of reduced cation uptake. Ammonium may depress water uptake (Quebedeaux & Ozbun, 1973), which, in turn, could restrict $Ca^{2+}$ uptake and distribution.

Magnesium inhibition of $Sr^{2+}$ uptake (in lieu of $Ca^{2+}$), and vice versa, has been attributed to uncompetitive effects on $Sr^{2+}$ absorption sites (Epstein & Leggett, 1954), i.e., $Ca^{2+}$ and $Mg^{2+}$ absorption sites are different and are separated from each other. Inhibitory effects of $Al^{3+}$ on $Ca^{2+}$ uptake have been attributed to $Al^{3+}$ replacement of $Ca^{2+}$ on binding or exchange sites of cell walls (Foy et al., 1972). This competition can in part explain the effect of Al toxicity in plants grown on acid soils. With $NO_3^-$ as the source of N, $Ca^{2+}$ uptake and transport in apple seedlings was greatly reduced by $Al^{3+}$ (Kotze, 1979; Kotze et al., 1976, 1977). When $NH_4^+$ was the dominant or sole source of N, $Ca^{2+}$ uptake was either enhanced or unaffected by $Al^{3+}$.

Inhibitory effects of $Mn^{2+}$ (Ramani & Kannan, 1976) and stimulatory effects of $BO_3^{3-}$ (Dixon et al., 1973) on $Ca^{2+}$ uptake have also been reported.

Calcium can influence the uptake of other cations; $Ca^{2+}$ often stimulates the uptake of other cations (called the *Viet's effect*) (Viets, 1944). Calcium generally enhances the absorption of the monovalent cations $K^+$ and $Rb^+$ and depresses the absorption of $Li^+$ and $Na^+$ (Hiatt & Leggett, 1974). However, reports on the depression of $K^+$ uptake by $Ca^{2+}$ have been noted (Hiatt & Leggett, 1974). Calcium appears to be essential for the integrity of the selective ion transport mechanism and for the permeability of cell membranes (Hiatt & Leggett, 1974). Since cell membrane characteristics are an integral part of the cation absorption process, $Ca^{2+}$ can affect ion uptake by modifying the membrane. Thus, $Ca^{2+}$ might influence its own absorption by inducing changes in the rate-limiting membrane. In addition, absorption rates for divalent cations are lower than those for monovalent cations.

Somewhat related to ion interactions are the differences in root cation exchange capacities (CEC) and the relationship of CEC to $Ca^{2+}$ and $Mg^{2+}$

uptake. Roots of dicotyledonous plants usually have higher CEC values and absorb more divalent cations than monocotyledonous plants. Plant species with high root CEC may respond more to added K than to added Ca (lime), thus causing inhibitory effects of $K^+$ on $Ca^{2+}$ uptake (Carolus, 1975). For example, tomato, with a relatively higher Ca concentration in the plant and a higher CEC than spinach, is often more responsive to $K^+$ and less responsive to $Ca^{2+}$.

**c. pH**—The effect of pH on $Ca^{2+}$ uptake by plants is relatively small (Moore et al., 1961a). Under field conditions where Ca disorders appear, increasing soil pH is usually accomplished by the addition of lime, which contains Ca. Thus, the effect of increasing soil pH is to increase Ca levels. In nearly all soil studies where pH effects on $Ca^{2+}$ uptake have been reported, $Ca^{2+}$ uptake by plants increased as pH increased (Mason, 1979). This increase would be expected because more Ca is added to the plant growth medium and because lime removes Al as an antagonist to $Ca^{2+}$ uptake. However, decreased $Ca^{2+}$ uptake with added $CaCO_3$ has been reported for some plants (Bhor et al., 1970). Since the effect of pH on $Ca^{2+}$ uptake is relatively small, the effect of adding lime is primarily to increase the Ca levels in the growth medium, which, in turn, increases $Ca^{2+}$ uptake by plants.

**d. Temperature**—Temperature has little effect on the rate of $Ca^{2+}$ uptake (Drew & Biddulph, 1971; Handley & Overstreet, 1961; Maas, 1969; Moore et al., 1961a, Wallace et al., 1963). No differences in $Ca^{2+}$ uptake were noted for bean roots at 5, 15, or 25°C (Drew & Biddulph, 1971), and $Ca^{2+}$ uptake was independent of temperature in several plant species (Smith & Wallace, 1956; Wallace, 1963, 1971a, b; Wallace et al., 1963). A 25-fold increase in $Ca^{2+}$ uptake due to increases in temperature (from 3 to 25°C) was reported for wheat (Johnson & Jackson, 1964). Translocation of $Ca^{2+}$ was also less at lower temperatures (Drew & Biddulph, 1971). Increases in $Ca^{2+}$ uptake were reported for wheat, maize, and soybean as temperatures increased, but the roots contained more Ca than the aerial organs (Saric & Curic, 1963). Calcium was stored mainly in the stems and roots of woody perennial plants and in bulbs and tubers of many other plants at lower temperatures (Wiersum, 1979). Total $Ca^{2+}$ uptake increased with higher temperatures (due to increased growth), but concentrations of Ca in plant tissue did not increase (Chu & Toop, 1975). Reduced $Ca^{2+}$ uptake has also been noted in plants grown at higher temperatures (24 vs. 4°C) because of the competitiveness of higher $K^+$ (Wallace & Hemaidan, 1963; Wiersum, 1979). The Ca/K ratios would be lower at lower temperatures because of the passive uptake of $Ca^{2+}$ and the active uptake of $K^+$.

**e. Light**—Most of the effects of light on the uptake and transport of $Ca^{2+}$ are indirect (Clark, 1981). Light has pronounced effects on metabolic processes, especially those that produce energy. Since $Ca^{2+}$ uptake is passive, energy-producing processes would not affect $Ca^{2+}$ uptake directly, but light effects on the transport of water and assimilates in the xylem and phloem would affect $Ca^{2+}$ uptake, transport, and fixation. Light absorption by plant parts and the surrounding medium usually increases the ambient and plant temperatures. Higher temperature increase transpiration and water movement that would directly affect $Ca^{2+}$ movement into roots and to the various plant parts. Likewise, shading would reduce $Ca^{2+}$ movement by decreasing temperatures and the rate of metabolic processes associated with light. Ion fluxes affected by light have no effect on $Ca^{2+}$ uptake (Jeschke, 1976).

**f. Water Stress**—Water stress effects on mineral element composition in plants has been reviewed (Clark, 1981). Calcium uptake and translocation in plants is probably more closely associated with water movement than most elements because of the passive movement of $Ca^{2+}$. In soils, water relations affect all factors that affect $Ca^{2+}$ availability. These include Ca concentrations in the soil solution, absorption by the root, translocation from root to shoots, and root extension to distant supply points. Light and temperature may also influence water stress by increasing or decreasing evaporation and transpiration and by influencing metabolic processes inside plants.

Water stress may reduce root permeability and transport capacity and thereby inhibit $Ca^{2+}$ uptake (Pitman et al., 1974; Wiersum, 1979). Water stress may not only decrease $Ca^{2+}$ uptake but may enhance the active uptake of $H_2PO_4^-$ and $K^+$ (Mattson, 1973; Wiersum, 1979), leading to imbalanced ratios of Ca/P and Ca/K. Under water stress, older roots surrounded by dry soil lose their ability to absorb nutrients and plants must be supplied almost exclusively by the more active root tips, which enhances uptake of divalent cations relative to monovalent cations (Mattson, 1973).

Plant species and genotypes within species vary in Ca nutrition, and these differences are important in water stress responses to Ca and other elements. For example, rye contained considerably less Ca than barley under water stress (Williams & Shapter, 1955). When plants were under water stress beyond the available moisture range, some had less $Ca^{2+}$ uptake than plants grown with adequate soil moisture (Brown et al., 1960).

External water stress increases internal plant water stress and increases competition among plant parts for available moisture and $Ca^{2+}$. Water stress appears to affect xylem water and $Ca^{2+}$ flow more than phloem movement (Wardlaw, 1967; Wiersum, 1979). This causes shifts in distribution of $Ca^{2+}$ into various plant parts. During periods of water stress, the fast-growing and lower-transpiring organs (fruit, pod, bud, tuber, seed) may not receive adequate $Ca^{2+}$ since they are fed mostly by phloem rather than by xylem transport. In addition, competition by leaves for water could divert $Ca^{2+}$ from these developing tissues as well as enhance withdrawal of water and $Ca^{2+}$ from them because of the greater suction of the more highly transpiring leaves. Water promotes greater xylem transport, which, in turn, enhances the supply of water and $Ca^{2+}$ to the developing and fruiting organs. Renewing water supplies abruptly after an extended water stress period may stimulate new vegetative growth, especially late in the season, which increases the competition between leaves and other tissues (Wiersum, 1979).

Mature maize plants grown under water stress had dry matter yield reductions of 56% but only 9% reductions in Ca concentrations (Jenne et al., 1958). These and other results (Gerakis et al., 1975) indicate that with some plants, water stress was more detrimental to dry matter accumulation than to $Ca^{2+}$ uptake and accumulation. Water stress affects some mineral elements more than others (Clark, 1981). Water stress enhanced Ca accumulation in several range and forage species (Abdel Rahman et al., 1971) but decreased Ca accumulation in other forage species (Gerakis et al., 1975).

**g. Environment and Cultural Practices**—Several articles on the effects of environment and cultural practices on $Ca^{2+}$ uptake and accumulation in plants, especially fruits, are presented in Shear (1979) and Atkinson et al. (1980). Special consideration is given to the article by Wiersum (1979). Some of the effects of environmental conditions and cultural practices on Ca nutrition are summarized briefly.

*1) Soil Volume*

The soil volume available to roots has little effect on Ca supply except where root growth is materially inhibited for some reason or where plants are grown in small containers. Under field conditions, soil volumes do not usually limit root growth, and the increased active surface area facilitates $Ca^{2+}$ uptake. Calcium has a relatively low influx rate, and $Ca^{2+}$ is one of the first elements to show reduced uptake because of limited root systems.

*2) Tillage*

In many acid soils, whether limed or not, the surface soil usually contains more Ca than the subsoil. Tillage operations may damage surface soil roots, forcing roots deeper in the soil profile to provide Ca. In this way, tillage may restrict $Ca^{2+}$ uptake, but such effects are seldom evident. In soils with ample supplies of subsoil Ca, tillage may enhance $Ca^{2+}$ uptake. The cooler temperatures of subsoils will inhibit $Ca^{2+}$ uptake relatively less than they do $K^+$ and $H_2PO_4^-$.

*3) Drainage and Irrigation*

A high water supply in soils will reduce Ca concentration in solution relatively more than monovalent cations because of cation exchange phenomena. However, total solution Ca will probably be greater. If allowed to persist, high water contents will decrease soil $O_2$ levels and increase $CO_2$ levels, which could raise Ca in the soil solution, especially if $HCO_3^-$ is present (Wiersum, 1969).

*4) Mulch*

Grass or straw mulches often contain considerable amounts of K, which may make $K^+$ predominant in uptake by roots growing in the mulch layers. The higher $K^+$ has the potential for restricting $Ca^{2+}$ uptake and causing improper Ca/K balances. On the other hand, mulches from dicotyledonous plants could conceivably enhance $Ca^{2+}$ uptake.

### *5) Changes in Transpiration*

Reducing transpiration rates may enhance Ca accumulation into some Ca-sensitive tissues (fruits and pods) by diverting it from the leaves. Since Ca moves with water, $Ca^{2+}$ could be directed to tissues in greater need of Ca. Reduced transpiration has been accomplished by shading and shelterbelts. Application of antitranspirant oil sprays have also helped overcome some Ca disorders (Wiersum, 1979). Under special conditions, reduced transpiration has resulted in development of root pressures that force xylem water into more Ca-sensitive tissues (Palzkill et al., 1976; Tibbitts & Palzkill, 1979).

Under greenhouse conditions, a reduction in transpiration may not benefit Ca supplies. The moist atmosphere and high light intensities often prevalent in greenhouses may cause excessive heating of plant parts, which reduces transpiration and, in turn, lowers the Ca supply. Ventilation and shading may be required to prevent this. Methods to reduce photosynthesis and reduce transport of assimilates in the phloem could enhance influx of xylem $Ca^{2+}$ to tissues that may otherwise be deficient.

### *6) Source/Sink Relationships*

Rapid growth usually results in increased phloem transport from leaves to sinks (fruit, seed, tuber). The increased phloem transport usually decreases $Ca^{2+}$ influx to the sink tissues. Avoiding conditions favorable to rapid growth should enhance the Ca supply to tissues that often suffer Ca deficiency. Density of plants may affect the growth rate; wide spacing favors rapid growth of individual plants. Pruning (leaves and roots) or removing buds can alter Ca relations by affecting source/sink ratios (Wiersum, 1979). Appropriate source/sink ratios may reduce Ca disorders in many plants.

### *7) Genotype Selection*

Some plant genotypes tolerate low Ca better than others or grow better under conditions that enhance Ca deficiency. Genotypes also have different Ca requirements (discussed later).

### *8) Calcium Applications*

Liming is the common method for adding Ca to benefit plants grown on deficient soils, but Ca sprays and dips are often used for such Ca-sensitive plants as apple and tomato (Atkinson et al., 1980; Shear, 1979). Lime and other Ca salts are added to enhance the Ca supply in soils, to increase soil pH, and to correct toxicities (primarily Al and Mn).

## B. Calcium Translocation

### 1. XYLEM TRANSPORT

Once $Ca^{2+}$ passes through the endodermal cells, it enters stele vessels (xylem and phloem) for longitudinal transport. Bowling (1973) determined electrochemical distributions of major ions ($K^+$, $Ca^{2+}$, $Mg^{2+}$, $NO_3^-$, and $SO_4^{2-}$) between the external solution and the epidermal cells and between the epidermal cells and the xylem root exudate of sunflower roots. The data provided no evidence to support active transport of either $Ca^{2+}$ or $Mg^{2+}$ during any stage of their movement across the root cells to the pericycle and their transfer into the xylem vessels.

From early experiments, the concept that $Ca^{2+}$ and other ions move with the transpiration stream by mass movement has prevailed (Hanger, 1979). More recent evidence abounds to indicate that $Ca^{2+}$ moves in the xylem by ion exchange (Bell & Biddulph, 1963; Biddulph et al., 1961; Ferguson & Bollard, 1976b; Isermann, 1970, 1971; Jacoby, 1967; Millikan and Hanger, 1964, 1967; Millthorpe and Moorby, 1969; Morrison & Heine, 1966; Shear & Faust, 1970; Van de Geijn et al., 1979). The xylem cylinder is visualized as an ion exchange column, and $Ca^{2+}$ and $Mg^{2+}$ are adsorbed onto negatively charged exchange sites on or in the cylinder walls. Although the ion exchange concept is generally accepted for alkali earth metal cation movement, especially $Ca^{2+}$ and $Mg^{2+}$ (Millthorpe & Moorby, 1969), it does not hold for all ions (Morrison & Heine, 1966). Calcium generally lags behind $K^+$ and $H_2PO_4^-$ in translocation, which supports the ion exchange concept of $Ca^{2+}$ movement.

As $Ca^{2+}$ moves in the xylem, considerable leakage to adjacent tissues occurs (Bell & Biddulph, 1963; Biddulph et al., 1961; Faust & Shear, 1973; Hanger, 1979; Wieneke, 1979). Adjacent tissues absorb $Ca^{2+}$ according to metabolic utilization (Bell & Biddulph, 1963; Faust & Shear, 1973). Young leaves and tissues are preferred. Leakage from the xylem can be rapid and occurs primarily in the apoplast, i.e., very little $Ca^{2+}$ enters these cells by symplastic movement. Calcium in the longitudinal veins of oat leaves did not move into mesophyll tissue because of the presence of suberized lamellae in the radial walls (Hanger, 1979). The suberized lamellae would not have restricted lateral flow of Ca if it had been in the symplasm. Lignin and oxalic acid, which have high affinities for $Ca^{2+}$ (Faust & Shear, 1973; Gallaher, 1975), may act as agents to bind $Ca^{2+}$ in older cells adjacent to the xylem. This binding of $Ca^{2+}$ may reduce the amount of $Ca^{2+}$ that is available to developing leaves or fruits.

The movement of $Ca^{2+}$ in the xylem from the roots is in the acropetal (toward tops) direction, with little or no $Ca^{2+}$ being moved basipetally (toward root tips) (Marschner & Richter, 1974). Calcium-45 was applied 6 to 9 cm from the root tip of maize seedlings and $^{45}Ca$ was measured in 3-cm segments of the roots as well as in the endosperm, leaves, and guttated vesicles. Little or no $^{45}Ca$ was detected in the 0- to 6-cm sections of the roots, indicating no movement of Ca to root tips. Except for the root section where $^{45}Ca$ was applied, the root segments contained similar amounts of $^{45}Ca$ until Ca accumulated in the leaves of the plant.

The movement of $Ca^{2+}$ in the xylem vessels is promoted by the presence of other divalent cations ($Sr^{2+}$, $Mg^{2+}$, $Mn^{2+}$, and $Zn^{2+}$) (Bell & Biddulph, 1963; Faust & Shear, 1973; Ferguson & Bollard, 1976b; Hanger, 1979; Millikan & Hanger, 1964, 1967; Shear & Faust, 1970) and by chelating agents, both synthetic (ethylenediaminetetraacetate [EDTA]) and natural (citrate, malate, and oxalate) (Bradfield, 1976; Ferguson & Bollard, 1976b; Hanger, 1979; Jacoby, 1967; Millikan & Hanger, 1964; Van De Geijn et al., 1979). The chelating agents compete for exchange sites and complex $Ca^{2+}$ into organic forms. Water and $K^+$ do not facilitate $Ca^{2+}$ movement in the xylem (Bell & Biddulph, 1963; Millikan & Hanger, 1967, 1969). The higher the concentration of divalent cations introduced, the faster and more extensive the movement of $Ca^{2+}$ in the xylem. However, high levels of $K^+$ and $H_2PO_4$-reduced $^{45}Ca$ translocation, but $K^+$ had no effect on total Ca in tomato plants (Chiu & Bould, 1976). Sodium also retarded the translocation of $Ca^{2+}$ from roots to shoots in scurvy grass (Le Saos, 1976). Translocation of $Ca^{2+}$ has also been inhibited by $Al^{3+}$, $Sc^{3+}$, and $Fe^{3+}$ (Johnson & Jackson, 1964; Clarkson & Sanderson, 1971).

Higher flow rates of water reduce the retention of $Ca^{2+}$ in the xylem (Emmert, 1969). The faster the water movement, the greater the distance $Ca^{2+}$ moved before readsorption on other exchange sites. Calcium moved in the free state and its use was not dependent on endogenous metabolism. Higher transport of $Ca^{2+}$ also occurred in slowly transpiring leaves of cabbage and carrot by enhancing root pressure flow (Tibbitts & Palzkill, 1979).

Plants grown with various sources of N have shown different patterns of $Ca^{2+}$ translocation. Ammonium reduced $Ca^{2+}$ uptake and translocation (Gysi, 1975) and increased Ca accumulation more in newer developing tissues than in older tissues (Shear & Faust, 1970). Nitrate enhanced the movement of $Ca^{2+}$ into older leaves of apple trees (Shear & Faust, 1970). Xylem exudate volumes and Ca concentrations in those exudates decreased in tomato plants transferred from $NO_3^-$ to $NH_4^+$ solutions (Wilcox et al., 1977). Nitrate-fed tomato plants accumulated higher $Ca^{2+}$ than $NH_4^+$- or urea-fed plants (Kirkby & Knight, 1977; Kirkby & Mengel, 1967). This increased $Ca^{2+}$ movement was attributed to the increased organic acid synthesis and enhanced cation uptake noted so frequently in plants supplied with $NO_3^-$ instead of $NH_4^+$ or other sources of N (Hiatt & Leggett, 1974; Kirkby, 1979).

## 2. PHLOEM TRANSPORT

Even though $Ca^{2+}$ has been reported in the phloem of many plant tissues (Bangerth, 1979; Eschrich et al., 1964; Faust & Shear, 1973; Ferguson, 1979; Hanger, 1979; Läuchli, 1968; Millikan & Hanger, 1969; Pate et al., 1975; Priestley, 1976; Ringoet et al., 1967, 1968; Stebbins & Dewey, 1972; Tammes & Van Die, 1966; Thomas, 1967; Wieneke, 1979; Wiersum et al., 1971), the immobility of $Ca^{2+}$ in the phloem has been confirmed by many investigators (Bangerth, 1979; Hanger, 1979; Ziegler, 1975). Under some circumstances, $Ca^{2+}$ can enter the sieve tubes of the phloem and be translocated like other substances (Millikan & Hanger, 1969; Ringoet et al., 1967, 1968; Wieneke, 1979). Ziegler (1975), therefore, proposed that $Ca^{2+}$ immobility in the phloem under normal conditions is not the inability of the phloem to transport $Ca^{2+}$ but the inability of $Ca^{2+}$ to enter into the sieve tubes of the phloem. On the other hand, Ferguson (1979) considered the concept that $Ca^{2+}$ is a "phloem-immobile" element to be a "somewhat misleading generalization." He suggested that $Ca^{2+}$ moves in the phloem and that the problem is one of $Ca^{2+}$ movement through nonvascular tissue and the loading and unloading of transport conduits.

Relative to $Ca^{2+}$ entry into and translocation in the sieve tubes, Peel (1972) found $^{45}Ca$ in sieve-tube exudates in willow to be sevenfold lower than for $^{137}Cs$. This difference was attributed to an unknown mechanism that discriminated between $Ca^{2+}$ and $Cs^{+}$. The excluded $Ca^{2+}$ was believed to be metabolically mediated and associated with a membrane boundaries. This fixed Ca was not recycled to new meristematic tissues during the time of Ca deficiency. The exclusion of $Ca^{2+}$ from sieve tubes may have caused cell membrane structures to disintegrate and become more functional for longitudinal conduction of solutes (Epstein, 1973). Since the plasmalemma is the only functional membrane in sieve tubes, the small amount of $Ca^{2+}$ normally found in sieve elements could satisfy that requirement. Without $Ca^{2+}$ in the sieve tubes of bean plants, assimilate translocation was inhibited (Geuy, 1972). Some proteinaceous filaments, which are characteristic of seive tubes, are Ca sensitive and may help regulate $Ca^{2+}$ in these structures (Koontz & Foote, 1966).

Potassium enhances phloem transport and could retard $Ca^{2+}$ influx via the xylem into sink tissues (Mengel & Haeder, 1977). In addition, the organic acids are located primarily in cell vacuoles that may act as strong sinks for $Ca^{2+}$. The high fixation capacity of the vacuoles could then prevent $Ca^{2+}$ from entering the transport stream. The fixation of $Ca^{2+}$ within vacuoles can be even stronger if oxalic acid is formed, because it binds $Ca^{2+}$ more tightly than most other prevalent organic acids. Thus, the Ca supply available for transport can be lowered dramatically (Brumagen & Hiatt, 1966).

Quantities of $Ca^{2+}$ are readily transported laterally out of the xylem and have been found to form Ca-oxalate crystals (Bell & Biddulph, 1963; Wieneke, 1979). Deposition of $Ca^{2+}$ in the sieve tubes as oxalate or $PO_4^{3-}$ has been suggested as a means of restricting $Ca^{2+}$ transport (Bangerth, 1979; Marschner, 1974; Ulrich & Mostafa, 1976; Van Goor & Wiersma, 1974). Even though oxalate is only sparingly soluble and does not recycle under most conditions (Gallaher, 1975), the evidence is inadequate to assume that oxalate crystals can immobilize all tissue $Ca^{2+}$ (Bangerth, 1979; Hanger, 1979). For example, $Ca^{2+}$ is also immobile in tissues free of Ca oxalate crystals (monocotyledons) even though oxalate crystal formation occurs in young leaf tissues. High amounts of water-soluble $Ca^{2+}$ have been found in older than in younger tissues (Millikan & Hanger, 1964). Phosphorus also has been implicated in the inhibition of $Ca^{2+}$ movement in tomato (Chiu & Bould, 1976) and sugar beets (Ulrich & Mostafa, 1976).

Immobilization of $Ca^{2+}$, particularly in woody plants, has been noted in the bark of trees, in pedicels of fruits (Leigel, 1970; Stebbins & Dewey, 1972; Terblanche et al., 1979; Thomas, 1970; Wieneke, 1979), and in the basal portions of trees during the growing season (Terblanche et al., 1979; Thomas, 1970; Wieneke & Führ, 1975). It has been assumed that this $Ca^{2+}$ immobilization or temporary fixation occurs mostly in the xylem (Bangerth, 1979).

The movement of $Ca^{2+}$ from xylem vessels to surrounding leaf tissue has received less attention than movement of $Ca^{2+}$ into roots and in the xylem. Ferguson (1979) noted that more Ca was found in the apical sections of maize leaves than in middle and basal sections. The portion of $Ca^{2+}$ transferred from the midrib to the surrounding lamina was also higher in apical than in basal sections of leaves. The restricted lateral movement in the basal portion of the leaf was attributed to reduced apoplastic movement because of suberized lamellae in the cells surrounding the bundle sheaths. The lack of $Ca^{2+}$ movement from the leaf to the translocation stream may be important. Since information on such $Ca^{2+}$ movement is limited, implications must be made from studies of water movement. Kuo et al. (1974) suggested that nearly all water (99%) flows out of the xylem in the major bundles in wheat leaves. If endodermal cells of roots are used as an analogy, the major xylem vessels should be more highly suberized than the minor xylem vessels, and $Ca^{2+}$ movement should be highly restricted in the major xylem vessels because $Ca^{2+}$ must move in the apoplast. Calcium immobility in leaf cells may be visualized as $Ca^{2+}$ binding to cell walls and cellular structures.

## 3. CALCIUM REDISTRIBUTION

Little of the initially absorbed $Ca^{2+}$ reaches the new aboveground growth (Biddulph et al., 1958; Green & Bukovac, 1968; Hanger, 1979; Levi, 1968; Millikan & Hanger, 1967; Rinne & Langston, 1960; Terblanche et al., 1979). Most $Ca^{2+}$ is either immobilized or greatly delayed in translocating to the growing leaf points or fruit. For example, less than 5% of the $^{45}C^{2+}$ absorbed by oat plants was recovered in the new top growth (Hanger, 1979). Older bean plants also redistributed more $Ca^{2+}$ than younger plants (Biddulph et al., 1958; Levi, 1968). Even though uptake of $^{45}Ca$ was fairly high during active growth of apple seedlings, very little $Ca^{2+}$ reached the fruit (Terblanche et al., 1979). Of that $Ca^{2+}$ that did, most remained in the pedicel. Redistribution of $Ca^{2+}$ in subterranean clover was enhanced by injections of a divalent cation through the petiole (Millikan & Hanger, 1964, 1967). Redistribution of $Ca^{2+}$ has also been greater in plants grown under low Ca conditions (Green & Bukovac, 1968; Millikan & Hanger, 1967).

In woody plants, the upward movement of $Ca^{2+}$ is slow, and its redistribution often continues over many years. Because of the Ca disorder bitter pit in apple, this plant has been studied extensively for $Ca^{2+}$ movement and redistribution. Therefore, several examples of Ca nutrition in apple are used here to describe $Ca^{2+}$ redistribution in plants. Calcium movement in herbaceous plants is similar to woody plants (Hanger, 1979).

Several investigators have reported slow upward movement of $Ca^{2+}$ in apple trees (Frasher, 1958; Hanger, 1979; Millikan, 1971, Shear & Faust, 1970; Wieneke & Führ, 1973). For example, $Ca^{2+}$ absorbed by roots of 30-cm tall apple seedlings required 3 d to reach the tip (Shear & Faust, 1970). Hanger (1979) reported that it took 14 d for $^{45}Ca^{2+}$ to travel 70 to 80 cm up single-stemmed apple trees early in the season. A considerable amount of $Ca^{2+}$ accumulated in the bark during the first months of the season before declining during the latter part of the season (Hanger, 1979). During dormancy, the bark of 2-year-old 'Golden Delicious' apple trees contained 59% of the total Ca in the tree (Terblanche et al., 1979), although the bark made up only 12% of the total biomass. During the period of rapid shoot extension, the $Ca^{2+}$ reserves in the tree contributed 23% of the Ca needed for new growth (shoots, leaves, and fruit). Both autumn and spring applied Ca were immobilized in these apple trees.

Xylem sap of apple may contain relatively high amounts of Ca (Bradfield, 1976; Wieneke, 1969). After bud burst (mouse-ear stage), Ca concentrations in apple sap rose to 4.5 m*M*, then decreased for the next 3 weeks to under 1 m*M* (Bradfield, 1976) and remained below 1 m*M* until flowering and fruit set. Citric acid concentrations followed similar trends. Peak Ca concentrations in apple sap coincided with flowering, fruit set, and during young fruit development (Hanger, 1979; Wieneke & Führ, 1973). Calcium was preferentially routed to the more metabolically active tissue (Millikan, 1971; Shear & Faust, 1970). Calcium moved into young apple fruits at linear rates until the fruits reached a fresh weight of about 30 g, at which time $Ca^{2+}$ influx decreased (Hanger, 1979). By midseason of fruit development, negligible amounts of $Ca^{2+}$ entered the fruit (Millikan, 1971; Wieneke & Führ, 1973), but it increased near fruit maturity (Askew et al., 1959, 1960). Even though total Ca in the fruit increased with age, its concentration decreased (Hanger, 1979; Wilkinson, 1968).

A high proportion of the Ca in mature apple fruits is deposited during early development (Führ & Wieneke, 1973; Hanger, 1979; Wieneke & Führ, 1973; Wilkinson, 1968). Calcium is distributed fairly uniformly in the newly developed fruit (Hanger, 1979; Wieneke, 1974), but as fruits age, $Ca^{2+}$ is deposited preferentially in the pedicel and proximal core tissue (Hanger, 1979; Millikan, 1971; Terblanche et al., 1979; Wieneke, 1969, 1974). Hanger (1979) suggests that the preferential accumulation of Ca in the pedicel could contribute to low fruit Ca, but long-term monitoring of $Ca^{2+}$ movement failed to show a barrier to Ca accumulation in the fruit. When Ca in the pedicel increased dramatically, more $Ca^{2+}$ entered the fruit and accumulated mainly in the peel or adjacent cells (Terblanche et al., 1979; Wieneke, 1974).

Calcium-45 injected into apple (Martin, 1967), as well as other woody plants (Ferrell & Johnson, 1956; Thomas, 1970), has been detected in buds, leaves, and fruit produced two and three seasons later. Continued mobility of $Ca^{2+}$ to meet Ca requirements of new tissue at the beginning of seasons is important in woody plants. Terblanche et al. (1979) and Wieneke and Führ (1975) noted that Ca in newly developing tissues was deposited the season before. The movement of this deposited Ca was highest early in the season and diminished as the season progressed (Wieneke & Führ, 1973). The $Ca^{2+}$ taken up during the latter part of the season was bound, and its mobility was noted in the following season (Führ & Wieneke, 1974; Wieneke & Führ, 1975). The major source for $Ca^{2+}$ redistribution was the roots, bark, and basal portions of the stem (Hanger, 1979; Terblanche et al., 1979). Improved distribution of $Ca^{2+}$ was attributed to its release from exchange sites and its maintenance in the free form (Bradfield, 1976).

In apple fruits, the peel and seeds have higher concentrations of Ca than the flesh. In addition, when Ca was applied to one side of apple fruits, the treated side was free of Ca deficiency symptoms (bitter pit), whereas the untreated side developed the disorder (Chittenden et al., 1969).

Calcium concentrations in seeds are much lower than in leaves and other vegetative tissues (Bollard, 1970; Clark, 1976). If germinating seeds must rely exclusively on seed Ca, the developing seedlings become Ca deficient in a relatively short time, because seedborne Ca is sufficient to maintain meristematic activity for only a few days. Most of the Ca for newly developing plants comes from cotyledons of dicotyledonous seeds (Ferguson & Bollard, 1976a; van Staden & Comins, 1976) and from aleurone layers in cereal grains (Eastwood & Laidman, 1971; Liu & Pomeranz, 1975). The movement of $Ca^{2+}$ from seed to newly developing tissue varies from almost nil (Okamoto, 1962) to nearly one third of the seed Ca (Eastwood & Laidman, 1971; Ferguson & Bollard, 1976a; Guardiola & Sutcliffe, 1972). Some $Ca^{2+}$ is translocated basipetally to the roots of new seedlings (Eastwood & Laidman, 1971; Liu et al., 1975), suggesting some phloem transport. Because of the high binding capacity of young tissue, it was proposed that the $Ca^{2+}$ must have been protected from fixation by organic complexes (Ferguson & Bollard, 1976).

Calcium accumulation in underground storage organs, e.g., potato and peanut, may occur by translocation from the roots to the storage organ (Beringer & Taha, 1976; Bledsoe et al., 1949; Wiersum, 1966). Little or no transpirational water passes through these underground organs, hence their $Ca^{2+}$ intake is greatly restricted. When potato tubers were exposed to air or buried in dry sand, they became infiltrated with a dye that had been injected into the xylem as a result of transpiration through the tubers (Wiersum, 1966). On the other hand, when tubers were buried in moist sand, they were not infiltrated with dye because of the lack of transpiration. Before young peanut pegs were buried, $Ca^{2+}$ was delivered to them by the transpiration stream, but once the pegs were buried, $Ca^{2+}$ from transpiration ceased (Beringer & Taha, 1976; Bledsoe et al., 1949). Calcium is absorbed by each storage organ from the external medium. If high water potential gradients were established between leaves and pegs, $Ca^{2+}$ will move out of peanut pegs to the vegetative tissue via transpiration (Bledsoe et al., 1949). In addition, young fruit or storage organs receive sufficient $Ca^{2+}$ via the transpiration stream during rapid cell division, but $Ca^{2+}$ intake decreases dramatically with the commencement of rapid cell growth and import of photosynthate. Two reasons for the decrease in $Ca^{2+}$ were proposed by Hanger (1979). First, sufficient water would be provided the storage tissue from phloem transport of photosynthate to satisfy the tissue needs. If phloem water delivery exceeded tissue demands, water could back flow in response to the water potential gradient in the xylem. Second, rapid growth would decrease the surface area to volume ratio and decrease transpiration potential to the storage organ. In turn, the transpiration would bypass the storage organ in favor of leaf tissues. Baker and Moorby (1969) noted $Sr^{2+}$ (in lieu of $Ca^{2+}$) moved to potato tubers only when foliar transpirational rates were lowered and water potential gradients favored flow to the tubers. Tomato fruits held at low transpiration rates also had low Ca concentrations (Wiersum, 1966).

Very little of the $Ca^{2+}$ applied to leaves or stems moves from the treated area to other parts of the plant (Hanger, 1979; Lewis & Martin, 1973). When $^{45}Ca$ was injected into a small apple tree branch, only 5% of the $^{45}Ca$ was translocated from the treated area into apple fruits, and about one fourth of the $Ca^{2+}$ reaching the fruit moved to the calyx flesh area (Lewis & Martin, 1973). Increased concentrations of $Ca^{2+}$, $Mg^{2+}$, and other polyvalent cations and chelating agents may overcome some of the $Ca^{2+}$ fixation in leaves and induce its translocation to other plant parts (Millikan & Hanger, 1964, 1967). Translocation of $Ca^{2+}$ in young yucca and in heat-girdled spiderwort stalks indicated that $Ca^{2+}$ moved primarily by the phloem (Penot et al., 1976; Wiersum et al., 1971). Since phloem movement of $Ca^{2+}$ is relatively small, the researchers concluded that $Ca^{2+}$ from Ca sprayed leaves must move by mechanisms other than phloem transport.

One of the likely pathways whereby $Ca^{2+}$ moves from Ca-treated leaves to other plant parts is by the back flow of water in the transpiration stream (Hanger, 1979). Calcium in leaves may readily reach the xylem vessels and be transported therein if the hydrostatic tension in the vessels is reduced by leaf injury, high atmospheric humidity, and application of aqueous solutions on leaves in amounts that exceed the volume transpired (Biddulph et al., 1959; Bukovac et al., 1956; Crafts & Crisp, 1971; Hanger, 1979; Zimmerman, 1960). Stem ringing of oat leaves and clover petioles did not restrict $^{45}Ca$ movement. In addition, movement of $^{45}Ca$ from tips of oat leaves coincided with cellular injury (to provide an entry point) in the uptake zone. Once an entry point appeared, $^{45}Ca$ moved readily down the leaf. Damage to bundle sheath cells around veins also increased $^{45}Ca$ movement in bean leaves treated with Ca (Biddulph et al., 1959).

## C. Calcium Functions in Plants

Considerable progress has been made within recent years to define the function of Ca in plants. As recently as the 1960s, Burstrom (1968) and Nason and McElroy (1963) reported that the function of Ca in plants was not entirely clear and questioned whether Ca even met the criteria of essentiality (Arnon, 1950). The functions of Ca in plants have been related to Ca effects on membranes, cell walls, enzymes and metabolism, phytohormones, and with calmodulin (Bangerth, 1979; Christiansen & Foy, 1979; Clarkson & Hanson, 1980; Cormier et al., 1982; Marmé, 1982, 1983; Marmé & Dieter, 1984; Wyn Jones & Lunt, 1967).

## 1. MEMBRANES

Considerable evidence indicates that an important function of Ca is to stabilize membranes (Bangerth, 1979; Christiansen & Foy, 1979; Clarkson & Hanson, 1980; Jackson, 1967). Evidence for this comes from experiments dealing with the effects of $Ca^{2+}$ on selective ion uptake, leakiness of cells, and electron microscope observations (Bangerth, 1979).

Early experiments showed the need for $Ca^{2+}$ to alleviate detrimental effects of distilled water on plant growth (True, 1914). The uptake of inorganic ions (notably monovalent cations) was enhanced when $Ca^{2+}$ was present in solution (Viets, 1944). Subsequent investigators have noted similar stimulatory effects of $Ca^{2+}$ on ion uptake in various kinds of plants (Jackson, 1967; Wyn Jones & Lunt, 1967). Other polyvalent cations ($Mg^{2+}$, $Sr^{2+}$, $Ba^{2+}$, and $Al^{3+}$) had similar effects, but they were usually less pronounced than those of $Ca^{2+}$. The internal Ca concentration of the plant had no influence on $K^+$ uptake. These results suggested that the controlling mechanism of $Ca^{2+}$ on $K^+$ was located in the outer cell areas, most likely the plasmalemma.

The influence of $Ca^{2+}$ on ion uptake is pH dependent (Jacobson et al., 1960, 1961a, b; Rains et al., 1964). When $H^+$ concentrations of external media were high (near pH 4.5), $Ca^{2+}$ had its greatest effects and counterbalanced the detrimental effect of high $H^+$ (Marschner et al., 1966). Two general effects appeared under acid conditions: (i) the competitive effects of $H^+$ on binding reactions of cations (Fried & Noggle, 1958) and (ii) the direct damage of $H^+$ on membranes (Rains et al., 1964). These and other investigations support the concept that $H^+$ altered cell membrane properties, including selective ion uptake, can be overcome by $Ca^{2+}$.

Closely associated with selective ion uptake by membranes is the leakiness, or increased permeability, of membranes at low $Ca^{2+}$ levels. In Ca-deficient tissues, cell membrane permeability is usually high, i.e., efflux of compounds is common but influx is not (Mengel & Kirkby, 1982). Mengel and Helal (1967) noted that $Ca^{2+}$ affected the efflux of $K^+$ and $H_2PO_4^-$ in barley roots but not their influx. Losses of inorganic P, soluble organic P and N compounds, $K^+$, and $Ca^{2+}$ were observed in barley roots grown at low pH and low $Ca^{2+}$ (Jacobson et al., 1950). Higher amounts of several simple sugars were noted in exudates from peanut plants grown at low $Ca^{2+}$ (0.25 m*M*) than at adequate $Ca^{2+}$ (1.25 m*M*) (Shay & Hale, 1973). Betacyanin and sucrose effluxes from red beet sections were noted when $Ca^{2+}$ was withheld from the bathing solution (Hawker et al., 1974; Siegel, 1970). Stress-induced leakiness (cold, $O_2$ deficiency) of cells can be prevented by the addition of $Ca^{2+}$ (Christiansen & Foy, 1979). Tomato and apple fruits have higher permeabilities when plants are grown with low Ca than with high Ca (Cooper & Bangerth, 1976; Rousseau et al., 1972; Van Goor, 1968). The less efficient absorption of sorbitol into fruit cells has also been associated with cell leakage (Bangerth et al., 1972). Christiansen and Foy (1979) proposed that many Ca deficiency disorders in plants may be due to poor membrane function in metabolic compartmentalization.

Leakiness of cells is partially prevented when $Ca^{2+}$ is substituted by $Mn^{2+}$ and $Sr^{2+}$ (Garrard & Humphreys, 1967; Siegel, 1970; van Steveninck, 1965), and $Sr^{2+}$ sprays reduced some Ca deficiency disorders such as blossom end rot (BER) in tomatoes, internal breakdown in apple, and blackheart in celery (Bangerth, 1973; Takatori et al., 1961; Wills et al., 1975). Other ions ($Mg^{2+}$, $K^+$, and $H^+$) were antagonistic to the $Ca^{2+}$ effects (Bangerth, 1974). Thus, the effects of competitive cations on Ca nutrition included not only $Ca^{2+}$ absorption but also membrane permeability caused by the replacement of $Ca^{2+}$ (Bangerth, 1979). In studies with chelating agents, removal of $Ca^{2+}$ from membranes has enhanced the concept that $Ca^{2+}$ is required to maintain membrane stability and integrity. Hanson (1960) noted that excised maize roots treated with EDTA, ribonuclease (RNase), or KCl had lower respiration rates and a loss of nucleotides. The addition of $Ca^{2+}$ or $Mg^{2+}$ reversed the effects of these agents (Foote & Hanson, 1964). The removal of $Ca^{2+}$ by EDTA also affected the uptake of $Cl^-$, $NO_3^-$, $Br^-$, and $H_2PO_4^-$ (Foote & Hanson, 1964; Hooymans, 1964; Hyde, 1966; Leggett et al., 1965; Pitman, 1964) and increased the leakiness of cells (van Steveninck, 1965).

Microscopic studies have indicated that Ca deficiency produces many cytological disorders (Bangerth, 1970, 1973; Bussler, 1962, 1963; Fuller, 1976; Hecht-Buchholz, 1979; Marinos, 1962; Marschner & Günther, 1964; Nilshammar et al., 1971). Microscopic studies showed initial Ca deficiency was characterized by swollen and brown-colored cell walls and, at later stages, an accumulation of brown substances in the vacuoles and intracellular spaces (Bussler, 1963). Cortex and pith cells of potato buds had brown-colored cellular contents (autolysis) before visual Ca deficiency symptoms were observed (Hecht-Buchholz, 1979). The xylem and phloem cells of Ca-deficient plants were intact, and only the walls of xylem cells appeared darker than cells of plants receiving adequate Ca. In electron microscope studies of barley (Marschner & Günther, 1964), the tonoplast of vacuolated cells disintegrated, followed by destruction of all cytomembranes and a total breakdown of compartmentalization. Potato bud cells, relatively rich in cytoplasm, showed initial Ca deficiency symptoms to be invagination and disintegration of the plasmalemma, accumulation of degenerative vesicles, and destruction of mitochondria (Hecht-Buchholz, 1979). In addition, the breakdown of the nuclear envelope was noted during the initial phases of Ca deficiency in barley (Marinos, 1962). In both of these latter studies, the proplastids were relatively persistent and cell walls were intact during the initial phases of Ca deficiency. Plastids of cells adjacent to Ca-deficient cells were deficient in starch, an abnormal condition of cells receiving adequate Ca (Hecht-Buchholz, 1979). In rye, a dark-stained material that accumulated in the space between invaginated plasmalemma and the wall of Ca-deficient cells was suggested to be denatured protein that had leaked from the cytoplasm because of the increased permeability (Hallam et al., 1973). Calcium-deficient conditions (loss of membrane integrity and accumulation of vesicles) were similar to changes noted in senescent cells (Hecht-Buchholz, 1979). The development of bitter pit in apples is characterized by the breakdown and fragmentation of the cuticle, abnormal compression or thicker than normal hypodermal cells, and highly perforated, thick-walled, and necrotic cortex cells (Simons & Chu, 1980). Hecht-Buchholz (1979) noted that the lack of $Ca^{2+}$ in plant cells did not appear to affect any special part of metabolism, but it caused a rapid autolysis of cells because of increased membrane permeabilities.

Little is known about the biochemical mechanisms involved in Ca-membrane responses. The binding of $Ca^{2+}$ with phospholipids (Ochiai, 1977; Oursel et al., 1973), cholesterol (Marfey & Chessin, 1974), proteins (Ochiai, 1977), or specially arranged carboxyl groups (Williams, 1976) could possibly alter pore radius or conformational structures in membranes (Christiansen & Foy, 1979; Garrard & Humphreys, 1967; Williams, 1976). Garrard and Humphreys (1967) suggested that $Ca^{2+}$ could bind with certain anions in the membrane to form bridges between structural components. Christiansen et al. (1970) suggested that $Ca^{2+}$ prevented cold-stressed leakiness of cells by preventing a change in membrane phase (liquid to solid). Later studies (Christiansen & Foy, 1979) indicated that the cold-induced leakiness was not dependent on metabolism and that the $Ca^{2+}$ reversal effects on cold-stressed membranes were due to nonmetabolic physical binding of $Ca^{2+}$ in the membranes. Thus, Ca deficiency appears to increase pore size, which then enhances permeability and autolysis of cells.

## 2. CELL WALLS

Considerable amounts of Ca are located in cell walls, especially as Ca pectate in the middle lamella, and various concepts on the functions of Ca in cell walls have been developed (Burstrom, 1968; Jackson, 1967; Wyn Jones & Lunt, 1967). They include formation of ionic bridges (Bennett-Clark, 1956; Jansen et al., 1960; Ochiai, 1977), increase in cell wall rigidity (Cormack, 1949; Tagawa & Bonner, 1957), increase in hydrophobicity and reduction in water permeability (Ochiai, 1977), and cell elongation (Burstrom, 1968). Explanations for the role of $Ca^{2+}$ in these various processes have not been fully resolved, and conflicting results are used to support the various concepts (Burstrom, 1968; Clarkson & Hanson, 1980; Christiansen & Foy, 1979; Marmé, 1982, 1983; Marmé & Dieter, 1984; Wyn Jones & Lunt, 1967).

Microscopic studies of plant tissues containing low Ca show considerable disorganization of cells (Bussler, 1962; Fuller, 1976, 1980; Marinos, 1962). However, this disorganization does not appear to be caused by disruption of the cell wall or the middle lamella (Bangerth, 1973; Hecht-Buchholz, 1979; Marinos, 1962; Marschner & Günther, 1964). Hecht-Buchholz (1979) noted that the cell walls of potato sprout cells undergoing Ca deficiency were relatively intact. Similar results were noted for barley tissues (Marinos, 1962). Even in potato sprouts with severe Ca deficiency autolysis, the cell walls showed dense lines that were considered to be Ca pectate and representative of the middle lamella (Hecht-Buchholz, 1979). The amount of Ca in the pectate fractions of rice leaf cells was not affected by the Ca level supplied (Ito & Fujiwara, 1967). In addition, Cleland (1960) noted that $Ca^{2+}$ was not removed from pectin and suggested that Ca was unaltered by auxin and that Ca-carboxyl linkages were not involved in cell elongation. These results would indicate that cell wall breakdown and disruption of Ca pectate as a cell wall binding agent are not causal events in Ca deficiency disorders.

The weakening of cell walls and middle lamella may occur under various Ca deficiency conditions. Many Ca deficiency–related disorders, e.g., fruit and vegetable mealiness, water core, internal breakdown, bitter pit, BER, corking, cracking, tipburn, and browning (Atkinson et al., 1980; Bangerth, 1973; Faust & Shear, 1968; Shear, 1975, 1979), can be associated with breakdown and disorganization of the cell walls of plants. The mechanisms for these effects of $Ca^{2+}$ appear to be less specific than the $Ca^{2+}$ effects on membranes (Bangerth, 1979) even though $Ca^{2+}$ is a major cation in the pectin-protein cement of the middle lamella (El Hinnawy, 1974; Ginzburg, 1961).

### 3. ENZYMES AND METABOLISM

Burstrom (1968) considered $Ca^{2+}$ to have no specific role in metabolic processes, but other investigators have noted several specific functions of Ca (Clarkson & Hanson, 1980; Wyn Jones & Lunt, 1967). Many enzymes have been reported to be stimulated or inhibited by $Ca^{2+}$, some of which are $\alpha$-amylase, esterase, pectinesterase, lipoxygenase, nucleases, protein kinase, pyruvate kinase, polygalacturonic transeleminase, glucose-6-phosphate dehydrogenase, and adenosine triphosphatase (ATPase). More extensive lists of $Ca^{2+}$-affected enzymes are given in Wyn Jones and Lunt (1967), Christiansen and Foy (1979), and Clarkson and Hanson (1980). Membrane-bound enzymes, particularly ATPases, require divalent cations for activity and have received considerable attention in recent years (Hodges, 1976). Alkaline phosphatase and $Mg^{2+}$-activated ATPase of cucumber have also been found to increase in activity under Ca-deficient conditions (Matsumoto & Kawasaki, 1981; Matsumoto & Yamaya, 1981; Yamaya & Matsumoto, 1981).

The action of $Ca^{2+}$ on membrane ATPases, which require divalent cations ($Ca^{2+}$, $Mg^{2+}$, and $Mn^{2+}$) for activation, appears to be as a cation–adenosine triphosphate (ATP) complex as substrate for the enzyme (Hodges, 1976). Unless $Ca^{2+}$ transport occurs, reason to believe that $Ca^{2+}$ is the active metal is unfounded. Some hydrolases e.g., $\alpha$-amylases, phosphorlyases, and nucleases that metabolize macromolecular substrates (starch) require bound $Ca^{2+}$ for activity (Scott & Peterson, 1979). Calcium may also form bridge complexes with starch. Thus, $Ca^{2+}$ may play a structural role to change enzyme conformation rather than act as a catalyst. In animals, $Ca^{2+}$ generally meets enzyme site requirements with enzyme-substrate-metal (E-S-M) type of enzymes ($Ca^{2+}$ does not act directly with the enzyme molecule but through the substrate), but not with enzyme-metal-substrate (E-M-S) type of enzymes (direct interaction between metal and protein) (O'Sullivan, 1973; Scrutton, 1973). Calcium usually inhibits E-M-S type of enzymes. Even though these types of enzyme-metal mechanisms have not been identified in plants, it is assumed that $Ca^{2+}$ will function similarly in plants as they do in animals. Thus, a major role of $Ca^{2+}$ appears to be its binding with proteins, nucleic acids, and lipids to affect cell adhesion, membrane and chromatin organization, and enzyme conformation (Clarkson & Hanson, 1980). These $Ca^{2+}$ functions appear to be related to Ca–calmodulin protein complexes.

Calcium can bind with many different proteins (Kretsinger, 1976, 1977), and recent findings have revealed that certain Ca-protein complexes (calmodulin) have specific regulatory functions in cells (Cheung, 1980; Cormier et al., 1982; Klee et al., 1980; Marmé, 1982, 1983; Marmé & Dieter, 1984). This Ca-binding protein appears to be ubiquitous in eukaryotes (Klee et al., 1980). Calmodulin-Ca complexes appear to act directly on effector systems (ATPases) and indirectly on regulatory systems (protein kinases) (Klee et al., 1980). Calmodulin-$Ca^{2+}$ is discussed in a later section.

Low Ca has been closely associated with high respiration rates in fruits (Bangerth et al., 1972; Bramlage et al., 1974; Cooper & Bangerth, 1976; Faust & Shear, 1972; Tingwa & Young, 1974) and leafy tissues (Pal et al., 1973). These higher respiration rates have been explained by the interaction of $Ca^{2+}$ with ATPase or pectin kinase on membrane permeability (Kylin & Kahr, 1973; Ralph et al., 1976) and by increased membrane permeability that allows respiratory substrates to diffuse from the vacuole to enzymes in the cytoplasm (Bangerth et al., 1972). However, Faust and Shear (1972) and Pal et al. (1973) did not attribute Ca effects on respiration as membrane-regulated processes; $Ca^{2+}$ appeared to have some direct effects. Pyruvate kinase, which is inhibited by $Ca^{2+}$ (Bangerth, 1979; Tomlinson & Turner, 1973), could be involved in such $Ca^{2+}$-affected processes as respiration or the formation of volatile substances (Bangerth, 1979). Low Ca in fruit delays ripening and lowers fruit firmness and storageability (Cooper & Bangerth, 1976; Sharples & Johnson, 1977; Tingwa & Young, 1974), increases such volatile substances as ethanol, ethylene, acetaldehyde, and acetic acid (Bangerth, 1973; Wills, 1972), causes chlorophyll degradation (Cooper & Bangerth, 1976), and enhances senescence and abscission (Ferguson et al., 1983; Leopold, 1977; Poovaiah, 1979; Sharples & Johnson, 1977).

## 4. PHYTOHORMONES

The phytohormone gibberellin, or gibberellic acid (GA), reduces $Ca^{2+}$ translocation to the aerial plant parts in many species (Bangerth, 1973; Wieneke et al., 1971; Wills et al., 1975) resulting in Ca disorders on the parts (vegetative and fruits) treated with GA. No correlation has been noted between $Ca^{2+}$ translocation and GA-induced growth processes (Wieneke et al., 1971). However, $Ca^{2+}$ enhanced GA-stimulated growth processes (Leopold, 1977) in oat coleoptile, cocklebur cotyledons, lettuce hypocotyls, and rice mesocotyls (Leopold et al., 1974). In the presence of GA, $Ca^{2+}$ inhibited senescence and abscission (Poovaiah & Leopold, 1973a, b). In contrast, $NH_4^+$ plus GA promoted the processes. The effect of $NH_4^+$ could be reversed by $Ca^{2+}$ (Poovaiah & Leopold, 1973a), possibly by altering the configuration of membranes or macromolecules and affecting the affinity of the hormone for its attachment site (Leopold, 1977).

Cytokinins (benzyladenine [BA] and kinetin) have enhanced $Ca^{2+}$ movement to aerial plant parts (Shear & Faust, 1970). The hypocotyls of mung beans had higher $Ca^{2+}$ uptake in the presence of cytokinins (Lau & Yang, 1975). Calcium enhanced BA-stimulated elongation of oat coleoptiles, lettuce hypocotyls, and cocklebur cotyledons and enhanced the development of lateral shoots of soybean (Leopold et al., 1974). On the other hand, $NH_4^+$ inhibited the development of lateral shoots of soybean (Poovaiah & Leopold, 1976b). Calcium plus BA inhibited leaf senescence in maize, and added $Ca^{2+}$ overcame the senescence-promoting effects of $NH_4^+$ (Poovaiah & Leopold, 1973a). The mechanism for cytokinin effects on $Ca^{2+}$ movement was suggested to be because of retarded senescence (Bangerth, 1979) or because transpiration rates were increased to enhance $Ca^{2+}$ influx (Isermann, 1970). Bound $Ca^{2+}$ was higher in normal tomato fruits than in Ca-deficient fruits (Poovaiah, 1979). Higher $Ca^{2+}$ uptake could also be explained by modified phosphorylation reactions of membrane proteins (Ralph et al., 1976).

Auxins inhibit $Ca^{2+}$ movement to aerial plant parts (Bangerth, 1973; Bangerth & Firuzeh, 1971; Oberly, 1973; Stahly & Benson, 1970, 1976; Wieneke et al., 1971). Spray applications of 2,3,5-triiodobenzoic acid (TIBA), which inhibits basipetal auxin transport, enhanced Ca deficiency disorders in tomato and apple fruits (Bangerth, 1973; Bangerth & Firuzeh, 1971; Oberly, 1973; Stahly & Benson, 1970, 1976). TIBA interfered with $Ca^{2+}$ transport into apical bud tips and youngest leaves of sunflower (Marschner & Ossenberg-Neuhaus, 1977). The negative effects of auxin on $Ca^{2+}$ translocation was considered to be due to growth rate increases caused by TIBA and indole-3-acetic acid (IAA) (Marschner & Ossenberg-Neuhaus, 1977), and decreased basipetal movement in favor of acropetal movement of $Ca^{2+}$ (Bangerth, 1976; Dela Fuente & Leopold, 1973). Increased lateral distribution of $Ca^{2+}$ in IAA-stimulated growth was also noted (Arslan-Cerim, 1966; Acharya Goswami & Audus, 1976). The latter findings helped explain the enhanced $Ca^{2+}$ translocation into parthenocarpic fruit of plants treated with auxins exhibiting high basipetal transport properties (indole auxins and $\alpha$-naphthaleneacetic acid [NAA]), but not with phenoxy type auxins (Bangerth, 1979). Auxin-stimulated growth was enhanced by $NH_4^+$ but inhibited by $Ca^{2+}$ (Poovaiah & Leopold, 1976a). The mechanism for $Ca^{2+}$ effects on reciprocal transport of auxins is not clear, but Leopold (1977) suggested that $Ca^{2+}$ effects on auxin-induced growth (as well as on other hormonal-induced responses) were on configuration changes in the membranes or macromolecules that could affect hormone affinity at its attachment site. Poovaiah and Leopold (1976a) found that $Ca^{2+}$ increased the specific binding of NAA to membrane pieces of maize coleoptiles.

Relationships between Ca-auxin growth effects and Ca deficiency disorders have been implied but not understood. Crisp et al. (1976) suggested that Ca-auxin processes in cell elongation (Burstrom, 1968) may be related to Ca deficiency disorders. These authors noted that the development of lettuce tipburn was associated with higher than optimum levels of IAA. If $Ca^{2+}$ were limiting or low during the auxin-mediated cell elongation phase, cell walls could collapse and necrosis develop. High IAA concentrations could arise because IAA oxidase was inactivated by chlorogenic acid, a compound that generally increases during Ca deficiency. A lettuce cultivar susceptable to tipburn had higher concentrations of chlorogenic acid than other lettuce cultivars tolerant to tipburn (Collier et al., 1979). Bangerth (1979) suggested that the effects of $Ca^{2+}$ on other auxin-mediated processes such as auxin binding (Poovaiah & Leopold, 1976a) and the acidification of auxin-treated tissues (Cohen & Nadler, 1976) did not appear to be closely related to Ca deficiency disorders. Whether Ca-auxin processes such as ultrastructure changes in plasma membranes (Moore & Bracker, 1976) are related to Ca deficiency disorders remains uncertain.

The effects of the phytohormone ethylene ($C_2H_4$) appear to mimic Ca deficiency disorders. For example, $C_2H_4$ increases membrane permeability and respiration rates, hastens ripening and senescence (Abeles, 1973; Suttle & Kende, 1978), and increases hydrogen peroxide ($H_2O_2$) in Ca-deficient tissue (Faust & Shear, 1968; Kevers et al., 1982). The localization of $C_2H_4$ synthesis was associated with cell wall membranes in apple (Mattoo & Lieberman, 1977). Cell walls are also the site for initial Ca deficiency symptoms to appear (Hecht-Buchholz, 1979). Calcium also inhibited or prevented $C_2H_4$ effects in the swelling of pea stems and the elongation of rice mesocotyles (Leopold et al., 1974) and cucumber cotyledons (Ferguson et al., 1983).

## 5. CALMODULIN

Recent information has greatly expanded and clarified many of the concepts on the function of $Ca^{2+}$ in plants (Cormier et al., 1982; Marmé, 1982, 1983; Marmé & Dieter, 1984). The discovery of calmodulin, its importance in $Ca^{2+}$ function, the similarities of plant and animal calmodulin properties, and the studies on calmodulin-dependent enzymes have broadened the scope and interpretation of the molecular mechanisms involved with or regulated by $Ca^{2+}$. Even though the information for plant systems is limited, these new concepts have caused extensive interest and investigation on $Ca^{2+}$. Calcium appears to function as a second messenger in plants and plays a major role in the regulation of cellular processes in conjunction with the primary messengers light, hormones, and gravity (Marmé & Dieter, 1984).

Calmodulin is a Ca-binding protein that is present in many cells. It binds with $Ca^{2+}$ in such a configuration that the Ca-protein complex stimulates many enzymes and physiological processes at Ca concentrations considered to be very low (10–1000 n*M*). In this respect, $Ca^{2+}$ stimulates processes at concentrations lower than those normally thought possible and in the concentration range usually reserved for such effectors as hormones.

Many biochemical and physiological processes are known to be regulated or influenced by $Ca^{2+}$. Information on the function of $Ca^{2+}$ in these processes has been mostly speculative. This has been most likely because of the difficulty in measuring free cytoplasmic $Ca^{2+}$ concentrations (Williamson, 1981) and the lack of evidence for the existence of Ca-containing molecules or compounds (Marmé, 1983; Marmé & Dieter, 1984). New techniques for $Ca^{2+}$ assays using metallochromic (Scarpa et al., 1978) and luminescence (Loewenstein & Rose, 1978) indicators and the discovery of the similarity between plant and animal calmodulin (Anderson & Cormier, 1978) have led to the search for $Ca^{2+}$-regulating processes in plants (Marmé, 1983; Marmé & Dieter, 1984).

Researchers have long recognized that the free cytoplasmic $Ca^{2+}$ concentration is very low (Bangerth, 1979; Marmé, 1983). Most of the $Ca^{2+}$ in cells is immobilized and may or may not be exchangeable or ionizable (Bygrave, 1978). Relatively high concentrations of $Ca^{2+}$ have been required for Ca-dependent enzyme activities in vitro. These relatively high in vitro $Ca^{2+}$ concentrations appear to be irrelevant for in situ conditions and have even been reported to inhibit some enzymes (Marmé, 1983). The discrepancy between in vitro $Ca^{2+}$ requirments for optimum enzyme activity and the very low free $Ca^{2+}$ in the cytoplasm can be explained by the lack of calmodulin in in vitro assays. When calmodulin is added in vitro, $Ca^{2+}$ concentration needed for optimum enzyme activity, or that which does not inhibit reactions, is of the same magnitude as is found in the cytoplasm (near 1 $\mu M$).

The low concentrations of free $Ca^{2+}$ in the cytoplasm are regulated by specific transport reactions (Marmé, 1983; Marmé & Dieter, 1984). Active transport of $Ca^{2+}$ into cellular compartments (mitochondria, chloroplasts, endoplasmic retriculum, and other microsomal fractions) and extrusion of $Ca^{2+}$ from the cytoplasm (pumps located at the plasma membrane) have been reported. These $Ca^{2+}$ transport systems appear to be similar in plants and animals. Plants have an additional $Ca^{2+}$ transport system not found in animals; the transport into and out of vacuoles with the control mechanism located at the tonoplast (Gross, 1982). Because of these, low free $Ca^{2+}$ concentrations can be controlled in the cytoplasm.

Plants have only a few enzymes and processes that are now known to be controlled by calmodulin. Those identified so far are nicotinamide adenine dinucleotide (NAD) kinase (Anderson & Cormier, 1978; Anderson et al., 1980; Dieter & Marmé, 1980b, c), microsomal $Ca^{2+}$-ATPase (Dieter & Marmé, 1980a, 1981), and microsomal $Ca^{2+}$ transport (Dieter & Marmé, 1980b). Other intracellular enzymes and processes identified as being affected by or requiring $Ca^{2+}$ and that may involve calmodulin are pyruvate kinase (Tomlinson & Turner, 1973), alkaline lipase (Muto & Beevers, 1974), phospholipid acyl-hydrolase (Hasson & Laties, 1976), and glutamic dehydrogenase (Joy, 1973; Nauen & Hartmann, 1980). The secretion of polysaccharides (Griffing & Ray, 1979; Morris & Northcote, 1977) and $\alpha$-amylase (Jacobson & Higgins, 1982; Jones & Jacobsen, 1983; Moll & Jones, 1982) from cell walls has also been noted at deficient levels of $Ca^{2+}$.

Many extracellular functions of $Ca^{2+}$ that may be similar to or have $Ca^{2+}$-calmodulin functions have also been identified. These include (i) plant growth hormone actions (Burstrom, 1968; Leopold, 1977; Thimann & Schneider, 1938), (ii) auxin transport (Day et al., 1978; Dela Fuente & Leopold, 1973; Janistyn, 1973), (iii) auxin-mediated $H^+$ secretion (Cohen & Nadler, 1976), (iv) $H_2O_2$ secretion (Kevers et al., 1982; Sticher et al., 1981), (v) abscission (Poovaiah & Leopold, 1973a), (vi) senescence (Ferguson et al., 1983; Provaiah & Leopold, 1973b), (vii) cation transport into plant cells (Epstein, 1961), (viii) cell plate formation (Becerra & Lopez-Saez, 1978; Paul & Goff, 1973), (ix) membrane fusion (Holz & Stratford, 1979; Portis et al., 1979, (x) membrane damage and leakiness (Poovaiah & Leopold, 1976), (xi) cell elongation (Cleland & Rayle, 1977), (xii) ultrastructural membrane alterations (Moore & Bracker, 1976), (xiii) osmoregulation (Kauss, 1974, 1981, 1983; Kauss et al., 1979), (xiv) pollen tube growth (Brewbaker & Kwack, 1963; Herth, 1978; Mascarenhas & Machlis, 1964; Reiss & Herth, 1978), (xv) gravimetric stimulation (Acharya Goswami & Audus, 1976; Arslan-Cerim, 1966; Biro et al., 1982), (xvi) sensitive plant leaf movement (Campbell & Thomson, 1977; Toriyama & Jaffe, 1972; Vanden Driessche, 1963, 1978), (xvii) *Chlamydomonas* flagellae movement (phototaxis), (Halldal, 1957; Hyams & Borisy, 1978; Nultsch, 1979; Nultsch et al., 1971; Schmidt & Eckert, 1976), (xviii) *Platymonas* rhizoplast contraction (Salisbury & Floyd, 1978), (xix) *Mougeotia* chloroplast movement (Dreyer & Weinenseel, 1979; Klein et al., 1980; Wagner & Klein, 1978), (xx) *Fucus* egg development (Jaffe, 1979; Nuccitelli & Jaffe, 1976), (xxi) *Poterioochromonas* cell shrinkage (Kauss, 1981, 1983; Van Belle, 1981), and (xxii) reversal of germination inhibition of sensitive fern by red light (Wayne & Hepler, 1982).

With the function of calmodulin in biochemical and physiological reactions becoming more apparent (Marmé, 1983; Marmé & Dieter, 1984), the search for its presence in various tissues has intensified. Calmodulin has been identified in garden pea, mung bean, wild carrot, and peanut (seeds) (Anderson & Cormier, 1978; Anderson et al., 1980), barley shoots and roots (Grand et al., 1980; Caldwell, 1982; Iverson et al., 1981), squash hypocotyls (Dieter & Marmé, 1980a), spinach leaves (Watterson et al., 1980), maize (Iverson et al., 1981), two basidiomycetes (Grand et al., 1980), the yeasts (*Candida* and *Saccharomyces*) (Hubbard et al., 1982), *Chlamydomonas* (Gitelman & Witman, 1980), *Mougeotia* (Marmé, 1983), *Neurospora* (Ortega-Perez et al., 1981), and *Blastocladiella* (Gomes et al., 1979). In many of the studies, the properties of plant calmodulin are similar to those in animals. In fact, calmodulin from animal sources has been used to activate reactions in plants and vice versa (Marmé, 1983).

The concept that hormones can act at very low doses with high specificity has been well established and recognized. This principle has been used in animal research to define the concept of second messenger compounds (Robinson et al., 1971). In animals, cyclic adenosine 5′-monophosphate (cAMP) has been established as a second messenger because it meets the designated criteria (Sutherland et al., 1968; Rasmussen, 1981). Marmé (1983) used the same criteria to suggest that $Ca^{2+}$ is a second messenger in plants. The criteria are:

1. The cytoplasmic, free $Ca^{2+}$ concentration is low and precisely controlled by $Ca^{2+}$ transport mechanisms that pump $Ca^{2+}$ actively out of the cytoplasm against an electrochemical gradient and allow passive leakage of $Ca^{2+}$ into the cytoplasm.
2. Hormones, neurotransmitters, and other substances (the primary signals) increase cytoplasmic $Ca^{2+}$ by enhancing $Ca^{2+}$ net influxes into the cell or the release from intracellular stores.
3. The intracellular $Ca^{2+}$ signal is coupled via calmodulin to biochemical reactions and physiological processes.
4. Calmodulin antagonists inhibit the effects of primary signals.
5. The effects of the primary signals can be mimicked by substances that increase free cytoplasmic $Ca^{2+}$ concentrations (e.g., the ionophore A 23 187).

Accumulated data on $Ca^{2+}$ functions in plants support the criteria that $Ca^{2+}$ is a second messenger in plants. That is, calmodulin has been found in cells (Anderson & Cormier, 1978), the $Ca^{2+}$ transport mechanisms have been idenified (Gross, 1982; Gross & Marmé, 1978; Hodges & Hanson, 1965), the modulation of the $Ca^{2+}$ fluxes by primary signals have been reported (Dieter & Marmé, 1981; Kubowicz et al., 1982), and $Ca^{2+}$ and calmodulin-dependent enzymes are present (Anderson & Cormier, 1978; Dieter & Marmé, 1981; Ranjeva et al., 1983). To substantiate this hypothesis, Marmé and Dieter (1984) note specific examples and outline a diagram to describe the reactions for this model.

## D. Calcium Requirements and Concentrations in Plants

### 1. REQUIREMENTS

Although many plants (especially dicotyledons) accumulate fairly high concentrations of Ca (20 000 to 30 000 $\mu g\ g^{-1}$ is not uncommon in leaves), the requirement for $Ca^{2+}$ in many metabolic reactions and physiological processes is relatively low (Burstrom, 1968; Marmé, 1983; Wallace & Soufi, 1975; Wyn Jones & Lunt, 1967). Wallace et al. (1966) noted that concentrations of Ca could be as low as 100 to 200 $\mu g\ g^{-1}$ in maize leaves and 800 $\mu g\ g^{-1}$ in tobacco leaves if the levels of Cu, Fe, Mn, Zn, and Mg were in proper balance. Wallace and Soufi (1975) pointed out that in many plants Ca may be required at concentrations similar to those of micronutrients. Many of the studies of Wallace and colleagues (Wallace, 1966, 1972, 1980) have indicated that 800 to 2000 $\mu g\ g^{-1}$ of Ca in the foliar parts of plants is sufficient for plant growth and reproduction. One reason that the apparent Ca requirement could be so low was attributed to reduced concentrations of heavy metals available to plants. One of the primary roles of Ca in plant nutrition has been suggested to be to detoxify the effects of heavy metals and other cation excesses in plants (Foy, 1974; Wallace & Saufi, 1975). Thus, the critical Ca requirement in a plant is variable and depends on the status of heavy metals (Wallace & Soufi, 1975) and other cations (Burstrom, 1968).

Much of the information on the apparently low Ca requirement of higher plants comes from root and seedling growth studies (Burstrom, 1968; Wyn Jones & Lunt, 1967). Calcium in the root growth medium at concentrations as low as 1 $\mu M$ was sufficient to allow maximum root elongation provided Cu and Fe were present at even lower levels (Wyn Jones & Lunt, 1967). Calcium concentrations at 10 $\mu M$ were sufficient to maintain cell elongation, and 0.1 to 1 $\mu M$ gave normal mitotic activity in the meristems of wheat roots (Burstrom, 1952). Root growth has been inhibited in many plant species when $Ca^{2+}$ was near or above 1 m$M$ (Burstrom, 1968). Inhibitions of shoot growth have been noted in various plant species with Ca in the root growth medium at 5 to 20 m$M$. Normal concentrations of Ca in oat coleoptiles were 3 m$M$ (1800 $\mu g\ g^{-1}$ fresh weight) for plants supplied with 10 to 100 m$M$ of $Ca^{2+}$. Pollen tube growth was enhanced at 0.1 to 5 m$M$ of $Ca^{2+}$ but inhibited by $Ca^{2+}$ concentrations near 10 m$M$ (Brewbaker & Kwack, 1963; Kwack, 1967; Kwack & Kim, 1967; Mascarenhas & Machlis, 1964).

Calcium requirements were examined by Loneragan et al. (1970) and defined as (i) solution Ca requirement, (ii) functional Ca requirement, and (iii) critical Ca concentration. *Solution Ca requirement* is the minimum level of Ca in the growth medium required by the plant to attain maximum growth. This definition does not consider internal Ca concentrations. *Functional Ca requirement* is the minimum Ca concentration at functional sites within plant cells that sustains maximum growth rates. High concentrations of Ca may accumulate in some cells and organelles but may not be involved in any growth process. Functional Ca concentrations remain relatively constant in a plant organ as dry matter increases. Calcium concentrations of 1000 to 2000 $\mu g\ g^{-1}$ have been noted for legumes and herbs and 500 to 1000 $\mu g\ g^{-1}$ in grasses and cereals (Loneragan & Snowball, 1969). *Critical Ca concentration* is the Ca concentration actually present in the plant organ at the time Ca becomes limiting to growth. These values may be relatively high. Higher concentrations were noted for legumes and herbs than for grasses and cereals (Loneragan et al., 1970).

## 2. CONCENTRATIONS

Concentrations of Ca found in many plants are given in Table 1. Calcium concentrations in plants vary according to plant part, plant age, location on the plant, and many other factors. In some plant tissues, critical or deficiency ranges of Ca have been noted, but these are dependent on so many factors that listing such would be of little value. Plant species vary extensively in the amount of Ca that they accumulate in their organs, especially leaves.

Plant age probably exerts as much influence on element concentration in plant tissues as any other factor. For example, Ca concentrations in maize plants varied extensively with age and plant part (Clark, 1975b). Maize leaves and tassels showed the greatest increases in Ca concentrations with age. Very little Ca accumulated in the kernels of maize plants. Over half of the $Ca^{2+}$ absorbed by maize plants remained in the leaves at maturity

Table 1—Ranges of Ca and Mg concentrations considered to be sufficient in plant tissues (Walsh & Beaton, 1973).

| Plant | Plant part | Stage of growth† | Range of concentration Ca | Mg |
|---|---|---|---|---|
| | | | μg g⁻¹, dry wt | |
| | | **Cereals** | | |
| Maize | Leaves | 3rd–4th leaf stage to 30–45 d after plant emergence | 9 000–16 000 | 3 000– 8 000 |
| | Ear leaf | Silk | 2 100–10 000 | 2 000– 4 000 |
| | Stalk | Silk | 1 000– 3 000 | 1 000– 3 000 |
| Grain sorghum | Leaves | 23–39 d after seeding | 9 000–13 000 | 3 500– 5 000 |
| | Leaves | YM, 37–56 d after seeding | 1 500– 9 000 | 2 000– 5 000 |
| | Leaf (third from top) | 66–97 d after seeding | 2 000– 6 000 | 1 000– 5 000 |
| Wheat and oat | Leaves | Head emergence | 2 000– 5 000 | 1 500– 5 000 |
| Barley | Leaves | NA | 3 000–12 000 | 1 200– 2 100 |
| Rice | Leaves | Midtillering to panicle differentiations | 1 600– 3 900 | 1 600– 3 900 |
| | | **Field and sugar crops** | | |
| Soybean | Leaves | NA | 3 600–20 000 | 2 600–10 000 |
| Cotton | Leaves | NA | 19 000–35 000 | 3 000– 7 500 |
| Peanut | Upper stem and leaves | Various | 7 500–20 000 | 3 000– 8 000 |
| Sugar beet | Petiole | NA | 2 000–25 000 | 1 000– 7 000 |
| | Blade | | 4 000–15 000 | 1 000–15 000 |
| Sugarcane | Sheaths 3–6 | NA | 1 000–20 000 | 1 500–40 000 |
| | Internodes 8–10 | NA | 500–20 000 | 1 000–10 000 |
| | | **Fruits and forest trees** | | |
| Apple | Leaves | NA | 14 000 | 4 100 |
| Cherry | Leaves | NA | 20 900 | 6 800 |
| Peach | Leaves | NA | 21 200 | 6 700 |
| Grape | Leaves | NA | 17 500 | 4 400 |
| Blueberry | Leaves | NA | 7 400 | 2 800 |
| Coniferous spp. | Leaves | NA | 1 200 | 7 000 |
| Hardwood spp. | Leaves | NA | 3 000–22 000 | |
| | | **Forage** | | |
| Alfalfa | Top | 15 cm | 5 000–30 000 | 3 000–10 000 |
| | | **Vegetables** | | |
| Asparagus, garden | Fern | 45–90 cm | 4 000– 5 000 | 1 500– 2 000 |
| Bean | Leaf or foliate | YM | 8 000–30 000 | 2 500– 7 000 |
| Beet, red | Leaf | Mature | 25 000–35 000 | 3 000– 8 000 |
| Broccoli | Leaf | YM, heading | 12 000–25 000 | 2 300– 4 000 |
| Brussels sprout | Leaf, upper | NA | 3 000–22 000 | 2 300– 4 000 |
| Cabbage | Leaf, wrapper | Half grown | 15 000–35 000 | 2 500– 4 500 |
| | Midrib, upper leaf | YM | 14 000–20 000 | 3 000– 5 000 |
| | Head, fourth section | Mature | 4 000– 6 000 | 1 400– 2 000 |
| Cantaloupe | Petiole | YM leaf | 30 000–50 000 | 7 000–10 000 |
| | Blade | YM | 50 000–70 000 | 10 000–15 000 |

(continued on next page)

Table 1—Continued.

| Plant | Plant part | Stage of growth† | Range of concentration Ca | Range of concentration Mg |
|---|---|---|---|---|
| | | | ——$\mu$g g$^{-1}$, dry wt—— | |
| Carrot | Leaf | YM, midgrowth | 20 000–35 000 | 4 300– 5 300 |
| Cauliflower | Leaf | YM, heading | 20 000–35 000 | 2 400– 2 600 |
| | Blade | Button | 7 200– 7 900 | |
| Celery | Leaf | YM, half grown | 4 000–30 000 | 2 000– 5 000 |
| Kale/collard | Leaf | YM, near maturity | 30 000–40 000 | |
| Lettuce | Leaf, wrapper | Half size | 14 000–20 000 | 5 000– 7 000 |
| Pea, garden | Leaves and stems | Bud to full bloom | 12 000–15 000 | 2 700– 3 500 |
| | Leaf | YM, midgrowth | 15 000–25 000 | 2 500– 4 000 |
| Pepper, green | Leaf | YM, midgrowth | 4 000– 6 000 | 10 000–17 000 |
| Potato | Leaf | YM, half grown | 20 000–40 000 | 5 000– 8 000 |
| | Leaf | Mature | 15 000–25 000 | 5 000–15 000 |
| | Leaf | Young, 30 cm | 6 000– 9 000 | 8 000–11 000 |
| Spinach | Leaf | YM, 30–50 d | 6 000–12 000 | 16 000–18 000 |
| | Leaf | YM, mature | 6 000–10 000 | 4 000– 8 000 |
| Corn, sweet | Ear leaf | After silk | 16 000–25 000 | 4 000– 8 000 |
| | Ear leaf | 80% silk | 9 000–13 000 | 3 000– 8 000 |
| Potato, sweet | Leaf | Mature | 7 300– 9 500 | 4 000– 8 000 |
| Turnip | Leaf | YM, midgrowth | 30 000–50 000 | |
| Tomato | Leaf | NA | | 6 000– 9 000 |
| | Leaf | YM, midgrowth | 14 000–18 000 | 4 000– 6 000 |
| | Leaf | YM, trellised | 5 000–20 000 | 6 000–10 000 |
| Watermelon | Leaf | YM, midgrowth | 25 000–35 000 | 6 000– 8 000 |

† Abbreviations: YM = young mature; NA = not available or given.

even though leaves represented only 10% of the dry matter of the plant. Only about 5% of the Ca was found in the reproductive parts (kernels, cob, husk, and tassel) even though these parts contributed over 40% of the dry matter. In young maize plants, Ca decreased in concentration from lower to upper leaves, but after development of about the 12th leaf, Ca concentrations in the newer developing leaves remained relatively constant (Clark, 1975a). These leaves increased uniformly in Ca concentration with age. Since $Ca^{2+}$ is essentially immobile in the phloem, $Ca^{2+}$ does not readily translocate to the reproductive parts of maize plants. Similar results have been noted for many plant species (Atkinson et al., 1980; Shear, 1979), and special treatments often have to be administered to such plants as apple and pear during growth or to their fruits to prevent Ca deficiency disorders from developing.

## 3. GENOTYPE DIFFERENCES

Plant species have been classified into two broad categories relative to Ca: calcifuges and calcicoles (Burstrom, 1968). Calcifuges are plants that generally grow well on acid, low-Ca soils, and calcicoles are plants that generally grow well on calcareous soils. Normal leaf Ca concentrations range from 4800 to 10 800 $\mu$g g$^{-1}$ in numerous calcifuges and from 15 600 to 22 700 $\mu$g g$^{-1}$ in calcicoles (Hou & Merkle, 1950). Calcifuge plants have generally been characterized by tolerance to low levels of Ca, low responses to added Ca, low $Ca^{2+}$ uptake rates, high K/Ca ratios in tissues, slow growth rates, tolerance to Al (and sometimes Mn), and susceptibility to Fe deficiency when grown on calcareous soils (Foy, 1974). The physiological factors associated with adaptation to acid or calcareous soils do not often follow the calcifuge-calcicole classification of plants. For example, tolerance to Al or other acid soil factor is often unrelated to the Ca requirements in plants, but other factors may be of equal or greater importance. Examples of differences among plants for some of these factors and tolerances to Al have been discussed by Foy (1974) and Kamprath and Foy (1984).

Genotype differences to Ca requirements, uptake, and yield responses have been noted for sheep fescue (Snaydon & Bradshaw, 1961), perennial ryegrass (Forbes & Gelman, 1981), white clover (Snaydon, 1962), red clover (Hunt et al., 1976), blue lupine and white lupine (Hocking & Pate, 1977), tobacco (McEvoy, 1963; Peeden & McCants, 1977), barley (Rasmusson et al., 1971; Young & Rasmusson, 1966), brussels sprout (Millikan & Hanger, 1969), cabbage (Maynard et al., 1965; Nieuwhof, 1960), cauliflower (Maynard et al., 1981), lettuce (Collier et al., 1976, 1979), maize hybrids (Andrade et al., 1975; Bradford et al., 1966; Clark, 1976; Furlani et al., 1977; Gorsline et al., 1961, 1968), maize inbreds (Clark, 1976, 1978; DeTurk, 1941; Porter & Moraghan, 1975; Sayre, 1955), tomato (English et al., 1981; Giordano et al., 1981; Greenleaf & Adams, 1969), snap bean (Shannon et al., 1967), peanut (Bathla et al., 1978; Beringer & Taha, 1976; Walker, 1975; Walker & Keisling, 1978), citrus and fruit tree rootstocks (Eaton & Robinson, 1977; Koksal, 1973; Poling & Oberly, 1979; Sharples, 1974), potato (Dyson & Digby, 1975), pearl millet (Oza et al., 1979), and garden strawberry (Guttridge et al., 1978).

Differences among plant genotypes (varieties, inbreds, or cultivars within species) have been associated with many factors. Maize genotypes responded differently to Ca and differed in production of dry matter per unit Ca and functional and critical concentrations of Ca (Clark, 1978). Maize and sorghum genotypes differed in $Ca^{2+}$ absorption properties so that one genotype developed Ca deficiency and another did not (Kawaski & Moritsugu, 1979). Differential Ca response among fescue genotypes have been attributed to differences in $Ca^{2+}$ absorption rather than to differences in $Ca^{2+}$ metabolism or internal $Ca^{2+}$ requirement (Snaydon & Bradshaw, 1961). Peanut genotypes differed in translocation of $Ca^{2+}$ from the shells to the meat (Beringer & Taha, 1976). Tomato genotypes differed in ability to deplete the growth medium of $Ca^{2+}$ and to use $Ca^{2+}$ already inside the plant (Giordano et al., 1981). Tomato tops had greater control over the uptake of $Ca^{2+}$ than did the roots (Gableman & Gerloff, 1982). Tomato genotypes that were tolerant to BER absorbed and accumulated more $Ca^{2+}$ than did BER susceptible genotypes (Greenleaf & Adams, 1969). Cabbage genotypes differed in $Ca^{2+}$ uptake and transport from the basal/wrapper leaves to the leaves inside the head (Maynard et al., 1965), and cauliflower leaf tipburn was related to the lack of $Ca^{2+}$ (Maynard et al., 1981).

## 4. BREEDING FOR Ca EFFICIENCY

Several assumptions and decisions are required before plant breeding approaches for solving Ca deficiency stress, or any mineral stress, in plants can be considered. Criteria for breeding plants for mineral stresses and some of the problems involved have been widely discussed (Devine, 1982; Duvick et al., 1981; Lafever, 1981; Saric, 1982). If the plant breeding approach is assumed to have merit, plants must be identified with differences in tolerance to Ca deficiency. Many plant species have genotypes that require or respond differently to Ca. Improvement of plants for Ca deficiency stresses should be feasible through plant breeding. Considerable progress has been made already by selecting and using Ca-efficient genotypes, but actual breeding for improved Ca efficiency has been limited (Saric, 1982).

The inheritance of Ca concentrations in the ear leaves of maize has been found to be under genetic control (Gorsline et al., 1961). Later studies noted that even though Ca and Mg concentrations in double-cross hybrids were not correlated with results found in single-cross parents (Craig & Thomas, 1970), the genetic control of $Ca^{2+}$ and $Mg^{2+}$ was associated with two or three genes (Gorsline et al., 1968). The locus for accumulation of a large number of mineral elements was suggested to be chromosome 9 (Barber, 1970), and this was verified for Ca, Mn, and P (Naismith et al., 1974). These elements did not have a common mechanism for control, however. The genetic control of Ca accumulation has also been noted in barley, wheat, and soybean (Fick & Rasmusson, 1967; Kleese et al., 1968; Rasmusson et al., 1964, 1971), but mechanisms for these differences are unknown. The frequency of resistant plants for tolerance to BER was low in segregating populations of tomato genotypes (Greenleaf & Adams, 1969). These authors suggested that tolerance to BER was recessive.

In a recent study on the inheritance of Ca deficiency traits in tomato, Giordano et al. (1981) noted 81% differences in dry matter yields between Ca-efficient and Ca-inefficient genotypes grown at low Ca (10 mg $plant^{-1}$). When these genotypes were grown with adequate Ca (400 mg $plant^{-1}$), no differences in dry matter yields were noted among genotypes. The Ca efficiency character in tomato appeared to be controlled by simple additive-dominance systems; the additive gene effects were highly significant. The genetic control of Ca in these tomato plants varied between families that indicated some kind of physiological control, e.g., more effective utilization of $Ca^{2+}$ in the tissue or a greater ability to absorb $Ca^{2+}$ from the growth medium at low levels of Ca. These authors concluded that based on the high broad-sense heritability values noted for the Ca efficiency characters, the selection of superior individuals for Ca efficiency in segregating populations should be successful.

## 5. DEFICIENCY

**a. Symptoms**—In vegetative tissue, Ca deficiency is characterized by growth reductions of the meristematic tissues. The deficiency is observed first in the growing tips or youngest leaves. Affected tissues usually become soft, and brown substances often accumulate. Leaves tend to curl toward the midrib, and margins are commonly serrated (Kawaski & Moritsugu, 1979). In sorghum, which may be typical of the cereal crops, Ca-deficient leaves are characterized by dead and tightly curled tips (swordlike) that tend to bend over and often stick to adjoining leaves (ladder effect). Leaves are usually sticky and gummy to the touch, are brittle, and commonly have small, sticky liquid vesicles on leaf margins and surfaces. Less severely affected leaves are speckled yellow and frequently show serrated breaks. Lower leaves are usually darker than normal leaves. Tips of old and new roots of Ca-deficient plants are stubby and dark brown, and root systems have slightly decreased fibrousness (Clark, 1982; Kawaski & Moritsugu, 1979). Tipburn is a common symptom in Ca-deficient leaves of many plants (Shear, 1975). Characteristic Ca deficiency symptoms on leaves of many plants are described by Sprague (1964).

In fleshy or fruit tissues, Ca deficiency is characterized by many types of disorders, e.g., cracking, watercore or soaking, pits, cavities, corking, internal breakdown, browning, and blackheart. Shear (1975) cited 35 disorders of fruits and vegetables that have been associated with Ca deficiency. Review articles on Ca deficiency disorders by Bangerth (1973), Faust and Shear (1968), Hewitt (1963, 1966), Shear (1975), and Simon (1978) are excellent references on the identification, symtomologies, characteristics, and causes for many Ca deficiency disorders. Other symposia (Atkinson et al., 1980; Shear, 1979) have assembled many reports on Ca deficiency disorders and methods for their prevention.

**b. Effects on Tissue Composition**—The role of $Ca^{2+}$ in metabolism might give clues to the types of compounds that could accumulate in Ca-deficient tissues. Hewitt (1963) reported changes in amino acids in Ca-deficient tissue. Amides (asparagine and glutamine) and proline were higher in Ca-deficient leaves of peppermint than in control plants (Steward et al., 1959). Changes in total organic acid concentrations in Ca-deficient tissue were negligible, and Ca deficiency did not change the type of organic acid present (Clark, 1968; Popp & Kinzel, 1981). Soluble sugars increased 10-fold in Ca-deficient cucumber leaves (Matsumoto & Teraoka, 1980). Changes in sucrose, fructose, glucose, maltose, and galactose occurred, but these changes were similar to changes in total sugars. Increases in starch, stachyose, raffinose, and cellibiose occurred in late stages of Ca deficiency. Calcium-deficient fruits had higher amounts of sorbitol and mannitol (Bangerth, 1973; Bangerth et al., 1972; Faust & Shear, 1968; Williams, 1966); fewer reducing sugars and sucrose (Williams, 1966); higher protein-N, total N, and soluble amino acids (Bangerth, 1970; Faust & Shear, 1968); fewer carbohydrates because of increased respiration (Bangerth, 1973); and more electrolytes ($K^+$, $Mg^{2+}$, and $H^+$) (Bangerth, 1970, 1973; Simon, 1978). Inbalances of $K^+$, $Mg^{2+}$, and $H^+$ relative to $Ca^{2+}$ (Bangerth, 1973; Faust & Shear, 1968; Marschner & Mengel, 1966; Simon, 1978) and higher amounts of volatile compounds such as acetaldehyde, acetic acid, ethanol, and $C_2H_4$

(Bangerth, 1973; Bangerth et al., 1972; Clijsters, 1965; Faust & Shear, 1968; Wills et al., 1970) have been reported in Ca-deficient tissues.

Deficiencies of Ca and other elements, as well as toxicities of mineral elements, have been described for maize and sorghum (Clark, 1970a, b, 1982; Clark et al., 1981). When Ca was deficient in maize leaves, P, S, Fe, B, Cu, and Zn increased and Ca and K decreased (Clark, 1970b). In sorghum, Ca-deficient leaves increased in K, Mg, Mn, and Fe and decreased in Ca, whereas their roots increased in P, Mn, and Fe and decreased in Ca (R. B. Clark, 1982, unpublished data).

## II. MAGNESIUM

Information on the chemistry of Mg in biological processes can be obtained in such references as Clarkson and Hanson (1980), Eichhorn (1973), Hughes (1972), and Sigel (1973–1978). Magnesium has a relatively small ionic radius (0.065 nm) compared with other divalent cations ($Ca^{2+}$ = 0.099 nm and $Mn^{2+}$ = 0.080 nm) that readily compete or substitute for it in many physiological processes and biochemical reactions. Its small ionic radius permits the formation of stable ionic complexes with "hard" (strongly electrophylic) electronegative donors such as $O^-$ in $PO_4^{3-}$ groups (Clarkson & Hanson, 1980). Magnesium is a hard Lewis acid. The coordination number of Mg is six, which allows it to form octohedral complexes, and $Mg^{2+}$ makes relatively strong geometrical demands on substrates and molecules with which it binds. Magnesium may bind with only one or two centers to maintain a regular octohedral structure with the remaining positions for water of hydration. As such, $Mg^{2+}$ forms ternary complexes with enzymes where a bridging metal is needed, and precise geometry is required. These complexes are shielded between electronegative groups and masked against oxidizing agents, which is typical of hard base-acid interactions. Magnesium also has high hydration energy, a low rate of water substitution, and a high rate constant in complex formation. These properties link themselves well to the reactions involved in $PO_4^{3-}$, nucleotide, and carboxyl transfers in which $Mg^{2+}$ is so commonly found.

### A. Magnesium Uptake

#### 1. MECHANISMS

Magnesium in plants is generally absorbed at lower quantities than either $Ca^{2+}$ or $K^+$ (Kirkby & Mengel, 1976; Mengel & Kirkby, 1982). The uptake of $Mg^{2+}$ is strongly reduced when metabolic energy is blocked (Pitman, 1976), a finding that suggests to some investigators that $Mg^{2+}$ is actively absorbed by plants. The evidence for active uptake of $Mg^{2+}$ is not conclusive. Magnesium is absorbed primarily from regions near the root apex (Russell & Clarkson, 1976), which supports the concept of passive uptake. Uptake processes for $Mg^{2+}$ are similar to those for $Ca^{2+}$.

## 2. FACTORS AFFECTING UPTAKE

The concentrations of Mg in soil solutions vary but are usually moderate (0.5–2.0 m*M*) compared with those of Ca (1.0–10.0 m*M*), which are high, and those of P (0.015–0.30 m*M*), which are low (Mengel et al., 1969). Magnesium concentrations are normally high in calcareous soils and relatively low in acid soils (Fried & Shapiro, 1961). Factors affecting $Mg^{2+}$ uptake are similar to those affecting $Ca^{2+}$ uptake.

**a. Cation Concentrations**—The competition of other cations with $Mg^{2+}$ is of particular importance in Mg nutrition (Mengel & Kirkby, 1982) and has caused Mg deficiency in some plants grown under field conditions. With near neutral soil pH and with adequate amounts of Mg, Ca, and K, alfalfa and maize plants tolerated relatively wide ranges of Ca/Mg ratios (0.8–5.0) before yields were reduced and deficiencies (or excesses) of Mg or Ca appeared (Simson et al., 1979). From the examination of data of various greenhouse and field experiments, Simson et al. (1979) noted that Ca/Mg ratios could be even wider than those noted for alfalfa and maize before plant yields would decrease.

Under controlled conditions, $Mg^{2+}$ uptake can be greatly affected by the level of Ca in the growth medium (Pitman, 1976). The rate and quantity of $Mg^{2+}$ absorbed by barley and maize roots were less with than without $Ca^{2+}$ (Moore et al., 1961b). The effect of $Ca^{2+}$ on $Mg^{2+}$ uptake was reported to be noncompetitive and at a site other than that involved in $Mg^{2+}$ uptake (Maas & Ogata, 1971). This effect of $Ca^{2+}$ is not general, because $Ca^{2+}$ and $K^+$ had little effect on $Mg^{2+}$ uptake by excised soybean roots (Leggett & Gilbert, 1969). Without $K^+$ in the solution, $Mg^{2+}$ in the vacuole of the roots increased over threefold. Shoot concentrations of $Mg^{2+}$ were unaffected by $K^+$ even though $Ca^{2+}$ increased by twofold. Moritsugu and Kawasaki (1980) noted that high rates of Ca caused high amounts of Mg to accumulate in roots of maize, sorghum, and barley but not in bean and rice, probably because the Ca caused a decrease in translocation of $Mg^{2+}$ from roots to shoots.

The uptake of $Mg^{2+}$ is usually decreased with increasing levels of $K^+$. This phenomenon has been noted in many plants and recognized for many years (Chu & Toop, 1975; Mengel & Kirkby, 1982). Even though this and other cation interactions occur, the total cation equivalents in plants remain relatively constant. High Mg concentrations have been reported in plants treated with low levels of K (Grimme et al., 1974; Leggett & Gilbert, 1969). Reduced plant growth could not explain the higher Mg concentrations in plants found at lower levels of $K^+$ (Kirkby & Mengel, 1976; Mengel & Kirkby, 1982). The antagonistic effects of $K^+$ on $Mg^{2+}$ uptake were related more to ionic balance effects than to competition for carrier at uptake sites (Kirkby & Mengel, 1976). Different plant parts may be affected differently by high $K^+$. In experiments on cation distribution in tomato leaves, roots, and fruits, high $K^+$ decreased Mg concentrations in both leaves and roots but not in fruits (Kirkby & Mengel, 1976). Fruit increases of Mg with high K were consistent over many harvests, which indicated that increases in Mg were real. Similar results have been reported for storage organs of other plants (Kirkby & Mengel, 1976). These increases in Mg were interpreted to mean that increased amounts of K in the growth medium could promote the translocation of $Mg^{2+}$ toward fruits and storage tissues. Magnesium deficiencies in leaves of apple trees have also been highly correlated with K concentrations of the leaf (Allen, 1980).

Competitive effects of $NH_4^+$ on $Mg^{2+}$ have also been noted for many plants (Hanada, 1980; Kirkby & Mengel, 1976; Mengel & Kirkby, 1982), and under some conditions Mg deficiencies have been induced in field and greenhouse grown plants. The mechanism for this competition is not clear, but Mengel and Kirkby (1982) suggested that it could be a direct effect of $NH_4^+$ on $Mg^{2+}$ or on $H^+$ release during $NH_4^+$ incorporation into amino acids.

**b. pH**—Above pH 5, $Mg^{2+}$ uptake by barley roots was independent of pH (Moore et al., 1961b). Below pH 5, $Mg^{2+}$ uptake decreased with decreasing pH, and at pH values below about 3.5, $Mg^{2+}$ diffused from the root tissue to the growth medium. Similar results were noted for lettuce (Arnon et al., 1942). In other studies (Christenson et al., 1973; John et al., 1972), $Mg^{2+}$ uptake was low at both very low and very high pH. In acid soils, decreased $Mg^{2+}$ uptake by plants was not attributed to low pH itself but to an interaction between $Al^{3+}$ and $Mn^{2+}$ (Farima et al., 1980a, b).

**c. Environment**—Temperature affects $Mg^{2+}$ uptake similar to that of $Ca^{2+}$, i.e., $Mg^{2+}$ is absorbed rather slowly by most plants, and temperature has relatively small effects (Chu & Toop, 1975; Moore et al., 1961b; Wallace, 1963, 1971a, b). Even though Mg concentrations in plants remain unchanged with temperature increases, Mg contents increase because of increased plant growth (Chu & Toop, 1975).

The effects of light and water stress on $Mg^{2+}$ uptake are mostly indirect (Clark, 1981). High light intensities decreased Mg concentrations in tomato plants but not Mg contents (Chu & Toop, 1975). High light intensity was suggested as a factor that might induce Mg deficiency. Magnesium contents of plants decreased more than those of Ca when water stresses were imposed on field-grown maize; Ca decreased by only 9%, but Mg decreased by 35% in mature plants (Jenne et al., 1958). In contrast, water stress enhanced Mg concentrations and accumulation in several grass and forage species (Abdel Rahman et al., 1971; Gerakis et al., 1975).

### B. Magnesium Translocation

From the limited data available on $Mg^{2+}$ transport in the xylem, the evidence supports the concept that $Mg^{2+}$ is transported similar to that of $Ca^{2+}$ (Kirkby & Mengel, 1976; Mengel & Kirkby, 1982). Much of the literature available on $Ca^{2+}$ transport in xylem should, therefore, be applicable to $Mg^{2+}$. More Mg usually accumulates in the older leaves than the younger leaves. When normal, intact sunflower roots were treated with $^{28}Mg$, older leaves contained more $^{28}Mg$ than the younger leaves (Schimansky, 1973).

Root-fed $Mg^{2+}$ was distributed more uniformly throughout the plant than leaf-fed $Mg^{2+}$. Upward translocation of $Mg^{2+}$ was directed by transpiration.

In contrast to $Ca^{2+}$, $Mg^{2+}$ is very mobile in the phloem (Schimansky, 1973; Steucek & Koontz, 1970) and is readily transported from one tissue to another. Magnesium readily moves from roots, stems, cotyledons, primary leaves, and secondary leaves to young newly developing leaves if plants are placed in solutions without Mg (Schimansky, 1973). When $^{28}Mg$ was fed to bean and barley leaves, it did not transport basipetally past a small segment of leaf that had been heat treated to block phloem transport (Steucek & Koontz, 1970). Under normal conditions, extensive amounts of $^{28}Mg$ moved basipetally from treated leaves to other leaves within 24 h. When the phloem tissue was separated from xylem tissue with thin plastic, $^{28}Mg$ moved from treated leaves to lower leaves. Autoradiographs clearly showed that $^{28}Mg$ readily moves to other plant tissues via the phloem (Schimansky, 1973; Steucek & Koontz, 1970). The direction and amount of transport of $Mg^{2+}$ was highly dependent on assimilated transport (Schimansky, 1973). Potassium enhanced $Mg^{2+}$ distribution into fruit and storage tissues. Since fruit and storage tissues are highly dependent on phloem transport for supplies of minerals, $Mg^{2+}$ is readily transported in the phloem.

## C. Magnesium Functions in Plants

### 1. ELECTROLYTE BALANCE

Most Mg salts are highly soluble, and more than 70% of $Mg^{2+}$ in plants is freely diffusible (Kirkby & Mengel, 1976). Magnesium is often associated with such organic anions as malate, citrate, pectate, and oxalate as well as with inorganic anions (Kirkby & Mengel, 1967). Of the elements found in the cytoplasm, $Mg^{2+}$ probably has the highest chemical activity (Clarkson & Hanson, 1980). In onion root cells, the activity of $Mg^{2+}$ in the cytoplasm was reported to be from 2 to 10 m*M* (Macklon & Sim, 1976); considerable $Mg^{2+}$ is bound to polyphosphates (ATP) and to anionic groups of proteins, nucleic acids, and phospholipids (Clarkson & Hanson, 1980). Under some circumstances, $Mg^{2+}$ may contribute to the electrical neutrality of organic compounds such as sugar phosphates, sugar nucleotides, and organic and amino acids.

### 2. CONSTITUENT OF CHLOROPHYLL

The most notable function of $Mg^{2+}$ is as a constituent of chlorophyll. The amount of Mg associated with chlorophyll is relatively low and is about 15 to 20% of total plant Mg (Neales, 1956). Even in Mg-deficient plants, the amount of $Mg^{2+}$ in chlorophyll does not usually exceed 30% (Michael, 1941). Although the role of $Mg^{2+}$ in chlorophyll is not completely understood, its specificity is attributed to its geometrical demands (Clarkson & Hanson, 1980). Coordination of $Mg^{2+}$ with four pyrrole-N rings gives a coordinately unsaturated electrophile that tends to form octohedral complexes as well as bridges between keto groups, carboxyl groups, or both (Fong et al., 1977; Katz et al., 1977).

### 3. ENZYME ACTIVITIES

The abundance of Mg in plants suggests that $Mg^{2+}$ is required in many enzyme reactions. In general terms, enzyme reactions that require $Mg^{2+}$ include $PO_4^{3-}$ or nucleotide transfer (phosphatases, kinases, ATPases, synthases, nucleotide transferases), carboxyl transfer (carboxylases, dicarboxylases), and activity with such enzymes as dehydrogenases, mutases, and lyases (Clarkson & Hanson, 1980). In many of its functions in ATP and adenosine diphosphate (ADP) reactions, $Mg^{2+}$ forms a bridge between the $PO_4^{3-}$ structure and the enzyme. A good example of this is the activation of ATPase by $Mg^{2+}$ (Balke & Hodges, 1975). The binding of $Mg^{2+}$ can be more complex than this; it may bind to both substrate and enzyme or to substrate alone (Scrutton, 1973).

Magnesium regulates ribosome integrity and stabilizes nucleic acids and membranes. The lack of $Mg^{2+}$ probably causes the dissociation of ribosomes into their subunits and destroys the ribosomal configuration that is necessary for protein synthesis (Ts'o, 1962; Watson, 1965). It is believed that $Mg^{2+}$ stabilizes the matrix of the nucleus by forming bridges with neighboring indiffusible anions (Wunderlich, 1978). The transfer of amino acyls from amino acyl transfer ribonucleic acid (tRNA) to a polypeptide chain appears to be activiated by $Mg^{2+}$.

Evidence indicates that $Mg^{2+}$ may regulate enzyme activity by binding to enzymes apart from their substrate sites (Portis & Heldt, 1976; Walker, 1974). The transport of $Mg^{2+}$ into and out of chloroplast membranes appears to control photosynthetic $CO_2$ fixation by ribulose diphosphate (RDP) carboxylase. Illuminated envelope-free chloroplasts efflux $Mg^{2+}$ (Dilley & Vernon, 1965; Nobel, 1967) most likely in response to proton uptake. This transport of $Mg^{2+}$ is light driven, and $Mg^{2+}$ moves from the cytoplasm or thylakoid (or both) to the stroma (Lin & Nobel, 1971; Walker, 1974). Total Mg in chloroplasts was 400 to 600 nmol $mg^{-1}$ of chlorophyll, and only small portions of $Mg^{2+}$ were transported across the thylakoid membrane during illumination (Portis & Heldt, 1976). Illumination of intact chloroplasts caused $Mg^{2+}$ to increase in the stroma, but this increase was small relative to the total $Mg^{2+}$ in chloroplasts. These $Mg^{2+}$ increases in the stroma during illumination could easily affect the affinity of RDP carboxylase for $CO_2$ (Bassham et al., 1968; Jensen, 1971; Walker, 1972; Walker et al., 1971). RDP carboxylase has been known to switch on in the light and switch off in the dark (Bassham et al., 1968; Walker, 1974), and changes in $Mg^{2+}$ could provide the basis for this light activation. Evidence for this concept is furnished by $O_2$ evolution from $CO_2$-dependent reactions, which are switched on either by increasing Mg concentration (from 1 to 5 m*M*) (Walker, 1973, Walker et al., 1971) or by increasing $CO_2$–$HCO_3^-$ concentrations (10-fold) (Walker et al., 1971) in reconstituted chloroplast systems (envelope-free chloroplasts supplemented with stromal proteins and coenzymes). In addition, $CO_2$ fixation was inhibited when the ionophore A 23 187 was added in the absence of external $Mg^{2+}$ (Portis & Heldt, 1976). If $Mg^{2+}$ was added, $CO_2$ fixation was restored to half restoration with 0.2 m*M* of $Mg^{2+}$ and to maximum restoration (but not full restoration) with 1 to 2 m*M* of $Mg^{2+}$. The ionophore caused a rapid and reversible light-dependent efflux of $Mg^{2+}$ from the thylakoid to the stroma.

The function of Mg in plants is summarized by Clarkson and Hanson (1980):

> $Mg^{2+}$ fills the need for small, strongly electropositive divalent cation which is readily mobile, which will coordinate with strongly nucleophilic liquids through largely ionic bonding and with limited geometric distortion. These properties have wide application in group transfer and some forms of macromolecular structure. These same properties prevent $Mg^{2+}$ from interfering with the formation by other metals of complexes which are more convalent and varied in character and geometry.

## D. Magnesium Requirements and Concentrations in Plants

### 1. REQUIREMENTS

Magnesium requirements and concentrations in plants vary with plant species, even within cultivars or genotypes of the same species. Normal Mg concentrations in many plants are given in Table 1. Critical and deficient concentrations of Mg in dry material are relatively low (usually $< 1000$ $\mu g$ $g^{-1}$) (Walsh & Beaton, 1973). Concentrations of Mg as low as 100 to 500 $\mu g$ $g^{-1}$ have been reported as critical or as causing deficiency symptoms in leaf tissue from sugar beets (Ulrich & Hills, 1973) and sugarcane (Humbert, 1973) and in various tree species (Leaf, 1973). Concentrations of Mg below about 2000 $\mu g$ $g^{-1}$ have been considered critical for the development of hypomagnesemia (grass tetany) in ruminants that consume the plant tissue (Grunes et al., 1970; Rendig & Grunes, 1979). Criteria for Mg requirements can be classified similar to that for Ca (Loneragan et al., 1970). Some plants accumulate relatively high concentrations of Mg. For example, tissues concentrations of over 10 000 $\mu g$ $g^{-1}$ have been reported for cantaloupe, green pepper, spinach, and potato (Geraldson et al., 1973).

### 2. CONCENTRATIONS

Many factors affect $Mg^{2+}$ uptake and accumulation, and plant age and plant part must be considered in evaluating Mg concentrations reported in plants (Walsh & Beaton, 1973). Magnesium remained relatively high in the different plant parts of maize (Clark, 1975b). During grain fill, leaves, tassels, and husks increased, roots and stalks decreased, and kernels and cobs decreased slightly in Mg from their initial concentrations in newly developing tissues. High amounts of Mg accumulated in maize kernels, but Ca did not. The relative concentrations of Mg and Ca in maize kernels were those predicted by Mg being phloem mobile and Ca being phloem immobile. Magnesium concentrations in leaves at various positions on maize plants remained relatively constant but increased with age (Clark, 1975a). As long as ample Mg is in the growth medium, most plants have the capacity to absorb, translocate, and accumulate sufficient $Mg^{2+}$ in the various plant parts to prevent Mg disorders, especially in harvested fruits and seeds.

### 3. GENOTYPE DIFFERENCES

Differences in genotype responses to Mg have been noted among plants grown on serpentine soils (high in Mg and low in Ca) (Kruckeburg, 1954; Proctor & Woodell, 1975; Shewry & Peterson, 1975; Walker, 1954; Walker et al., 1955) and among forage plants associated with hypomagnesemia in ruminants (Brown & Sleper, 1980; Currier et al., 1981; Fairbourn & Batchelder, 1980; Gladstone & Loneragan, 1970; Gross & Jung, 1978; Grunes et al., 1970; Hill & Guss, 1976; Hill & Jung, 1975; Lane & Lamp, 1977; McNaught, 1970; Nguyen & Sleper, 1981; Rendig & Grunes, 1979; Stratton & Sleper, 1979; Todd, 1966; Walker & Graffis, 1979). Many grass and legume genotypes have different capabilities to accumulate Mg or other cations that interact with $Mg^{2+}$ (particularly $K^+$ and $Ca^{2+}$). In addition, genotype differences in responses to Mg have been noted in celery (Pope & Munger, 1953), hops (Marocke et al., 1979), maize hybrids (Agarwala & Mehrotra, 1979; Andrade et al., 1975; Broersma & van Ryswyk, 1979; Gallaher et al., 1981), maize inbreds (Clark, 1975c; Foy & Barber, 1958; Kovacevic, 1980), and sunflower (Madhok & Walker, 1969).

Some plant species require higher levels of Mg for growth than others; tobacco and citrus trees require high Mg (Adams & Pearson, 1967), and maize is more tolerant of low soil Mg than cotton (Adams, 1969). A certain *Silene* species accumulated higher Mg than other *Silene* species when grown on serpentine soils (Shewry & Peterson, 1975). On acid soils, weeping lovegrass accumulated more Mg than fescue (Fleming et al., 1974).

Two sunflower genotypes, when grown in nutrient solutions with different levels of Mg, showed marked differences in Mg accumulation and the amount of Mg required for maximum dry matter production (Madhok & Walker, 1969). At Mg levels above those required for normal growth (2–10 m*M*), the cultivated sunflower genotype (*Helianthus annuus* L.) decreased in dry matter production, and the sunflower genotype endemic to serpentine soils (*H. bolanderi* Gray ssp. *exilis*) increased in dry matter. Similar responses were noted for two maize inbreds (Clark, 1975c). The Mg-efficient maize inbred transported $Mg^{2+}$ more readily to the tops than did the Mg-inefficient inbred. The Mg-inefficient inbred had higher concentrations of P, Ca, Mn, Fe, and Cu that could possibly interact with Mg. The dry matter produced per unit Mg was higher for the Mg-efficient inbred than for the Mg-inefficient inbred at lower levels of Mg, but they reversed at higher levels of Mg (1–4 m*M*).

## 4. BREEDING FOR Mg EFFICIENCY

Improvement of plant tolerances for low Mg levels appears feasible through plant breeding. Early experiments with celery indicated that Mg deficiency was controlled by a single gene (Pope & Munger, 1953). Inheritance of Mg, as well as that for many other elements, was noted in maize (Gorsline et al., 1964). The inheritance traits in maize were ascribed to additive gene action in the grain and to nonadditive gene action in the leaves. Additive gene action was also important for the inheritance of Mg accumulation in alfalfa (Hill & Jung, 1975). Genetic differences were noted for the accumulation of Mg in wheat and barley genotypes (Kleese et al., 1968). Open-pollinated progeny of tall fescue showed differential Mg accumulation (Haaland et al., 1978). Heritability estimates for Mg in perennial ryegrass, tall fescue, and reed canarygrass were found to be sufficiently high that breeding plants for improved Mg concentrations in plants should be feasible (Cooper, 1973; Hovin et al., 1978; Sleper, 1979; Sleper et al., 1977). Although specific combining ability was nonsignificant, genetic combining ability was significant, broad-sense and narrow-sense heritabilities were high, and additive gene action was important in Mg accumulation of tall fescue (Nguyen & Sleper, 1981). Nearly all of the evidence on inheritance studies points to the feasibility of breeding plants for higher Mg concentrations without having to sacrifice plant dry matter yields.

## 5. MAGNESIUM DEFICIENCY AND EXCESS

**a. Symptoms**—Magnesium deficiencies appear first in the older tissues and leaves, indicating that it is readily remobilized. Initial and moderate symptoms in maize and sorghum are characterized by light coloring (loss of chlorophyll) between the veins. For more severe Mg deficiency, dark necrotic lesions appear between the veins, and leaves usually turn reddish purple. Severe Mg deficiency causes leaves to turn brown and become brittle and margins to curl. The pale, diffuse symptoms are fairly uniform over most of the affected leaf, whereas necrosis appears first on new tips or along margins of leaves, then progresses toward the base and midrib. Magnesium deficient roots of sorghum plants were red and stunted. Sprague (1964) showed pictorial characteristics of Mg deficiencies on many plants, and Kirkby and Mengel (1976), Mengel and Kirkby (1982), and Embleton (1966) describe Mg deficiency symptoms for many plant species.

**b. Physiological Effects**—Since $Mg^{2+}$ is a constituent of chlorophyll, lighter-colored leaves might be expected in Mg-deficient plants, but other changes occur in plants. The mitochondria of apex barley root cells formed granules and cell membranes were disrupted (Marinos, 1963). The grana in chloroplasts of beans were reduced in number, were irregularly shaped, and were less compartmentalized (Thomson & Weier, 1962). Lamellar structures were deformed in chloroplasts, and mitochondria cristae were underdeveloped in Mg-deficient apple leaves (Chevalier & Huguet, 1975). Chloroplasts of Mg-deficient maize leaves had more bound water and were more rounded and swollen than normal chloroplasts (Dekov et al., 1979). If water stress and Mg deficiency occurred, the destruction of chloroplasts was even greater. When leaf Mg concentrations decreased from 200 to 50 mg $kg^{-1}$ of dry matter, photosynthetic $CO_2$ uptake decreased by one third and respiration increased twofold in sugar beet leaves (Terry & Ulrich, 1974). Magnesium deficiency had less effect on leaf stomatal diffusion resistance than on mesophyll resistance. New types of lactic and glucose-6-phosphate dehydrogenases appeared, and the intensity of alcohol dehydrogenase increased in Mg-deficient maize roots and leaves (Kudrev & Babalukova, 1975). Total lipids decreased more than phospholipids in Mg-deficient maize leaves (Kudrev & Georgieva, 1976). The lipid fraction that decreased the most was phosphatidylglycerol, and lesser decreases occurred in diphosphatidylglycerol, phosphatidylcholine, and phosphatidylethanolamine. The degree of phospolipid and fatty acid saturation also decreased (Georgieva et al., 1979). Magnesium deficiency decreased the rate of transpiration by leaves (Kudrev & Georgieva, 1976) and had little effect on relative proportions of storage proteins in pea seed (Randall et al., 1979).

The accumulation of a particular compound in Mg-deficient tissue is unpredictable because Mg is required in so many different metabolic reactions. Without photosynthesis or energy transfer, the initial building blocks for metabolism are not available. Many specific changes in metabolic compounds caused by Mg deficiency are given by Hewitt (1963).

**c. Effects on Growth**—Magnesium deficiency decreases vegetative growth and delays the onset of the reproductive phase (Mengel & Kirkby, 1982). Roots are affected more than tops (Bouma et al., 1979; Clark, 1975c). Dry matter top/root ratios of normal maize plants (2.5–3.0) increased to 7.0 to 8.0 in Mg-deficient plants (Clark, 1975c).

Reduced root growth caused by Mg deficiency may be related in some way to Al toxicity. Root concentrations of Mg decreased more than any other element in Al-treated maize plants (Clark, 1977). At a high Al treatment level, Mg concentration in roots of an Al-susceptible genotype was only 2% that in normal plants. Leaf tissue was not affected as severely (25% of normal plants). In subsequent studies, maize and sorghum plants grown on Al toxic soils showed similar decreases in Mg concentration. These decreases in Mg were more dramatic than other elements usually associated with Al toxicity (R. B. Clark, 1982, unpublished data).

**d. Effects on Tissue Composition**—Magnesium deficiency affects the accumulation of other mineral elements in leaves and roots (Clark, 1970a, b, 1975c, 1982). Maize leaves deficient in Mg had decreased N, Ca, Mg, and S but increased K, P, B, Cu, and Zn. Sorghum leaves deficient in Mg had decreased Mg but increased P, Ca, Mn, and Fe, whereas Mg-deficient roots had decreased Mg and increased P, K, Ca, Mn, Fe, and Ca (R. B. Clark, 1982, unpublished data). Other elements usually affect Mg accumulation less than Ca, K, and $NH_4^+$-N. Excess Al was especially detrimental to Mg in maize and sorghum (Clark, 1977; R. B. Clark, 1982, unpublished data) and in potato (Lee, 1971). High $PO_4^{3-}$ induced Mg deficiency in apple trees (Allen, 1980).

# III. MOLYBDENUM

Molybdenum has the highest atomic weight of any element essential to plant growth. Molybdenum has six readily removable valence electrons (d orbital), which makes it highly oxyphilic and allows it to undergo redox activity at high oxidation states [Mo(VI)/Mo(V)] (Frausto da Silva & Williams, 1976). Molybdenum can readily pass into six or eight coordination and be reduced in one electron step. The Mo(VI) molecule forms highly charged centers, but on reduction to Mo(V), the element has high affinity for N and S ligands. The most stable state of Mo in aqueous solutions is molybdate ($MoO_4^{2-}$), which is Mo(VI) and the form available to plants (Clarkson & Hanson, 1980; Mengel &Kirkby, 1982). Molybdenum has properties that resemble nonmetals (Frausto da Silva & Williams, 1976), and $MoO_4^{2-}$ is a weak acid that can form complex polyanions such as phosphomolybdate.

The valency states of Mo in reactions where it is required in plant growth are uncertain (Bowden, 1975; Bray & Swann, 1972). The state of Mo in oxidized enzymes is nearly always Mo(VI) and is usually reduced to Mo(V). However, lower oxidation states may occur under special conditions (Frausto da Silva & Williams, 1976).

The Mo flavoprotein complexes have redox potentials in the appropriate range so that the polarizing effect of Mo on coordinated N or O may cause Mo to change oxidation states by release of electrons and protons with a subsequent change in pH (Stieffel, 1973). When Mo complexes release electrons, they also release protons. This property is significant for reduction of $NO_3^-$ to $NO_2^-$ by $NO_3^-$ reductase (NR). Complexes of Mo are relatively labile, except for complexes with cyanide, and this lability appears to be important to the type of reactions that are characteristic of Mo complexes (Spence, 1976). For example, oxybridged dimer units may readily form on the reduction of Mo(VI) to MO(V), which could explain the two electron transfer in the reduction of $NO_3^-$ to $NO_2^-$.

## A. Molybdenum Uptake

### 1. MECHANISMS

As previously mentioned, the form of Mo absorbed by plants is $MoO_4^{2-}$ (Clarkson & Hanson, 1980; Mengel & Kirkby, 1982). Evidence that Mo is absorbed by active or passive mechanisms is indirect (Moore, 1972). Molybdate absorption is decreased by added $SO_4^{2-}$ (Bolle-Jones, 1957; Reisenauer, 1963; Stout et al., 1951) and by increased pH (Stout et al., 1951), but it is enhanced by added $PO_4^{3-}$ (Stout & Meagher, 1948; Stout et al., 1951). These data are often interpreted to mean that Mo absorption is metabolically controlled. Many investigators believe that $MoO_4^{2-}$ is absorbed into plant roots by active processes. Chemical potential data calculated for anion absorption strongly support the theory of active absorption for anions, and since $MoO_4^{2-}$ is an anion, it is interpreted that $MoO_4^{2-}$ uptake is an active process (Lüttge & Higinbotham, 1979).

### 2. FACTORS AFFECTING UPTAKE

Many factors affect $MoO_4^{2-}$ absorption by plants (Moore, 1972). The uptake and transport of $MoO_4^{2-}$ are affected by its interactions with other elements. Although the role of P in enhancing $MoO_4^{2-}$ absorption has not been fully eluciated, it was suggested that P increases the release of Mo from root cells into the translocation system (Stout et al., 1951). Stimulated $MoO_4^{2-}$ absorption by P has been attributed to the formation of a phosphomolybdate complex that is more readily absorbed by plants (Barshad, 1951). Confirmation of P enhancement effects on $MoO_4^{2-}$ absorption have been noted in other studies (Greenwood & Hallsworth, 1960; Gupta & Munro, 1969).

The inhibitory effects of $SO_4^{2-}$ on $MoO_4^{2-}$ absorption have been attributed to the competition of two anions of similar size (Stout et al., 1951). The inhibition of $MoO_4^{2-}$ utilization inside the plant by $SO_4^{2-}$ has also been suggested (Reisenauer, 1963). A strong inhibition of $SO_4^{2-}$ on $MoO_4^{2-}$ absorption was not attributed to a pH change in the root growth medium of brussels sprout but to an interaction between $SO_4^{2-}$ and $MoO_4^{2-}$ at the site of $MoO_4^{2-}$ absorption (Gupta & Munro, 1969). Reduction of $MoO_4^{2-}$ absorption by $SO_4^{2-}$ has been attributed to the lowering of soil pH, which would release more $SO_4^{2-}$ for competition with $MoO_4^{2-}$ (Gubler et al., 1982). The lowering of pH also enhances the absorption of $MoO_4^{2-}$ by soil particles (Mengel & Kirkby, 1982).

Even though strong Fe-Mo interactions have been noted in plants (Olsen, 1972; Olsen & Watanabe, 1979), the reported effects are mostly Mo deterrents on Fe uptake rather than the reverse. Mutual antagonisms between Mo and Cu have been noted (MacKay et al., 1966). Excess Cu caused Mo deficiency, and excess Mo caused Cu deficiency. Extensive Mn-Mo interactions have been noted (Jackson, 1967).

## B. Molybdenum Translocation

The predominant form of Mo in xylem transport, assuming no association with other compounds, appears to be $MoO_4^{2-}$ (Tiffin, 1972). Molybdenum complexes with S amino acids (cysteine) have been reported (Huang & Haight, 1970), and Mo-S forms of translocation cannot be overlooked (Tiffin, 1972). Molybdate binds with many sugars and other polyvalent compounds (Bayer & Voelter, 1966; Bourne et al., 1960) that may serve as translocation agents for Mo. Considerable Mo has been located in the phloem and vascular parenchyma of plant tissues (Hewitt & Agarwala, 1952), indicating some phloem movement of Mo. However, Mo is only moderately mobile in plants (Mengel & Kirkby, 1982).

## C. Molybdenum Functions in Plants

The only identifiable function of Mo in higher plants is in the enzyme NR (Clarkson & Hanson, 1980). Even though Mo is a constituent of the enzyme nitrogenase, this enzyme is located in symbiotic bacteria and in actinomycetes but has not been found in higher plants. Microorganisms have other Mo-containing enzymes, e.g., $SO_3^{2-}$ oxidase, aldehyde oxidase, xanthine dehydrogenese, and xanthine oxidase, but their presence in plants have not been confirmed. Because NR is the only established function for Mo in higher plants, this will be the only reaction discussed relative to Mo function in plants. The literature abounds with articles on nitrogenase and other Mo-containing enzymes (Bowden, 1975; Brill, 1980; Dilworth, 1974; Eady & Postgate, 1974; Hardy, 1978; Ljones, 1979; Mortenson & Throneley, 1979; Roberts & Brill, 1981; Stewart & Gallon, 1980).

Two metalloprotein enzymes convert $NO_3^-$ to $NH_4^+$ in the assimilatory $NO_3^-$ reducing system in plants; these are NR and $NO_2^-$ reductase (Guerrero et al., 1981). Nitrate reductase contains Mo, and $NO_2^-$ reductase contains Fe. Two main types of NR have been identified: (i) ferredoxin-dependent NR found in the procaryote cyanobacteria (blue-green algae) and presumably in chemorganic and photosynthetic bacteria, and (ii) pyridine nucleotide [NAD(P)H]–dependent NR found in eukaryotic organisms (including higher plants). The ferredoxin-dependent NR appears to be tightly bound to photosynthetic membranes and the NAD(P)H-dependent NR is soluble (Losada & Guerrero, 1979; Vennesland & Guerrero, 1979). Only NAD(P)H NR is discussed here, because it is the form found in higher plants. The ferredoxin-dependent NR reaction has been described extensively by Guerrero et al. (1981).

Nitrate reductase of fungi, green algae, and higher plants is catalyzed by reduced NAD(P)H (Garrett & Amy, 1978; Hewitt, 1975; Hewitt & Notton, 1980; Losada & Guerrero, 1979; Vennesland & Guerrero, 1979). The overall equation for this reaction is

$$NO_3^- + NAD(P)H + H^+ \xrightarrow{2\,e^-} NO_2^- + NAD(P)^+ + H_2O.$$

Specificity of NR reactions for NADH or NAD(P)H have been classified into (i) NADH-specific, (ii) NAD(P)H-dependent, and (iii) NADPH-specific enzymes (Guerrero et al., 1981). The type of NR present in most plants is NADH specific (Guerrero et al., 1981), but both NADH- and NAD(P)H-dependent enzymes have been reported in soybean leaves (Campbell, 1976) and rice seedlings (Shen et al., 1976).

Purified sources of NR have identified flavin adenine dinucleotide (FAD), cytochrome b-557, and Mo as ubiquitous prosthetic groups (Guerrero et al., 1981). Depending on the NR source, reports of one Mo atom per molecule of NR (Jacob & Orme-Johnson, 1980) or two Mo atoms per molecule of NR (Giri & Ramadoss, 1979) have been documented. Although information on Mo and other prosthetic groups contained in NR of higher plants is limited, it is believed that spinach NR contains one Mo atom (and two Fe atoms) per molecule of NR (Notton & Hewitt, 1979). The NR enzyme contains subunits, and each source of NR appears to be made up of different numbers of subunits (Guerrero et al., 1981). The NR molecule from various sources appears to have different molecular weights.

The location of Mo in the NR enzyme of higher plants has not been established, but it is thought to be similar to that in fungi (Guerrero et al., 1981). In fungi, the Mo atom appears to be associated with a low molecular weight polypeptide, called a *Mo cofactor* (Garrett & Amy, 1978; Hewitt, 1975; Notton & Hewitt, 1979). Except for the nitrogenase enzyme (Johnson, 1980), this Mo cofactor may be common to all molybdoenzymes (Guerrero et al., 1981). The active Mo cofactor has been proposed to be composed of Mo and a reduced form of pterin (Johnson et al., 1980).

Evidence supports the concept that Mo participates in the terminal moity of the NR reaction (Garrett & Amy, 1978; Hewitt & Notton, 1980; Losada & Guerrero, 1979; Vennesland & Guerrero, 1979). The sequence of events in the NR reaction are outlined as follows (Mengel & Kirkby, 1982):

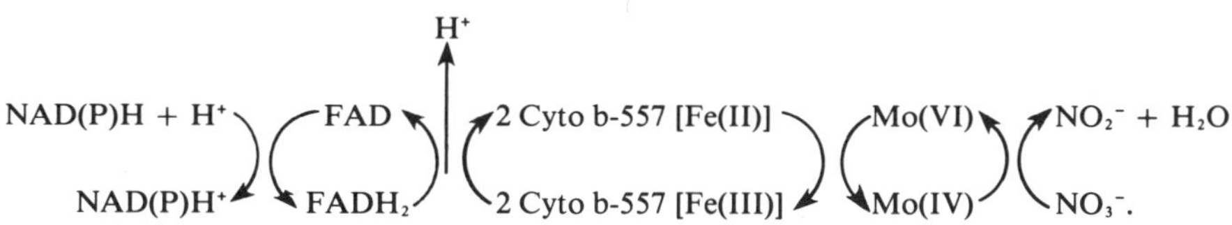


Nitrate appears to bind and be reduced at the Mo end of the reaction (Guerrero et al., 1981). The oxidation state of Mo has not been adequately determined but is thought to be Mo(VI)/Mo(IV) (Hewitt & Notton, 1980; Jacob & Orme-Johnson, 1980; Losada & Guerrero, 1979; Vennesland & Guerrero, 1979). However, electron spin resonance spectra data indicated that both Mo(V) and Mo(IV) appear under certain conditions (Jacob & Orme-Johnson, 1980; Solomonson, 1979). Active participation of FAD (Garrett & Amy, 1978; Notton & Hewitt, 1979; Vennesland & Guerrero, 1979) and cytochrome b-557 (Losada & Guerrero, 1979; Vennesland & Guerrero, 1979) in the NR reaction have been shown, and the position of these compounds in the sequence is shown in the above reaction.

In plants, NR appears to be located primarily in the leaves and, more specifically, in or with the chloroplasts (Guerrero et al., 1981). Even though NR is reportedly located in or with the chloroplast, NR has also been located intracellularly and is frequently referred to as a cytoplasmic enzyme (Beevers & Hageman, 1972; Dalling et al., 1972; Wallsgrove et al., 1979). Because NR has been found in different locations (Guerrero et al., 1981), the intracellular location of NR in green tissue of higher plants remains a controversy. If NR is located exterior to the chloroplast, the shuttling of energy from the chloroplast (photosynthetic organelle) to the outside has been proposed (Krause & Heber, 1976). Nitrate reductase has been found in roots, embryos, cotyledons, scutella, aleurone cells, pollen grains, and various other tissues (Beevers & Hageman, 1972; Hewitt et al., 1976). The properties of NR in nonchlorophyllous tissues have not been fully characterized, but root tissue NR appears to be similar to that in leaves (Beevers & Hageman, 1972; Ida et al., 1974; Oaks, 1979).

Nitrate reductase activity is affected by many environmental factors (Srivastava, 1980). Molybdenum-deficient plant tissues have lower NR activity, and this activity can be increased rapidly with Mo added exogenously (Afridi & Hewitt, 1964, 1965; Hewitt & Afridi, 1959; Mulder et al., 1959; Randall, 1969). When Mo was removed from NR by cyanide, NR activity ceased, but it recovered with the addition of Mo (Nicholas & Nason, 1955). Nitrate reductase activity was positively related to Mo in citrus (Shaked & Bar-Akiva, 1967). Tungsten (W) is a competitive inhibitor of Mo in NR, and when Mo is replaced by W in NR reactions, activity is low (Hewitt, 1975; Notton & Hewitt, 1971). The activity of NR has also been found to be lower in Ca-deficient tissues (Prasil, 1975). Since NR is also a S protein (Beevers & Hageman, 1969), S deficiency reduces NR activity (Friedrich & Schrader, 1978). Azide and cyanide competitively inhibit NR, and their effects appear to be associated with the terminal portion of the reaction complex (Vennesland & Guerrero, 1979), i.e., the portion of the reaction involving Mo and $NO_3^-$. Chlorate may replace $NO_3^-$ as a substrate in purified preparations of NR and is a competitive inhibitor in the reaction (Notton & Hewitt, 1979; Vennesland & Guerrero, 1979). Nitrite, the product of the NR reaction, has been reported to be an inhibitor of NR in some plant sources (Guerrero & Gutierrez, 1977; McDonald & Coddington, 1974; Vennesland & Guerrero, 1979).

## D. Molybdenum Requirements and Concentrations in Plants

### 1. CONCENTRATIONS

The concentration of Mo in plant materials is usually fairly low, normally below 1.0 $\mu g\ g^{-1}$ of dry tissue (Mengel & Kirkby, 1982). Table 2 lists some Mo concentrations reported for normal plant tissues. Critical Mo concentrations (Mo deficiency symptoms appeared) were reported to be about 0.1 $\mu g\ g^{-1}$ for cauliflower (Mengel & Kirkby, 1982), melon (Gubler et al., 1982), and tomato (Johnson, 1966), and below about 0.2 $\mu g\ g^{-1}$ in maize (Brown & Clark, 1974) and alfalfa (James et al., 1968). Plants from maize seeds containing $< 0.02$ $\mu g$ of Mo seed$^{-1}$ developed severe Mo deficiency, those with 0.03 to 0.06 $\mu g$ of Mo seed$^{-1}$ developed mild to moderate Mo deficiency, and those with 0.08 $\mu g$ of Mo seed$^{-1}$ or more developed no Mo deficiency (Wier & Hudson, 1966). Adequate supplies of Mo in most plant

Table 2—Molybdenum concentrations considered to be sufficient in healthy plant tissues (Walsh & Beaton, 1973; Johnson, 1966).

| Plant | Plant part | Range of concentration |
|---|---|---|
| | | $\mu g\ g^{-1}$, dry wt |
| Peanut | Upper stems and leaves (early peg) | 1.0 - 5.0 |
| Sugar beet | Blade and tops | 0.20–20.0 |
| Alfalfa | Tops | 0.5 -10.0 |
| Forest trees | (Many species) | 0.05- 0.15 |
| Maize | Stems, leaves | >0.11 |
| Bean | Tops | 0.40 |
| Spinach | Leaves | 1.60 |
| Tomato | Leaves | 0.68 |

tissues appear to be near or above 0.5 $\mu g\ g^{-1}$ (Mengel & Kirkby, 1982). Physiological requirements for Mo were found to be below 1.0 $\mu g\ g^{-1}$ (Stout & Meagher, 1948). Excess Mo can be taken into plants without causing toxicity problems (Mengel & Kirkby, 1982). For example, cauliflower plants could vary 100-fold in Mo concentrations without deleterious effects (Agarwala & Hewitt, 1954), and Mo accumulated as high as 1500 $\mu g\ g^{-1}$ in cotton leaves without disorders appearing (Joham, 1953). Excess Mo disorders appeared in tomato leaves when Mo concentration reached 1000 to 2000 $\mu g\ g^{-1}$ (Johnson, 1966).

## 2. PLANT SPECIES DIFFERENCES

Plant species differ considerably in their Mo requirements (Clark, 1976; Mengel & Kirkby, 1982). Members of the genus *Brassica* (cauliflower and cabbage) require high levels of Mo, as do legumes because of their association with symbiotic microorganisms and their involvement in $N_2$ fixation. In the United States, a survey of crop species showing Mo deficiency followed the order of alfalfa > cauliflower > broccoli > soybean > clover > citrus (Berger, 1962). In other studies, *Brassica* sp., lettuce, tomato, and sugar beet were highly susceptible to Mo deficiency (Hewitt & Bolle-Jones, 1952a, b), clover and alfalfa were moderately susceptible to Mo deficiency (Hewitt & Bolle-Jones, 1952a, b), and large-seeded legumes, grasses, barley, most cereals, and potato were relatively tolerant to low levels of Mo (Hewitt & Bolle-Jones, 1952a, b; Johnson et al., 1952; Plant, 1952). In general, monocotyledonous plants require less Mo and are less sensitive to Mo deficiency than dicotyledonous plants (Clark, 1976; Mengel & Kirkby, 1982). Seedborne Mo may be largely responsible for many differences among plants for susceptibility to Mo deficiency (Chipman et al., 1970; Dios Vidal & Broyer, 1965; Harris et al., 1965; Hewitt et al., 1954; Meagher et al., 1952; Peterson & Purvis, 1961; Robinson & Campbell, 1956; Wier & Hudson, 1966). Seed from different sources or locations may also help explain plant differences in occurrence of Mo deficiency (Robinson & Campbell, 1956; Wier & Hudson, 1966). For Mo deficiencies to be induced in some plants, seeds may need to be produced under low Mo conditions for one or two generations to reduce seedborne Mo (Dios Vidal & Broyer, 1965; Hewitt et al., 1954; Meagher et al., 1952; Peterson & Purvis, 1961). Some seeds have 10 times the amount of Mo needed for their growth cycle (Meagher et al., 1952).

Since Mo is adsorbed strongly by acid soils, Mo deficiencies often appear in plants grown under these conditions. For example, Mo deficiencies appeared when maize (Brown & Clark, 1974; Lutz et al., 1972) and alfalfa (James et al., 1968) were grown on strongly acid soils. In addition to the low Mo availability in acid soils, the high availability of Mn and Al in acid soils appear to accentuate Mo deficiencies (Mengel & Kirkby, 1982).

Very little Mo was absorbed during early growth stages of soybean plants (Ishizuka, 1982), but during later growth stages (reproductive), Mo was absorbed and its concentration increased in the plant parts. Seed Mo was believed to supply early plant growth needs. From initial pod filling to maturity, Mo concentrations increased in seeds, decreased in roots and pod shells, and remained relatively constant in leaves, stems, and nodules. Early maturing soybean varieties contained less Mo than later maturing varieties. Concentrations of Mo decreased extensively in leaves, stems, and roots throughout the growth cycle of cotton plants (Bhatt & Appukuttan, 1982). Relative Mo concentrations in the plant parts were leaves > roots > stems.

## 3. GENOTYPE DIFFERENCES

Differential responses of plants to low Mo have been recognized for a long time. Ever since "whiptail" symptoms in cauliflower and other *Brassica* sp were diagnosed as Mo deficiency (Davies, 1945; Mitchell, 1945), this deficiency has been reported for plants from many parts of the world (Chipman et al., 1970; Millikan, 1961). Differences among genotypes for susceptibility to Mo deficiency at low levels of Mo have been noted in cauliflower (Chipman et al., 1970; Gammon et al., 1954; Robinson & Campbell, 1956), maize (Brown & Clark, 1974; Dios Vidal & Broyer, 1965; Lutz et al., 1972; Noonan, 1953; Wier & Hudson, 1966), *Brassica* sp. (Hewitt & Bolle-Jones, 1952a; Neenan & Goodman, 1954; Plant, 1952), grasses (Kretschmer & Allen, 1956), lettuce (Plant, 1952), citrus (Stewart & Leonard, 1952), alfalfa (Younge & Takahashi, 1953), and pasture legumes (Johansen, 1978).

Reasons for differential responses of genotypes within species is not understood, but examples showing some of these differences in maize have been given by Clark (1976). For example, genotypes from both Spanish and American origin were susceptible to Mo deficiency, whereas another American genotype was not (Dios Vidal & Broyer, 1965). Differences among genotypes for leaf and seed Mo were noted when plants were grown with added Mo, and these relative differences remained for plants grown without added Mo. Uptake of Mo could not be attributed to Mo differences in seeds, since one genotype with low seed Mo produced plants with high leaf Mo, and a genotype with high seed Mo produced plants with low leaf Mo. Seed Mo differed with location where plants were grown, but the seed Mo had a marked effect on whether plants grown from these seeds developed Mo deficiency symptoms (Wier & Hudson, 1966). Certain genotypes consistently had lower Mo than others when grown in a soil at various pH values (Lutz et al., 1972).

Maize inbred lines were grown on various soils to induce mineral deficiency stresses (Clark & Brown, 1974a). During these studies, one genotype (Pa36) grew poorly on an acid, low P, high Al soil, whereas another genotype (WH) grew normally. This was unexpected since Pa36 was more efficient for absorbing P (Clark & Brown, 1974b) and was more Al tolerant (Clark, 1977) than WH. Spectographic data showed that WH consistently contained more Mo than Pa36 (Brown & Clark, 1974). When sufficient Mo was added to the soil, Pa36 grew as well as WH and other genotypes that were tolerant of low Mo. Seed Mo concentrations in Pa36 and WH could not explain their differential responses to Mo.

## 4. DEFICIENCIES

**a. Symptoms**—Molybdenum deficiencies resemble N deficiencies in many plants (Hagstrom & Berger, 1965) because Mo is closely related to N nutrition. Older leaves become pale or chlorotic first, and with increasing severity of deficiency, symptoms will show on other leaves (Mengel & Kirkby, 1982). Middle leaves may show severe symptoms of Mo deficiency because they are not as fully developed as older leaves. Leaves may wither, gray-green markings in the interveinal tissue of leaves may appear, leaves may be limp, and leaves may curl in on themselves. The middle lamella of cell walls may not completely form in leaves, leaving only the leaf rib (Bussler, 1970). Because of this, leaves often develop the whiptail appearance that is characteristic of Mo deficiency in cauliflower. Plant yields decrease and flower formation is restricted (Hewitt et al., 1954). Molybdenum-deficient plants often show necrotic symptoms on the leaf margins in response to the accumulation of $NO_3^-$ (Maynard, 1979). Pictorial presentations of Mo deficiencies on plant leaves can be observed in Hewitt (1963, 1966), Mengel and Kirkby (1982), and Sprague (1964).

**b. Effects on Plant Constituents**—The major compound to accumulate in Mo-deficient tissues is $NO_3^-$ (Brown & Clark, 1974; Hewitt & Bolle-Jones, 1952a, b; Maynard, 1979; Wier & Hudson, 1966; Wilson, 1949). Molybdenum-deficient plants often contain low levels of $NH_4^+$-soluble organic N, and protein N (Hewitt et al., 1957; Possingham, 1956). The inorganic P/organic P ratio is usually high (Possingham, 1954). Decreased ascorbic acid (Agarwala & Hewitt, 1954) and altered amino acid accumulations occur (Hewitt, 1963). The uptake of $MoO_4^{2-}$ is higher in plants fed $NO_3^-$ than in those fed $NH_4^+$ (Giordano et al., 1966). Sterile cauliflower plants did not develop Mo deficiency if plants were fed $NH_4^+$ only (Hewitt & Gundry, 1970). Under nonsterile conditions, a Mo requirement was noted because $NH_4^+$ was nitrified by contaminating organisms.

## REFERENCES


Abdel Rahman, A. A., A. F. Shalaby, and M. O. El Monayeri. 1971. Effect of moisture stress on metabolic products and ions accumulation. Plant Soil 34:65–90.

Abeles, F. B. 1973. Ethylene in plant biology. Academic Press, Inc., New York.

Acharya Goswami, K. K., and L. J. Audus. 1976. Distribution of calcium, potassium and phosphorus in *Helianthus annuus* hypocotyls and *Zea mays* coleoptiles in relation to tropic stimuli and curvatures. Ann. Bot. (London) 40:49–64.

Adams, F. 1969. Response of corn to lime in field experiments on Alabama soils. Alabama Agric. Exp. Stn. Bull. 391.

Adams, F. 1974. Soil solution. p. 441–481. *In* E. W. Carson (ed.) The plant root and its environment. Univ. Press of Virginia, Charlottesville.

Adams, F., and R. W. Pearson. 1967. Crop response to lime in the southern United States and Puerto Rico. *In* R. W. Pearson and F. Adams (ed.) Soil acidity and liming. Agronomy 12: 161–206. Am. Soc. of Agron., Madison, Wis.

Afridi, M. M. R. K., and E. J. Hewitt. 1964. The inducible formation and stability of nitrate reductase in higher plants: I. Effects of nitrate and molybdenum on enzyme activity in cauliflower (*Brassica oleracea* var. *botrytis*). J. Exp. Bot. 15:251–271.

Afridi, M. M. R. K., and E. J. Hewitt. 1965. The inducible formation and stability of nitrate reductase in higher plants: II. Effects of environmental factors, anti-metabolites, and amino acids on induction. J. Exp. Bot. 16:628–645.

Agarwala, S. C., and E. J. Hewitt. 1954. Molybdenum as a plant nutrient: IV. The interrelationships of molybdenum and nitrate supply in chlorophyll and ascorbic acid fractions in cauliflower plants grown in sand culture. J. Hortic. Sci. 29:291–300.

Agarwala, S. C., and N. K. Mehrotra. 1979. Relative susceptibility of some maize varieties to magnesium deficiency in sand culture. Indian J. Plant Physiol. 22:9–13.

Allen, M. 1980. Relationships between soil factors and magnesium deficiency in apple. p. 279–283. *In* D. Atkinson et al. (ed.) Mineral nutrition of fruit trees. Butterworth (Publ.) Inc., Woburn, Mass.

Anderson, J. M., H. Charbonneau, H. P. Jones, R. O. McCann, and M. J. Cormier. 1980. Characterization of the plant nicotinamide adenine dinucleotide kinase activator protein and its identification as calmodulin. Biochemistry 19:3113–3120.

Anderson, J. M., and M. J. Cormier. 1978. Calcium-dependent regulator of NAD kinase in higher plants. Biochem. Biophys. Res. Commun. 84:595–602.

Anderson, W. P. (ed.) 1973. Ion transport in plants. Academic Press, Inc., New York.

Andrade, A. G., H. P. De Haag, G. D. Oliveira, and J. R. De Sarruge. 1975. The differential accumulation of nutrients by five cultivars of maize (*Zea mays* L.): I. Accumulation of macronutrients. An. Esc. Super. Agric. Luiz de Queiroz Univ. Sao Paulo 32:115–149.

Arnon, D. I. 1950. Criteria of essentiality of inorganic micronutrients for plants. p. 31–39. *In* Trace elements in plant physiology. The Chronica Botanic Co., Waltham, Mass.

Arnon, D. I., W. E. Fratzke, and C. M. Johnson. 1942. Hydrogen ion concentration in relation to absorption of inorganic nutrients by higher plants. Plant Physiol. 17:515–524.

Arslan-Cerim, N. 1966. The redistribution of radioactivity in geotropically stimulated hypocotyls of *Helianthus annuus* pretreated with radioactive calcium. J. Exp. Bot. 17:236–240.

Askew, H. O., E. T. Chittenden, R. J. Monk, and J. Watson. 1959. Chemical investigations on bitter pit of apples: I. Physical and chemical changes in leaves and fruits of Cox's orange variety during the season. N.Z. J. Agric. Res. 2:1167–1186.

Askew, H. O., E. T. Chittenden, R. J. Monk, and J. Watson. 1960. Chemical investigations on bitter pit of apples: III. Chemical composition of affected and neighboring healthy tissues. N.Z. J. Agric. Res. 3:169–178.

Atkinson, D., J. E. Jackson, R. O. Sharples, and W. M. Waller (ed.) 1980. Mineral nutrition of fruit trees. Butterworth (Publ.) Inc., Woburn, Mass.

Baker, D. A., and J. L. Hill (ed.) 1975. Ion transport in plants cells and tissues. Elsevier North-Holland, Inc., New York.

Baker, D. A., and J. Moorby. 1969. The transport of sugar, water and ions into developing potato tubers. Ann. Bot. (London) 33:729–741.

Balke, N. E., and T. K. Hodges. 1975. Plasma membrane adenosine triphosphatase of oat roots. Plant Physiol. 55:83–86.

Bangerth, F. 1970. Bitter pit of apples, an unsolved problem of fruit physiology up to day. Gartenbauwissenschaft 35:91–120.

Bangerth, F. 1973. Investigations upon Ca related physiological disorders. Phytopathol. Z. 77: 20–37.

Bangerth, F. 1974. Antagonism between calcium and other elements in apple fruit. Acta Hortic. 45:49–52.

Bangerth, F. 1976. A role for auxin and auxin transport inhibitors on the Ca content of artificially induced parthenocarpic fruits. Physiol. Plant. 37:191–194.

Bangerth, F. 1979. Calcium-related physiological disorders of plants. Annu. Rev. Phytopathol. 17:97–122.

Bangerth, F., D. R. Dilley, and D. H. Dewey. 1972. Effect of postharvest calcium treatments on internal breakdown and respiration of apple fruits. J. Am. Soc. Hortic. Sci. 97:679–682.

Bangerth, F., and P. Firuzeh. 1971. The effect of 2,3,5-triiodobenzoic acid (TIBA) on the mineral content and bitter pit disease of "Boskoop" apple fruits. Z. Pflanzenkr. Pflanzenschutz. 78:93–97.

Barber, W. D. 1970. An investigation of the gentic control of the accumulation of phosphorus and other elements by corn leaves using chromosomal translocations. Ph.D. Diss. Pennsylvania State Univ., University Park. Diss. Abstr. 30B:3033–3034.

Barshad, I. 1951. Factors affecting the molybdenum content of pasture plants: II. Effect of soluble phosphates, available nitrogen, and soluble sulfates. Soil Sci. 71:387–398.

Bassham, J. A., P. Sharp, and I. Morris. 1968. The effect of $Mg^{2+}$ concentration on the pH optimum and Michaelis constants of the spinach chloroplast ribulosediphosphate carboxylase (carboxydesmutase). Biochim. Biophys. Acta 153:898–900.

Bathla, R. N., S. M. Virmani, and M. L. Chaudhry. 1978. Pattern of calcium and magnesium accumulation in some varieties of groundnut (*Arachis hypogaea* L.). J. Indian Soc. Soil Sci. 26:290–296.

Bayer, E., and W. Voelter. 1966. Molybdate complexes of sugars and alcohols. Ann. Chem. (Warsaw) 696:194–208.

Becerra, J., and J. F. Lopez-Saez. 1978. Effects of caffeine, calcium and magnesium on plant cytokinesis. Exp. Cell Res. 111:301–308.

Beevers, L., and R. H. Hageman. 1969. Nitrate reduction in higher plants. Annu. Rev. Plant Physiol. 20:495–522.

Beevers, L., and R. H. Hageman. 1972. The role of light in nitrate metabolism in higher plants. p. 85–113. *In* A. C. Giese (ed.) Photophysiology, Vol. 7. Academic Press, Inc., New York.

Bell, C. W., and O. Biddulph. 1963. Translocation of calcium. Exchange versus mass flow. Plant Physiol. 38:610–614.

Bennett-Clark, T. A. 1956. Salt accumulation and the mode of action of auxin: A preliminary hypothesis. p. 284–291. *In* R. L. Wain and F. Wightman (ed.) The chemistry and mode of action of plant growth substances. Butterworth (Publ.) Inc., Woburn, Mass.

Berger, K. C. 1962. Micronutrient deficiencies in the United States. J. Agric. Food Chem. 10: 178–181.

Beringer, H., and M. A. Taha. 1976. $^{45}$Calcium absorption by two cultivars of groundnut (*Arachis hypogaea*). Exp. Agric. 12:1–7.

Bhatt, J. G., and E. Appukuttan. 1982. Manganese and molybdenum contents of the cotton plant and different stages of growth. Commun. Soil Sci. Plant Anal. 13:463–471.

Bhor, S. M., M. M. Kibe, and G. K. Zende. 1970. Interrelationship between free lime status of soils and the uptake of Mn, P, and Ca by paddy and *jowar.* J. Indian Soc. Soil Sci. 18: 479–484.

Biddulph, O., and S. Biddulph, and R. Cory, and H. Koontz. 1958. Circulation patterns for phosphorus, sulphur, and calcium in the bean plant. Plant Physiol. 33:293–300.

Biddulph, O., R. Cory, and S. Biddulph. 1959. Translocation of calcium in the bean plant. Plant Physiol. 34:512–519.

Biddulph, O., F. S. Nakayama, and R. Cory. 1961. Transpiration stream and ascension of calcium. Plant Physiol. 36:429–436.

Biro, R. L., C. C. Hale, O. F. Wiegand, and S. J. Roux. 1982. Effects of chlorpromazine on gravitropism in *Avena* coleoptiles. Ann. Bot. (London) 50:735–747.

Bledsoe, R. W., C. L. Comer, and H. C. Harris. 1949. Absorption of radioactive calcium by the peanut fruit. Science (Washington, D.C.) 109:329–330.

Bollard, E. G. 1970. The physiology and nutrition of developing fruits. p. 387–425. *In* A. C. Hulme (ed.) The biochemistry of fruits and their products, Vol. 1. Academic Press, Inc., New York.

Bolle-Jones, E. W. 1957. Molybdenum: Effects on the growth and composition of *Hevea.* J. Rubber Res. Inst. Malya. 15:141–157.

Bouma, D., E. J. Dowling, and H. Wahjoedi. 1979. Some effects of potassium and magnesium on the growth of subterranean clover (*Trifolium subterraneum*). Ann. Bot. (London) 43:529–538.

Bourne, E. J., D. H. Hutson, and H. Weigel. 1960. Paper ionophoresis of sugars and other cyclic polyhydroxy compounds in molybdate solution. J. Am. Chem. Soc. 82:4252–4256.

Bowden, F. L. 1975. The biochemical function of molybdenum. p. 205–267. *In* C. A. McAuliffe (ed.) Techniques and topics in bioinorganic chemistry. John Wiley & Sons, Inc., New York.

Bowling, D. J. F. 1966. Active transport of ions across sunflower roots. Planta 69:377–382.

Bowling, D. J. F. 1973. The origin of the trans-root potential and the transfer of ions to the xylem of sunflower roots. p. 483–491. *In* W. P. Anderson (ed.) Ion transport in plants. Academic Press, Inc., New York.

Bowling, D. J. F., A. E. S. Macklow, and R. M. Spanswick. 1966. Active and passive transport of the major nutrient ions across the root of *Ricinus communis.* J. Exp. Bot. 17:410–416.

Bradfield, E. C. 1976. Calcium complexes in the xylem sap of apple shoots. Plant Soil 44: 496–499.

Bradford, R. R., D. E. Baker, and W. I. Thomas. 1966. Effect of soil treatments on chemical element accumulation of four corn hybrids. Agron. J. 58:614–617.

Bramlage, W. J., M. Drake, and J. H. Baker. 1974. Relationships of calcium content to respiration and postharvest condition of apples. J. Am. Soc. Hortic. Sci. 99:376–378.

Bray, R. C., and J. C. Swann. 1972. Molybdenum-containing enzymes. p. 107–144. *In* J. D. Dunitz et al. (ed.) Structure and bonding, Vol. 11. Springer-Verlag New York, New York.

Brewbaker, J. L., and B. H. Kwack. 1963. The essential role of calcium ion in pollen germination and pollen tube growth. Am. J. Bot. 50:859–865.

Brill, W. J. 1980. Biochemical genetics of nitrogen fixation. Microbiol. Rev. 44:449–467.

Broersma, K., and A. L. van Ryswyk. 1979. Magnesium deficiencies observed in forage corn varieties. Can. J. Plant Sci. 59:541–544.

Brown, D. A., G. A. Place, and J. V. Pettiet. 1960. The effect of soil moisture upon cation exchange in soils and nutrient uptake by plants. Int. Congr. Soil Sci., Trans. 7th (Madison, Wis.) 3:443–449.

Brown, J. C., and R. B. Clark. 1974. Differential response of two maize inbreds to molybdenum stress. Soil Sci. Soc. Am. Proc. 38:331–333.

Brown, J. R., and D. A. Sleper. 1980. Mineral concentration in two tall fescue genotypes grown under variable soil nutrient levels. Agron. J. 72:742–745.

Brumagen, D. M., and A. J. Haitt. 1966. The relationship of oxalic acid to the translocation and utilization of calcium in *Nicotiana tabacum.* Plant Soil 24:239–249.

Bukovac, M. J., S. H. Wittwer, and H. B. Tukey. 1956. Anesthetization by di-ethyl ether and the transport of foliar applied radiocalcium. Plant Physiol. 31:254–255.

Burstrom, H. G. 1952. Studies on growth and metabolism of roots: VIII. Calcium as a growth factor. Physiol. Plant 5:391–402.

Burstrom, H. G. 1968. Calcium and plant growth. Biol. Rev. 43:287–316.

Bussler, W. 1962. Ca-deficiency symptoms in sunflowers. Z. Pflanzenernaehr. Dueng. Bodenkd. 99:207–215.

Bussler, W. 1963. The development of Ca-deficiency symptoms. Z. Pflanzenernaehr. Dueng. Bodenkd. 100:53–58.

Bussler, W. 1970. The development of molybdenum deficiency symptoms in cauliflower. Z. Pflanzenernaehr. Bodenkd. 125:36–50.

Bygrave, F. L. 1978. Calcium movements in cells. Trends Biochem. Sci. (Pers. Ed.) 3:175–182.

Caldwell, C. R. 1982. Kinetics, temperature dependence, and cation inhibition of barley root plasma membrane-bound ATPase in relation to biophysical properties of membrane lipids. Ph.D. Diss. Michigan State Univ., East Lansing. Diss. Abstr. 42B:4638.

Campbell, N. A., and W. W. Thomson. 1977. Effects of lanthanum and ethylenediaminetetraacetate on leaf movements of *Mimosa.* Plant Physiol. 60:635–639.

Campbell, W. H. 1976. Separation of soybean leaf nitrate reductases by affinity chromatography. Plant Sci. Lett. 7:239–247.

Carolus, R. 1975. Calcium relationships in vegetable nutrition and quality. Commun. Soil Sci. Plant Anal. 6:285–298.

Cheung, W. Y. 1980. Calmodulin plays a pivotal role in cellular regulation. Science (Washington, D.C.) 207:19–27.

Chevalier, S., and C. Huguet. 1975. Magnesium deficiency effects on apple trees: Ultrastructural evolution in deficient leaves of apple trees. Ann. Agron. 26:351–362.

Chipman, E. W., D. C. Mackay, H. C. Gupta, and H. B. Cannon. 1970. Response of cauliflower cultivars to molybdenum deficiency. Can. J. Plant Sci. 50:163–167.

Chittenden, E. T., D. J. Stanton, and J. Watson. 1969. Bitter pit in Cox's orange apples. N.Z. J. Agric. Sci. 12:240–247.

Chiu, T.-F., and C. Bould. 1976. Effects of shortage of calcium and other cations on $^{45}$Ca mobility, growth and nutritional disorders of tomato plants (*Lycopersicon esculentum*). J. Sci. Food Agric. 27:969–977.

Christenson, D. R., R. P. White, and E. C. Doll. 1973. Yields and magnesium uptake by plants as affected by soil pH and calcium levels. Agron. J. 65:205–206.

Christiansen, M. N., H. R. Carns, and D. J. Slyter. 1970. Stimulation of solute loss from radicals of *Gossypium hirsutum* L. by chilling, anaerobiosis, and low pH. Plant Physiol. 46:53–56.

Christiansen, M. N., and C. D. Foy. 1979. Fate and function of calcium in tissues. Commun. Soil Sci. Plant Anal. 10:427–442.

Chu, C. B., and E. W. Toop. 1975. Effects of substrate potassium, substrate temperature and light intensity on growth and uptake of major cations by greenhouse tomato plants. Can. J. Plant Sci. 55:121–126.

Clark, R. B. 1968. Organic acids of maize (*Zea mays* L.) as influenced by mineral deficiencies. Crop Sci. 8:165–167.

Clark, R. B. 1970a. Mineral deficiencies of corn: Their effects on growth and interactions between nutrients. Agrichem. Age 13:4, 8.

Clark, R. B. 1970b. Effects of mineral nutrient levels on the inorganic composition and growth of corn (*Zea mays* L.). Ohio Agric. Res. Dev. Center Res. Circ. no. 181.

Clark, R. B. 1975a. Mineral element concentrations in corn leaves by position on the plant and age. Commun. Soil Sci. Plant Anal. 6:439–450.

Clark, R. B. 1975b. Mineral element concentrations of corn plant parts with age. Commun. Soil Sci. Plant Anal. 6:451–464.

Clark, R. B. 1975c. Differential magnesium efficiency in corn inbreds: I. Dry-matter yields and mineral element composition. Soil Sci. Soc. Am. Proc. 39:488–491.

Clark, R. B. 1976. Plant efficiencies in the use of calcium, magnesium and molybdenum. p. 175–191. *In* M. J. Wright (ed.) Plant adaptation to mineral stress in problem soils. Cornell Univ. Agric. Exp. Stn., Ithaca, N.Y.

Clark, R. B. 1977. Effect of aluminum on growth and mineral elements of Al-tolerant and Al-intolerant corn. Plant Soil 47:653–662.

Clark, R. B. 1978. Differential response of corn inbreds to calcium. Commun. Soil Sci. Plant Anal. 9:729–744.

Clark, R. B. 1981. Effect of light and water stress on mineral element composition of plants. J. Plant Nutr. 3:853–885.

Clark, R. B. 1982. Plant response to mineral element toxicity and deficiency. p. 71–142. *In* M. N. Christiansen and C. F. Lewis (ed.) Breeding plants for less favorable environments. John Wiley & Sons, Inc., New York.

Clark, R. B., and J. C. Brown. 1974a. Differential mineral uptake by maize inbreds. Commun. Soil Sci. Plant Anal. 5:213–227.

Clark, R. B., and J. C. Brown. 1974b. Differential phosphorus uptake by phosphorus-stressed corn inbreds. Crop Sci. 14:505–508.

Clark, R. B., P. A. Pier, D. Knudsen, and J. W. Maranville. 1981. Effect of trace element deficiencies and excesses on mineral nutrients in sorghum. J. Plant Nutr. 3:357–374.

Clarkson, D. T. 1974. Ion transport and cell structure in plants. John Wiley & Sons, Inc., New York.

Clarkson, D. T., and J. B. Hanson. 1980. The mineral nutrition of higher plants. Annu. Rev. Plant Physiol. 31:239–298.

Clarkson, D. T., and J. Sanderson. 1970. Further observations on the uptake of ions by different portions of the intact root system of barley (var. Maris Badger). Annu. Rep. Agric. Res. Counc. Letcombe Lab. 20:17–19.

Clarkson, D. T., and J. Sanderson. 1971. Inhibition of the uptake and long-distance transport of calcium by aluminum and other polyvalent cations. J. Exp. Bot. 22:837–851.

Clarkson, D. T., J. Sanderson, and R. S. Russell. 1968. Ion uptake and root age. Nature (London) 220:805–806.

Cleland, R. 1960. Effect of auxin upon loss of calcium from cell walls. Plant Physiol. 35:581–584.

Cleland, R. E., and D. L. Rayle. 1977. Reevaluation of the effect of calcium ions on auxin-induced elongation. Plant Physiol. 60:709–712.

Clijsters, H. 1965. Malic acid metabolism and initiation of the internal breakdown of "Jonathan" apples. Physiol. Plant. 18:85–94.

Cohen, J. D., and K. D. Nadler. 1976. Calcium requirement for indoleacetic acid-induced acidification by *Avena* coleoptiles. Plant Physiol. 57:347–350.

Collier, G. F., V. C. Huntington, and E. F. Cox. 1979. A possible role for chlorogenic acid in calcium-related disorders of vegetable crops with particular reference to lettuce tipburn. Commun. Soil Sci. Plant Anal. 10:481–490.

Collier, G. F., M. A. Scaife, and V. C. Huntington. 1976. Nutritional aspects of physiological disorders. Wellsbourne Natl. Vegetable Res. Stn. Rep., Wellsbourne, Warwick, England, p. 43–44.

Cooper, J. P. 1973. Genetic variation of herbage constituents. p. 379–417. *In* G. W. Butler and R. W. Bailey (ed.) Chemistry and biochemistry of herbage, Vol. 2. Academic Press, Inc., New York.

Cooper, T., and F. Bangerth. 1976. The effect of Ca and Mg treatments on the physiology chemical composition and bitter pit development of "Cox's orange" apples. Sci. Hortic. (Amsterdam) 5:49–57.

Cormack, R. G. H. 1949. The development of root hairs in angiosperms. Bot. Rev. 15:583–612.

Cormier, M. J., H. W. Jarrett, and H. Charbonneau. 1982. Role of $Ca^{++}$-calmodulin in metabolic regulation in plants. p. 125–139. *In* S. Kakiuchi et al. (ed.) Calmodulin and intracellular $Ca^{++}$ receptors. Plenum Press, New York.

Crafts, A. S., and C. E. Crisp. 1971. Phloem transport in plants. W. H. Freeman & Co. Publ., San Francisco.

Craig, W. F., and W. I. Thomas. 1970. Prediction of chemical accumulation in maize double-cross hybrids from single-cross data. Crop Sci. 10:609–610.

Crisp. P., G. F. Collier, and T. H. Thomas. 1976. The effect of boron on tipburn and auxin activity in lettuce. Sci. Hortic. (Amsterdam) 5:215–226.

Currier, C. G., C. G. Williams, C. S. Hoveland, and C. B. Elkins. 1981. The relationship of grass tetany potential with root growth in tall fescue. Agron. Abstr., p. 136.

Dalling, M. J., N. E. Tolbert, and R. H. Hageman. 1972. Intracellular location of nitrate reductase and nitrite reductase: I. Spinach and tobacco leaves. Biochim. Biophys. Acta 283:505–512.

Davies, E. B. 1945. A case of molybdenum deficiency in New Zealand. Nature (London) 156: 392.

Day, D. A., B. L. Bertagnolli, and J. B. Hanson. 1978. The effect of calcium on the respiratory responses of corn mitochondria. Biochem. Biophys. Acta 502:289–297.

Dekov, I., T. Kudrev, and L. Petrova. 1979. Influence of magnesium deficiency and water stress on ultrastructural changes of chloroplasts and some parameters of water regime in maize plants. p. 133–138. *In* T. Kudrev et al. (ed.) Mineral nutrition of plants, Vol. 1. Bulgarian Acad. of Sci., Sofia.

Dela Fuente, R. K., and A. C. Leopold. 1973. A role for calcium in auxin transport. Plant Physiol. 51:845–847.

DeTurk, E. E. 1941. Plant nutrient deficiency symptoms. Physiological basis. Ind. Eng. Chem. 33:648–653.

Devine, T. E. 1982. Genetic fitting of crops to problem soils. p. 143–173. *In* M. N. Christiansen and C. F. Lewis (ed.) Breeding plants for less favorable environments. John Wiley & Sons, Inc., New York.

De Wit, C. T., W. Dijkshoorn, and J. C. Noggle. 1963. Ionic balance and growth of plants. Versl. Landbouwkd. Onderz. 69:1–68.

Dieter, P., and D. Marmé. 1980a. $Ca^{2+}$ transport in mitochondrial and microsomal fractions form higher plants. Planta 150:1–8.

Dieter, P., and D. Marmé. 1980b. Calmodulin activation of plant microsomal $Ca^{2+}$ uptake. Proc. Natl. Acad. Sci. USA 77:7311–7314.

Dieter, P., and D. Marmé. 1980c. Partial purification of plant NAD kinase by calmodulin-sepharose affinity chromatography. Cell Calcium 1:279–286.

Dieter, P., and D. Marmé. 1981. A calmodulin-dependent, microsomal ATPase from corn (*Zea mays* L.). FEBS Lett. 125:245–248.

Dilley, R. A., and L. P. Vernon. 1965. Ion and water transport processes related to the light-dependent shrinkage of spinach chloroplasts. Arch. Biochem. Biophys. 111:365–375.

Dilworth, M. J. 1974. Dinitrogen fixation. Annu. Rev. Plant Physiol. 25:81–114.

Dios Vidal R., and T. C. Broyer. 1965. Deficiency symptoms and essentiality of molybdenum in corn hybrids. Agrochimica 9:273–284.

Dixon, B., G. R. Sager, and V. M. Shorrocks. 1973. Effect of calcium and boron on the incidence of tree and storage pit in apples of the cultivar Egremont russet. J. Hortic. Sci. 48: 403–411.

Drew, M. C., and O. Biddulph. 1971. Effect of metabolic inhibitors and temperature on uptake and translocation of 45-Ca and 42-K by intact bean plants. Plant Physiol. 48:426–432.

Dreyer, E. M., and M. H. Weinenseel. 1979. Phytochrome-mediated uptake of calcium in *Mougeotia* cells. Planta 146:31–39.

Duvick, D. N., R. A. Kleese, and N. M. Frey. 1981. Breeding for tolerance of nutrient imbalances and constraints to growth in acid, alkaline and saline soils. J. Plant Nutr. 4:111–129.

Dyson, P. W., and J. Digby. 1975. Effects of calcium on sprout growth of ten potato cultivars. Potato Res. 18:363–377.

Eady, R. R., and J. R. Postgate. 1974. Nitrogenase. Nature (London) 249:805–810.

Eastwood, D., and D. L. Laidman. 1971. The mobilization of macronutrient elements in the germinating wheat grain. Phytochemistry 10:1275–1284.

Eaton, G. W., and M. A. Robinson. 1977. Interstock effects upon apple leaf and fruit mineral content. Can. J. Plant Sci. 57:227–234.

Eichhorn, G. L. (ed.) 1973. Inorganic biochemistry, Vol. 1 and 2. Elsevier North-Holland, Inc., New York.

El Hinnawy, E. 1974. Chelating compounds as cell wall-loosening agents in cell suspension culture of *Melilotus alba.* Z. Pflanzenphysiol. 71:207–219.

Embleton, T. W. 1966. Magnesium. p. 225–263. *In* H. D. Chapman (ed.) Diagnostic criteria for plants and soils. Univ. of California Press, Berkeley.

Emmert, F. H. 1969. Retention and passage of calcium and strontium in stems of *Phaseolus vulgaris* as mediated by xylem stream flow rate and dinitrophenol. Physiol. Plant. 22: 246–252.

English, J., A. V. Barker, and D. V. Maynard. 1981. Calcium nutrition among tomato strains. HortScience 16:39.

Epstein, E. 1961. The essential role of calcium in selective cation transport by plant cells. Plant Physiol. 36:437–444.

Epstein, E. 1972. Mineral nutrition of plants: Principles and perspectives. John Wiley & Sons, Inc., New York.

Epstein, E. 1973. Flow in the phloem and the immobility of calcium and boron: A new hypothesis in support of an old one. Experientia 29:133–134.

Epstein, E., and J. E. Leggett. 1954. The absorption of alkaline earth cations by barley roots: Kinetics and mechanism. Am. J. Bot. 41:785–791.

Eschrich, W., B. Eschrich, and H. B. Currier. 1964. Historadiographic demonstration of calcium-45 in the phloem of *Cucurbita maxima.* Planta 63:2146–2154.

Fairbourn, M. L., and A. R. Batchelder. 1980. Factors influencing magnesium in high plains forage. J. Range Manage. 33:435–438.

Farima, M. P. W., M. E. Sumner, C. O. Plank, and W. S. Letzsch. 1980a. Effect of pH on soil magnesium and its absorption by corn. Commun. Soil Sci. Plant Anal. 11:981–992.

Farima, M. P. W., M. E. Sumner, C. O. Plank, and W. S. Letzsch. 1980b. Exchangeable aluminum and pH as indicators of lime requirement for corn. Soil Sci. Soc. Am. J. 44: 1036–1041.

Faust, M., and C. B. Shear. 1968. Corking disorders of apples: A physiological and biochemical review. Bot. Rev. 34:441–469.

Faust, M., and C. B. Shear. 1972. The effect of calcium on respiration of apples. J. Am. Soc. Hortic. Sci. 97:437–439.

Faust, M., and C. B. Shear. 1973. Calcium translocation patterns in apples. Trans. 3rd Symp. Accumulation and translocation of nutrients and regulators in plant organisms. Pr. Inst. Sadow. Skierniewica Ser. E 3:423–436.

Ferguson, I. B. 1979. The movement of calcium in non-vascular tissue of plants. Commun. Soil Sci. Plant Anal. 10:217–224.

Ferguson, I. B., and E. G. Bollard. 1976a. The movement of calcium in germinating pea seeds. Ann. Bot. (London) 40:1047–1055.

Ferguson, I. B., and E. G. Bollard. 1976b. The movement of calcium in woody stems. Ann. Bot. (London) 40:1057–1065.

Ferguson, I. B., C. B. Watkins, and J. E. Harman. 1983. Inhibition by calcium of senescence of detached cucumber cotyledons: Effect of ethylene and hydroperoxide production. Plant Physiol. 71:182–186.

Ferrell, W. K., and F. D. Johnson. 1956. Mobility of calcium-45 after injection into western white pine. Science (Washington, D.C.) 124:364–365.

Fick, G. N., and D. C. Rasmusson. 1967. Heritability of Sr-59 and Ca-35 accumulation in barley seedlings. Crop Sci. 7:315–317.

Fleming, A. L., J. W. Schwartz, and C. D. Foy. 1974. Chemical factors controlling the adaptation of weeping lovegrass and tall fescue to acid mine spoils. Agron. J. 66:715–718.

Fong, F. K., V. J. Koester, and L. Galloway. 1977. Endo and exo carbomethoxy carbonyl bonding in hydrated chlorophyll a dimers. Experimental criteria for the determination of the P700 structure. J. Am. Chem. Soc. 99:2372–2375.

Foote, B. D., and J. B. Hanson. 1964. Ion uptake by soybean root tissues depleted of calcium by EDTA. Plant Physiol. 39:450–460.

Forbes, J. C., and A. L. Gelman. 1981. Copper and other minerals in herbage species and varieties on copper deficient soils. Grass Forage Sci. 35:25–30.

Foy, C. D. 1974. Effects of soil calcium availability on plant growth. p. 565–600. *In* E. W. Carson (ed.) The plant root and its environment. Univ. Press of Virginia, Charlottesville.

Foy, C. D., and S. A. Barber. 1958. Magnesium absorption and utilization by two inbred lines of corn. Soil Sci. Soc. Am. Proc. 22:57–62.

Foy, C. D., A. L. Fleming, and G. C. Gerloff. 1972. Differential aluminum tolerance in two snapbean varieties. Agron. J. 64:815–818.

Frasher, D. A. 1958. The translocation of rubidium-86 and calcium-45 in trees. p. 347–363. *In* K. V. Thiman (ed.) The physiology of forest trees. Ronald Press, New York.

Frausto da Silva, J. J. R., and R. J. P. Williams. 1976. The uptake of elements by biological systems. p. 67–121. *In* J. D. Dunitz et al. (ed.) Structure and bonding, Vol. 29. Springer-Verlag New York, New York.

Fried, M., and J. C. Noggle. 1958. Multiple site uptake of individual cations by roots as affected by hydrogen ion. Plant Physiol. 33:139–144.

Fried, M., and R. E. Shapiro. 1961. Soil-plant relationships in ion uptake. Annu. Rev. Plant Physiol. 12:91–112.

Friedrich, J. W., and L. E. Schrader. 1978. Sulfur deprivation and nitrogen metabolism in maize seedlings. Plant Physiol. 61:900–903.

Führ, R., and J. Wieneke. 1974. Secondary translocation of Ca-45 to the fruit of apple trees the year after dormancy. p. 171–175. *In* R. L. Bieleski et al. (ed.) Mechanisms of regulation of plant growth. Royal Soc. New Zealand Bull. no. 12, Wellington.

Fuller, M. M. 1976. The ultrastructure of the outer tissues of cold-stored apple fruits of high and low calcium content in relation to cell breakdown. Ann. Appl. Biol. 83:299–304.

Fuller, M. M. 1980. Cell ultrastructure in apple fruits in relation to calcium concentration and fruit quality. p. 51–55. *In* D. Atkinson et al. (ed.) Mineral nutrition of fruit trees. Butterworth (Publ.) Inc., Woburn, Mass.

Furlani, P. R., R. Hiroce, O. C. Bataglia, and W. J. Silva. 1977. Accumulation of micronutrients, silicon and dry matter in two simple maize hybrids. Brigantia 36:223–229.

Gableman, W. H., and G. C. Gerloff. 1982. The search for, and interpretation of, genetic controls that enhance plant growth under deficiency levels of macronutrient. p. 301–312. *In* M. R. Saric (ed.) Genetic specificity of mineral nutrition of plants, Serbian Acad. Sci. Arts, Belgrade, Yugoslavia.

Gallaher, R. N. 1975. The occurrence of calcium in plant tissue as crystals of calcium oxalate. Commun. Soil Sci. Plant Anal. 6:315–330.

Gallaher, R. N., and M. D. Jellum, and J. B. Jones, Jr. 1981. Leaf magnesium concentration efficiency versus yield efficiency of corn hybrids. Commun. Soil Sci. Plant Anal. 12:345–354.

Gammon, N. Jr., G. M. Volk, E. N. McCubbin, and A. H. Eddins. 1954. Soil factors affecting molybdenum uptake by cauliflower. Soil Sci. Soc. Am. Proc. 18:302–305.

Garrard, L. A., and T. E. Humphreys. 1967. Effect of divalent cations on the leakage of sucrose from corn scutellum slices. Phytochemistry 6:1085–1095.

Garrett, R. H., and N. K. Amy. 1978. Nitrate assimilation in fungi. Adv. Microb. Physiol. 18: 1–65.

Georgieva, M., T. Kudrev, and P. Stanchev. 1979. Influence of magnesium deficiency on phospholipid content and fatty acid composition in maize plant roots. p. 147–163. *In* T. Kudrev et al. (ed.) Mineral nutrition of plants, Vol. 1. Bulgaria Acad. of Sci., Sofia.

Gerakis, P. A., F. P. Guerrero, and W. A. Williams. 1975. Growth, water relations and nutrition of three grassland annuals as affected by drought. J. Appl. Ecol. 12:125–135.

Geraldson, C. M., G. R. Klacan, and O. A. Lorenz. 1973. Plant analysis as an aid in fertilizing vegetable crops. p. 365–379. *In* L. M. Walsh and J. D. Beaton (ed.) Soil testing and plant analysis. Soil Sci. Soc. of Am., Madison, Wis.

Geuy, B. 1972. The leaf uptake of $^{14}CO_2$ and translocation of $^{14}C$-assimilates in bean plants grown under conditions of P-, Ca- or K-deficiency. Bull. Acad. Pol. Sci. Ser. Sci. Biol. 2:803–808.

Ginzburg, B. Z. 1961. Evidence for a protein gel structure crosslinked by metal cations in the intercellular cement of plant tissue. J. Exp. Bot. 12:85–107.

Giordano, L. B., W. H. Gableman, and G. C. Gerloff. 1981. Inheritance of differences in calcium utilization by tomatoes under low-calcium stress. J. Am. Soc. Hortic. Sci. 107:664–669.

Giordano, P. M., H. V. Koontz, and E. J. Rubins. 1966. $^{14}C$ distribution in photosynthate of tomato as influenced by substrate copper and molybdenum level and nitrogen source. Plant Soil 24:437–447.

Giri, L., and C. S. Ramadoss. 1979. Physical studies on assimilatory nitrate reductase from *Chlorella vulgaris.* J. Biol. Chem. 254:11703–11712.

Gitelman, S. E., and G. B. Witman. 1980. Purification of calmodulin from *Chlamydomonas*: Calmodulin occurs in cell bodies and flagella. J. Cell Biol. 87:764–770.

Gladstones, J. S., and J. F. Loneragan. 1970. Nutrient elements in herbage plants in relation to soil adaptation and animal nutrition. p. 350–354. *In* M. J. T. Norman (ed.) Proc. 11th Int. Grassland Congr., Surfers Paradise, Queensland, Australia. 13–23 April. Univ. of Queensland Press, St. Lucia, Australia.

Gomes, S. L., L. Mennucci, and J. C. Maia. 1979. A calcium-dependent protein activator of mammalian cyclic nucleotide phosphodiesterase from *Blastocladiella emersonii.* FEBS Lett. 99:39–42.

Gorsline, G. W., R. L. Ragland, and W. I. Thomas. 1961. Evidence for inheritance of differential accumulation of calcium, magnesium and potassium in maize. Crop Sci. 1:154–156.

Gorsline, G. W., W. I. Thomas, and D. E. Baker. 1964. Inheritance of P, K, Mg, Ca, B, Zn, Mn, Al, and Fe concentrations by corn (*Zea mays* L.) leaves and grain. Crop Sci. 4:207–210.

Gorsline, G. W., W. I. Thomas, and D. E. Baker. 1968. Major gene inheritance of Sr-Ca, Mg, K, P, Zn, Cu, B, Al-Fe, and Mn concentrations in corn (*Zea mays* L.). Pennsylvania State Agric. Exp. Stn. Bull. 746.

Grand, R. J. A., A. C. Nairn, and S. V. Perry. 1980. The preparation of calmodulins from barley (*Hordeum* sp.) and basidiomycete fungi. Biochem. J. 185:755–760.

Green, D. W., and M. J. Bukovac. 1968. Redistribution of calcium in *Phaseolus vulgaris* L. Proc. Am. Soc. Hortic. Sci. 93:368–378.

Greenleaf, W. H., and F. Adams. 1969. Genetic control of blossom end rot disease in tomatoes through calcium metabolism. J. Am. Soc. Hortic. Sci. 94:248–250.

Greenwood, E. A. N., and E. G. Hallsworth. 1960. Studies on the nutrition of forage legumes: II. Some interactions of calcium, phosphorus, copper, and molybdenum on the growth and chemical composition of *Trifolium subterraneum* L. Plant Soil 12:97–127.

Griffing, L. R., and P. M. Ray. 1979. Dependence of cell wall secretion of calcium. Plant Physiol. 63(Suppl.):51.

Grimme, H., L. C. von Braunschweig, and K. Nemeth. 1974. Potassium, calcium and magnesium interactions in relation to uptake and yield. Landwirtsch. Forsch. Sonderh. 30(2): 93–100.

Gross, C. F., and G. A. Jung. 1978. Magnesium, Ca, K concentration in temperate-origin forage species as affected by temperature and Mg fertilization. Agron. J. 70:397–403.

Gross, J. 1982. Oxalate-enhanced active calcium uptake in membrane fractions from zucchini squash. p. 369–376. *In* D. Marmé et al. (ed.) Plasmalemma and tonoplast: Their functions in the plant cell. Elsevier North-Holland, Inc., New York.

Gross, J., and D. Marmé. 1978. ATP-dependant $Ca^{2+}$ uptake into plant membrane vesicles. Proc. Natl. Acad. Sci. USA 75:1232–1236.

Grunes, D. L., P. R. Stout, and J. R. Brownell. 1970. Grass tetany of ruminants. Adv. Agron. 22:331–374.

Guardiola, J. L., and J. F. Sutcliffe. 1972. Transport of materials from the cotyledons during germination of seeds of the garden pea (*Pisum sativum* L.). J. Exp. Bot. 23:322–337.

Gubler, W. D., R. G. Grogan, and P. P. Osterli. 1982. Yellows of melons caused by molybdenum deficiency in acid soil. Plant Dis. 66:449–451.

Guerrero, M. G., and M. Gutierrez. 1977. Purification and properties of the NAD(P)H:nitrate reductase of the yeast *Rhodotorula gultinis.* Biochim. Biophys. Acta 482:272–285.

Guerrero, M. G., J. M. Vega, and M. Losada. 1981. The assimilatory nitrate-reducing system and its regulation. Annu. Rev. Plant Physiol. 32:169–204.

Gupta, U. C., and D. C. Munro. 1969. Influence of sulfur, molybdenum, and phosphorus on chemical composition and yields of brussels sprouts and of molybdenum on sulfur contents of several plant species grown in the greenhouse. Soil Sci. 107:114–118.

Guttridge, C. G., R. F. Hughes, C. R. Stinchcombe, and W. W. George. 1978. Nutrient effects on tipburn of strawberry. Exp. Hortic. 30:36–41.

Gysi, C. 1975. The calcium, potassium and phosphorus uptake of apple seedlings with respect to the form of nitrogen. Eidg. Anst. Forstl. Versuchswes. Mitt. 51:41–47.

Haaland, R. L., C. B. Elkins, and C. S. Hoveland. 1978. A method for detecting genetic variability for grass tetany potential in tall fescue. Crop Sci. 18:339–340.

Hagstrom, G. R., and K. C. Berger. 1965. Molybdenum deficiencies of Wisconsin soils. Soil Sci. 100:52–56.

Hallam, N. D., B. E. Roberts, and D. J. Osborne. 1973. Embryogenesis and germination in rye (*Secale cercale* L.): III. Fine structures and biochemistry of the non-viable embryo. Planta 110:279–290.

Halldal, P. 1957. Importance of calcium and magnesium ions in phototaxis of motile green algae. Nature (London) 179:215–216.

Hanada, K. 1980. Studies of nitrogen nutrition for muskmelon: 1. The influence of the form and the level of nitrogen supplied on the growth of melon. Kyushu Daigaku Nogakubu Gakugei Zasshi 34:67–80. (Biol. Abstr. 71:76310.)

Handley, R., and R. Overstreet. 1961. Uptake of calcium and chlorine in roots of *Zea Mays.* Plant Physiol. 36:766–769.

Hanger, B. C. 1979. The movement of calcium in plants. Commun. Soil Sci. Plant Anal. 10: 171–193.

Hanson, J. B. 1960. Impairment of respiration, ion accumulation and ion retention in root tissue treated with ribonuclease and ethylenediamine tetraacetic acid. Plant Physiol. 35: 372–379.

Hardy, R. W. F. (ed.) 1978. A treatise on dinitrogen fixation. J. Wiley & Sons, Inc., New York.

Harris, H. B., M. B. Parker, and B. J. Johnson. 1965. Influence of molybdenum content of soybean seed and other factors associated with seed source on progeny response to applied molybdenum. Agron. J. 57:397–399.

Hasson, E. P., and G. G. Laties. 1976. Purification and characterization of an A type phospholipase from potato and its effect on potato mitochondria. Plant Physiol. 57:148–152.

Hawker, J. S., H. Marschner, and W. J. S. Downton. 1974. Effects of sodium and potassium on starch synthesis in leaves. Aust. J. Plant Physiol. 1:491–501.

Hecht-Buchholz, Ch. 1979. Calcium deficiency and plant ultrastructure. Commun. Soil Sci. Plant Anal. 10:67–81.

Herth, W. 1978. Ionophore A 23187 stops tip growth, but not cytoplasmic streaming, in pollen tubes of *Lilium longiflorum.* Protoplasma 96:275–282.

Hewitt, E. J. 1963. The essential nutrient elements: Requirements and interactions in plants. p. 137–360. *In* F. C. steward (ed.) Plant physiology: A treatise, Vol. 3. Inorganic nutrition of plants. Academic Press, Inc., New York.

Hewitt, E. J. 1966. Sand and water culture methods used in the study of plant nutrition. Tech. Commun. no. 22, Commonwealth Agric. Bureau, Farnham Royal, England.

Hewitt, E. J. 1975. Assimilatory nitrate-nitrite reduction. Annu. Rev. Plant Physiol. 26:73–100.

Hewitt, E. J., and M. M. R. K. Afridi. 1959. Adaptive synthesis of nitrate reductase in higher plants. Nature (London) 183:57–58.

Hewitt, E. J., and S. C. Agarwala. 1952. Reduction of triphenyltetrazolium chloride by plant tissues and its relation to the molybdenum status. Nature (London) 169:545–546.

Hewitt, E. J., S. C. Agarwala, and A. H. Williams. 1957. Molybdenum as a plant nutrient: VIII. The effects of different molybdenum levels and nitrogen supplies on the nitrogen fractions in cauliflower plants grown in sand cultures. J. Hortic. Sci. 32:34–48.

Hewitt, E. J., and E. W. Bolle-Jones. 1952a. Molybdenum as a plant nutrient: I. The influence of molybdenum on growth of some *Brassica* crops in sand culture. J. Hortic. Sci. 27: 245–256.

Hewitt, E. J., and E. W. Bolle-Jones. 1952b. Molybdenum as a plant nutrient: II. The effects of molybdenum deficiency on some horticultural and agricultural crop plants in sand cultures. J. Hortic. Sci. 27:257–265.

Hewitt, E. J., and E. W. Bolle-Jones, and P. Miles. 1954. The productions of copper, zinc and molybdenum deficiencies in crop plants grown in sand culture with special reference to some effects of water supply and seed reserves. Plant Soil 5:205–222.

Hewitt, E. J., and C. S. Gundry. 1970. The molybdenum requirement of plants in relation to nitrogen supply. J. Hortic. Sci. 45:351–358.

Hewitt, E. J., D. P. Hucklesby, and B. A. Notton. 1976. Nitrate metabolism. p. 633–681. *In* J. Bonner and J. E. Varner (ed.) Plant biochemistry. Academic Press, Inc., New York.

Hewitt, E. J., and B. A. Notton. 1980. Nitrate reductase systems in eukaryotic and prokaryotic organisms. p. 273–325. *In* M. Coughlan (ed.) Molybdenum and molybdenum-containing enzymes. Pergamon Press, Inc., Elmsford, N.Y.

Hiatt, A. J., and J. E. Leggett. 1974. Ionic interactions and antagonisms in plants. p. 101–134. *In* E. W. Carson (ed.) The plant root and its environment. Univ. Press of Virginia, Charlottesville.

Higinbotham, N., B. Etherton, and R. J. Foster. 1967. Mineral ion contents and cell transmembrane electropotentials of pea and oat seedling tissue. Plant Physiol. 42:37–46.

Hill, R. R., Jr., and S. B. Guss. 1976. Genetic variability for mineral concentration in plants related to mineral requirements of cattle. Crop Sci. 16:680–685.

Hill, R. R., Jr., and G. A. Jung. 1975. Genetic variability for chemical composition of alfalfa: I. Mineral elements. Crop Sci. 15:652–657.

Hocking, P. J., and J. S. Pate. 1977. Mobilization of minerals to developing seeds of legumes. Ann. Bot. (London) 41:1259–1278.

Hodges, T. K. 1976. ATPases associated with membranes of plant cells. p. 260–283. *In* U. Lüttge and M. G. Pitman (ed.) Transport in plants II. Part A: Cells. Encyclopedia of plant physiology, New Series, Vol. 2A. Springer-Verlag New York, New York.

Hodges, T. K., and J. B. Hanson. 1965. Calcium accumulation by maize mitochondria. Plant Physiol. 40:101–108.

Holz, R. W., and C. A. Stratford. 1979. Effects of divalent ions on vesicle-vesicle fusion studied by a new luminescence assay for fusion. J. Membr. Biol. 46:331–358.

Hooymans, J. J. M. 1964. The role of calcium in the absorption of anions and cations by excised barley roots. Acta Bot. Neerl. 13:507–540.

Hou, H.-Y., and F. G. Merkle. 1950. Chemical composition of certain calcifugous and calcicolous plants. Soil Sci. 69:471–486.

Hovin, A. W., T. L. Tew, and R. E. Stucker. 1978. Genetic variability for mineral elements in reed canarygrass. Crop Sci. 18:423–427.

Huang, T. J., and G. P. Haight, Jr. 1970. A paramagnetic monomeric molybdenum(V)-cysteine complex as a model for molybdenum-enzyme interaction. J. Am. Chem. Soc. 92:2336–2342.

Hubbard, M., M. Bradley, M. P. Sullivan, M. Shepherd, and I. Forrester. 1982. Evidence for the occurrence of calmodulin in the yeasts *Candida albicans* and *Saccharomyces cerevisiae.* FEBS Lett. 137:85–88.

Hughes, M. N. 1972. The inorganic chemistry of biological processes. John Wiley & Sons, Inc., New York.

Humbert, R. P. 1973. Plant analysis as an aid in fertilizing sugar crops: Part II. Sugarcane. p. 289–298. *In* L. M. Walsh and J. D. Beaton (ed.) Soil testing and plant anlaysis. Soil Sci. Soc. of Am., Madison, Wis.

Hunt, I. V., J. Frame, and R. D. Harkess. 1976. Removal of mineral nutrients by red clover varieties. J. Br. Grassl. Soc. 31:171–179.

Hyams, J. S., and G. C. Borisy. 1978. Isolated flagellar apparatus of *Chlamydomonas*: Characterization of forward swimming and alteration of wave form and reversal of motion of calcium ions *in vitro.* J. Cell Sci. 33:235–253.

Hyde, A. H. 1966. Nature of the calcium effect in phosphate uptake by barley roots. Plant Soil 24:328–332.

Ida, S., E. Mori, and Y. Morita. 1974. Purification, stabilization and characterization of nitrate reductase from barley roots. Planta 121:213–224.

Isermann, K. 1970. The effect of absorption in the xylem on calcium distribution in higher plants. Z. Pflanzenernaehr. Bodenkd. 126:191–203.

Isermann, K. 1971. Effect of chelating agents on calcium uptake and distribution in higher plants. Z. Pflanzenernaehr. Bodenkd. 128:195–207.

Ishizuka, J. 1982. Characteristics of molybdenum absorption and translocation in soybean plants. Soil Sci. Plant Nutr. 28:63–77.

Ito, A., and A. Fujiwara. 1967. Functions of calcium in the cell wall of rice leaves. Plant Cell Physiol. 8:409–422.

Iverson, D. B., M. Schleicher, J. G. Zendegui, and D. M. Watterson. 1981. Structure and function of calmodulin from higher plants. Fed. Proc. Fed. Am. Soc. Exp. Biol. 40:1738.

Jackson, W. A. 1967. Physiological effects of soil acidity. *In* R. W. Pearson and F. Adams (ed.) Soil acidity and liming. Agronomy 12:43–124. Am. Soc. of Agron., Madison, Wis.

Jacob, G. S., and W. H. Orme-Johnson. 1980. Molecular properties and mechanism of action of nitrate reductase from *Neurospora crassa.* p. 327–344. *In* M. Coughlan (ed.) Molybdenum and molybdenum-containing enzymes. Pergamon Press, Inc., Elmsford, N.Y.

Jacobson, J. V., and T. J. V. Higgins. 1982. Characterization of the $\alpha$-amylase synthesized by aleurone layers of Himalaya barley in response to gibberellic acid. Plant Physiol. 70: 1647–1653.

Jacobson, L., R. J. Hannapel, D. P. Moore, and M. Schaedle. 1961a. Influence of calcium on the selectivity of ion absorption process. Plant Physiol. 36:58–61.

Jacobson, L., R. J. Hannapel, D. P. Moore, and M. Schaedle. 1961b. Effect of root to solution ratio in ion absorption experiments. Plant Physiol. 36:62–65.

Jacobson, L., D. P. Moore, and R. J. Hannapel. 1960. Role of calcium in absorption of monovalent cations. Plant Physiol. 35:352–358.

Jacobson, L., R. Overstreet, H. M. King, and R. Handley. 1950. A study of potassium absorption by barley roots. Plant Physiol. 25:639–647.

Jacoby, B. 1967. The effect of the roots on calcium ascent in bean stems. Ann. Bot. (London) 31:725–730.

Jaffe, L. F. 1979. Control of development of ionic currents. p. 199–231. *In* R. A. Cone and J. E. Dowling (ed.) Membrane transduction mechanisms. Raven Press, New York.

James, D. W., T. L. Jackson, and M. E. Harward. 1968. Effect of molybdenum content of alfalfa grown on acid soils. Soil Sci. 105:397–402.

Janistyn, B. 1973. Indole-3-acetic acid-induced efflux of calcium ions by corn coleoptile sections. (In German.) Z. Naturforsch. C: Biosci. 28:777–778.

Jansen, E. F., R. Jang, P. Albersheim, and J. Bonner. 1960. Pectic metabolism of growing cell walls. Plant Physiol. 35:87–97.

Jenne, E. A., H. F. Rhoades, C. H. Yien, and O. W. Howe. 1958. Change in nutrient element accumulation by corn with depletion of soil moisture. Agron. J. 50:71–74.

Jensen, R. G. 1971. Activation of $CO_2$ fixation in isolated spinach chloroplasts. Biochim. Biophys. Acta 234:360–370.

Jeschke, W. D. 1976. Ionic relations of leaf cells. p. 160–194. *In* U. Lüttge and M. G. Pitman (ed.) Transport in plants II. Part B: Tissues and organs. Encyclopedia of plant physiology, New Series, Vol. 2B. Springer-Verlag New York, New York.

Joham, H. J. 1953. Accumulation and distribution of molybdenum in the cotton plant. Plant Physiol. 28:275–280.

Johansen, C. 1978. Comparative molybdenum concentrations in some tropical pasture legumes. Commun. Soil Sci. Plant Anal. 9:1009–1017.

John, M. K., G. W. Eaton, V. W. Case, and H. H. Chuah. 1972. Timing of alfalfa (*Medicago sativa* L.): II. Effect on mineral composition. Plant Soil 37:363–374.

Johnson, C. M. 1966. Molybdenum. p. 286–301. *In* H. D. Chapman (ed.) Diagnostic criteria for plants and soils. Univ. of California Press, Berkeley, Calif.

Johnson, C. M., G. A. Pearson, and P. R. Stout. 1952. Molybdenum nutrition of crop plants: II. Plant and soil factors concerned with molybdenum deficiencies in crop plants. Plant Soil 4:178–196.

Johnson, J. L. 1980. Characterization of the molybdenum cofactor of sulfite oxidase, xanthine oxidase and nitrate reductase identification of pterdine as a structural component. J. Biol. Chem. 255:1783–1786.

Johnson, J. L., B. E. Hainline, and K. V. Rajagopalan. 1980. The molybdenum cofactor common to nitrate reductase, xanthine dehydrogenase and sulfate oxidase. p. 345–383. *In* M. Coughlan (ed.) Molybdenum cofactor of sulfite oxidase and nitrate reductase. Pergamon Press, Inc., Elmsford, N.Y.

Johnson, R. E., and W. A. Jackson. 1964. Calcium uptake and transport by wheat seedlings as affected by aluminum. Soil Sci. Soc. Am. Proc. 28:381–386.

Jones, R. L., and J. V. Jacobsen. 1983. Calcium regulation of the secretion of $\alpha$-amylase isoenzymes and other proteins from barley aleurone layers. Planta 158:1–9.

Joy, K. W. 1973. Control of glutamate dehydrogenase from *Pisum sativum* roots. Phytochem. 12:1031–1040.

Kamprath, E. J., and C. D. Foy. 1984. Lime-fertilizer-plant interactions in acid soils. *In* O. Engelstad (ed.) Fertilizer technology and use. Soil Sci. Soc. of Am., Madison, Wis. (In press.)

Katz, J. J., J. R. Norris, and L. L. Shipman. 1977. Models for reaction-center and antenna chlorophyll. p. 16–55. *In* J. M. Olsen and H. Hind (ed.) Chlorophyll-proteins, reaction centers, and photosynthetic membranes. Brookhaven Symp. of Biology no. 28, Brookhaven Natl. Lab., Upton, N.Y.

Kauss, H. 1974. Osmoregulation of *Ochromonas*. p. 90–94. *In* U. Zimmermann and J. Dainty (ed.) Membrane transport in plants. Springer-Verlag New York, New York.

Kauss, H. 1981. Sensing of volume changes by *Poterioochromonas* involves a $Ca^{2+}$-regulates system which controls activation of isofloridoside-phosphate synthase. Plant Physiol. 68: 420–424.

Kauss, H. 1983. Volume regulation in *Poterioochromonas*: Involvement of calmodulin in the $Ca^{2+}$-stimulated activation of isofloridoside-phosphate synthase. Plant Physiol. 71:169–172.

Kauss, H., K. S. Thomson, M. Thomson, and W. Jeblick. 1979. Osmotic regulation: Physiological significance of proteolytic and nonproteolytic activation of isofloridoside-phosphate synthase. Plant Physiol. 63:455–459.

Kawaski, T., and M. Moritsugu. 1979. A characteristic symptom of calcium deficiency in maize and sorghum. Commun. Soil Sci. Plant Anal. 10:41–56.

Kevers, C., L. Sticher, C. Penel. H. Greppin, and Th. Gasper. 1982. Calcium-controlled peroxidase secretion by sugarbeet cell suspensions in relation to habituation. Plant Growth Regul. 1:61–66.

Kirkby, E. A. 1968. Influence of ammonium and nitrate nutrition on the cation-anion balance and nitrogen and carbohydrate metabolism of white mustard plants grown in dilute nutrient solutions. Soil Sci. 105:133–141.

Kirkby, E. A. 1979. Maximizing calcium uptake by plants. Commun. Soil Sci. Plant Anal. 10:89–113.

Kirkby, E. A., and A. H. Knight. 1977. Influence of the level of nitrate nutrition on ion uptake and assimilation, organic acid accumulation, and cation-anion balance in whole tomato plants. Plant Physiol. 60:349–353.

Kirkby, E. A., and K. Mengel. 1967. Ionic balance in different tissues of the tomato plant in relation to nitrate, urea, or ammonium nutrition. Plant Physiol. 42:6–14.

Kirkby, E. A., and K. Mengel. 1976. The role of magnesium in plant nutrition. Z. Pflanzenernaehr. Bodenkd. 2:209–222.

Klee, C. B., T. H. Crouch, and P. G. Richman. 1980. Calmodulin. Annu. Rev. Biochem. 49: 489–575.

Kleese, R. A., D. C. Rasmusson, and L. H. Smith. 1968. Genetic and environmental variation in mineral element accumulation in barley, wheat, and soybeans. Crop Sci. 8:591–593.

Klein, K., G. Wagner, and M. R. Blatt. 1980. Heavy-mereomysin-decoration of microfilaments from *Mougeotia* protoplasts. Planta 150:345–356.

Koksal, A. J. 1973. Interrelation between varieties, rootstocks, and interstocks of apple: II. Nutrient status of leaves. Gartenbauwissenschaft 38:287–310.

Koontz, H. V., and R. E. Foote. 1966. Transpiration and calcium deposition by unifoliate leaves of *Phaseolus vulgaris* differing in maturity. Physiol. Plant. 19:313–321.

Kotze, W. A. G. 1979. Ionic interactions in the uptake and transport of calcium by apple seedlings. Commun. Soil Sci. Plant Anal. 10:115–127.

Kotze, W. A. G., C. B. Shear, and M. Faust. 1976. Effect of nitrogen source and the presence or absence of aluminum on the growth and calcium nutrition of apple seedlings. J. Am. Soc. Hortic. Sci. 102:279–282.

Kotze, W. A. G., C. B. Shear, and M. Faust. 1977. Effect of nitrogen source and aluminum in nutrient solution on the growth and mineral nutrition of apple and peach seedlings. J. Am. Soc. Hortic. Sci. 101:305–309.

Kovacevic, V. 1980. The role of the genotype in the mineral nutrition of inbred corn lines: 4. Control and accumulation of magnesium. Zb. Rad. Poljopr. Inst. Osijek 10:88–100.

Krause, G. H., and U. Heber. 1976. Energetics of intact chloroplasts. p. 171–214. *In* J. Barber (ed.) The intact chloroplasts. Elsevier North-Holland, Inc., New York.

Kretschmer, A. E., Jr., and R. J. Allen, Jr. 1956. Molybdenum in Everglades soils and plants. Soil Sci. Soc. Am. Proc. 20:253–257.

Kretsinger, R. H. 1976. Calcium-binding proteins. Annu. Rev. Biochem. 45:239–266.

Kretsinger, R. H. 1977. Evolution of the informational role of calcium in eukaryotes. p. 63–72. *In* R. H. Wassermann et al. (ed.) Calcium-binding proteins and calcium function. Elsevier North-Holland, Inc., New York.

Kruckeburg, A. R. 1954. The ecology of serpentine soils: III. Plant species in relation to serpentine soils. Ecology 35:267–274.

Kubowicz, B. D., L. N. Vanderhoef, and J. B. Hanson. 1982. ATP-dependent calcium transport in plasmalemma preparations from soybean hupocotyls: Effect of hormone treatments. Plant Physiol. 69:187–191.

Kudrev, T., and N. Babalukova. 1975. The influence of magnesium insufficiency on the fraction of easily soluble protein and on the isoenzyme composition of certain dehydrogenases. Agrochimica 19:336–347.

Kudrev, T., and M. Georgieva. 1976. The influence of magnesium deficiency on the change of certain phospholipids in the leaves of maize. Agrochimica 20:222–232.

Kuo, J., T. P. O'Brien, and M. J. Canny. 1974. Pit-field distribution, plasmodesmatal frequency, and assimilate flux in the mesotome sheath cells of wheat leaves. Planta 121: 97–118.

Kwack, B. H. 1967. Studies on cellular site of calcium action in promoting pollen growth. Physiol. Plant. 20:825–833.

Kwack, B. H., and I.-H. Kim. 1967. Effects of calcium ion and the protective action on survival and growth inhibition of pollen. Physiol. Plant. 20:73–82.

Kylin, A., and M. Kahr. 1973. The effect of magnesium and calcium ions on adenosine triphosphatases from wheat and oat roots at different pH. Physiol. Plant. 28:452–457.

Lafever, H. N. 1981. Genetic differences in plant responses to soil nutrient stress. J. Plant Nutr. 4:89–109.

Lane, D. W. A., and C. A. Lamp. 1977. Preliminary selection of ryegrass plants for Mg accumulating ability. p. 385–387. *In* E. Wojahn and H. Thöns (ed.) Proc. 13th Grassland Congr., Leipzig, G.D.R. 18–27 May. Akademie-Verlag, Berlin.

Lau, O.-L., and S. F. Yang. 1975. Interaction of kinetin and calcium in relation to their effect on stimulation of ethylene production. Plant Physiol. 55:738–740.

Läuchli, A. 1968. Distribution and transport of ions in plant tissues studied with the electron probe x-ray microanalyzer: II. Long distance transport of ions to pea fruits. Planta 83: 137–149.

Leaf, A. L. 1973. Plant analysis as a aid in fertilizing forests. p. 427–454. *In* L. M. Walsh and J. D. Beaton (ed.) Soil testing and plant analysis. Soil Sci. Soc. of Am., Madison, Wis.

Lee, C. R. 1971. Influence of aluminum on plant growth and mineral nutrition of potatoes. Agron. J. 63:604–608.

Leggett, J. E., R. A. Galloway, and H. G. Gauch. 1965. Calcium activation of orthophosphate absorption by barley roots. Plant Physiol. 40:897–902.

Leggett, J. E., and W. A. Gilbert. 1969. Magnesium uptake by soybeans. Plant Physiol. 44: 1182–1186.

Leigel, W. 1970. Calcium oxalate depositions in pedicels of some apple varieties. Angew. Bot. 44:223–232.

Leopold, A. C. 1977. Modification of growth regulatory action with inorganic solutes. p. 33–41. *In* C. A. Stutte (ed.) Plant growth regulators: Chemical activity, plant responses, and economic potential. Adv. Chem. Series 159, Am. Chemical Soc., Washington, D.C.

Leopold, A. C., B. W. Poovaiah, R. K. Dela Fuente, R. J. Williams. 1974. Regulation of growth with inorganic solutes. p. 780–788. *In* Plant growth substances. Hirokawa Publ. Co., Tokyo.

Le Saos, J. 1976. Calcium translocation to aenial parts in the halophyte *Cochlearis anglica* L.: Sodium chloride effects. Physiol. Veg. 14:381–390.

Levi, E. 1968. The distribution of mineral elements following leaf and root uptake. Physiol. Plant. 21:213–226.

Lewis, T. L., and D. Martin. 1973. Longitudinal distribution of applied calcium, and of naturally occurring calcium, magnesium, and potassium, in Menton apple fruits. Aust. J. Agric. Res. 24:263–271.

Lin, D. C., and P. S. Nobel. 1971. Control of photosynthesis by $Mg^{2+}$. Arch. Biochem. Biophys. 145:622–632.

Liu, D. J., and Y. Pomeranz. 1975. Distribution of mineral elements in barley at the cellular level by x-ray analysis. Cereal Chem. 52:620–629.

Liu, D. J., Y. Pomeranz, and G. S. Robbins. 1975. Mineral content of developing and malted barley. Cereal Chem. 52:678–686.

Ljones, T. 1979. Nitrogen fixation and bioenergetics: The role of ATP in nitrogenase catalysis. FEBS Lett. 98:1–8.

Loewenstein, W. R., and B. Rose. 1978. Calcium in (junctional) intercellular communication and a thought on its behaviour in intracellular communication. Ann. N.Y. Acad. Sci. 307:285–307.

Loneragan, J. F., I. C. Rowland, A. D. Robson, and K. Snowball. 1970. The Ca nutrition of plants. p. 358–362. *In* M. J. T. Norman (ed.) Proc. 11th Int. Grassland Congr., Surfers Paradise, Queensland, Australia. 13–23 April. Univ. of Queensland Press, St. Lucia, Australia.

Loneragan, J. F., and K. Snowball. 1969. Calcium requirements of plants. Aust. J. Agric. Res. 20:465–478.

Losada, M., and M. G. Guerrero. 1979. The photosynthetic reduction of nitrate and its regulation. p. 365–408. *In* J. Barber (ed.) Photosynthesis in relation to model systems. Elsevier North-Holland, Inc., New York.

Lüttge, U., and N. Higinbotham. 1979. Transport in plants. Springer-Verlag New York, New York.

Lüttge, U., and M. G. Pitman (ed.) 1976. Transport in plants II. Part B: Tissues and organs. Encyclopedia of plant physiology, New Series, Vol. 2B. Springer-Verlag New York, New York.

Lutz, J. A., Jr., C. F. Genter, G. W. Hawkins. 1972. Effect of soil pH on element concentration and uptake by maize: II. Cu, B, Zn, Mn, Mo, Al, and Fe. Agron. J. 64:583–585.

Maas, E. V. 1969. Calcium uptake by excised maize roots and interactions with alkali cations. Plant Physiol. 44:985–989.

Maas, E. V. and G. Ogata. 1971. Absorption of magnesium and chloride by excised corn roots. Plant Physiol. 47:357–360.

MacKay, D. C., E. W. Chipman, and U. C. Gupta. 1966. Copper and molybdenum nutrition of crops grown on acid sphagnum peat soil. Soil Sci. Soc. Am. proc. 30:755–759.

Macklon, A. E. S. 1975. Cortical cell fluxes and transport to the stele in excised root segments of *Allium cepa* L.: II. Calcium. Planta 122:131–141.

Macklon, A. E. S., and A. Sim. 1976. Cortical cell fluxes and transport to the stele in excised root segments of *Alluim cepa* L.: III. Magnesium. Planta 128:5–9.

Madhok, O. P., and R. B. Walker. 1969. Magnesium nutrition of two species of sunflower. Plant Physiol. 44:1016–1022.

Marfey, P., and H. Chessin. 1974. Cholesterol: Complexes with metalic salts. Biochim. Biophys. Acta 337:136–144.

Marinos, N. G. 1962. Submicroscopic aspects of mineral deficiency: I. Calcium deficiency in the shoot apex of barley. Am. J. Bot. 49:834–841.

Marinos, N. G. 1963. Studies on microscopic aspects of mineral deficiencies: II. Nitrogen, potassium, sulfur, phosphorus, and magnesium deficiencies in the shoot apex of barley. Am. J. Bot. 50:998–1005.

Marmé, D. 1982. The role of $Ca^{2+}$ and calmodulin in plants. What's New Plant Physiol. 13:37–40.

Marmé, D. 1983. Calcium transport and function. p. 599–625. *In* A. Läuchli and R. L. Bieleski (ed.) Inorganic plant nutrition. Encyclopedia of plant physiology, New Series, Vol. 15. Springer-Verlag New York, New York.

Marmé, D., and P. Dieter. 1984. The role of $Ca^{2+}$ and calmodulin in plants. *In* W. Y. Cheung (ed.) Calcium and cell function. (In press.)

Marocke, R., G. Correge, and R. Trendel. 1979. Varietal differences in the response of hops to magnesium deficiency. C. R. Hebd. Sean. Acad. Sci. 289:97–100.

Marschner, H. 1974. Calcium nutrition of higher plants. Neth. J. Agric. Sci. 22:275–282.

Marschner, H., and I. Guenther. 1964. Ion uptake and cell structure of barley roots depending on calcium supply. Z. Planzenernaehr. Dueng. Bodenkd. 107:118–136.

Marschner, H., R. Handley, and R. Overstreet. 1966. Potassium loss and changes in fine structure of corn root tips induced by $H^+$-ion. Plant Physiol. 41:1725–1735.

Marschner, H., and K. Mengel. 1966. Effect of Ca- and H-ions on the membrane permeability of young barley roots under various metabolic conditions. Z. Pflanzenernaehr. Dueng. Bodenkd. 112:39–49.

Marschner, H., and H. Ossenberg-Neuhaus. 1977. Effect of 2,3,5-triiodobenic acid (TIBA) on calcium translocation and cation exchange capacity in sunflower. Z. Pflanzenphysiol. 85: 29–44.

Marschner, H., and C. Richter. 1974. Calcium transport in roots of maize and bean plants. Plant Soil 40:193–210.

Martin, D. 1967. $^{45}Ca$ movement in apple trees. Experiments in Tasmania 1960–1965. Field Stn. Rec. (Aust. CSIRO Div. Plant Ind.) 6:49–54.

Mascarenhas, J. P., and L. Machlis. 1964. Chemotropic response of the pollen of *Antirrhinum majus* to calcium. Plant Physiol. 39:70–77.

Mason, J. L. 1979. Increasing calcium content of calcium-sensitive tissues. Commun. Soil Sci. Plant Anal. 10:349–371.

Matsumoto, H., and T. Kawasaki. 1981. Changes of membrane-associated $Mg^{2+}$-activated ATPase of cucumber roots during calcium starvation. Physiol. Plant. 52:442–448.

Matsumoto, H., and K. Teraoka. 1980. Accumulation of sugars in cucumber leaves during calcium starvation. Plant Cell Physiol. 21:1505–1513.

Matsumoto, H., and T. Yamaya. 1981. Increase in alkaline phosphatase activity in cucumber roots during calcium starvation. Plant Cell Physiol. 22:1137–1140.

Mattoo, A. K., and M. Lieberman. 1977. Localization of the ethylene-synthesizing system in apple tissue. Plant Physiol. 60:794–799.

Mattson, S. 1973. Ionic relationships of soil and plant: IV. Ion uptake in relation to membrane activity and soil moisture. Acta Agric. Scand. 23:11–16.

Maynard, D. N. 1979. Nutritional disorders of vegetable crops: A review. J. Plant Nutr. 1:1–23.

Maynard, D. N., B. Gersten, and H. F. Vernell. 1965. The distribution of calcium as related to internal tipburn, variety, and calcium nutrition of cabbage. Proc. Am. Soc. Hortic. Sci. 86:392–396.

Maynard, D. N., D. C. Warner, and J. C. Howell. 1981. Cauliflower (*Brassica oleracea*) leaf tipburn, a calcium deficiency disorder. Hortic. Sci. 16:193–195.

McDonald, D. W., and A. Coddington. 1974. Properties of the assimilatory nitrate reductase from *Aspergillus nidulans*. Eur. J. Biochem. 46:169–178.

McEvoy, E. T. 1963. Varietal differences in calcium level in leaves of flue cured tobacco. Can. J. Plant Sci. 43:141–145.

McNaught, K. J. 1970. Diagnosis of mineral deficiencies in grass-legume pastures by plant analysis. p. 334–338. *In* M. J. T. Norman (ed.) Proc. 11th Int. Grassland Congr., Surfers Paradise, Queensland, Australia. 13–23 April. Univ. of Queensland Press, St. Lucia, Australia.

Meagher, W. R., C. M. Johnson, and P. R. Stout. 1952. Molybdenum requirement of leguminous plants supplied with fixed nitrogen. Plant Physiol. 27:223–230.

Mengel, K., H. Grimme, and K. Nemeth. 1969. Potential and actual availability of plant nutrients in the soils. Landwirtsch. Forsch. Sonderh. 23(1):79–80.

Mengel, K., and H. E. Haeder. 1977. Effect of potassium supply on the rate of phloem sap exudation and the composition of phloem sap of *Ricinus communis*. Plant Physiol. 59: 282–284.

Mengel, K., and M. Helal. 1967. Influence of the exchangeable $Ca^{++}$ of young barley roots on the fluxes of $K^{+}$ and phosphate, an interpretation of the Viets effect. Z. Pflanzenphysiol. 57:223–234.

Mengel, K., and E. A. Kirkby. 1982. Principles of plant nutrition. 3rd ed. Int. Potash Inst., Bern, Switzerland.

Michael, G. 1941. Uptake and distribution of magnesium and its role in higher plants. Z. Pflanzenernaehr. Bodenkd. 25:65–120.

Millikan, C. R. 1961. Plant varieties and species in relation to the occurrence of deficiencies and excesses of certain nutrient elements. J. Aust. Inst. Agric. Sci. 27:220–233.

Millikan, C. R. 1971. Mid-season movement of $^{45}Ca$ in apple trees. Aust. J. Agric. Res. 22: 923–930.

Millikan, C. R., and B. C. Hanger. 1964. Effect of calcium level in the substrate on the distribution of $^{45}Ca$ in subterranean clover (*Trifolium subterraneum* L.). Aust. J. Biol. Sci. 17: 823–844.

Millikan, C. R., and B. C. Hanger. 1967. Redistribution of $^{45}Ca$ in *Trifolium subterraneum* L. and *Antirrhinum majus* L. Aust. J. Biol. Sci. 20:1119–1130.

Millikan, C. R., and B. C. Hanger. 1969. Movement of foliar-applied $^{45}Ca$ in brussels sprouts. Aust. J. Biol. Sci. 22:545–558.

Millthorpe, F. L., and J. Moorby. 1969. Vascular transport and its significance in plant growth. Annu. Rev. Plant Physiol. 20:117–138.

Mitchell, K. J. 1945. Preliminary note on the use of ammonium molybdate to control whiptail in cauliflower and broccoli crops. N.Z. J. Sci. Technol. Sect. A 27:287–293.

Moll, B. A., and R. L. Jones. 1982. $\alpha$-Amylase secretion by single barley aleurone layers. Plant Physiol. 70:1149–1155.

Moore, D. J., and C. E. Bracker. 1976. Ultrastructural alteration of plant plasma membranes induced by auxin and calcium ions. Plant Physiol. 58:544–547.

Moore, D. P. 1972. Mechanisms of micronutrient uptake by plants. p. 171–198. *In* J. J. Mortvedt et al. (ed.) Micronutrients in agriculture. Soil Sci. Soc. of Am., Madison, Wis.

Moore, D. P. 1974. Physiological effects of pH on roots. p. 135–151. *In* E. W. Carson (ed.) The plant root and its environment. Univ. Press of Virginia, Charlottesville.

Moore, D. P., L. Jacobson, and R. Overstreet. 1961a. Uptake of calcium by excised barley roots. Plant Physiol. 36:53–57.

Moore, D. P., R. Overstreet, and L. Jacobson. 1961b. Uptake of magnesium and its interaction with calcium in excised barley roots. Plant Physiol. 36:290–295.

Moritsugu, M., and T. Kawasaki. 1980. Effect of calcium on magnesium uptake by plants: I. The apparent accelerative effect of calcium on magnesium uptake. Ber. Ohara Inst. Landwirtsch. Biol. Okayama Univ. 18:55–64. (Biol. Abstr. 70:74303.)

Morris, M. R., and D. H. Northcote. 1977. Influence of cations at the plasma membrane in controlling polysaccharide secretion from sycamore suspension cells. Biochem. J. 166: 603–618.

Morrison, T. M., and R. W. Heine. 1966. Transport of minerals in tree xylem. Ann. Bot. (London) 30:807–819.

Mortenson, L. E., and R. N. F. Thorneley. 1979. Structure and function of nitrogenase. Annu. Rev. Biochem. 48:387–418.

Mulder, E. G., R. Boxma, and W. L. Van Veen. 1959. The effect of molybdenum and nitrogen deficiencies on nitrate reduction in plant tissues. Plant Soil 10:335–355.

Muto, S., and H. Beevers. 1974. Lipase activities in castor bean endosperm during germination. Plant Physiol. 54:23–28.

Naismith, R. W., M. W. Johnson, and W. I. Thomas. 1974. Genetic control of relative calcium, phosphorus, and manganese accumulation on chromosome 9 in maize. Crop Sci. 14:845–849.

Nason, A., and W. D. McElroy. 1963. Modes of action of the essential mineral elements. p. 451–536. *In* F. C. Steward (ed.) Plant physiology: A treatise, Vol. 3. Inorganic nutrition of plants. Academic Press, Inc., New york.

Nauen, W., and T. Hartmann. 1980. Glutamate dehydrogenase from *Pisum sativum* L. Planta 148:7–16.

Neales, T. F. 1956. Components of the total magnesium content within the leaves of white clover and perennial rye grass. Nature (London) 177:388–389.

Neenan, M., and O. G. Goodman. 1954. Varietal susceptibility of brassicas to molybdenum deficiency. Nature (London) 74:792–793.

Nguyen, H. T., and D. A. Sleper. 1981. Genetic variability of mineral concentrations in *Festuca arundinacea* Schreb. Theor. Appl. Genet. 59:57–63.

Nicholas, D. J. D., and A. Nason. 1955. Role of molybdenum as a constituent of nitrate reductase from soybean leaves. Plant Physiol. 30:135–138.

Nieuwhof, W. 1960. Internal tipburn in white cabbage: I. Variety trials. Euphytica 9:203–208.

Nilshammar, M., B. Walles, and A. Kylin. 1971. Effect of calcium deficiency on the ultrastructure of the green alga *Scenedesmus.* Z. Pflanzenphysiol. 66:197–205.

Nobel, P. S. 1967. Relation of swelling and photophosphorylation to light-induced ion uptake by chloroplasts *in vitro*. Biochim. Biophys. Acta 131:127–140.

Noonan, J. B. 1953. Molybdenum deficiency in maize and other crops in the Taree District. Agric. Gaz. NSW 64:422–424.

Notton, B. A., and E. J. Hewitt. 1971. The role of tungsten in the inhibition of nitrate reductase activity in spinach (*Spinacea oleracea* L.) leaves. Biochem. Biophys. Res. Commun. 44:702–710.

Notton, B. A. and E. J. Hewitt. 1979. Structure and properties of higher plant nitrate reductase especially *Spinacea oleracea*. p. 227–244. *In* E. J. Hewitt and C. V. Cutting (ed.) Nitrogen assimilation of plants. Academic Press, Inc., New York.

Nuccitelli, R., and L. F. Jaffe. 1976. The ionic components of the current pulses generated by developing fucoid eggs. Dev. Biol. 49:518–531.

Nultsch, W. 1979. Effect of external factors on phototaxis of *Chlamydomonas reinhardtii*: III. Cations. Arch. Microbiol. 123:93–99.

Nultsch, W., G. Throm, and I. von Rimscha. 1971. Phototactic investigations in *Chlamydomonas reinhardii* Dangeard in homocontinuous culture. Arch. Mikrobiol. 80:487–490.

Oaks, A. 1979. Nitrate reductase in roots and its regulation. p. 217–226. *In* E. J. Hewitt and C. V. Cutting (ed.) Nitrogen assimilation of plants. Academic Press, Inc., New York.

Oberly, G. H. 1973. Effect of 2,3,5-triiododbenzoic acid on the bitter pit and calcium accumulation in "Northern spy" apples. J. Am. Soc. Hortic. Sci. 98:269–271.

Ochiai, E.-L. 1977. Bioinorganic chemistry. Allyn & Bacon, Boston, Mass.

Okamoto, H. 1962. Transport of cations from cotyledon to seedling of the embryonic plants of *Vigua sesquipedalis.* Plant Cell Physiol. 3:83–94.

Olsen, S. R. 1972. Micronutrient interactions. p. 243–264. *In* J. J. Mortvedt et al. (ed.) Micronutrients in agriculture. Soil Sci. Soc. of Am., Madison, Wis.

Olsen, S. R., and F. S. Watanabe. 1979. Interactions of added gypsum in alkaline soils with uptake of iron, molybdenum, and zinc by sorghum. Soil Sci. Soc. Am. J. 43:125–130.

Ortega-Perez, R., D. van Tuinen, D. Marme, J. A. Cox, and G. Turian. 1981. Purification and identification of calmodulin from *Neurospora crassa.* FEBS Lett. 133:205–208.

O'Sullivan, W. J. 1973. Kinases. p. 582–607. *In* G. L. Eichhorn (ed.) Inorganic biochemistry, Vol. 1. Elsevier North-Holland, Inc., New York.

Oursel, A., A. Lamant, L. Salsac, and P. Mazliak. 1973. Comparative study of lipids and the passive binding of calcium in the roots and subcellular fractions of *Lupinus luteus* and *Vicia faba.* Phytochemistry 12:1865–1874.

Oza, A. M., S. Singh, and M. B. Kamath. 1979. Differential accumulation of phosphorus and calcium by pearl millet cultivars from applied sources. J. Nucl. Agric. Biol. 8:54–57.

Pal, R. N., V. K. Rai, and M. M. Laloraya. 1973. Calcium in relation to respiratory activity of peanut and linseed plants. Biochem. Physiol. Pflanz. 164:258–265.

Palzkill, D. A., T. W. Tibbitts, and P. H. Williams. 1976. Enhancement of calcium transport to inner leaves of cabbage for prevention of tipburn. J. Am. Soc. Hortic. Sci. 101:645-648.

Pate, J. S., P. J. Sharkey, and O. A. M. Lewis. 1975. Xylem to phloem transfer of solutes in fruiting shoots of legumes, studies by a phloem bleeding technique. Planta 122:11-26.

Paul, D. C., and Ch. W. Goff. 1973. Comparative effects of caffeine, its analogs and calcium deficiency on cytokinesis. Exp. Cell Res. 78:399-413.

Peeden, G. F., and C. B. McCants. 1977. Influence of variety and soil applications of Ca on development of Ca deficiency in tobacco. Agron. J. 69:71-76.

Peel, A. J. 1972. The control of solute movement into sieve elements. Pestic. Sci. 3:631-641.

Penot, M., J. Floc'h, and M. Penot. 1976. Comparitive study of $^{45}$Ca absorption and translocation in various groups of plants. Planta 129:7-14.

Peterson, N. K., and E. R. Purvis. 1961. Development of molybdenum deficiency symptoms in certain crop plants. Soil Sci. Soc. Am. Proc. 25:111-117.

Pitman, M. G. 1963. The effect of divalent cations on the uptake of salt by beet root tissue. J. Exp. Bot. 15:444-456.

Pitman, M. G. 1976. Ion uptake by plant roots. p. 95-128. *In* U. Lüttge and M. G. Pitman (ed.) Transport in plants II. Part B: Tissues and organs. Encyclopedia of plant physiology, New Series, Vol. 2B, Springer-Verlag New York, New York.

Pitman, M. G. 1977. Ion transport into the xylem. Annu. Rev. Plant Physiol. 28:71-88.

Pitman, M. G., U. Lüttge, A. Läuchli, and E. Ball. 1974. Effect of previous water stress on ion uptake and transport in barley seedlings. Aust. J. Plant Physiol. 1:377-385.

Plant, W. 1952. The molybdenum content of some brassicae and lettuce crops in acid soils. Int. Soc. Soil Sci., Trans. Comm. II, IV (Dublin) 2:176-177.

Poling, E. B., and G. H. Oberly. 1979. Effects of rootstock on mineral composition of apple leaves. J. Am. Soc. Hortic. Sci. 104:799-801.

Poovaiah, B. W. 1979. Role of calcium in ripening and senescence. Commun. Soil Sci. Plant Anal. 10:83-88.

Poovaiah, B. W., and A. C. Leopold. 1973a. Deferral of leaf senescence with calcium. Plant Physiol. 52:236-239.

Poovaiah, B. W., and A. C. Leopold. 1973b. Inhibition of abscission by calcium. Plant Physiol. 51:848-851.

Poovaiah, B. W. and A. C. Leopold. 1976a. Effects of inorganic solutes on the binding of auxin. Plant Physiol. 58:783-785.

Poovaiah, B. W., and A. C. Leopold. 1976b. Effects of inorganic salts on tissue permeability. Plant Physiol. 58:182-185.

Pope, D. T., and H. M. Munger. 1953. Heredity and nutrition in relation to magnesium deficiency chlorosis in celery. Proc. Am. Soc. Hortic. Sci. 61:472-480.

Popp, M., and H. Kinzel. 1981. Changes in the organic acid content of some cultivated plants induced by mineral ion deficiency. J. Exp. Bot. 32:1-8.

Porter, O. A., and J. T. Moraghan. 1975. Differential response of two corn inbreds to varying root temperature. Agron. J. 67:515-518.

Portis, A. R., Jr., and H. W. Heldt. 1976. Light-dependent changes of the $Mg^{2+}$ concentration in the stroma in relation to the $Mg^{2+}$ dependency of $CO_2$ fixation in intact chloroplasts. Biochim. Biophys. Acta 449:434-446.

Portis, A., C. Newton, W. Pangborn, and D. Papahadjopoulos. 1979. Studies on the mechanism of membrane fusion: Evidence for an intermembrane $Ca^{2+}$-phospholipid complex, synergism with $Mg^{2+}$ and inhibition by spectrin. Biochemistry 18:780-790.

Possingham, J. V. 1954. The effect of molybdenum on the organic and inorganic phosphorus of plants. Aust. J. Biol. Sci. 7:221-224.

Possingham, J. V. 1956. The effect of mineral nutrition on the content of free amino acids and amides in tomato plants: I. A comparison of the effects of deficiencies of copper, zinc, manganese, iron, and molybdenum. Aust. J. Biol. Sci. 9:539-551.

Prasil, I. 1975. The effect of calcium on nitrate absorption and assimilation in pumpkin plants. Biol. Plant. 17:223-225.

Preistley, C. A. 1976. Some effects of ringing and deblossoming braces of young apple trees in leaf composition. J. Exp. Bot. 27:1325-1332.

Procter, J., and S. R. J. Woodell. 1975. The ecology of serpentine soils. Adv. Ecol. Res. 9: 255-366.

Quebedeaux, B., Jr., and J. L. Ozbun. 1973. Effect of ammonium nutrition on water stress, water uptake, and root pressure in *Lycopersicon esculentum* Mill. Plant Physiol. 52: 677–679.

Rains, D. W., W. E. Schmid, and E. Epstein. 1964. Absorption of cation by roots: Effects of hydrogen ions and essential role of calcium. Plant Physiol. 39:274–278.

Raju, V. K., H. Marschner, and Romheld. 1972. Effect of iron nutritional status on ion uptake, substrate pH and production and release of organic acids and riboflavin by sunflower plants. Z. Pflanzenernaehr. Bodenkd. 132:177–190.

Ralph, R. K., S. Bullivant, S. J. Wojcik. 1976. Effects of kinetin on phosphorylation of leaf membrane proteins. Biochim. Biophys. Acta 412:319–327.

Ramani, S., and S. Kannan. 1976. Inhibition of $Ca^{2+}$ and $Zn^{2+}$ uptake by $Mn^{2+}$ in excised rice roots. Z. Naturforsch. 31C:12–14.

Randall, P. J. 1969. Changes in nitrate and nitrate reductase levels on restoration of molybdenum to molybdenum-deficient plants. Aust. J. Agric. Res. 20:635–642.

Randall, P. J., J. A. Thomsom, and H. E. Schroeder. 1979. Cotyledonery storage proteins in *Pisum sativum*: IV. Effects of sulfur, phosphorus, potassium and magnesium deficiencies. Aust. J. Plant Physiol. 6:11–24.

Ranjeva, R., G. Refeno, A. M. Boudet, and D. Marme. 1983. Plant quinate:$NAD^+$ oxidoreductase is activated by $Ca^{2+}$-calmodulin-dependent phosphorylation. Proc. Natl. Acad. Sci. USA 80:5222–5224.

Rasmussen, H. 1981. Calcium and cAMP as synarchic messengers. Wiley & Sons, Inc., New York.

Rasmusson, D. C., A. J. Hester, G. N. Fick, and I. Byrne. 1971. Breeding for mineral content in wheat and barley. Crop Sci. 11:623–626.

Rasmusson, D. C., L. H. Smith, and R. A. Kleese. 1964. Inheritance of Sr-89 accumulation in wheat and barley. Crop Sci. 4:586–589.

Raven, J. A. 1977. Hydrogen ion and calcium (2+) ion in phloem and symplast: Relation of relative immobility of the ions to the cytoplasmic nature of the transport path. New Phytol. 79:465–480.

Reisenauer, H. M. 1963. The effect of sulfur on the absorption and utilization of molybdenum by peas. Soil Sci. Soc. Am. Proc. 27:553–555.

Reiss, H. D., and W. Herth. 1978. Visualization of the $Ca^{2+}$ gradient in growing pollen tubes of *Lilium longiflorum* with chlorotetracycline fluorescence. Protoplasma 97:373–377.

Rendig, V. V., and D. L. Grunes. 1979. Grass tetany. ASA Spec. Pub. no. 35, Am. Soc. of Agron., Crop Sci. Soc. of Am., and Soil Sci. Soc. of Am., Madison, Wis.

Ringoet, A., R. V. Rechenmann, and H. Veen. 1967. Calcium movement in oat leaves measured by semi-conductor detectors. Radiat. Bot. 7:81–90.

Ringoet, A., G. Sauer, and A. J. Gielink. 1968. Phloem transport of calcium in oat leaves. Planta 80:15–20.

Rinne, R. W., and R. Langston. 1960. Effect of growth on redistribution of some mineral elements in peppermint. Plant Physiol. 35:210–215.

Roberts, G. P., and W. J. Brill. 1981. Genetics and regulation of nitrogen fixation. Annu. Rev. Microbiol. 35:207–235.

Robinson, D. B., and J. E. Campbell. 1956. Whiptail of cauliflower in Prince Edward Island. Can. J. Agric. Sci. 36:241–244.

Robinson, G. A., R. W. Butcher, and E. W. Sutherland. 1971. Cyclic AMP. Academic Press, Inc., New York.

Rousseau, G. G., F. J. Haasbroek, and C. K. Visser. 1972. Bitter pit in apples: The effect of calcium on permeability changes in apple fruit tissue. Agroplantae 4:73–83.

Russell, R. S., and D. T. Clarkson. 1976. Ion transport in root systems. p. 401–411. *In* P. S. Savies and N. Sunderland (ed.) Perspectives in experimental biology, Vol. 2. Botany. Pergamon Press, Inc., New York.

Salisbury, J. L., and G. L. Floyd. 1978. Calcium-induced contraction of the rhizoplast of a quadriflagellate green alga. Science (Washington, D.C.) 202:975–976.

Saric, M. R. (ed.) 1982. Genetic specificity of mineral nutrition of plants. Serbian Acad. Sci. Arts, Belgrade, Yugoslavia.

Saric, M. R., and R. Curic. 1963. The uptake of P, Ca, S, and Fe by the plants in dependence of temperature. Agrochimica 7:173–184.

Sayre, J. D. 1955. Mineral nutrition of corn. p. 293–314. *In* G. F. Sprague (ed.) Corn and corn improvement. Academic Press, Inc., New York.

Scarpa, A., F. L. Brinely, T. Tiffert, and G. R. Dubyak. 1978. Metallochromic indicators of ionized calcium. Ann. N.Y. Acad. Sci. 307:86–112.

Schimansky, C. 1973. Investigations on translocation of magnesium ($^{28}$Mg) in sunflower plants. Z. Pflanzenernaehr. Bodenkd. 136:68–82.

Schmidt, J. A., and R. Eckert. 1976. Calcium couples flagellar reversal to photostimulation in *Chamydomonas reinhardtii.* Nature (London) 262:713–715.

Scott, M. G., and R. L. Peterson. 1979. The root endodermis in *Ranunculus acris*: I. Structure and ontogeny. Can. J. Bot. 57:1040–1062.

Scrutton, M. C. 1973. Metal enzymes. p. 381–437. *In* G. L. Eichhorn (ed.) Inorganic biochemistry, Vol. 1. Elsevier North-Holland, Inc., New York.

Shakad, A., and A. Bar-Akiva. 1967. Nitrate reductase activity as an indication of molybdenum level and requirement of citrus plants. Phytochemistry 6:347–350.

Shannan, S., J. J. Natti, and J. D. Atkin. 1967. Relation of calcium nutrition to hypocotyl necrosis of snap bean (*Phaseolus vulgaris* L.). Proc. Am. Soc. Hortic. Sci. 90:180–190.

Sharples, R. O. 1974. Number and size of fruits. Acta Hortic. 45:21–28.

Sharples, R. O., and D. S. Johnson. 1977. The influence of calcium on senescence changes in apple. Ann. Appl. Biol. 85:450–453.

Shay, F. J., and M. G. Hale. 1973. Effects of low levels of calcium on exudation of sugars and sugar derivatives from intact peanut roots under axenic conditions. Plant Physiol. 51: 1061–1063.

Shear, C. B. 1975. Calcium-related disorders of fruits and vegetables. Hortic. Sci. 10:361–365.

Shear, C. B. (ed.) 1979. International symposium on calcium nutrition of economic crops. Commun. Soil Sci. Plant Anal. 10:1–501.

Shear, C. B., and M. Faust. 1970. Calcium transport in apple trees. Plant Physiol. 45:670–674.

Shen, T. C., E. A. Funkhouser, and M. G. Guerrero. 1976. NADH- and NAD(P)H-nitrate reductase in rice seedlings. Plant Physiol. 58:292–294.

Shewry, P. R., and P. J. Peterson. 1975. Calcium and magnesium in plants and soil from a serpentine area on Unst, Shetland. J. Appl. Ecol. 12:381–391.

Siegel, S. M. 1970. Further studies on regulation of betacyanin efflux from beetroot tissue: Ca-ion-reversible effects of hydrochloric acid and ammonia water. Physiol. Plant. 23: 251–257.

Sigel, H. (ed.) 1973–1978. Metal ions in biological systems, Vol. 1–7. Marcel Dekker, Inc., New York.

Simon, E. W. 1978. The symptoms of calcium deficiency in plants. New Phytol. 80:1–15.

Simons, R. K., and M. C. Chu. 1980. Scanning electron microscopy and electron microprobe studies of bitter pit in apples. p. 57–69. *In* D. Atkinson et al. (ed.) Mineral nutrition of fruit trees. Butterworth (Publ.) Inc., Woburn, Mass.

Simson, C. R., R. B. Cory, and M. E. Sumner. 1979. Effect of varying Ca:Mg ratios on yield and composition of corn (*Zea mays*) and alfalfa (*Medicago sativa*). Commun. Soil Sci. Plant Anal. 10:153–162.

Sleper, D. A. 1979. Plant breeding, selection, and species in relation to grass tetany. p. 63–77. *In* V. V. Rendig and D. L. Grunes (ed.) Grass tetany. ASA Spec. Pub. no. 35, Am. Soc. of Agron., Crop Sci. Soc. of Am., and Soil Sci. Soc. of Am., Madison, Wis.

Sleper, D. A., G. B. Garner, K. H. Asay, R. Boland, and E. E. Picket. 1977. Breeding for Mg, Ca, K, and P content in tall fescue. Crop Sci. 17:433–438.

Smith, R. L., and A. Wallace. 1956. Influence of cation ratio, temperature, and time on adsorption and absorption of calcium and potassium by citrus and other plant species. Soil Sci. 82:9–19.

Snaydon, R. W. 1962. The growth and competitive ability of contrasting populations of *Trifolium repens* on calcareous and acid soils. J. Ecol. 50:439.

Snaydon, R. W., and A. D. Bradshaw. 1961. Differential response to calcium within the species *Festuca ovina* L. New Phytol. 60:219–234.

Solomonson, L. P. 1979. Structure of *Chlorella* nitrate reductase. p. 199–205. *In* E. J. Hewitt and C. V. Cutting (ed.) Nitrogen assimilation of plants. Academic Press, Inc., New York.

Spence, J. T. 1976. Reactions of molybdenum coordination compounds as models for biological systems. p. 279–324. *In* H. Sigel (ed.) Metal enzymes in biological systems, Vol. 5. Marcel Dekker, Inc., New York.

Sprague, H. B. (ed.) 1964. Hunger signs in crops. ed. 3. David McKay Co., Inc., New York.

Srivastava, H. S. 1980. Regulation of nitrate reductase activity in higher plants. Phytochemistry 19:725–733.

Stahly, E. A., and N. R. Benson. 1970. Calcium levels of "Golden Delicious" apples sprayed with 2,3,5-triiodobenzoic acid. J. Am. Soc. Hortic. Sci. 95:726–727.

Stahly, E. A., and N. R. Benson. 1976. Calcium levels of "Golden Delicious" apples as influenced by calcium sprays, 2,3,5-triiodobenzoic acid and other plant growth regulator sprays. J. Am. Soc. Hortic. Sci. 101:120–122.

Stebbins, R. L., and D. H. Dewey. 1972. Role of transpiration and phloem transport in accumulation of $^{45}$Calcium in leaves of young apple trees. J. Am. Soc. Hortic. Sci. 97:471–474.

Steucek, G. L., and H. V. Koontz. 1970. Phloem mobility of magnesium. Plant Physiol. 46: 50–52.

Steward, F. C., F. Crane, K. Millar, R. M. Zacharias, R. Rabson, and D. Margolis. 1959. Nutritional and environmental effects on the nitrogen metabolism of plants. p. 148–176. *In* H. K. Porter (ed.) Utilization of nitrogen and its compounds by plants. Symp. Soc. Exp. Biol. no. 13. Academic Press, Inc., New York.

Stewart, I., and C. D. Leonard. 1952. Molybdenum deficiency in Florida citrus. Nature (London) 170:714–715.

Stewart, W. D. P., and J. R. Gallon (ed.) 1980. Nitrogen fixation. Academic Press, Inc., New York.

Sticher, L., C. Penel, and H. Greppin. 1981. Calcium requirement for the secretion of peroxidases by plant cell suspensions. J. Cell Sci. 48:345–353.

Stieffel, E. J. 1973. Proposed molecular mechanism for the action of molybdenum in enzymes: Coupled proton and electron transfer. Proc. Natl. Acad. Sci. USA 70:988–992.

Stout, P. R., and W. R. Meagher. 1948. Studies of the molybdenum nutrition of plants with radioactive molybdenum. Science (Washington, D.C.) 108:471–473.

Stout, P. R., W. R. Meagher, G. A. Pearson, and C. M. Johnson. 1951. Molybdenum nutrition of crop plants: I. The influence of phosphate and sulfate on the absorption of molybdenum from soils and solution cultures. Plant Soil 3:51–87.

Stratton, S. D., and D. A. Sleper. 1979. Genetic variation and interrelationships of several minerals in orchardgrass herbage. Crop Sci. 19:477–481.

Sutherland, E. W., G. S. Robinson, and R. W. Butcher. 1968. Some aspects of the biological role of adenosine 3′-5′-monophosphate (cyclic AMP). Circulation 37:279–306.

Suttle, J. S., and H. Kende. 1978. Ethylene and senescence in petals of *Tradescantia.* Plant Physiol. 62:267–271.

Tagawa, T., and J. Bonner. 1957. Mechanical properties of the *Avena* coleoptile as related to auxin and to ionic interactions. Plant Physiol. 32:207–212.

Takatori, F. H., O. A. Lorenz, and G. H. Cannell. 1961. Strontium and calcium for the control of blackheart of celery. Proc. Am. Soc. Hortic. Sci. 77:406–414.

Tammes, P. M. L., and J. Van Die. 1966. Studies on phloem exudation from *Yucca flaccida*: IV. Translocation of macro and micronutrients by the phloem sap-stream. Acta Bot. Neerl. 13:76–83.

Terblanche, J. H., L. H. Wooldridge, I. Hesebeck, and M. Joubert. 1979. The redistribution and immobilization of calcium in apple trees with special reference to bitter pit. Commun. Soil Sci. Plant Anal. 10:195–215.

Terry, N., and A. Ulrich. 1974. Effects of magnesium deficiency on the photosynthesis and respiration of leaves of sugar beet. Plant Physiol. 54:379–381.

Thimann, K. V., and C. L. Schneider. 1938. The role of salts, hydrogen-ion concentration and agar in the response of the *Avena* coleoptile to auxin. Am. J. Bot. 25:270–280.

Thomas, W. A. 1967. Dye and calcium ascent in dogwood trees. Plant Physiol. 42:1800–1802.

Thomas, W. A. 1970. Retention of calcium-45 by dogwood trees. Plant Physiol. 45:510–511.

Thomson, W. W. and T. E. Weier. 1962. An electron microscope study of chloroplasts from leaves deficient in nitrogen, phosphorus, magnesium and zinc. Plant Physiol. 37(Suppl.): xi.

Tibbitts, T. W., and D. A. Palzkill. 1979. Requirement for root-pressure flow to provide adequate calcium to low-transpiring tissue. Commun. Soil Sci. Plant Anal. 10:251–257.

Tiffin, L. O. 1972. Translocation of micronutrient in plants. p. 199–229. *In* J. J. Mortvedt et al. (ed.) Micronutrients in agriculture. Soil Sci. Soc. of Am., Madison, Wis.

Tingwa, P. O., and R. E. Young. 1974. The effect of calcium on the ripening of avocado (*Persea americana* Mill.) fruits. J. Am. Soc. Hortic. Sci. 99:540–542.

Todd, J. R. 1966. Methods of increasing magnesium contents of herbage with particular reference to the prevention of hypomagnesaemic tetany in ruminants. p. 178–180. *In* A. G. G. Hill (ed.) Proc. 10th Int. Grassland Congr., Helsinki, Finland. 7–16 July. Valtioneuvoston, Kirjapaino, Helsinki, Finland.

Tomlinson, J. D., and J. F. Turner. 1973. Pyruvate kinase of higher plants. Biochim. Biophys. Acta 329:128–139.

Toriyama, H., and M. J. Jaffe. 1972. Migration of calcium and its role in the regulation of siesmonasty in the moter cell of *Mimosa pudica* L. Plant Physiol. 49:72–81.

True, R. H. 1914. The harmful action of distilled water. Am. J. Bot. 1:255–273.

Ts'o, P. O. P. 1962. The ribosomes-ribonucleoprotein particles. Annu. Rev. Plant Physiol. 13: 45–80.

Ulrich, A., and F. J. Hills. 1973. Plant analysis as an aid in fertilizing sugar crops: Part 1. Sugar beets. p. 271–288. *In* L. M. Walsh and J. D. Beaton (ed.) Soil testing and plant analysis. Soil Sci. Soc. of Am., Madison, Wis.

Ulrich, A., and M. A. E. Mostafa. 1976. Calcium nutrition of the sugarbeet. Commun. Soil Sci. Plant Anal. 7:483–495.

Van Belle, H. 1981. R23471: A potent inhibitor of calmodulin-activated enzymes. Cell Calcium 2:483–494.

Van de Geijn, S. C., C. M. Petit, and H. Roejofsen. 1979. Measurement of the cation exchange capacity of the transport system in intact plant stems: Methodology and preliminary results. Commun. Soil Sci. Plant Anal. 10:225–236.

Vanden Driessche, T. 1963. Implications of contractile proteins in the leaf movement in *Mimosa pudica* L. Ann. Physiol. Veg. Univ. Bruxelles 8:101–112.

Vanden Driessche, T. 1978. The molecular mechanism of *Mimosa* leaf seismonastic movement: A re-evaluation. Arch. Biol. 89:435–449.

Van Goor, B. J. 1968. The role of calcium and cell permeability in the disease blossom-end rot of tomatoes. Physiol. Plant 21:1110–1121.

Van Goor, B. J., and D. Wiersma. 1974. Redistribution of potassium, calcium, magnesium, and manganese in the plant. Physiol. Plant. 31:163–168.

van Staden, J., and N. R. Comins. 1976. Energy dispersive x-ray analysis of protein bodies in *Protea compecta* cotyledons. Planta 130:219–221.

van Steveninck, R. F. M. 1965. The significance of calcium on the apparent permeability of cell membranes and the effects of substitution with other divalent ions. Physiol. Plant. 18:54–69.

Vennesland, B., and M. G. Guerrero. 1979. Reduction of nitrate and nitrite. p. 425–444. *In* M. Gibbs and E. Latzko (ed.) Photosynthesis II. Photosynthesis carbon metabolism and related processes. Encyclopedia of plant physiology, New Series, Vol. 6. Springer-Verlag New York, New York.

Viets, F. G., Jr. 1944. Calcium and other polyvalent cations as accelerators of ion accumulation by excised barley roots. Plant Physiol. 19:466–480.

Wagner, G., and K. Klein. 1978. Differential effects of calcium on chloroplast movement on *Mougeotia*. Photochem. Photobiol. 27:137–140.

Walker, D. A. 1972. The affinity of ribulose diphosphate carboxylase for $CO_2$-bicarbonate. p. 1773–1778. *In* G. Forti et al. (ed.) Photosynthesis two centuries after its discovery by Joseph Priestley, Vol. 3. Dr. W. Junk Publ., The Hague.

Walker, D. A. 1973. Photosynthetic induction of the light activation of ribulose diphosphate carboxylase. New Phytol. 72:209–235.

Walker, D. A. 1974. Chloroplast and cell—the movement of certain key substrates, etc. across the chloroplast envelope. p. 1–49. *In* D. H. Northcote (ed.) Plant biochemistry, Vol. 11. Biochemistry Series 1, MTC international review science. Butterworth (Publ.) Inc., Woburn, Mass.

Walker, D. A., A. V. McCormick, and D. M. Stokes. 1971. $CO_2$-dependent oxygen evolution by envelope-free chloroplasts. Nature (London) 233:346–347.

Walker, M. E. 1975. Calcium requirements for peanuts. Commun. Soil Sci. Plant Anal. 6: 299–313.

Walker, M. E., and T. C. Keisling. 1978. Response of five peanut cultivars to gypsum fertilization on soils varying in calcium content. Peanut Sci. 5:57–60.

Walker, R. B. 1954. The ecology of serpentine soils: II. Factors effecting plant growth on serpentine soils. Ecology 35:259–266.

Walker, R. B., W. H. Walker, and P. R. Ashworth. 1955. Calcium-magnesium nutrition with special reference to serpentine soils. Plant Physiol. 30:214–221.

Walker, W. M., and D. W. Graffis. 1979. Grass species and cultivar differences in magnesium concentration. Commun. Soil Sci. Plant Anal. 10:1239–1248.

Wallace, A. 1963. Solute uptake by intact plants. Edwards Bros., Inc., Ann Arbor, Mich.

Wallace, A. (ed.) 1966. Current topics in plant nutrition. Edwards Bros., Inc., Ann Arbor, Mich.

Wallace, A. 1971a. Effect of root temperature on mineral content of *Salsola kali* shoots. p. 202–206. *In* A. Wallace (ed.) Regulation of the micronutrient status of plants by chelation agents and other factors. Edwards Bros., Inc., Ann Arbor, Mich.

Wallace, A. 1971b. Low root temperature, calcium, and nitrate ion interactions on nonexchangeable rubidium, cesium, and sodium absorption by bush beans. Plant Soil 34:121–131.

Wallace, A. (ed.) 1972. Regulation of the micronutrient status of plants by chelating agents and other factors. Edwards Bros., Inc., Ann Arbor, Mich.

Wallace, A. (ed.) 1980. Trace element excesses in plants. J. Plant Nutr., Vol. 2, Issues 1 and 2.

Wallace, A., E. Frolich, and O. R. Lunt. 1966. Calcium requirements of higher plants. Nature (London) 209:634.

Wallace, A., and N. Hemaidan. 1963. Effect of temperature on the ratio of Ca to K absorbed by bush beans in short-time studies. p. 120–121. *In* A. Wallace (ed.) Solute uptake by intact plants. Edwards Bros., Inc., Ann Arbor, Mich.

Wallace, A., N. Hemaidan, C. Carmack, R. T. Mueller, and G. A. Wallace. 1963. Uptake of $Ca^{47}$ by bush bean plants. p. 142–146. *In* A. Wallace (ed.) Solute uptake by intact plants. Edwards Bros., Inc., Ann Arbor, Mich.

Wallace, A., and S. M. Soufi. 1975. Low and variable critical concentrations of calcium in plant tissues. Commun. Soil Sci. Plant Anal. 6:331–337.

Wallsgrove, R. M., P. J. Lea, and B. J. Miflin. 1979. Distribution of the enzymes of nitrogen assimilation within the pea leaf cell. Plant Physiol. 63:232–236.

Walsh, L. M., and J. D. Beaton (ed.) 1973. Soil testing and plant analysis. Soil Sci. Soc. of Am., Madison, Wis.

Wardlaw, I. 1967. The effect of water stress on translocation in relation to photosynthesis and growth: I. Effect during grain development in wheat. Aust. J. Biol. Sci. 20:25–39.

Wardlaw, I. F., and J. B. Passioura (ed.) 1976. Transport and transfer processes in plants. Academic Press, Inc., New York.

Watson, J. D. 1965. Molecular biology of the gene. W. A. Benjamin, Inc., Menlo Park, Calif.

Watterson, D. M., D. B. Iverson, and L. J. Van Eldik. 1980. Spinach calmodulin: Isolation, characterization, and comparison with vertebrate calmodulins. Biochemistry 19:5762–5768.

Wayne, R., and P. K. Hepler. 1982. The role of calcium in phytochrome-mediated fern spore germination. Plant Physiol. 69(Suppl.):25.

Wieneke, J. 1969. Calciumtranslöckation und Atiologie der Strippikeit bein Apfel. Erw. Obstb. 11:225–231.

Wieneke, J. 1974. Translocation of $^{45}$Ca in apple trees: II. Transport to the fruit and distribution. Gartenbauwissenschaft 39:57–67.

Wieneke, J. 1979. Calcium transport and its microautoradiographic localization in the tissue. Commun. Soil Sci. Plant Anal. 10:237–250.

Wieneke, J., O. Biddulph, C. G. Woodbridge. 1971. Influence of growth regulating substances on absorption and translocation of calcium in pea and bean. J. Am. Soc. Hortic. Sci. 96: 721–724.

Wieneke, J., and F. Führ. 1973. Investigations on translocation of $^{45}$Ca in apple trees: I. Transport and distribution in relation to time of uptake. Gartenbauwissenschaft 38:91–108.

Wieneke, J., and F. Führ. 1975. Investigations on translocation of $^{45}$Ca in apples trees: IV. Secondary calcium translocation after dormancy. Gartenbauwissenschaft 40:105–111.

Wier, R. G., and A. Hudson. 1966. Molybdenum deficiency in maize in relation to seed reserves. Aust. J. Exp. Agric. Anim. Husb. 6:35–41.

Wiersum, L. K. 1966. Calcium content of fruits and storage tissue in relation to the mode of water supply. Acta Bot. Neerl. 15:406–418.

Wiersum, L. K. 1969. Soil water content in relation to nutrient uptake by the plant. Versl. Meded. Commun. Hydrol. Onderz TNO 15:74–89.

Wiersum, L. K. 1979. Effects of environment and cultural practices on calcium nutrition. Commun. Soil Sci. Plant Anal. 10:259–278.

Wiersum, L. K., C. A. Vonk, and P. M. L. Tammes. 1971. Movement of $^{45}$Ca in the phloem of *Yucca*. Naturwissenschaften. 58:99.

Wilcox, G. E., J. E. Hoff, and C. M. Jones. 1973. Ammonium reduction of calcium and magnesium content of tomato and sweet corn leaf tissue and influence on incidence of blossom end rot of tomato fruit. J. Am. Soc. Hortic. Sci. 98:86–89.

Wilcox, G. E., C. A. Mitchell, and J. E. Hoff. 1977. Influence of nitrogen form on exudation rate, and ammonium, amide, and cation composition of xylem exudate in tomato. J. Am. Soc. Hortic. Sci. 102:192–196.

Wilkinson, B. G. 1968. Mineral composition of apples: IX. Extra uptake of calcium by the fruit. J. Sci. Food Agric. 19:646–647.

Williams, M. W. 1966. Relationship of sugars and sorbital to watercore in apples. Proc. Am. Soc. Hortic. Sci. 88:67–75.

Williams, R. F., and R. E. Shapter. 1955. A comparative study of growth and nutrition in barley and rye as affected by low-water treatment. Aust. J. Biol. Sci. 8:435–466.

Williams, R. J. P. 1974. Calcium ions: Their ligands and their functions. p. 133–138. *In* R. M. S. Smellie (ed.) Calcium and cell regulation. Biochem. Soc. Symp. no. 39, Biochemistry Soc., London.

Williams, R. J. P. 1976. Calcium chemistry and its relation to biological function. p. 1–17. *In* C. J. Duncan (ed.) Calcium in biological systems. Symp. Soc. Exp. Biology, no. 30. Cambridge Univ. Press, New York.

Williamson, R. W. 1981. Free $Ca^{2+}$ concentration in the cytoplasm: A regulator of plant cell function. What's New Pltn Physiol. 12:45–48.

Wills, R. B. H. 1972. Effect of calcium on production of volatiles by apples. J. Sci. Food Agric. 23:1131–1134.

Wills, R. B. H., K. J. Scott, E. T. Carroll. 1975. Use of alkaline earth metals to reduce the incidence of storage disorders of apples. Aust. J. Agric. Res. 26:169–171.

Wills, R. B. H., K. J. Scott, and W. B. McGlasson. 1970. Role for acetate in the development of low-temperature breakdown in apples. J. Sci. Food Agric. 21:42–44.

Wilson, R. D. 1949. A field response of rock melons to molybdenum. J. Aust. Inst. Agric. Sci. 15:118–121.

Wunderlich, F. 1978. The nucleus matrix: Dynamic protein-structure in cell nucleus. Naturwiss. Rundsch. 31:282–288.

Wyn Jones, R. G. and O. R. Lunt. 1967. The funtion of calcium in plants. Bot. Rev. 33:407–426.

Yamaya, T., and H. Matsumato. 1981. Properties of alkaline phosphatase in cucumber roots induced by calcium starvation. Plant Cell Physiol. 22:1355–1365.

Young, W. I., and D. C. Rasmusson. 1966. Variety differences in strontium and calcium accumulation in seedlings of barley. Agron. J. 58:481–483.

Younge, O. R., and M. Takahashi. 1953. Response of alfalfa to molybdenum in Hawaii. Agron. J. 45:420–428.

Ziegler, H. 1975. Nature of transported substances. p. 59–100. *In* M. H. Zimmerman and J. A. Milburn (ed.) Transport in plants I. Phloem transport. Encyclopedia of plant physiology, New Series, Vol. 1. Springer-Verlag New York, New York.

Zimmerman, H. M. 1960. Transport in the phloem. Annu. Rev. Plant Physiol. 11:167–190.

# 4 Liming Materials and Practices

**STANLEY A. BARBER**

*Purdue University*
*West Lafayette, Indiana*

Liming soil is an ancient agricultural practice. Cato and Varro (about 200 B.C.) and Dickson (1788) referred to the use of lime in the first and second centuries B.C. The Romans tested for soil acidity by tasting fresh water percolated through a basket of soil. The early settlers in England used marl. Use of lime in the United States was promoted by the writings of Ruffin (1821, 1852) in the *American Farmer* and in his books on liming in which he reported that marl applications improved crop yields on his farms (Ihde, 1952). Agricultural experiment stations in several states began research on lime between 1880 and 1902 (Linsley, 1954; Patterson, 1906; Latta, 1885). They used either burned lime, gas lime, or marl since crushed agricultural limestone was not readily available.

An *agricultural liming material* is defined as a material whose Ca and Mg compounds are capable of neutralizing soil acidity. These materials include quicklime, hydrated lime, limestone (both calcitic and dolomitic), marl, shells, and byproducts such as slag.

Limestone is the main liming material used. It may be calcite ($CaCO_3$), dolomite ($CaCO_3 \cdot MgCO_3$), or a mixture of these two minerals. Pure dolomite has equal molecular ratios of $CaCO_3$ and $MgCO_3$, and on a weight basis, it consists of 54.3% $CaCO_3$ and 45.7% $MgCO_3$, or 21.7% Ca and 13.1% Mg, since the molecular weight of Mg is less than that of Ca.

The total use of agricultural limestone in the United States is shown in Figure 1. Between 1935 and 1950, the use of agricultural limestone increased almost 10-fold. In the 1930s, federal programs developed to conserve the soil included financial assistance for the purchase of lime. This undoubtedly stimulated its use. The use of limestone decreased after 1950, reaching a low in 1954. Use between 1954 and 1980 increased about 75%, although there was a significant, albeit temporary, decline in use in the early 1970s. The amount of lime used is affected by changes in cropping programs and also by the way in which federal subsidy funds are used. Agronomists estimate that farmers of the United States could profitably use 80 million t of lime $yr^{-1}$. This is three times the amount used in 1981.

The use of hydrated lime, burnt lime, marl, and miscellaneous liming materials has decreased in recent years. The total of these materials, al-


*Soil Acidity and Liming*—Agronomy Monograph no. 12 (2nd Edition).

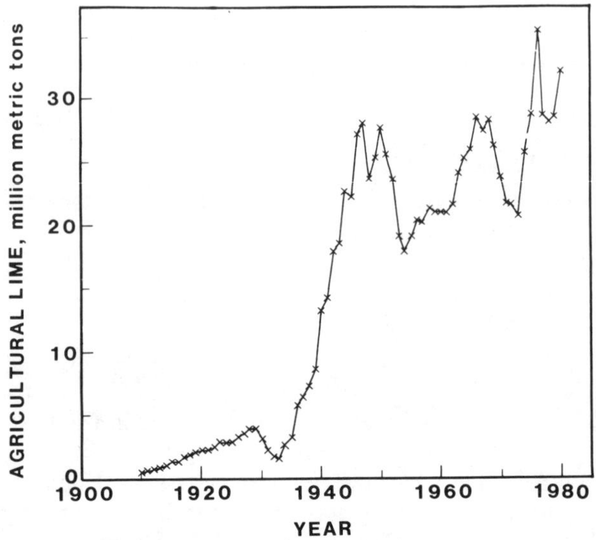


Fig. 1—Annual consumption of agricultural limestone in the United States (Nal. Limestone Inst., Fairfax, Va.).

though important in local areas, is not over 0.5 million t $yr^{-1}$ in the United States. Few recent statistics are available on their use.

Agricultural limestone is applied to reduce the levels of $H^+$ and soluble $Al^{3+}$ in the soil and to make a more favorable soil environment for root growth.

There are several ideas about liming soils. One is that all of the limestone should react with the soil as rapidly as possible to replace the exchangeable acidity. Since $Ca^{2+}$ and $Mg^{2+}$ are exchngeable cations, they are relatively resistant to loss by leaching from soils. Thus, little is gained by delaying the rate at which the lime reacts with the soil.

Where soils are low in cation exchange capacity (CEC), it is important to know how fast the limestone reacts so that the desired soil pH level will be reached and maintained. On soils very low in CEC, some slow release may extend the time the limestone will last before reliming is necessary.

Another belief is that the use of larger particles of limestone in a soil will result in the neutralization of islands of soil about the particles, and these will supply the $Ca^{2+}$ and $Mg^{2+}$ needed (Albrecht, 1946; Linsley, 1954). Proponents of this idea believe that acid soil between limestone particles may help keep micronutrients in greater supply. These acid areas may also supply excess $Al^{3+}$ and $Fe^{3+}$, which can be deleterious to plant growth. However, this hypothesis has not been supported by experimental evidence.

Calcium and $Mg^{2+}$ availability to the plant depends on both the level of exchangeable $Ca^{2+}$ and $Mg^{2+}$ in the soil and the level of $Ca^{2+}$ and $Mg^{2+}$ in the soil water that transports these ions to plant roots by mass flow (Al-Abbas & Barber, 1964). When we consider a given soil with a constant anion concentration in the soil solution, the availability of these nutrients increases as more $Ca^{2+}$ and $Mg^{2+}$ moves from the limestone particle onto exchange sites and into solution. This reaction of limestone with the acid soil causes both an increase in available $Ca^{2+}$ and $Mg^{2+}$ and a decrease in the amounts of $Fe^{3+}$ and $Al^{3+}$ that reach the plant root. It is generally desirable to use lime that reacts almost completely with the soil within a 2- or 3-yr period. Since the rate of reaction is related to fineness and pulverization adds to the cost of the product, the fineness of the limestone must necessarily represent a balance between the cost of manufacture and the increased benefits of fine pulverization.

## I. LIMING MATERIALS

### A. Agricultural Limestone

Agricultural limestone is usually a mixture of calcite and dolomite that has been crushed so that it is sufficiently fine to react readily with the soil. The rock from which the product is obtained may contain silica and other impurities that have no effect on soil acidity. These impurities merely dilute the limestone and make it less effective as an agricultural liming material.

#### 1. MEASUREMENT OF QUALITY

**a. Calcium and Mg Contents**—Pure calcite ($CaCO_3$) contains 400 g $kg^{-1}$ (40.0%) of Ca. Dolomite [$CaMg(CO_3)_2$] contains 217 g $kg^{-1}$ (21.7%) of Ca and 131 g $kg^{-1}$ (13.1%) of Mg. The dividing point between dolomitic limestone and calcitic limestone is not clearly defined in the agricultural liming trade. Usually, calcitic limestones refer to those containing very little Mg, whereas dolomitic limestones may vary from pure dolomite to limestones with only a few grams per kilogram of Mg. Pettyjohn (1957) has suggested a classification scheme as follows:

| Material | Mg content | |
|---|---|---|
| | g $kg^{-1}$ | (%) |
| Dolomite | 117–131 | (11.7–13.1) |
| Calcitic dolomite | 65–117 | (6.5–11.7) |
| Dolomitic limestone | 13–65 | (1.3–6.5) |
| Limestone, Mg | 6–13 | (0.6–1.3) |
| Limestone, high Ca | 0–6 | (0–0.6) |

The chemical effectiveness of an agricultural limestone is measured by its $CaCO_3$ equivalence. This is defined as the acid-neutralizing capacity of the material by weight in relation to $CaCO_3$. A dolomitic limestone has a higher neutralizing capacity than a calcitic limestone because of the lower atomic weight of Mg. Pure dolomite has a $CaCO_3$ equivalence of 1.08.

Standard procedures have been developed for the determination of $CaCO_3$ equivalence by the Association of Official Analytical Chemists. The most recent revisions can be obtained by referring to the *Journal of the*

*Association of Official Analytical Chemists.* Presently, a known amount of limestone is dissolved in a known amount ofacid, and the excess acid is back-titrated with a base. The results are calculated to give a percent $CaCO_3$ equivalence.

**b. Particle Size**—Agricultural limestone produced by regular crushing procedures has a range of particle sizes. The upper size limit has usually been determined by the screen used in the crushing plant. Crushed limestone is usually fine enough for most of it to pass a U.S. Standard no. 8 sieve. This is a sieve that has eight wires per inch and openings of 2.36 mm on each side. The U.S. Standard, ASTM-specified sieves, and their sieve openings are as follows:

| U.S. sieve designation, no. | Opening size, mm | U.S. sieve designation, no. | Opening size, mm |
|---|---|---|---|
| 4 | 4.75 | 40 | 0.425 |
| 8 | 2.36 | 60 | 0.25 |
| 10 | 2.00 | 80 | 0.18 |
| 20 | 0.85 | 100 | 0.15 |
| 30 | 0.60 | 200 | 0.075 |

The terms *fine limestone* and *coarse limestone* are qualitative terms that have little meaning unless the amounts passing various sieves are given. Some confusion has also resulted in the method of reporting the fineness of limestone according to sieve size. For example, limestones labeled 10-mesh size can be interpreted to mean that all particles are of a size that just passes through a 10-mesh sieve. However, it usually means that all limestone particles pass through a 10-mesh sieve, including those that are much finer than 10-mesh, and that the 10-mesh sieve has been used to control the upper limit of particle size. This alone is not a very informative measure of particle size. The best description of limestone particle size can be obtained by sieving through a series of sieves so that the amount of each particle-size fraction is known.

Limestone used for agricultural liming should be fine enough for at least 80% to pass through a U.S. Standard no. 8 sieve. Some limestones are pulverized so that all particles pass through a no. 100 sieve. Whittaker and Chichilo (1964) analyzed 193 limestone samples obtained from 35 states and found that most of the limestones were crushed so that 20 to 55% passed through a no. 100 sieve.

Agricultural limestone is best evaluated by determining the amount of various particle-size fractions, e.g., $>$ 8, 8 to 20, 20 to 60, 60 to 100, and $<$ 100, where the numbers refer to the mesh size of the sieves. However, measuring each particle-size fraction is very inconvenient and has led to efforts to gain similar information from single measurements. A fairly successful approach to this has taken advantage of the fact that a plot of the cumulative percentage of limestone passing through each sieve vs. the sieve size gives an S-shaped curve. In addition, the cumulative percentage of limestone passing through each sieve plotted on a probability scale against the

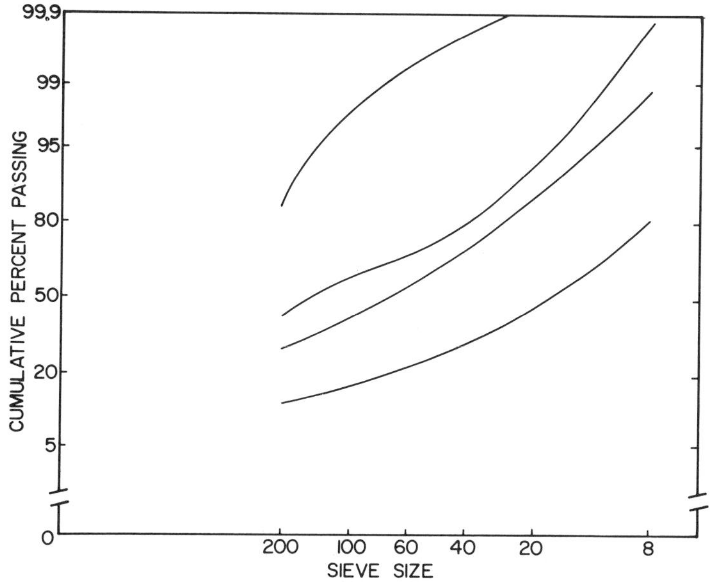


Fig. 2—Log normal plot of several limestones showing the distribution of various size fractions.

sieve opening on a log scale gives a relationship that approaches linearity. Schollenberger and Salter (1943) used this to evaluate limestones and called it a *log normal plot.* Examples of log normal plots for several limestones are shown in Figure 2. Some samples are atypical and are curvilinear toward the extremes. Variations from linearity may occur at the ends of the curve; however, since this is plotted on a probability scale, the variation from linearity of values at the ends of the curves may reflect a difference of only 1 or 2%. The log normal plot has been found most useful for characterizing limestones because the lines approach linearity, and the slope of the line does not change appreciably when a coarse limestone is compared with a finer one. The log normal plot is also useful for estimating the amount passing through a sieve that is different from those used for characterization.

Schollenberger and Salter (1943) used the log normal plot for describing limestones to develop a chart for limestone evaluation. Schollenberger and Whittaker (1962) showed that these plots sometimes departed significantly from linearity. However, the log normal plot usually approaches linearity when limestones are prepared by the normal crushing processes with no fine particles removed and a truly representative sample obtained.

Particle-size distribution of agricultural limestone may be determined by either dry or wet sieving. Almost all specifications for limestone in use in the United States in 1964 were based on dry sieving. Amounts passing through the various size sieves obtained by wet sieving may be larger than those obtained by dry sieving. Some limestones form aggregates when wet. These aggregates are retained after drying and are measured as coarser particles. Gehrke and Johnson (1963) found that the addition of clay caused aggregates to form. P. Chichilo and C. W. Whittaker (1958, personal communication) compared wet with dry sieving and found that the percent of particles < 60- to 100-mesh size was appreciably greater with wet sieving. Wiese (1956) and Elphick (1954) also obtained a higher percentage of fine particles by wet sieving.

G. P. Risk at Purdue University (1958, unpublished data) found that rolling dry limestone between a hard rubber roller and a rubber mat disintegrated the aggregates, resulting in little difference between wet and dry sieving.

The greatest differences between wet and dry sieving occur for particles passing through sieves finer than 80-mesh size, whereas there is generally little difference in the amounts passing through sieves coarser than 80-mesh size. It appears that if an accurate measure of the finest particle size is desired, wet sieving is preferable. However, since an aggregate occupies only one locus in the soil, the values obtained by wet sieving may give the limestone too high a rating. Wet sieving may also be preferable if the aggregates are artifacts resulting from the sampling procedure.

The most common method of measuring fineness of agricultural limestone is to determine the percentage that passes through a specific size sieve or set of sieves. State lime laws use the percentage passing through one or more sieves as the basis of measuring grades according to fineness. Calcium carbonate equivalence is also measured, and sometimes the product of $CaCO_3$ equivlence and the percentage passing through a specific sieve are used to describe the minimum limit for a particular grade of limestone.

There has been little research to determine the number or size of the sieves that should be used to adequately characterize an agricultural limestone. Barber (1957) investigated the relationship between the percentage passing through various sieves and the quality of the limestone. The quality parameter was determined by multiplying the amount of each particle-size fraction by a factor related to the diameter of the particle and then adding the values as described by Bear and Allen (1932). Hence, this gives a single parameter that accurately described the particle-size distribution of a limestone in terms of rate of dissolution. Using 99 samples collected from Indiana limestone quarries, Barber found that the correlation between the 3-month value for limestone decomposition and the amount passing through a 40-mesh sieve gave an $r^2$ value of 0.99. Correlation between decomposition in soil and the amounts passing through 8-, 20-, 60-, 80-, 100-, and 200-mesh sieves gave $r^2$ values of 0.47, 0.93, 0.96, 0.94, 0.92, and 0.73, respectively. The sieves near the midrange, the no. 40 or 60 sieves, gave the highest correlation values.

Results of similar calculations for other limestone samples reported in the literature are shown in Table 1. In the regression equations of Table 1, $Y$ refers to the quality of limestone as measured by equal diameter reduction, and $X$ refers to the percentage of the limestone sample that passed through a no. 60 sieve. A high correlation was obtained in all cases with the regression equations being similar. These data indicate that the use of just the no.

Table 1—Correlation between the percentage of limestone passing through a U.S. Standard no. 60 sieve and the quality of the limestone.

| Origin of samples | No. of samples | $r^2$ | Regression equation | Reference |
|---|---|---|---|---|
| Indiana (1955) | 99 | 0.96 | $Y = 0.77X + 26.7$ | Barber (1957) |
| Indiana (1958) | 75 | 0.91 | $Y = 0.86X + 23.4$ | S. A. Barber (1958, unpubl. data, Purdue Univ.) |
| Tennessee | 20 | 0.97 | $Y = 0.77X + 27.8$ | Shaw & Robinson (1959) |
| Maryland | 15 | 0.96 | $Y = 0.80X + 26.0$ | Thomas & Gross (1952) |
| Kentucky | 29 | 0.94 | $Y = 0.87 + 24.6$ | J. A. Harrison (1958, unpubl. data, Univ. of Kentucky) |
| South Carolina | 6 | 0.99 | $Y = 0.85X + 23.1$ | Plant (1960) |

60 sieve adequately characterizes the fineness characteristics of most limestones. The use of an additional sieve, possibly a no. 8 or 20, would prevent atypical samples from being incorrectly measured.

Whittaker and Chichilo (1964) used a somewhat different approach. They investigated samples from 193 limestone quarries in 35 states. They fractionated the limestone with sieves ranging from no. 8 to 270 and plotted the data on log normal graph paper. They found the graphs linear in the central portion (sieves no. 30 to 140) but departing from linearity frequently at the lower and upper ends. These departures were more than random variations. For each sample, Whittaker and Chichilo determined the median particle size, i.e., the sieve size that passed through 50% of the sample. They considered median particle size the most important single statistic of particle-size distribution because it indicates an average particle size. They proposed that characterization of agricultural limestone be made by using two sieves—one that passes 95% of the sample and one that passes through 50% of the sample. The average of the median particle sizes from the limestone samples analyzed by Whittaker and Chichilo ranged from 0.166 mm for the Middle Atlantic region to 0.450 mm for the west north-central region of the United States.

**c. Chemical Methods for the Determination of Limestone Quality—** The rate of reaction of a limestone with the soil depends on both its fineness and its chemical and physical composition. Between 1947 and 1965, considerable research effort was expended in an attempt to get a chemical method that gives a single numerical rating of limestone quality. Ideally, this rating would combine the effects of the different limestone properties and would be correlated with the rate at which the limestone reacts with the soil. Reviews of research on this problem have been presented by Schollenberger and Whittaker (1953, 1956) and Shaw and Robinson (1959).

Balks and Wehrmann (1938) in Germany investigated the rate of reaction between limestone of various particle sizes and sodium oxalate. A procedure for measuring the reaction between oxalate and the surface of agricultural limestone was first proposed in the United States by Barnes (1947). Since oxalate reacts to form calcium oxalate on the limestone-particle surface, the amount of oxalate used is a measure of Ca on the surface and, hence, of limestone reactivity. Barnes' procedure was investigated by

Thomas and Gross (1952) and Webster et al. (1953), who found that the oxalate procedure sometimes gave excessively high values for dolomitic limestones relative to values for calcitic limestones. Modifications of the procedure have been proposed by Elphick (1954) and by Schollenberger and Whittaker (1956).

Dilute acetic acid has also been used to estimate limestone reactivity. Beacher and Markle (1949) obtained a fair correlation between this measure of reactivity and the change in soil pH after 10 weeks of incubation of limestone added to an acid soil.

An extensive study of chemical reagents for rating limestones was reported by Thomas and Gross (1952). They evaluated oxalic, hydrofluoric, phosphoric, acetic, phthalic, stearic, oleic, and formic acids as well as potassium hydrogen phthalate, sodium fluoride, and potassium and sodium oxalates. However, none of these was equally effective for both calcitic and dolomitic limestones.

Schollenberger and Whittaker (1953) determined reactivity of agricultural limestone by comparing the solubility of limestone in an $NH_4Cl$ solution with the rate of reaction as determined by the equal-diameter reduction concept described later in this chapter. Shaw and Robinson (1959), in an extensive study of the $NH_4Cl$ procedure, found that the coarser particles dissolved somewhat more rapidly than would be expected on the basis of equal-diameter reduction. They also found that the evlauation of calcitic and dolomitic limestone by the $NH_4Cl$ procedure was more in accord with the reaction of limestone with acid soils than was evaluation by equal-diameter reduction.

The use of ethylenediaminetetraacetic acid (EDTA) to evaluate reactivity of limestone with soil was proposed by El Gibably and Axley (1955). They obtained a correlation of $r^2 = 0.62$ between EDTA ratings and changes in soil pH produced in 24 d by nine commercial limestones applied to field plots. A laboratory study gave a correlation of $r^2 = 0.92$ between the EDTA rating and the change in pH caused by mixing the limestone in a soil-water suspension. This procedure created interest in the Northeast, and a collaborative study among several states was initiated to evaluate the procedure. Hunter et al. (1963) reported that the correlation of EDTA ratings of 39 limestones with changes in soil pH brought about by shaking the limestone with a soil-water suspension gave an $r^2$ of 0.72. In this same collaborative study, Whittaker et al. (1963) reported that the trisodium salt of EDTA was superior to the disodium salt. However, neither salt accounted for all limestone reactivity. The percent of $CaCO_3$ in the limestone was as highly correlated with limestone reactivity as was the reaction between EDTA and limestone.

Schollenberger and Whittaker (1962) compared reactivity values obtained by the $NH_4Cl$, oxalate, and EDTA methods with values computed from (i) particle-size distribution, (ii) change in soil pH, and (iii) reaction of the limestone with acid soil. Using 24 limestone samples, they found that none of the methods gave a satisfactory correlation with any of the latter three standards of reference; neither did these standards correlate well with one another, indicating the need for a better standard of reference. Some of the methods correlated highly with the quality ratings for either calcitic or dolomitic limestones separately, but seldom was there a high correlation when both types of limestone were included in the same regression analysis.

Reactivity of limestone should be related to its exposed surface area. Love and Whittaker (1954), using adsorption of Kr, found that measured surface area was much greater than that calculated from a geometric surface, indicating considerable internal surface. In addition, particles of agricultural limestone of the same size but from different sources varied widely in measured surface areas.

Limestone reactivity is also affected by its contents of calcite and dolomite. Skinner et al. (1959) developed a simple, convenient method for determining the amounts of calcite and dolomite in a limestone as well as its $CaCO_3$ equivalence. They determined the amounts of calcite and dolomite from the measurement of their relative rates of reaction with HCl. Calcite reacts in 30 to 90 s, whereas dolomite reacts over a period of several minutes. Since the logarithm of the remaining amount of unreacted dolomite in excess HCl gives an approximate linear relationship with time, extrapolation of this line to time zero gives a measure of the original dolomite content. Results from this method have been shown by Turner (1960), Turner and Skinner (1960), and Brydon and Rice (1965) to agree with results obtained by regular chemical analysis.

None of the methods discussed above has provided a satisfactory procedure for obtaining a single numerical rating for limestone effectiveness. The total capacity of the limestone to react is related to its $CaCO_3$ equivalence, whereas the rate at which limestone reacts is related to particle size, chemical composition, and the physical nature of the limestone particle. The relative quantities of calcite and dolomite in the limestone particle and the size of the particles are the two most important factors affecting reaction rate. Although some procedures work well for evaluating either calcitic or dolomitic limestones singly, they seldom give acceptable ratings when both types are combined. As of 1982, the preferred method for rating agricultural limestone in the United States is still the determination of $CaCO_3$ equivalence plus a measure of fineness.

**d. State Specifications of Liming Materials**—Most states in the United States either have lime laws that regulate the quality of agricultural limestone sold within the state or have minimum specifications that liming materials must meet to qualify for economic assistance in federally sponsored conservation programs. Specifications vary from state to state, and the number and size of sieves used for characterizing limestone also vary.

An agricultural lime subcommittee of the American Society for Testing and Materials has developed specifications for uniform characterization of liming materials. These specifications were published as ASTM Designation C 602-69 (ASTM, 1969). Use of no. 8 and 60 sieves were recommended for measuring limestone fineness. Officials of the Association of American Plant Food Control Officials (1982, p. 34) prepared a Model Agricultural

Liming Materials Bill and have published it in their proceedings as an aid in standardizing the lime laws of the various states. They specify quality by the minimum amounts passing through various sieves, by the CaO and MgO or $CaCO_3$ and $MgCO_3$ contents, and by the neutralizing power expressed as weight percent of $CaCO_3$.

**e. Moisture Content**—Agricultural limestone as sold from the quarry may contain up to 10% moisture. Since the product is sold by weight, the percent of moisture is an important consideration. S. A. Barber (1960, unpublished data) found the moisture content of limestone samples collected from 50 Indiana quarries from February through May varied from 0.1 to 10.4%, with an average of 4.5%. A similar set of 50 samples collected from July through September gave a range in moisture content of 0.0 to 8.7%, with an average of 3.6%.

**f. Sampling Procedure**—Agricultural limestone offered for sale may be in large stockpiles, bins, trucks, or railroad cars. A representative sample must be obtained to get a true estimate of the limestone quality. The representative sample is important for the determination of $CaCO_3$ equivalence, particle-size distribution, and moisture content. Since more than one ledge of limestone may occur in a single quarry, the chemical composition can vary. Furthermore, fine particles within a pile may have different $CaCO_3$ content than coarse particles.

When agricultural limestone is stored in stockpiles, segregation of particles according to size will occur. Although I found no published research on agricultural limestone sampling, samplers generally recognize that samples should not be taken from near the base of the pile (coarse particles tend to roll to this position) or from near the top of the pile (particles may be finer than average).

To sample a stockpile, bin, or truck, a probe should be used. A probe 39 cm long and 2.5 cm in diam with a solid point and a flared opening on one side has been found suitable for sampling in Indiana. The probe is inserted horizontally with the opening at the bottom, given a one-half turn, and removed. Probes from 10 to 20 locations are mixed by riffling and recombining before subsampling to obtain the final sample.

**g. Calculation of Effectiveness**—A single number evaluation of limestone has been calculated by some (Murphy & Follett, 1978) to facilitate varying rate of application according to fineness and $CaCO_3$ equivalent. One method, where 8- and 60-mesh sieves were used, multiplies the percent on the 8-mesh sieve by zero, the percent passing through an 8-mesh sieve and held on a 60-mesh sieve by 0.50, and the percent passing through a 60-mesh sieve by 1.00. The three values are added to give the fineness factor. This factor is multiplied by the $CaCO_3$ equivalent to obtain an effective $CaCO_3$ content of the limestone. Where moisture content is appreciable, the limestone is corrected for moisture content. Recommended rates of lime may then be given according to effective $CaCO_3$, and rates of particular agricultural liming materials may be adjusted accordingly.

## 2. QUALITY OF AGRICULTURAL LIMESTONE AS A FACTOR IN THE RESPONSE OF CROPS TO LIMING

A large number of papers on limestone quality as related to crop response have appeared in the literature since 1900. Most of these papers were published between 1920 and 1960. Few investigations of this nature have been reported since 1960, presumably because the effect of limestone particle size on rate of dissolution and hence effectiveness as a liming material had been well characterized by 1960, and additional research was unnecessary. The relationship between particle size and quality has been studied empirically by many investigators, and some have developed theories for limestone dissolution in the soil. The results of many of these investigations have been reviewed by Frear (1921), MacIntyre and Shaw (1925), Pierre (1930), Lyon (1931), Walker et al. (1932), Fippin (1939), and Elphick (1954, 1955a, b).

The research on evaluation of agricultural limestone can be divided into (i) those experiments concerned with the rate of dissolution of limestone particles by acid soils and (ii) those concerned with the effect of various particle sizes on plant growth. The dissolution of limestone particles by acid soils has been studied both by measuring the amount of limestone remaining in the soil and by studying the effect of limestone on soil pH and on exchangeable Ca and Mg.

**a. Particle Size and Reaction Rate of Agricultural Limestone with Soil**—Upon incorporation into an acid soil, limestone particles immediately begin to dissolve. The dissolved $CaCO_3$ (and $MgCO_3$ in the case of dolomite) react with $H^+$ ions to give Ca soil, $Ca(HCO_3)_2$, and $H_2CO_3$. The $H_2CO_3$ equilibrates with $H_2O$ and $CO_2$, and the excess $CO_2$ is lost by volatilization. Hence, the reaction approaches completion, and exchangeable $H^+$ and $Al^{3+}$ are replaced by $Ca^{2+}$. Neutralization of the $H^+$ ions next to the limestone-particle surfaces causes an increase in Ca concentration at this point; thus, a concentration gradient is established that causes diffusion of Ca away from the surface. The $Ca^{2+}$ will either be accompanied by anions, e.g., $HCO_3^-$, or a counter diffusion of $H^+$ and $Al^{3+}$ will occur toward the particle to maintain the balance of charges. Dissolution of limestone particles is affected both by the rate of solution of the particle, which is highly sensitive to pH, and by the rate of diffusion of $Ca^{2+}$ away from the particle surface.

The rate at which the soil mass is affected by a lime application depends on the distance the diffusing Ca must move before the zones of neutralization around individual limestone particles overlap. The rate of Ca diffusion in the soil is very slow; hence, finely ground limestone uniformly mixed with the soil provides more particles per unit volume of soil, less distance between particles, and, therefore, more rapid neutralization of the soil.

A reasonable diffusion coefficient, $D_e$, for Ca in a silt loam soil is $7 \times 10^{-9}$ cm$^2$ s$^{-1}$. The mean diffusion distance of an ion is $(2D_e t)^{1/2}$, where $t$ is time in seconds. Using this relation, Ca would diffuse 0.35 cm in 100 d and 0.78 cm in 500 d. If all particles were 0.60 mm in diam (30-mesh size) and had a specific gravity of 2.54, there would be 1820 particles $g^{-1}$. If limestone were applied at a rate of 6 t $ha^{-1}$ or 3 g $kg^{-1}$ of soil, the average distance between particles (assuming a soil with a specific gravity of 1.25) would be 1.04 cm, or a half distance of 0.52 cm. This illustrates why limestone particles need to be fine to react with the soil. This calculation assumes solution of the particle does not limit reaction of the limestone particle with the soil.

Little research has been done on Ca diffusion from limestone in soils. The maximum distance that Ca diffused into an acid soil from a large piece of limestone was found by DeTurk (1938) to be 0.6 cm after 260 d and 1.7 cm after 528 d. These values are for maximum rather than mean diffusion distances, and they compare favorably with the calculations made above with estimated values. In addition to diffusion, Ca may be moved away from limestone particles through mass transport by water. In most acid soils of low salt content, however, the amount of Ca in solution is a very small percentage of the total exchangeable $Ca^{2+}$ present. Barber et al. (1963) found the Ca content of this water was frequently about 30 mg $L^{-1}$. If such a soil contained 2000 mg $kg^{-1}$ of exchangeable $Ca^{2+}$ and a moisture content of 33%, 30 mg $L^{-1}$ of Ca in solution would be only 0.5% of the Ca in exchangeable and soluble forms. Also, there is experimental evidence (Longnecker & Sprague, 1940) that applied Ca does not move as rapidly in soil as it would if movement were primarily by mass flow.

A relationship between particle size and rate of limestone dissolution, based on the hypothesis of equal-diameter reduction, has been investigated by Kriege (1929), Bear and Allen (1932), Salter and Schollenberger (1940), Schollenberger and Salter (1943), and Schollenberger and Whittaker (1962). This hypothesis states that the rate of reduction in diameter of limestone particles is the same for all particle sizes; hence, if particle A has a diameter twice that of particle B, it will take particle A twice as long to completely dissolve.

Swartzendruber and Barber (1965) investigated the concept of equal-diameter reduction and presented a complete mathematical development as well as an examination of the experimental validity of the hypothesis. Derivation of the dissolution equation from the equal-diameter reduction hypothesis was based on the assumptions that (i) the initial mass of limestone was present as spheres of uniform size, density, and composition; and (ii) the rate of loss of mass was directly proportional to the instantaneous surface area of the spheres. The mathematical formulation of the second assumption is

$$dm/dt = kS \qquad [1]$$

where $m$ is the mass of limestone that has dissolved from the surface after time $t$ that the limestone spheres have been uniformly mixed with the soil; $k$ is a proportionality constant; and $S$ is the total area of the spheres. From Eq. [1], they showed that

$$u = 1 - (1 - ct)^3 \qquad [2]$$

where $u$ is the fractional mass dissolved and is equal to $M/M_o$, $M_o$ is the initial mass; $M$ is the dissolved mass; and $c$ is equal to $2k/pD_o$ where $p$ is the density, and $D_o$ is the initial diameter of the spherical particle.

These equations are based on spherical particles, but limestone particles are not spherical. However, since differences due to particle shape are likely to be similar for various sizes of limestone, the errors should tend to cancel each other when relative comparisons are made among sizes. Swartzendruber and Barber (1965) further investigated the validity of Eq. [2] from experimental data of Elphick (1955a) by first rearranging equation [2] to give

$$(1 - u)^{1/3} = 1 - ct. \qquad [3]$$

A plot of $(1 - u)^{1/3}$ vs. $t$ should give a straight line with a slope of $-c$ and an intercept of 1. Figure 3 shows the type of fit obtained with the experimental data. The dissolution data from two coarser materials fit a straight line reasonably well, whereas data from the finer materials indicate a curved relationship in which the slope of the line decreases with time. Departure from linearity may be due to a progressive change in the concentration of H and Ca in solution at the particles' surface. This would, of course, affect the rate of dissolution.

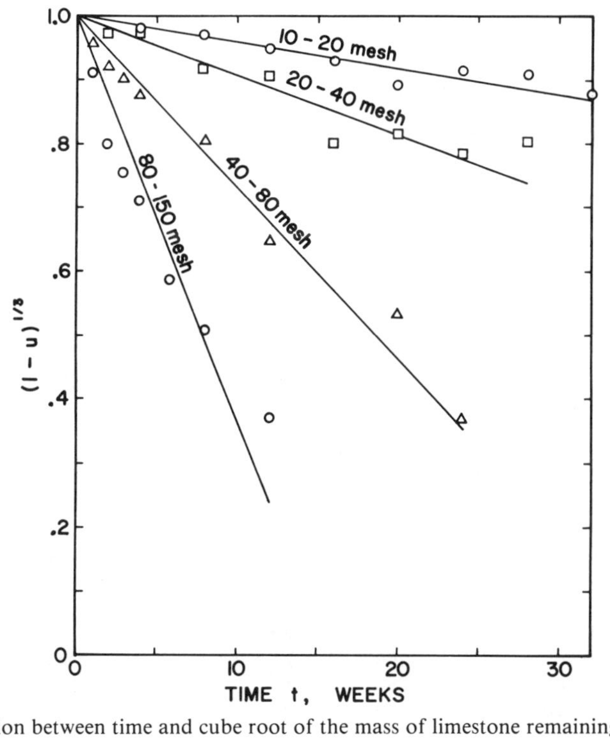


Fig. 3—Relation between time and cube root of the mass of limestone remaining in the soil as affected by particle size.

An alternate derivation can be developed by expressing the undissolved limestone mass as $1 - u$. Swartzendruber and Barber (1965) showed that

$$1 - u = (D_o - a)^3/D_o \quad [4]$$

where $a$ is the diameter reduction after time $t$. This equation is similar to the equation of Elphick (1955a) in which $d$ was diameter reduction.

The rate of disappearance of limestone particles in soil has been measured by many investigators. Usually they have measured carbonate remaining in the soil. Type of limestone, time, particle size, rate of application, and soil have varied with investigators. The equal-diameter reduction hypothesis, presented above, was used in an effort to place the different research results in the literature on a common basis. The rate of diameter reduction, $dD/dt$, will hereafter be referred to as Δ. Values are calculated for several different experiments (Table 2) according to the equation of Elphick (1955a) where

$$\Delta = \frac{D_o\{1 - [(M_o - M)/M_o]^{1/3}\}}{t}. \quad [5]$$

The rate of diameter reduction, shown in Table 2, varied from 0.07 to 12.2 μm week$^{-1}$. Observing that shorter time periods generally gave larger values, I plotted the Δ values of calcitic limestones that fell within the particle-size range of 20- to 60-mesh size against time to give the relationship shown in Figure 4. There is an obvious change in Δ with time. This clearly shows that

Table 2—Rate of solution of limestone particles in soil calculated from data reported in the literature.

| Particle size, mesh | Initial soil pH | Type of limestone† | Time | Soil type | Reduction in particle diameter | Reference |
|---|---|---|---|---|---|---|
| | | | Weeks | | μm week$^{-1}$ | |
| 40–50 | 5.6 | C | 5 | | 12.20 | Morgan & Salter |
| 40–50 | 5.6 | D | 5 | | 5.08 | (1923) |
| 4–10 | 4.9 | C | 180 | | 5.48 | Stewart & Wyatt |
| 4–10 | 4.9 | D | 180 | | 3.04 | (1919) |
| 10–20 | 5.4 | C | 12 | Wakanui | 5.74 | Elphick (1955a) |
| 20–40 | 5.4 | C | 12 | silt loam | 6.37 | |
| 80–150 | 5.4 | C | 12 | | 8.28 | |
| 4–20 | 4.9 | D | 78 | Crosby silt | 1.65 | S. A. Barber (1960, |
| 4–20 | 5.5 | D | 78 | loam (Aeric | 1.45 | unpubl. data, |
| 4–20 | 6.0 | D | 78 | Ochraqualfs; | 1.47 | Purdue Univ.) |
| 4–20 | 6.5 | D | 78 | B horizon) | 0.71 | |
| 4–6 | 4.6 | C | 104 | Mahoning | 3.07 | Bear & Allen (1932) |
| 8–10 | 4.6 | C | 104 | silt loam | 2.42 | |
| 14–20 | 4.6 | C | 104 | (Aeric | 1.17 | |
| 28–35 | 4.6 | C | 104 | Ochraqualfs) | 1.61 | |
| 4–5 | Acid soil | C | 260 | Cisne silt loam | 1.24 | DeTurk (1938) |
| 4–5 | Acid soil | D | 260 | (Mollic | 0.18 | |
| 8–10 | Acid soil | C | 260 | Albaqualfs) | 1.32 | |

(continued on next page)

Table 2—Continued.

| Particle size, mesh | Initial soil pH | Type of lime-stone† | Time | Soil type | Reduction in particle diameter | Reference |
|---|---|---|---|---|---|---|
| | | | Weeks | | μm week⁻¹ | |
| 8–10 | Acid soil | D | 260 | | 1.67 | |
| 8–10 | Limed soil | C | 260 | | 1.12 | |
| 8–10 | Limed soil | D | 260 | | 0.61 | |
| 8–12 | -- | C | 156 | | 0.66 | White & Gardner (1918) |
| 20–40 | -- | C | 156 | | 0.71 | |
| 8–12 | -- | D | 156 | | 0.25 | |
| 20–40 | -- | D | 156 | | 0.50 | |
| 60–80 | -- | D | 156 | | 0.50 | |
| 10–20 | 6.2 | C | 208 | Cumberland clay loam | 0.11 | McIntyre & Shaw (1930) |
| 10–20 | 6.2 | D | 208 | | 0.07 | |
| 20–40 | 6.2 | C | 208 | | 1.32 | |
| 20–40 | 6.2 | D | 208 | | 0.58 | |
| 40–80 | 6.2 | C | 208 | | 0.89 | |
| 40–80 | 6.2 | D | 208 | | 0.41 | |
| 4–8 | 5.0 | C | 78 | Canfield silt loam | 2.16 | Meyer & Volk (1952) |
| 40–50 | 5.0 | C | 78 | | 1.60 | |
| 100+ | 5.0 | C | 78 | | 0.76 | |
| 20–40 | 4.0 | C | 26 | Westmore-land silt loam | 3.56 | Beacher et al. (1952) |
| 60–80 | 4.0 | C | 26 | | 2.36 | |
| 100–200 | 4.0 | C | 26 | | 1.50 | |
| 20–40 | 4.0 | D | 26 | | 1.34 | |
| 60–80 | 4.0 | D | 26 | | 1.09 | |
| 100–200 | 4.0 | D | 26 | | 0.94 | |
| 20 | 5.3 | C | 26 | Grundy silt loam (Aquic Argiudolls) | 3.63 | Walker et al. (1932) |
| 40 | 5.3 | C | 26 | | 4.42 | |
| 100 | 5.3 | C | 26 | | 2.11 | |
| 20–40 | 4.6–5.0 | D | 12 | Avg of 3 different soils | 2.50 | Dawson et al. (1939) |
| 40–60 | 4.6–5.0 | D | 12 | | 2.53 | |
| 60–80 | 4.6–5.0 | D | 12 | | 2.95 | |
| 80–100 | 4.6–5.0 | D | 12 | | 3.03 | |
| 100–200 | 4.6–5.0 | D | 12 | | 2.94 | |
| 20–40 | 5.4 | D | 5–7 | Dunbar very fine sandy loam (Aeric Paleaquults) | 4.76 | Collins & Speer (1939) |
| 40–60 | 5.4 | D | 5–7 | | 2.45 | |
| 60–80 | 5.4 | D | 5–7 | | 5.73 | |
| 80–100 | 5.4 | D | 5–7 | | 4.76 | |
| 100–200 | 5.4 | D | 5–7 | | 4.70 | |
| 3–4 | Neutral | | 65 | Garden soil | 1.40 | Greiner (1950) |
| 6–8 | Neutral | | 65 | | 1.42 | |
| 10–14 | Neutral | | 65 | | 1.64 | |
| 10–12 | 4.7 | C | 52 | Hartsells fine sandy loam (Typic Hapludults) | 6.05 | Shaw (1960) |
| 18–20 | 4.7 | C | 52 | | 4.99 | |
| 35–40 | 4.7 | C | 52 | | 4.75 | |
| 10–12 | 4.7 | D | 52 | | 4.74 | |
| 18–20 | 4.7 | D | 52 | | 2.76 | |
| 35–40 | 4.7 | D | 52 | | 1.21 | |
| 8–20 | -- | C | 100 | | 1.17 | Ames & Schollen-berger (1916) |
| 20–80 | -- | C | 74 | | 1.48 | |
| 8–20 | -- | D | 100 | | 0.20 | |
| 20–80 | -- | D | 74 | | 0.29 | |

† C = calcitic; D = dolomitic.

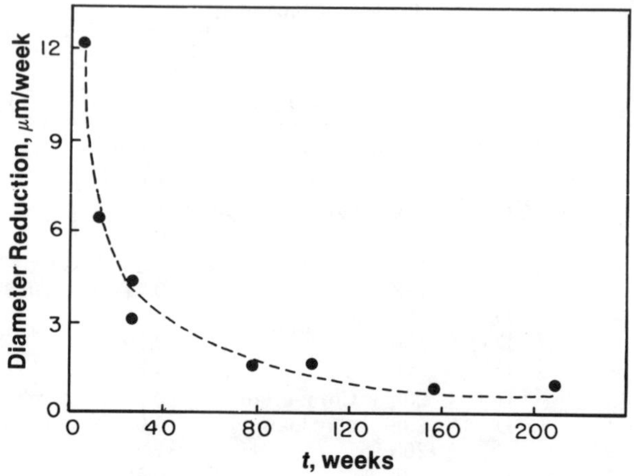


Fig. 4—Relation between rate of limestone-particle diameter reduction and time.

the dissolution rate of limestone particles is not merely a function of particle diameter but that other factors, e.g., diffusion, may also affect the rate of dissolution.

The data in Table 2 also show a variability in the $\Delta$ values with particles of different sizes. There was a tendency for $\Delta$ to increase as particle size decreased when time periods exceeded 150 weeks, whereas $\Delta$ tended to decrease as particle size decreased when time periods were 26 to 78 weeks.

The data in Table 2 further show that the rate of particle disappearance was less for dolomitic limestone than for calcitic. In experiments where the two limestones were compared, the average $\Delta$ was 2.83 $\mu$m week$^{-1}$ for calcitic limestones and 1.39 $\mu$m week$^{-1}$ for dolomitic limestones.

There have been numerous experiments in which the effect of particle size on the reaction rate of lime in the soil was determined by measuring changes in soil pH caused by adding different particle-size fractions of limestone to acid soil. Beacher et al. (1952) determined the relative value of different particle-size fractions by measuring the decrease in H, Al, and Mn and the increase in Ca, Mg, P, and pH. Their results are shown in the following table:

| Particle-size fraction, mesh | Relative effectiveness | |
|---|---|---|
| | Calcitic | Dolomitic |
| 20–60 | 19 | 11 |
| 60–110 | 50 | 25 |
| 100–200 | 85 | 39 |
| 200+ | 99 | 60 |

Motto and Melstead (1960) studied the effect of particle size of limestone on the change in pH of three acid soils and found that the 10- to 28-mesh fraction was 14% as effective as the < 100-mesh fraction, whereas particles > 10-mesh size essentially had no effect on soil pH.

Table 3—Effect of limestone fineness and type of limestone on soil pH (T. J. Heeg, 1958, personal communication, Guelph Univ., Guelph).

| Limestone fineness† | Type of limestone | Initial soil pH | Soil pH 2 yr after application of limestone | |
|---|---|---|---|---|
| | | | 5.6 t $ha^{-1}$ | 11.2 t $ha^{-1}$ |
| % | | | | |
| 45 | Calcitic | 5.7 | 6.0 | 6.4 |
| 80 | Calcitic | 5.4 | 6.5 | 6.8 |
| 45 | Dolomitic | 5.3 | 5.7 | 6.2 |

† Percent passing through a 60-mesh sieve.

The dissolution rate of limestone is reduced by low temperatures. Shaw (1960) found dissolution slower at 10°C than at 30°C. MacIntyre and Shaw (1930) found slower rates of dissolution in winter than in summer in Tennessee. Rates of dissolution also vary with moisture level. In one study, limestone incubated with soil at moisture levels of 60, 80, and 100% saturation gave lower dissolution rates at the lower moisture levels (Shaw, 1960). Jorgenson found that the rate of reaction of limestone was much faster on an acid peat soil than on an acid sandy loam (Jorgenson, 1957). There was not as much difference between the reaction rates of dolomitic and calcitic limestone with the peat soil as there was with the sandy loam soil.

Many researchers have studied the effect of particle-size fraction on the rate of soil pH change. In general, the pH of acid soils increased more rapidly with the use of finer limestone. Rost and Fieger (1927) compared limestone screenings (71% coarser than 20-mesh size) with < 100-mesh limestone at eight field locations. The finer lime gave larger increases in soil pH in all cases. Similar results are shown in Table 3 for another field experiment in which soil pH was measured 2 yr after application of three different limestones (T. J. Heeg, 1958, personal communication). Hoyert and Axley (1952) compared three limestones applied at two rates on seven field locations. Soils were sampled for pH four times over a 3-yr period. Application rates varied according to lime requirements. At the low rate of lime, which was used at five locations only, increase in soil pH was greater for the finer limestone than for the coarser material.

**b. Particle Size in Relation to Crop Response to Liming**—If researchers are to obtain experiments that relate limestone particle size to increases in yield, they must satisfy several conditions:

1) Crop growth must increase due to lime application.
2) The relation between rate of limestone application and yield must be known.
3) The most reactive particle-size fraction of limestone must be applied at a rate that is somewhat less than that required for maximum yield.
4) The time allowed for the reaction between limestone and soil must be reasonable since even the largest particles will eventually dissolve completely.

In greenhouse experiments, limestone is usually mixed more uniformly with the soil than it is in field experiments. Differences in mixing could definitely lead to differences between otherwise similar experiments.

### *1. Greenhouse Research*

In the United States, among the early experiments on the effect of limestone fineness on plant growth were those of White (1917) and Thomas and Frear (1915) in Pennsylvania and Kopeloff (1917) in New Jersey. White grew nine different crops in succession over a 3-yr period on a highly acid, silty clay loam soil limed with different particle-size fractions of both calcitic and dolomitic limestones. He found little difference between calcitic and dolomitic lime for all particle sizes except for the 20- to 40-mesh size, where the dolomitic was much less effective. However, he found a large difference for particle size. The average relative yields for all crops and limestones were 100, 69, 22, and 5 for particle sizes of < 100-, 60- to 80-, 20- to 40-, and 8- to 12-mesh sizes, respectively.

Thomas and Frear (1915) grew medium red clover in greenhouse pots on the same soil that White had used and treated it with five different particle-size fractions of limestone. The experiment lasted for 8.5 months. The relative yields were 100, 95, 76, 58, and 12 for particle sizes of < 100-, 80- to 100-, 60- to 80-, 40- to 60-, and 20- to 40-mesh sizes, respectively.

Kopelof (1917) applied four different particle-size fractions of limestone particles to acid soils from four different regions of the United States and grew four successive crops of crimson clover. The soils were Norfolk sandy loam (Typic Paleudults), Carrington silt loam (inactive series), Cumberland silt loam (Rhodic Paleudalfs), and Sierra sandy loam (Ultic Haploxeral fs). The average relative yields obtained were 100, 97, 95, and 77% for particle sizes of > 200-, 100- to 200-, 60- to 80-, 20- to 40-mesh sizes, respectively. The relative yield was 50% where no lime was applied.

In the 1950s, Meyer and Volk (1952) in Ohio and Beacher et al. (1952) in Pennsylvania conducted greenhouse pot experiments on particle size of limestone. Meyer and Volk evaluated a wide range of particle-size fractions from both calcitic and dolomitic limestones. They applied the different particle-size fractions at the rate of 6.7 t $ha^{-1}$ to a Canfield silt loam (Aquic Fragiudalfs) of pH 5.0, grew alfalfa for 10 months, and obtained five harvests. A crop of soybean was grown following the last alfalfa harvest. The effects of particle size and type of limestone on crop yields are shown in Table 4. Yield increased with increasing fineness of both materials.

Beacher et al. (1952) conducted a greenhouse experiment to evaluate the coarse fractions of limestone. They used a Westmoreland silt loam (Ultic Hapludalfs) of pH 4.0 and, after lime application, grew crimson clover for 6 months, followed by alfalfa. The results, presented in Table 5, show that yield of both crops increased with increasing fineness of both calcitic and dolomitic limestone.

Perkins (1961) evaluated several particle-size fractions of both calcitic and dolomitic limestones in a greenhouse experiment with alfalfa using a

Table 4—Effect of size of limestone particle and type of limestone on yield of alfalfa and soybean in the greenhouse (Meyer & Volk, 1952).

| Particle size, mesh | Plant yield, oven-dry wt pot$^{-1}$ | | | |
|---|---|---|---|---|
| | Alfalfa | | Soybean | |
| | Calcitic | Dolomitic | Calcitic | Dolomitic |
| | g | | | |
| 5–8 | 5.3 | 4.9 | 7.8 | 8.9 |
| 8–20 | 5.6 | 5.0 | | |
| 20–30 | 7.5 | 6.4 | 10.8 | 11.9 |
| 30–40 | 8.0 | 6.5 | | |
| 40–50 | 8.2 | 7.9 | 12.2 | 12.1 |
| 50–60 | 8.8 | 7.9 | | |
| 60–80 | 9.5 | 7.7 | 13.3 | 14.1 |
| 80–100 | 9.5 | 8.7 | | |
| <100 | 9.5 | 8.3 | 11.9 | 14.7 |
| No lime | 4.6 | 4.6 | | |

Table 5—Effect of size of limestone particle and type of limestone on yield of crimson clover and alfalfa in the greenhouse (Beacher et al., 1952).

| Particle size, mesh | Plant yield, oven-dry wt pot$^{-1}$ | | | |
|---|---|---|---|---|
| | Crimson clover | | Alfalfa | |
| | Calcitic | Dolomitic | Calcitic | Dolomitic |
| | g | | | |
| 20–40 | 0 | 0 | 1.2 | 0.1 |
| 60–80 | 10.9 | 1.4 | 4.7 | 6.3 |
| 100–200 | 12.9 | 12.7 | 5.9 | 7.1 |
| <200 | 15.1 | 15.9 | 6.5 | 7.6 |

Cecil sandy loam (Typic Hapludalts) of pH 5.3. The finer fractions were more effective than the coarse fractions. Fractions coarser than 50-mesh size of calcitic limestone were more effective in producing increased alfalfa yields than were similar fractions of dolomitic limestone. There was little difference between fractions of dolomitic and calcitic limestones when they were finer than 50-mesh size.

The results of the greenhouse experiments reported in this section show that the finer limestone fractions are more effective than the coarser fractions. They also show that the coarser fractions of dolomitic limestone are sometimes less effective than similar particle-size fractions of calcitic limestone. The experiments were conducted for periods varying from 6 months to 3 yr after lime application. Material coarser than 8-mesh size was shown to have very little value as a liming material. The 8- to 20-mesh material had a small effect for time periods of 6 months or more. The 40- to 60-mesh fraction gave responses to 60 to 80% of that obtained from limestone finer than 100-mesh size. Although dolomitic limestone was usually less effective than calcitic limestone, the difference between them tended to diminish as particle size was reduced.

### 2. *Field Research*

Almost since the first use of limestone in the United States, people have been concerned about the fineness of crushing of the product. Hence, many field experiments have been conducted to evaluate the effect of particle size. These experiments have been of two types: (i) sieving the limestone into several particle-size fractions and evaluating each particle-size fraction separately, and (ii) using regular limestones and making comparisons on the basis of their fineness as determined by the percentage of the sample passing through various size sieves (e.g., 50% passing through a 60-mesh sieve). When the latter type of experiment was used to make comparisons among experiments for this review, the fineness used is based on the percentage that passed through a 60-mesh sieve. Where the value for 60-mesh material was not given, it was estimated by interpolation from the log normal relationship of the values that were given.

The data for each type of experiment were evlauated separately; then the average results from the two methods were compared. In some experiments, several crops were used in a rotation. The following equivalents for crops were used to obtain a single measurement value: 100 kg of corn grain = 84 kg of wheat grain = 54 kg of soybean = 99 kg of oat grain = 188 kg of hay. Where field experiments were conducted for a long period, only the data from the first 6 or 8 yr were used.

The results of nine field experiments or summaries of field experiments with different particle-size fractions of limestone are shown in Table 6. The increase in yield produced by limestone is reported on a relative basis where the largest yield increase is given a value of 100. Where several rates of application were used in the research, either the average for all rates or only the lowest rate was used. The average of all rates was used if differences between materials were evident at the higher rates. The data are separated according to calcitic and dolomitic limestones where both were used. Where the type of limestone was unknown, it is listed in the calcitic column.

A few of the results in Table 6, e.g., those from Kentucky and Ohio, were quite variable. They indicate a coarser fraction as being more effective than a finer one. In most of these cases, the yield response was rather low and the degree of error associated with the data was undoubtedly large. By 1982 yield standards, many of these experiments had low yields, and many had yield variabilities that were attributable to such things as poor stands and weeds. All data were summarized since their averages show the same relationship as that shown by the experiments having the highest yields and greatest response to lime. The inclusion of the lower yielding experiments increased the range of relative values found. Hence, this range is not necessarily due entirely to real differences in relative effectiveness of the different particle-size fractions of limestone.

The averages and ranges for the data in Table 6 are shown in Table 7. The wide spread in the ranges indicates variability because of soils, crops, and limitations of experimental designs. Nevertheless, the 4- to 10-mesh

Table 6—Summary of the influence of discrete limestone particle-size fractions on relative crop response to limestone in field experiments.

| Location and soil type | Particle size, mesh | Limestone rate | Crop | Relative yield Calcitic | Relative yield Dolomitic | Reference |
|---|---|---|---|---|---|---|
| | | t ha$^{-1}$ | | | | |
| Maryland, unclassified sandy loam | 10–20 | -- | Mangel and carrots | 11 | 36 | Broughton et al. (1916) |
| | 20–40 | -- | | 34 | 75 | |
| | 40–80 | -- | | 61 | 79 | |
| | <80 | -- | | 100 | 100 | |
| Rhode Island, gravelly loam | 10–20 | -- | Alfalfa (5-yr avg) | 66 | | Hartwell & Damon (1919) |
| | 20–40 | -- | | 64 | | |
| | 40–80 | -- | | 100 | | |
| | <80 | | | 80 | | |
| Wooster, Ohio | 3–8 | -- | Soybean, wheat, and clover | 50 | 0 | Ames & Schollenberger (1916) |
| | 8–20 | | | 96 | 33 | |
| | 20–80 | | | 100 | 95 | |
| | <80 | | | | | |
| Kentucky, Decatur (like) silt loam (Rhodic Paleudults), pH 5.2 | 10–20 | -- | Corn, wheat, and hay (11 12-yr avg) | 78 | | Roberts et al. (1939) |
| | 40–50 | | | 100 | | |
| | 70–80 | | | 85 | | |
| | <100 | | | 78 | | |
| Alabama, Norfolk loamy sand (Typic Paleudults), pH 5.9 | 4–10 | 1.68 | Sudangrass and crimson clover (7-yr avg) | 38 | 24 | Davis (1951) |
| | 10–20 | | | 42 | 49 | |
| | 20–60 | | | 100 | 79 | |
| | 60–100 | | | 75 | 100 | |
| | 100–200 | | | 80 | 83 | |
| Alabama, Norfolk loamy sand, pH 5.9 | 4–10 | 3.4 | Sudangrass and crimson clover (7-yr avg) | 36 | 31 | |
| | 10–20 | | | 63 | 46 | |
| | 20–60 | | | 78 | 99 | |
| | 60–100 | | | 100 | 100 | |
| | 100–200 | | | 91 | 91 | |
| Iowa, Tama silt loam (Typic Argiudolls) | 10–20 | -- | Second-yr alfalfa | 33 | | Firkins & Pierre (1944) |
| | 20–40 | | | 63 | | |
| | 40–60 | | | 100 | | |
| | 10–20 | -- | Seventh-yr alfalfa | 64 | | |
| | 20–40 | | | 92 | | |
| | 40–60 | | | 100 | | |
| Cass County, Michigan | 10–20 | 4.5 | Alfalfa (4-yr avg) | 33 | | D. Longnecker (1958, personal comm., Michigan State Univ.) |
| | 40–60 | | | 71 | | |
| | <80 | | | 100 | | |
| Wisconsin, Withee silt loam (Aeric Glossaqualfs), pH 5.0 | 8–20 | 13.4 | Alfalfa (first harvest yr) | 55 | | Love et al. (1960) |
| | 20–40 | | | 74 | | |
| | 40–60 | | | 90 | | |
| | 60–100 | | | 94 | | |
| | <100 | | | 100 | | |
| England, 8 sites | 20–40 | -- | Mostly acid-sensitive crops (avg of 8 experiments for 4 yr) | 89 | | Walker (1952) |
| | 40–100 | | | 95 | | |
| | <100 | | | 100 | | |

Table 7—Summary of the effect of particle-size fraction of limestone on relative crop yield from nine field experiments reported in Table 6.

| Particle size, mesh† | Relative yield | |
|---|---|---|
| | Avg | Range |
| 4–10 | 33 | 0–50 |
| 10–20 | 54 | 11–96 |
| 20–40 | 84 | 34–100 |
| 40–80 or 40–100 | 99 | 61–100 |
| 80+ or 100+ | 100 | 78–100 |

† Some samples had only a portion of the category shown.

fraction gave the lowest yields, with a progressive increase in relative yield as particle size became smaller.

A summary of all the available published results of field experiments on the effect of fineness of commercial limestone on crop yield are reported in Table 8. The amount passing through the 60-mesh sieve was used for the purpose of presenting diverse data on a common basis. The limestone treatment resulting in the greatest yield increase was given a rating of 100, with the yield increases from other treatments in the experiment being given relative ratings. Where experiments were conducted for many years, only the data from the first 6 or 8 yr were used. A summary of these data (Table 9) shows clearly that the effectiveness of limestone for increasing crop yield was lower for materials having less than about 50% passing through a 60-mesh screen.

Table 9 also provides a comparison of field experiment results obtained with limestone fractions and those with unfractionated commercial limestone. The calculated values of relative yield were obtained by multiplying the amount of each limestone fraction (Table 10) in a material of given fineness (left-hand column of Table 9) by the relative yield of that fraction as shown in Table 7. The resulting values for the several fractions were then summed to give the figures listed in the right-hand column of Table 9. The results computed in this way from experiments that compared specific particle-size fractions of limestone are strikingly similar to those from experiments that compared different percentages passing through a 60-mesh sieve (Tables 8 and 9). Although no distinct mesh size separates "effective" from "ineffective" limestones, material coarser than 20-mesh size was relatively ineffective, whereas material finer than 60-mesh size was highly effective. The data also indicate that limestone with 50% passing through a 60-mesh sieve was essentially as effective as limestone with 100% passing through a 60-mesh sieve.

The rate vs. yield curves for finely pulverized limestones have a steeper slope at lower rates of lime than the curves for coarse limestones. In Figure 5, the assumption is made that 6 t $ha^{-1}$ of limestone having 100% passing through a 60-mesh sieve will give a relative yield increase of 100. A typical response curve was then drawn for the rates of limestone for this material using data reported in the literature. The curves for the coarser limestones

Table 8—Summary of the influence of limestone fineness on crop response to limestone in field experiments.

| Location and soil type | Limestone fineness† | Limestone rate | Crop | Relative yield Calcitic | Relative yield Dolomitic | Reference |
|---|---|---|---|---|---|---|
| | % | t ha$^{-1}$ | | | | |
| Illinois, Cisne silt loam | 28 | Applied triennially | Corn, soybean, wheat, and clover (6-yr avg) | 73 | 43 | Hopkins et al. (1919) |
| | 40 | | | 88 | 67 | |
| | 95 | | | 100 | 81 | |
| Illinois (2-yr avg) | 28 | 10.4 | Sweetclover | 91 | 81 | Mumford (1928, p. 31) |
| | 91 | 8.4 | | | | |
| | 95 | | | 100 | 92 | |
| Wooster, Ohio | 28 | 4.5 | Corn, oat, wheat, and hay | 64 | | Thorne (1919) |
| | 55 | | | 100 | | |
| | 28 | 9.0 | | 71 | | |
| | 55 | | | 100 | | |
| Wooster, Ohio, Canfield silt loam, pH 5.3 | 30 | -- | Corn, oat, wheat, and hay (4-yr avg) | 61 | | Ohio Agric. Exp. Stn. (1951) |
| | 50 | | | 100 | | |
| | 65 | | | 84 | | |
| Ohio, Canfield silt loam, pH 5.6 | 21 | -- | Timothy and alfalfa | 57 | | Ohio Agric. Stn. (1951) |
| | 38 | | | 95 | | |
| | 62 | | | 100 | | |
| Indiana, Newton fine sandy loam (Typic Humaquepts) | 28 | 9.0 | Corn, oat, and hay (8-yr avg) | 74 | 94 | Wiancho et al. (1929) |
| | 42 | | | 100 | | |
| | 95 | | | 94 | | |
| Indiana, Clermont silt loam (Typic Ochraqualfs) | 28 | 6.7 | Corn, wheat, and hay (6-yr avg) | 86 | 78 | Wiancho et al. (1929) |
| | 42 | | | 99 | | |
| | 95 | | | 100 | | |
| Indiana, Zanesville silt loam (Typic Fragiudalfs), pH 5.2 | 10 | -- | Alfalfa (3-yr avg) | 69 | | Barber (1962) |
| | 24 | | | 100 | | |
| | 45 | | | 96 | | |
| | 63 | | | 88 | | |
| Missouri, Putnam silt loam (Mollic Albaqualfs), pH 5.5 | 33 | 4.5 | Corn (8-yr avg) | 100 | | Albrecht (1946) |
| | 100 | | | 76 | | |
| Ontario, Canada, Oneida clay | 45 | 5.6 | Alfalfa (1-yr avg) | 62 | | Richards (1958) |
| | 98 | | | 100 | | |
| | 45 | 11.2 | | 91 | 42 | |
| | 98 | | | 100 | 67 | |
| England, 8 sites | 30 | -- | Mostly acid-sensitive crops (11 crops summarized for 5 experimental sites) | 94 | | Crowther & Walker (1952) |
| | | | | 92 | | |
| | 62 | | | 96 | | |
| | 100 | | | 100 | | |
| Ohio, Mahoning clay loam | 16 | 4.5 | Alfalfa-timothy (2-yr avg) | | 30 | Volk et al. (1952) |
| | 33 | | | | 86 | |
| | 60 | | | | 100 | |

† Percent passing through a 60-mesh sieve.

were drawn to cross the 6 t ha$^{-1}$ lime rate at relative yield increases equal to the average of those shown in Table 9. A horizontal line drawn at a relative yield of 80 shows the amount of each particle-size fraction required to give

Table 9—Summary of effect of limestone fineness on crop yield in field experiments.

| Fineness of limestone† | No. of tests | Relative yield from field experiments | | Relative yield calculated from limestone fraction data of Table 7 |
|---|---|---|---|---|
| | | Avg | Range | Avg |
| <20 | 2 | 50 | 30–69 | 58 |
| 20–30 | 11 | 77 | 53–100 | 71 |
| 30–40 | 5 | 84 | 62–100 | 75 |
| 40–50 | 9 | 92 | 62–100 | 83 |
| 50–60 | 4 | 100 | 100 | 89 |
| 60–80 | 5 | 94 | 84–100 | 94 |
| >80 | 10 | 97 | 76–100 | 100 |

† Percent passing through a 60-mesh sieve.

Table 10—Average particle-size distribution of commercial limestones ground to different fineness.†

| Percent passing through 60-mesh sieve | Percent distribution among size fractions | | | | |
|---|---|---|---|---|---|
| | 4–10 mesh | 10–20 mesh | 20–40 mesh | 40–90 mesh | 90+ mesh |
| 15 | 40 | 35 | 7 | 8 | 10 |
| 25 | 20 | 25 | 21 | 14 | 20 |
| 35 | 15 | 25 | 18 | 17 | 25 |
| 45 | 5 | 23 | 17 | 20 | 35 |
| 55 | 2 | 14 | 16 | 20 | 48 |
| 70 | 0 | 8 | 10 | 20 | 62 |

† Data presented are calculated from data of 244 limestone samples of Table 1.

the equivalent effectiveness of 6 t $ha^{-1}$ of limestone having 31 to 40% passing through a 60-mesh sieve. With the use of these curves, the relative amounts of limestone of different degrees of fineness required for equal effect on crop yield are shown in the first two columns of Table 11. Data of this type are needed to determine the relative value of the limestone. These data reflect the average of numerous field experiments on limestone fineness. Consequently, there was a range of values about each mean with individual experiments sometimes giving values that varied considerably from the average. The period of time considered in all of these evaluations varied from 1 to 6 yr.

**c. Relationship of Limestone Fineness to Its Economic Value**—The values for metric tons of each limestone fineness to give equal crop response are shown in Table 11. These values are then used to show what quarry prices would result in equal per-hectare costs for each limestone fineness. The assumed costs for hauling and spreading were \$7 $t^{-1}$. The reference limestone was that with 41 to 50% passing through a 60-mesh sieve and was given a value of \$6 $t^{-1}$ at the quarry. Since 6.1 t $ha^{-1}$ was required, the total liming cost per hectare was \$79.30. Since less limestone is required as the material is crushed finer, these calculations show that the

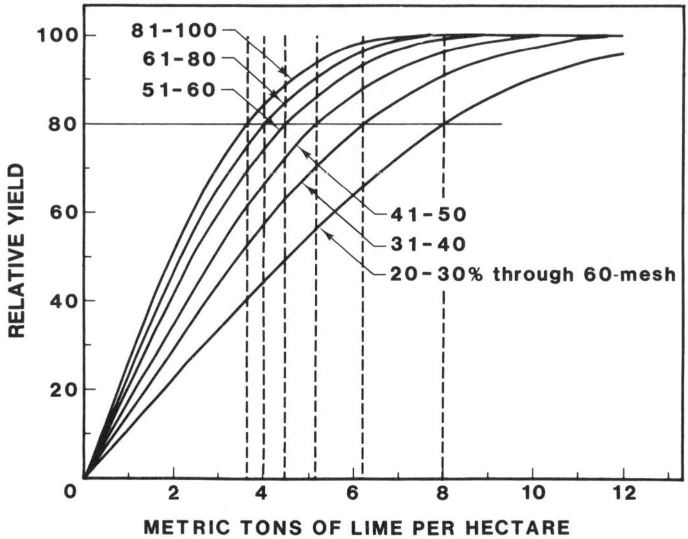


Fig. 5—Relation between limestone fineness, rate of limestone application, and crop yield based on average data from field experiments.

price that may be paid for lime at the quarry increases with increasing lime fineness. Where costs of hauling and spreading are greater than those used here, the difference in quarry prices of limestone for equal per-hectare costs would be even greater between fine and coarse materials.

Unfortunately, it is difficult to get reliable crushing costs for agricultural limestone. Crushing is usually done in conjunction with the production of aggregate for roads, construction, or steel mills. However, the production of very fine agricultural limestone would probably entail an additional crushing operation that undoubtedly varies with the type of stone and the type of plant operation. However, the additional crushing costs are usually low compared with hauling and spreading costs.

Table 11—Relative quarry prices necessary to provide equal cost per-hectare applications of limestones of varying degrees of fineness.

| Limestone fineness† | Limestone required for equal crop response | Hauling and spreading cost‡ | Quarry price for equal cost of \$79.30 $ha^{-1}$ |
|---|---|---|---|
| % | t $ha^{-1}$ | \$ $ha^{-1}$ | \$ $t^{-1}$ |
| 20–30 | 8.7 | 60.90 | 2.11 |
| 31–40 | 6.7 | 46.90 | 4.84 |
| 41–50 | 6.1 | 42.70 | 6.00 |
| 51–60 | 5.2 | 36.40 | 8.25 |
| 61–80 | 4.7 | 32.90 | 9.87 |
| 81+ | 4.0 | 28.00 | 12.83 |

† Percent passing through a 60-mesh sieve.
‡ Hauling and spreading cost \$7 $t^{-1}$. Values at 1982 prices.

Charles (1957, 1958) has discussed the relationship of energy and size reduction in crushing. He developed a general differential equation relating energy and size reduction of brittle and semibrittle materials.

**d. Calcium and Mg Content**—The range in agricultural limestones varies from almost pure calcitic to almost pure dolomitic limestone. The rate of dissolution and, hence, limestone effectiveness is affected by the Mg content of the limestone.

### *1. Effect on Rate of Dissolution*

Dolomitic limestone generally reacts more slowly with acid soils than does calcitic. The rate of limestone disappearance in a number of experiments was shown for both types of limestone in Table 2. A summary of 18 comparisons between dolomitic and calcitic limestone showed that the calcitic disappeared at a rate approximately twice that of the dolomitic. The ratio of the disappearance rates of calcitic/dolomitic limestones varied from 0.8 to 6.9, but most of the values were near the average of 2.0. These rates of dissolution were for limestone uniformly mixed with soil.

This information can be used to calculate the fineness needed for a dolomitic limestone to react with the soil at a rate equivalent to that of a calcitic limestone of a given fineness. For equal reaction rate, the particles of dolomitic limestone need to average approximately one-half the diameter of the calcitic limestone. If a calcitic limestone has 50% passing through a 60-mesh sieve and a particle distribution following the log normal relationship, the dolomitic limestone should have 68% passing through a 60-mesh sieve if it is to react with the soil at about the same rate as the calcitic limestone. However, the difference between the two materials increases as coarser materials are compared. The data in Table 11 were assumed to represent the average of both dolomitic and calcitic limestones and were calculated both from experiments showing crop response and from those investigating the

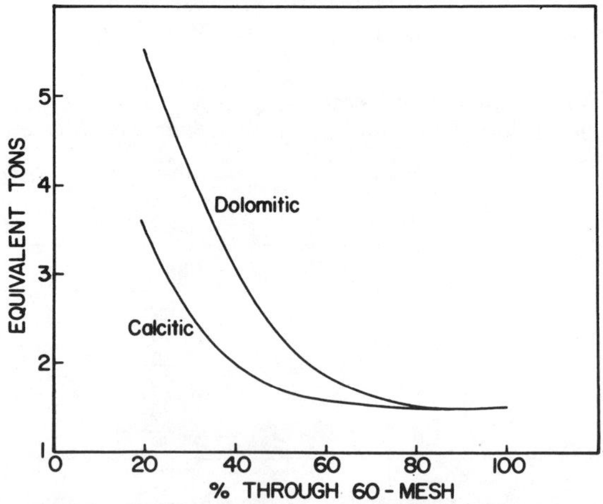


Fig. 6—Influence of fineness of limestone on the relative effectiveness of calcitic and dolomitic limestones for increasing crop yield.

rate of disappearance of limestone particles. If we separate these into calcitic and dolomitic and assume that the rate of disappearance of calcitic limestone is twice that of dolomitic, we get the curves shown in Figure 6. These curves show the effect of fineness on the relative amounts of each type of limestone required for equal results in a 2- to 4-yr period. Since many limestones are mixtures of dolomite and calcite, their relative effectiveness would be somewhere between the two curves shown in Figure 6.

### 2. *Effect on Crop Production*

The field experiments comparing dolomitic and calcitic limestones showed calcitic to be more effective than dolomitic in most cases when a period of $< 3$ yr was used and the same size particles were compared. The experimental error in many of the experiments appeared to be large, and yields were frequently low.

The effect of dolomitic limestone on yield compared with the effect of calcitic limestone may reflect responses to Mg application on soils low in Mg. If dolomitic limestone is superior, it is probably because of a response to Mg. If calcitic limestone is superior, it is probably because it reacted with the soil more rapidly and reduced the acid soil yield limiting factors. Nine of the field experiments reported in Table 6 and 8 compared dolomitic with calcitic limestone. In three of these experiments, dolomitic limestone gave the larger yield increase. In the remaining six, calcitic limestone of the same particle-size range gave considerably larger yield responses. Although dolomitic limestone reacts more slowly with the soil, it has the advantage of supplying Mg as well as Ca.

**e. Grindability**—The power required for crushing some limestones is greater than that required for others. Elphick (1956) studied the grindability of 23 calcitic limestones by grinding the 10- to 20-mesh fraction in a ball mill for 20 min. He then measured the percentage of stone that had been ground finer than 80-mesh size. It varied from 7.6% for a very hard, flinty limestone to 80.6% for a soft, shelly lime. The physical appearance of the limestone did not necessarily indicate its ease of grinding.

S. A. Barber (1958, unpublished data) made a similar determination of 16 Indiana limestones. He ground 100 g of 8- to 20-mesh limestone for 20 min in a ball mill, after which 25.8 to 78.5% passed through a 60-mesh sieve. The dolomitic limestone was ground as readily as the calcitic. The easiest stone to grind was a dolomitic limestone, whereas the most resistant limestone to grinding had a $CaCO_3/MgCO_3$ ratio of 26:1. Although grinding in a ball mill may not give the same relative values that would be obtained with a jaw crusher, the values do show that there is a variation among stones, and it is not necessarily related to the Mg content of the limestone.

**f. Effect of Physical Properties of the Limestone**—Limestones vary from dense, highly crystalline materials to soft, porous ones. Elphick (1954) determined the chemical and physical properties of 32 New Zealand limestones and found a wide range in ease of crushing. When the 20- to 40-mesh separates were investigated, he found a wide range in relative surface area as measured by its reaction with oxalate. He later measured what he termed *equivalent gross volume,* which was described as the gross volume (in milliliters) of limestone that contained 1 g of pure $CaCO_3$ (Elphick, 1955a, b). This value was inversely proportional to the density of Ca in calcitic limestones. The rates of solution of the 20- to 40-mesh separate of 25 limestones were highly correlated with their equivalent gross volumes ($r^2$ of 0.90).

Morgan and Salter (1923) investigated the effect of hardness, porosity, specific gravity, crystalline composition, and chemical analysis on the rate of dissolution of limestone particles. Of these properties, only the relative amounts of Ca and Mg had a large effect on rate of dissolution. Stukenholtz (1958) compared the effect of several Nebraska limestones in changing soil pH. The limestone sources were from the Cretaceous, Pennsylvanian, and Permian formations. The Cretaceous limestones reacted more rapidly than the others did when similar particle-size fractions were compared.

**g. Content of Nutrients Other Than Ca and Mg**—Agricultural limestones contain small amounts of minerals other than $CaCO_3$ and $MgCO_3$. Since limestone is sometimes applied at rates up to 20 t $ha^{-1}$, a low concentration of a micronutrient in the limestone may result in a significant application of the micronutrient. Chichilo and Whittaker (1961) analyzed 194 agricultural limestones from all over the United States, with the number of samples from each state roughly proportional to the amount of limestone used in that state. A summary of their data, presented in Table 12, gives the average and the range for each element determined.

Table 12—Chemical analysis of 194 samples of agricultural limestones from the United States (Chichilo & Whittaker, 1961).

| | Content | |
|---|---|---|
| Element | Avg | Range |
| | g $kg^{-1}$ | |
| Ca | 302.1 | 179.8–397.6 |
| Mg | 49.2 | 0.4–128.6 |
| Si | 23.6 | 0.3–155.3 |
| Al | 4.5 | 0.1–21.5 |
| Fe | 4.3 | 0.1–31.1 |
| K | 2.3 | <0.001–18 |
| S | 1.1 | <0.1–13.5 |
| Na | 0.3 | <0.01–1.5 |
| | mg $kg^{-1}$ | |
| Mg | 330 | 20–3000 |
| F | 230 | <10–1410 |
| P | 210 | 10–5660 |
| Zn | 31 | <1–425 |
| V | 11 | <1–106 |
| B | 4 | <1–21 |
| Cu | 2.7 | <0.3–89 |
| Mo | 1.1 | <0.1–92 |
| Co | <1 | <1–6 |

There was a positive correlation between the amount of Si present and the content of elements other than Ca and Mg, suggesting that some of the cations were present as silicates. There was enough Mn in some limestones to be a significant factor in Mn fertilization of the soil, if the Mn were largely available. However, Mn is much less available at high soil pH levels. The quantities of Mn in the limestones were not greatly different from those in many soils; thus, unless it is much more available to plants than the average soil Mn, it will not add greatly to Mn availability. Other nutrients present in some of the limestones in significant amounts were Mo and Zn.

Warren and Delavault (1961) analyzed 112 limestone samples from Ontario, Canada, for Pb, Cu, Zn, and Mo. The range of values they obtained were Pb, 3 to 65 mg $kg^{-1}$; Cu, 5 to 20 mg $kg^{-1}$; Zn, 11 to 520 mg $kg^{-1}$; Mo, 0.4 to 7 mg $kg^{-1}$. The values for Cu, Zn, and Mo were similar to those obtained by Chichilo and Whittaker (1961). Some values for Pb were surprisingly high, and Warren and Delavult (1961) found that Pb content was high in crops growing on soils limed with a limestone high in Pb.

**h. Fluid Lime**—Finely ground limestone is difficult to spread. Alley et al. (1980) found spreading problems due to bridging over when limestone was finer than 55% passing through a 100-mesh sieve. This problem can be circumvented by making fluid lime by adding water and 1 to 2% clay to form a suspension. The fluid is about 50 to 60% solids. This material can then be spread with fluid-spreading equipment and can result in an even distribution without loss of material. Fluid lime is a satisfactory material when compared on a limestone-material basis with agricultural ground limestone of similar particle size. Fertilizer materials such as N and K as liquid fertilizer may be mixed with fluid lime so that both materials are applied in one application.

## B. Agricultural Marl

Marl is a soft $CaCO_3$ deposit usually found in poorly drained, low lying areas. Its use in agriculture was reviewed by Berquist et al. (1932). Marl, an amorphous material that is powdery when dry, is very finely divided and should react rapidly with the soil. However, it may not react any more rapidly than limestone of larger particle size because it is usually spread as lumps that cannot be mixed uniformly with the soil.

Marl usually contains only a small amount of $MgCO_3$. Shickluna et al. (1958) found the Mg content of a number of marls from Michigan varied from 8 to 31 g $kg^{-1}$. Marl is usually handled on a volume rather than weight basis. On the average, a cubic meter of the marl in Michigan contained 24 kg of $MgCO_3$ and 666 kg of $CaCO_3$.

The $CaCO_3$ equivalence of marl varies depending on the impurities that it contains. Some states require a minimum $CaCO_3$ equivalence of 0.80, but others require only 0.70. In a few states where considerable quantities of marl are used, 2 $m^3$ of marl are considered the equivalent of 1 t of ground limestone.

## C. Agricultural Slag

Three different forms of slag are used as agricultural liming materials:

1) Blast furnace slag, used mainly to neutralize soil acidity and supply Ca and Mg.
2) Open-hearth slag, used because of its high Mn content in addition to its liming effect.
3) Basic slag, which was used in the past as P fertilizer because it contained 50 to 100 g of available P $kg^{-1}$. However, basic slag produced since 1975 has only about 10 g of available P $kg^{-1}$ and thus is no longer used as a P fertilizer.

### 1. BLAST FURNACE SLAG

Blast furnace slag is a byproduct of pig iron production. Limestone, iron ore, and coke are heated to about 1900°C in a blast furnace. This causes the iron to separate out while the silicates and alumina in the ore and coke combined with the Ca and Mg from the limestone. Small amounts of other materials in the ore also go into the slag. The molten slag is tapped from the furnaces and either is allowed to air-cool into a crystalline, cellular substance or is cooled rapidly in cold water that causes it to form lightweight, amorphous, porous granules.

Granulated, water-quenched, blast furnace slag weighs from 640 to 1000 kg $m^{-3}$, and air-cooled slag weighs about 1150 kg $m^{-3}$. The weight of pulverized limestone of comparable particle size is approximately 1380 kg $m^3$. Less water-quenched slag is made today because of pollution problems.

Slag is mainly Ca and Mg aluminosilicates, but it also contains smaller amount of other elements. The range of the chemical composition of a number of samples of blast furnace slag is shown in Table 13.

When slag is added to an acid soil, the $CaSiO_3$ probably hydrolyzes and neutralizes soil acidity in a manner similar to that of limestone. The capacity of the slag to neutralize soil acidity is measured by its $CaCO_3$

Table 13—Range in chemical composition of slags.

| Element | Blast furnace slag† | Open hearth slag‡ | Basic slag§ |
|---|---|---|---|
| | g $kg^{-1}$ | | |
| Ca | 260–320 | 160 | 270–420 |
| Mg | 20–70 | 50 | 10–50 |
| Si | 150–200 | 80 | 14–60 |
| Al | 54–85 | 15 | 10–20 |
| Fe | 2–20 | 240 | 40–170 |
| Mn | 2–12 | 100 | 60–40 |
| S | 10–30 | 1 | 2–110 |
| P | <10 | 10 | 50–100¶ |

† Whittaker (1955).
‡ Data supplied by United States Steel Corp.
‡ Waggaman (1952).
¶ Values since 1975 are < 10 g $kg^{-1}$.

equivalence, as is limestone. The $CaCO_3$ equivalence of blast furnace slag usually ranges between 0.80 and 1.00.

The use of blast furnace slag as a soil liming material in the United States was reviewed by Whittaker (1955). When compared with limestone, the amounts of slag used in the United States for liming soils are not great. Although early patents for agricultural slag date back to 1881, it has been used in this country as a liming material since about 1925. A number of experiments, many of which compared slag with limestone or $CaCO_3$, have been conducted to evaluate the use of slag as a liming material.

Ames (1916) found granulated slag was less effective than precipitated $CaCO_3$ for the reduction of soil acidity. However, pulverized limestone is also less effective than precipitated $CaCO_3$. Williams (1926) found in greenhouse experiments that it took 4.7 t of granulated slag to equal the effect of 1 t of finely ground limestone, but grinding the slag more than doubled its effectiveness.

Crane (1930) compared granulated blast furnace slag with limestone in greenhouse experiments in Illinois. The slag was as effective as the limestone when the two materials of equivlent fineness (all passing through a 100-mesh sieve) were compared on the basis of their neutralizing power as determined by acid titration to a phenolphthalein end point. It required up to four times as much coarse slag (11% passed through a 60-mesh sieve) to give plant growth equal to that obtained from fine slag (all passed through a 100-mesh sieve).

The first field experiments with slag, reported by White et al. (1937) in Pennsylvania, compared granulated slag, pulverized slag, air-cooled slag, and limestone. Two granulated slags, one with 12.8% passing through a 60-mesh sieve and one with 98% passing through a 60-mesh sieve, were compared with two fine limestones (80 and 98% passing through a 60-mesh sieve). The finely pulverized slag gave crop responses equal to those from limestone of equal fineness. The coarser granulated slag was from one-third to one-half as effective as the finely ground slag when evaluated over a period of 9 yr. Although the air-cooled slag was not compared at equivalent particle-size fractions of the other materials, it was somewhat less effective than the other slags.

In an experiment with crimson clover, Naftel (1942) compared blast furnace slag with dolomitic and calcitic limestones and found that slag gave much higher yields because of the B contained in the slag.

MacIntyre (1946), working with phosphate reduction electric furnace slag from Wilson Dam, Alabama, showed that granulated slag was as effective in its ability to raise soil pH as limestone of equal fineness and $CaCO_3$ equivalence. Greenhouse experiments with sudangrass showed that slag, because of its P content, was much superior to limestone on P-deficient soil.

Experiments in Alabama by Carter et al. (1951) compared air-cooled slag with limestone at three locations over a 3-yr period. Three slags of different degrees of fineness were compared with limestone. The amounts passing through a 60-mesh sieve were 62 to 72% for the finest slag, 38 to 50% for the medium slag, and 21 to 30% for the coarsest slag. The limestone had 63% passing through a 60-mesh sieve. They found no difference in yield response to comparable rates and degrees of fineness of slag and limestone, provided B applications were made to prevent its deficiency.

In experiments in Ohio, Volk et al. (1952) compared blast furnace slag with limestone. He used granulated blast furnace slag, air-cooled slag, and limestone in three degrees of fineness. The materials were applied at either two or three rates, and field experiments were conducted at three locations. The three degrees of fineness had approximately 20, 40, and 60% passing through a 60-mesh sieve. The field experiments were supplemented with greenhouse experiments where both crop response and the change in soil pH were measured. Effect on hay yields was no different among the coarsest granulated slag, air-cooled slag at all degrees of fineness, and dolomitic limestone at all degrees of fineness. When the three materials were compared at a fineness of 20% passing through a 60-mesh sieve, the granulated slag was definitely superior to air-cooled slag or dolomitic limestone. Air-cooled slag and dolomitic limestone were about equally effective when compared at equal degrees of fineness.

Chichilo et al. (1954) compared air-cooled, blast furnace slag with calcitic limestone. They used two degrees of fineness for each on two soils at three rates of application. The slag and limestone were equally effective for increasing soil pH and increasing sweetclover growth on a Norfolk sandy loam, but on a Fallsington loam (Typic Ochraquults), the limestone was more effective. They analyzed the top growth and found in comparing the slag with a similar limestone treatment that the slag gave a greater increase in Mn content of plants grown on the loam soil than did the limestone. The finer slag, 100% passing through an 80-mesh sieve, caused a greater increase in soil pH and higher sweetclover yields than the coarser slag that had 20% passing through a 60-mesh sieve.

The research with blast furnace slag indicates that its effectiveness usually increases as the degree of fineness to which it is crushed increases. The granulated slag was more effective than air-cooled slag of equal fineness. Air-cooled slag and limestone of equal fineness were equally effective. In some cases, blast furnace slag gave much larger yields than limestone, but this has usually been attributed to inclusions of micronutrients in the slag.

## 2. OPEN-HEARTH SLAG

Open-hearth slag comes from the open-hearth furnaces where pig iron is made into steel. This slag may contain approximately 200 g of Fe and 100 g of Mn kg$^{-1}$. The chemical analysis of one sample of open-earth slag is given in Table 13. This material has been investigated mainly for its value as a source of Mn or Fe in soils deficient in these nutrients. Experiments in Indiana (S. A. Barber, 1963, unpublished data) indicated that a 2 or 4 t ha$^{-1}$ application of this slag corrected Mn deficiency in soybean growing on Mn-deficient soil. Because of the high amounts of Fe and Mn in this slag, it has a lower $CaCO_3$ equivalence.

### 3. BASIC SLAG

Prior to 1975, basic slag (or Bessemer slag) contained 50 to 100 g of P $kg^{-1}$ (Waggaman, 1952) and was a byproduct from the refining of pig iron to steel where highly phosphatic iron ores are used. Much more of this material is produced in Europe than in the United States. In recent years, basic slag produced in the United States contains only about 10 g of available P $kg^{-1}$. The range of elements usually found in this slag is shown in Table 13. Experiments have shown that much of the P in this slag is readily available to crops.

## D. Miscellaneous Materials for Liming

Miscellaneous materials used for liming include various shells, flue dust from cement plants, lime from pollution control systems, refuse lime from sugar beet factories, paper mills, calcium carbide plants, rock wool plants, water softening plants, and byproduct lime from Pb mines. These materials are used in local areas in the vicinity of the source of the material. Most of the materials contain a high percentage of $CaCO_3$, and their value is proportional to their $CaCO_3$ equivalence and fineness. Little research has been published on these materials because they are used only in local areas and their total use is relatively small. They are generally finely pulverized so that they react rapidly with the soil.

Burned lime and hydrated lime are also used as liming materials. These were the first liming materials used. Their value is proportional to their $CaCO_3$ equivalence, which is usually much greater than that of $CaCO_3$.

# II. APPLICATION METHODS

Lime is usually broadcast on the soil surface at rates of 2 to 12 t $ha^{-1}$. The lime is then mixed with the soil during the course of tillage operations that follow, e.g., plowing, chiseling, disking, cultivating, and harrowing. Undoubtedly, the mixing is not uniform throughout the plow layer. Hulburt and Menzel (1953), in experimenting with the soil mixing characteristics of tillage implements, broadcast radioactive P on the surface and studied the effect of the tillage implement in mixing the $^{32}P$ throughout the soil. In separate experiments, they used sorghum seed pellets to measure mechanical mixing. They found that (i) tilling twice with a rotary tiller approximated uniform mixing; (ii) plowing placed most of the materials at the bottom of the plow layer; (iii) splitting the applications and applying one half before plowing and one half after plowing and then disking gave fair vertical distribution, but horizontal distribution was poor; and (iv) cultivators, harrows, or disks resulted in mixing with only the top 3 to 5 cm of soil.

Because of nonuniformity of mixing, the effectiveness of limestone in raising the average soil pH is somewhat less than that found in the laboratory where the limestone is more uniformly mixed with the soil. Walker (1952) found that two to three times as much limestone was required to raise the pH of the soil in the field as would be calculated from the results of uniform mixing in the laboratory.

Liming of the subsoil has been investigated because many subsoils are acid. This research ranges from thorough mixing of the lime with the subsoil by removing the soil and mixing it with lime and fertilizer (Fehrenbacker et al., 1958) to applying the lime in channels with a subsoil chisel (DeRoo, 1956). Liming of subsoil in the United States has frequently had no effect on subsequent crop yields (Fehrenbacher et al., 1958; Giddens et al., 1962; Woodruff & Smith, 1947); however, Doss et al. (1979) obtained increases in cotton and corn yields from liming to the 30-cm depth in studies on two soils (Rhodic Paleudults and Typic Paleudults) in Alabama. Gonzalez-Erico et al. (1979) found liming of Oxisol subsoils (15–30 cm) increased corn yields. The soil had a pH of 4.5 and 75% Al saturation before liming. The unlimed subsoil restricted root growth.

Liming has also been applied as a band near the seed of legume crops. This type of application enables the seedling to start growth, but unless the roots can contact a favorable soil pH at lower depths as the roots extend through the soil, it usually is not a satisfactory substitute for liming the whole plow layer. MacLeod and Bradfield (1964) started alfalfa in plugs of soil at pH 7.0 and then placed them in acid soil. The growth was less than where the entire acid soil (pH 4.8) was limed to pH 5.6 or 6.4.

Liming the subsoil, as well as the topsoil, has been investigated in greenhouse experiments. Hourigan et al. (1961) found that when the surface soil was adequately limed, liming the subsoil had little effect on yield. Longnecker and Merkle (1952) found that root extension into acid soil (pH 4.0) was limited unless the acid soil was limed, but they obtained maximum yields by liming only the surface layers. However, moisture deficiency due to shallow rooting should not be as much of a factor in plant growth in pot experiments as it would be under field conditions.

Lathwall and Peech (1965) studied the effect of depth of liming in field experiments. At low rates (2 and 4 t $ha^{-1}$), liming the uppermost 7.5-cm layer of soil was superior to applying the same amount of lime in the top 15-cm layer for growth of alfalfa on Mardin silt loam (Typic Fragiochrepts) at pH 4.8. When 8 and 16 t $ha^{-1}$ were used, depth of liming had no effect.

Woodhouse (1956) compared surface, mixed, and plow sole applications of lime to an acid Cecil sandy clay loam (Typic Hapludults). The mixed appliction gave the highest yields of alfalfa, followed by the plow sole treatment. The surface application gave very little more growth than the unlimed treatments.

When large applications of limestone are applied, split applications are sometimes recommended. One half of the lime is applied and worked into the soil before plowing, and the other half is applied and worked into the soil after plowing. In this way, the lime is mixed with more of the acid soil within the plow layer.

The rate of application of lime is also influenced by plowing. Where the soil is plowed to a 22.5-cm depth, the lime application should be 50% larger than where the soil is plowed to a 15-cm depth (Shoemaker, 1964).

## REFERENCES


Al-Abbas, A. J., and S. A. Barber. 1964. Effect of root growth and mass-flow on the availability of soil calcium and magnesium to soybeans in a greenhouse experiment. Soil Sci. 97:103–107.

Albrecht, W. A. 1946. Plant nutrition and the hydrogen ion: V. Relative effectiveness of coarsely ground and finely pulverized limestone. Soil Sci. 61:265–271.

Alley, M. M., G. D. McCart, D. D. Wolf, and E. S. Smith. 1980. Particle size effects on stockpiling and spreading agricultural limestone. Soil Sci. Soc. Am. J. 44:438–440.

American Society for Testing Materials. 1969. Book of standards. Am. Soc. for Testing Materials, Philadelphia.

Ames, J. W. 1916. Blast furnace slag as a source of bases for acid soils. Ohio Agric. Exp. Stn. Mon. Bull. 1:359–362.

Ames, J. W., and C. J. Schollenberger. 1916. Liming and lime requirements of soils. Ohio Agric. Exp. Stn. Res. Bull. 306:300–303.

Balks, R., and O. Wehrmann. 1938. Comparative studies on the solubility of carbonate liming materials. Z. Pflanzenernaehr. Bodenkd. 9:313–321.

Barber, S. A. 1957. Measurement of the fineness of agricultural limestone. Proc. Indiana Acad. Sci. 66:242–245.

Barber, S. A. 1962. Lime requirements for forage crops, Southern Indiana Forage Farm. Purdue Univ. Agric. Exp. Stn. Res. Progress Rep. 24.

Barber, S. A., J. M. Walker, and E. H. Vasey. 1963. Mechanisms for the movement of plant nutrients from the soil and fertilizer to the plant root. J. Agric. Food Chem. 11:204–207.

Barnes, E. F. 1947. A new method for estimating the surface of liming materials and the insoluble calcium compounds. Soil Sci. 63:285–289.

Beacher, R. L., D. Longnecker, and F. G. Merkle. 1952. Influence of form, fineness, and amount of limestone on plant development and certain soil characteristics. Soil Sci. 73: 75–82.

Beacher, R. L., and F. G. Merkle. 1949. The influence of form and fineness of lime compounds upon the correction of acidity and upon the nutrient status of soils. Soil Sci. Soc. Am. Proc. 13:391–393.

Bear, F. E., and L. Allen. 1932. Relation between fineness of limestone particles and their rates of solution. Ind. Eng. Chem. 24:998–1001.

Bergquist, S. G., H. H. Musselman, and C. E. Millar. 1932. Marl, its formation, excavation, and use. Michigan Agric. Exp. Stn. Spec. Bull. 224.

Broughton, L. B., R. C. Williams, and G. S. Frazee. 1916. Tests of availability of different grades of ground limestone. Maryland Agric. Exp. Stn. Bull. 193.

Brydon, J. E., and H. M. Rice. 1965. Use of the intercept method in determining the calcite-dolomite ratio for rating agricultural limestones. Agron. J. 57:283–285.

Carter, O. R., B. L. Collier, and F. L. Davis. 1951. Blast furnace slags as agricultural liming material. Agron. J. 43:430–433.

Cato and Varro. ca. 200 B.C. Roman farm management. Translated by a Virginia farmer. 1913. MacMillan, Inc., New York.

Charles, R. J. 1957. Energy-size reduction relationships in comminution. Min. Eng. (Littleton, Colo.) 9:80–88.

Charles, R. J. 1958. Collected discussion of energy-size reduction relationships in comminution. Min. Eng. (Littleton, Colo.) 10:481–484.

Chichilo, P. P., W. H. Armiger, A. W. Specht, and C. W. Whittaker. 1954. Plant nutrients from slag—furnace slag as a source of plant nutrients and its liming effectiveness relative to limestone. J. Agric. Food Chem. 2:458–462.

Chichilo, P., and C. W. Whittaker. 1961. Trace elements in agricultural limestones of the United States. Agron. J. 53:139–144.

Collins, E. R., and F. R. Speer. 1939. Decomposition of dolomite limestone in soils when used as a neutralizing agent in complete fertilizers. J. Assoc. Off. Agric. Chem. 22:142–147.

Crane, F. H. 1930. A comparison of some effects of blast furnace slag and of limestone on an acid soil. J. Am. Soc. Agron. 22:968–973.

Crowther, E. M., and T. W. Walker. 1952. The relative values of alternative liming materials. Agriculture (London) 59:251–260.

Davis, F. L. 1951. Effect of fineness of agricultural lime upon crop response. Agron. J. 43: 251–255.

Dawson, P. R., E. F. Snyder, W. L. Leighty, and F. R. Reid. 1939. Decomposition of dolomitic limestone in soils when used as a neutralizing agent in complete fertilizers. J. Assoc. Off. Agric. Chem. 22:137–141.

DeRoo, H. C. 1956. Subsoiling, plowing and deep placement of lime or fertilizer in one operation. Agron. J. 48:476–477.

DeTurk, E. E. 1938. Properties of Illinois soils which are related to their need for limestone and factors controlling effectiveness of limestone. Circ. Ill. State Geol. Surv. 23:191–203.

Dickson, A. 1788. The husbandry of the ancients, Vol. 1. Dickson and Greech, Edinburgh.

Doss, B. D., W. T. Damas, and Z. F. Lund. 1979. Depth of lime incorporation for correction of subsoil acidity. Agron. J. 71:541–544.

El Gibaly, H., and J. H. Axley. 1955. A chemical method for the rating of agricultural limestones used as soil amendments. Soil Sci. Soc. Am. Proc. 19:301–302.

Elphick, B. L. 1954. Studies in use of agricultural limestone: I. Chemical and physical properties of South Island (New Zeal.) agricultural limestones. N.Z. J. Sci. Tech. Sect. A 36A: 134–166.

Elphick, B. L. 1955a. Studies in use of agricultural limestone: II. Solubility of limestone in acid soil as influenced by particle size. N.Z. J. Sci. Tech. Sect. A 37A:156–173.

Elphick, B. L. 1955b. Studies in use of agricultural limestone. III. Solubility of limestone in acid soil as influenced by type of stone and particle size. N.Z. J. Sci. Tech. Sect. A 37A: 241–258.

Elphick, B. L. 1956. The reaction of crushed limestone with an acid soil. Int. Congr. Soil Sci., Trans. 6th (Paris) B:439–445.

Fehrenbacker, J. B., J. P. Vavra, and A. L. Lang. 1958. Deep tillage and deep fertilization experiment on a claypan soil. Soil Sci. Soc. Am. Proc. 22:553–557.

Fippin, E. O. 1939. Soil acidity and liming, and some factors that should determine the standards for lime for the soil. p. 57–70. *In* Symp. on Lime. Am. Soc. for Testing Materials, Philadelphia.

Firkins, B. J., and W. H. Pierre. 1944. Finer, better lime please. Farm Sci. Rep. Jan.:14–15.

Frear, W. 1921. The fineness of lime and limestone application as related to crop production. J. Am. Soc. Agron. 13:171–184.

Gehrke, C. W., and E. J. Johnson. 1963. Agricultural liming materials: Mechanical analysis and EDTA titration. J. Assoc. Off. Agric. Chem. 46:611–616.

Giddens, J., H. F. Perkins, E. R. Beaty, and H. D. Morris. 1962. Deep soil preparation and deep placement of fertilizer, lime. Ga. Agric. Res. 4:10–11.

Gonzalez-Erico, E., E. J. Kamprath, G. C. Naderman, and W. V. Soares. 1979. Effect of depth of lime incorporation on the growth of corn on an Oxisol of central Brazil. Soil Sci. Soc. Am. J. 43:1155–1158.

Greiner, E. W. 1950. Availability by particle sizes. Rock Prod. 53:123–125.

Hartwell, B. L., and S. C. Damon. 1919. A field comparison of hydrated lime with limestone of different degrees of fineness. Rhode Island Agric. Exp. Stn. Bull. 180.

Hopkins, C. G., F. W. Garrett, J. E. Whitchurch, and H. F. T. Fahrnkopf. 1919. Illinois crop yields from experimental fields. Ill. Agric. Exp. Stn. Bull. 219:452–454.

Hourigan, W. R., R. E. Franklin, Jr., E. O. McLean, and D. R. Bhumbla. 1961. Growth and Ca uptake by plants as affected by rate and depth of liming. Soil Sci. Soc. Am. Proc. 25: 491–494.

Hoyert, J. H., and J. H. Axley. 1952. Influence of liming materials on pH values of six Maryland soils. Soil Sci. 73:61–70.

Hulburt, W. C., and R. G. Menzel. 1953. Soil mixing characteristics of tillage implements. Agric. Eng. 34:702–708.

Hunter, A. S., H. Kinney, C. W. Whittaker, J. H. Axley, M. Peech, and J. E. Steckel. 1963. A collaborative study of the ethylenediaminetetraacetate (EDTA) method of rating agricultural limestones: I. Reproducibility of ratings and correlations with chemical and physical characteristics of materials. Agron. J. 55:351–354.

Ihde, A. J. 1952. Edmund Ruffin, soil chemist of the Old South. J. Chem. Educ. 29:407–414.

Jorgenson, J. R. 1957. The effect of kind, amount and particle size of lime on reaction, residual carbonates and exchangeable Ca, Mg, and K in an organic and a mineral soil. M.S. Thesis. Michigan State Univ., East Lansing.

Kopeloff, N. 1917. The influence of fineness of division of pulverized limestone on crop yields as well as the chemical and bacteriological factors in soil fertility. Soil Sci. 4:19–76.

Kriege, H. F. 1929. Factors affecting the dissolution rates of limestone in acid media with particular reference to the industrial and agricultural values of different limestones. Rock Prod. 32:73–80.

Lathwell, D. J., and M. Peech. 1965. Interpretation of chemical soil tests. New York (Cornell Univ.) Agric. Exp. Stn. Bull. 995.

Latta, W. C. 1885. Experiments with nitrogenous, phosphatic and other fertilizers. Indiana Agric. Exp. Stn. Bull. 2.

Linsley, C. M. 1954. Limestone, how to use it, when it use it, where to use it. Univ. of Illinois College of Agric. Ext. Circ. 721.

Longnecker, D., and F. G. Merkle. 1952. Influence of placement of lime compounds on root development and soil characteristics. Soil Sci. 73:71–74.

Longnecker, T. C., and H. B. Sprague. 1940. Rate of penetration of lime in soils under permanent grass. Soil Sci. 50:277–288.

Love, J. R., R. B. Corey, and C. C. Olsen. 1960. Effect of particle size and rate of application of dolomitic limestone on soil pH and growth of alfalfa. Int. Congr. Soil Sci., Trans. 7th (Madison, Wis. 3:293–301.

Love, K. S., and C. W. Whittaker. 1954. Surface area and reactivity of limestones. J. Agric. Food Chem. 2:1268–1272.

Lyon, T. L. 1931. Relative effectiveness of limestone particles of different sizes. New York (Cornell Univ.) Agric. Exp. Stn. Bull. 531.

MacIntyre, W. H. 1946. Certain "glassy" and crystal line calcium silicate materials. Their distinctive behavior and liming effectiveness as registered by plant response and soil pH. Soil Sci. Soc. Am. Proc. 10:71–80.

MacIntyre, W. H., and W. M. Shaw. 1925. The disintegration of limestone and dolomite separates as influenced by zone of incorporation. Soil Sci. 20:403–416.

MacIntyre, W. H., and W. M. Shaw. 1930. The influence of season upon the disintegration of limestone in soil. J. Am. Soc. Agron. 22:272–276.

MacLeod, L. B., and R. Bradfield. 1964. Greenhouse evaluation of the relative importance of lime and potassium for the establishment and maintenance of alfalfa and orchardgrass seedlings. Agron. J. 56:381–386.

Meyer, T. A., and G. W. Volk. 1952. Effect of particle size of limestones on soil reaction, exchangeable cations, and plant growth. Soil Sci. 73:37–52.

Association of American Plant Food Control Officials. 1982. Model agricultural liming materials bill. Official pub. no. 34, Assoc. of Am. Plant Food Control Officials, Lexington, Ky.

Morgan, M. F., and R. M. Salter. 1923. Solubility of limestones as related to their physical properties. Soil Sci. 15:293–305.

Motto, H. L., and S. W. Melstead. 1960. The efficiency of various particle-size fractions of limestone. Soil Sci. Soc. Am. Proc. 24:488–490.

Mumford, H. W. 1928. Annual report. Illinois Agric. Exp. Stn.

Murphy, L. S., and H. Follett. 1978. Liming, taking another look at the basics. Agrichem. Age 22:22–26.

Naftel, J. A. 1942. Soil liming investigations: IV. Response of crimson clover to boron with and without lime on Coastal Plain soils. J. Am. Soc. Agron. 34:975–985.

Ohio Agricultural Experiment Station. 1951. Handbook of Ohio experiments in agronomy. Book series B-1, Ohio Agric. Exp. Stn., Wooster.

Patterson, H. J. 1906. Results of experiments on the liming of soils. Maryland Agric. Exp. Stn. Bull. 110.

Perkins, H. F. 1961. Particle size influences effectiveness of limestone. Ga. Agric. Res. 3:3–4.

Pettyjohn, F. J. 1957. Sedimentary rocks. 2nd ed. Harper & Row, Publishers, Inc., New York.

Pierre, W. H. 1930. Neutralizing value of liming materials. Soil Sci. 29:137–158.

Plant, A. N. 1960. Factors contributing to the poor agreement between laboratory established lime requirements and pH values obtained under field conditions on four South Carolina soils. M.S. Thesis. Clemson Univ., Clemson, S.C.

Richards, N. R. 1958. Progress report. Dep. of Soils, Res., and Advisory Service, Ontario Agric. College, Guelph.

Roberts, C., J. F. Freeman, and E. J. Kinney. 1939. Soil management experiments. Kentucky Agric. Exp. Stn. Bull. 397, p. 337-343.

Rost, C. D., and F. A. Fieger. 1927. Effects of different kinds and amounts of liming materials upon the hydrogen-ion concentration of the soil. Int. Congr. Soil Sci., Trans. 1st (Washington, D.C.) 2:45-65.

Ruffin, E. 1821. On the composition of soils, and their improvement by calcareous manures. Am. Farmer 3:313-320.

Ruffin, E. 1852. Essays on calcareous manures. 5th ed. J. W. Randolph, Richmond, Va.

Salter, R. M., and C. J. Schollenberger. 1940. Criteria for the evaluation of agricultural ground limestone. p. 71-82. *In* Symp. on Lime. Am. Soc. for Testing Materials, Philadelphia.

Schollenberger, C. J., and R. M. Salter. 1943. A chart for evaluating agricultural limestone. J. Am. Soc. Agron. 35:955-956.

Schollenberger, C. J., and C. W. Whittaker. 1953. The ammonium chloride liming materials reaction. J. Assoc. Off. Agric. Chem. 36:1130-1146.

Schollenberger, C. J., and C. W. Whittaker. 1956. The oxalate-limestone reaction in the activity evaluation of agricultural limestones. J. Assoc. Off. Agric. Chem. 39:750-763.

Schollenberger, C. J., and C. W. Whittaker. 1962. A comparison of models of evaluating activities of agricultural limestone. Soil Sci. 93:161-171.

Shaw, W. M. 1960. Rate of reaction of limestone with soils. Univ. of Tennessee Agric. Exp. Stn. Bull. 319.

Shaw, W. M., and B. Robinson. 1959. Chemical evaluation of neutralizing efficiency of agricultural limestone. Soil Sci. 89:262-272.

Shickluna, J. C., E. D. Longnecker, and J. A. Porter. 1958. A rapid chemical determination of the magnesium carbonate content of marl coupled with a survey of this material in Michigan marl deposits. Mich. Agric. Exp. Stn. Q. Bull. 41:198-203.

Shoemaker, H. 1964. Fit lime to plowing depth. Better Crops Plant Food 48(1):24-27.

Skinner, S. I. M., R. L. Halstead, and J. E. Brydon. 1959. Quantitative manometric determination of calcite and dolomite in soils and limestones. Can. J. Soil Sci. 39:197-204.

Stewart, R., and F. H. Wyatt. 1919. The comparative value of various forms of limestone. Soil Sci. 7:272-278.

Stukenholtz, D. D. 1958. Relative effectiveness of nine limestones. M.S. Thesis. Univ. of Nebraska, Lincoln.

Swartzendruber, D., and S. A. Barber. 1965. The dissolution of limestone particles in the soil. Soil Sci. 100:287-291.

Thomas, R. P., and H. M. Gross. 1952. A method for estimating the reacting rate of different particle sizes of limestone. Soil Sci. 73:53-60.

Thomas, W., and W. Frear. 1915. Experiments to determine the influence of the fineness of subdivision and richness in magnesium carbonate of crushed limestone used for amendment of acid soils. J. Ind. Eng. Chem. 7:1041-1042.

Thorne, C. E. 1919. The maintenance of soil fertility. Ohio Agric. Exp. Stn. Res. Bull. 336: 607-608.

Turner, R. C. 1960. An investigation of the intercept method for determining the proportion of dolomite and calcite in mixtures of the two: I. Theoretical aspects of the rate of solution of dolomite when a number of crystals are present. Can. J. Soil Sci. 40:219-231.

Turner, R. C., and S. I. M. Skinner. 1960. An investigation of the intercept method for determining the proportion of dolomite and calcite in mixtures of the two: II. Experimental rates of solution of dolomite and calcite in samples consisting of a number of crystals. Can. J. Soil Sci. 40:232-241.

Volk, G. W., R. B. Harding, and C. E. Evans. 1952. A comparison of blast furnace slag and limestone as a soil amendment. Ohio Agric. Exp. Stn. Res. Bull. 708:3-19.

Waggaman, W. H. 1952. Basic slag, degreased bone and dicalcium phosphate. p. 356-375. *In* W. H. Waggaman (ed.) Phosphoric acid phosphates and phosphatic fertilizers. Van Nostrand Reinhold Co., New York.

Walker, R. H., P. E. Brown, and A. W. Yound. 1932. Some chemical and bacteriological effects of various kinds and amounts of lime on certain Iowa soils: Part I. Laboratory and greenhouse experiments. Iowa Agric. Exp. Stn. Res. Bull. 148.

Walker, T. W. 1952. The estimation of the lime requirements of soils, soil and crop yield data from the Hayes Adams and other liming experiments. J. Soil Sci. 3:261-278.

Warren, H. V., and R. E. Delavault. 1961. The lead, copper, zinc, and molybdenum contents of some limestones and related rocks in southern Ontario. Econ. Geol. 56:1265–1272.

Webster, R. H., C. E. Evans, G. W. Volk, and P. F. Pratt. 1953. Surface areas and relative solubility of agricultural limestone. Agron. J. 45:39–42.

White, J. W. 1917. Relative value of limestone of different degrees of fineness for soil improvement. Pennsylvania Agric. Exp. Stn. Bull. 149.

White, J. W., and F. D. Gardner. 1918. Relative value of limestone of different degrees of fineness for soil improvement. Pennsylvania Agric. Exp. Stn. Bull. 152.

White, J. W., F. J. Halben, and C. D. Jeffries. 1937. The agricultural value of specially prepared blast furnace slag. Pennsylvania Agric. Exp. Stn. Bull. 341.

Whittacker, C. W. 1955. Blast furnace slag in agriculture. Pit Quarry 62:1–6.

Whittacker, C. W., J. H. Axley, M. Peech, J. E. Steckel, E. O. McLean, and A. S. Hunter. 1963. A collaboration study of the ethylenediaminetetraacetate (EDTA) method for rating agricultural limestones: II. Relationship of ratings and of calcium carbonate content of limestones to their reactivity in the soil. Agron. J. 55:355–358.

Whittaker, C. W., and P. Chichilo. 1964. Interpretation of sieve analysis of agricultural limestones and suggested particle size specifications. Agron. J. 56:561–565.

Wiancho, A. T., G. P. Walker, and S. D. Conner. 1929. Comparison of soil liming materials. Purdue Univ. Agric. Exp. Stn. Bull. 329.

Wiese, R. A. 1956. Quality of Nebraska agricultural limestones. Unpublished report. Lincoln, Nebr.

Williams, C. G. 1926. Action of granulated slag on acid soils. Ohio Agric. Exp. Stn. Res. Bull. 402:18–20.

Woodhouse, W. W., Jr. 1956. Effect of placement and rate of phosphorus, potash, and limestone on the growth of alfalfa and lespedeza. Soil Sci. Soc. Am. Proc. 20:15–18.

Woodruff, C. M., and D. D. Smith. 1947. Subsoil shattering and subsoil liming for crop production in claypan soils. Soil Sci. Soc. Am. Proc. 11:539–542.

# 5 Crop Response to Lime in the Southern United States

**FRED ADAMS**

*Auburn University*
*Auburn, Alabama*

Although liming has long been a practice in some areas of the southern United States, it did not become a general practice until the second half of the 20th century. Liming is now recognized by most growers of the region as an essential component of profitable crop production on acid soils.

Extensive use of liming materials began during the "conservation era" of the 1930s under a subsidy payment program sponsored by the federal government. This program did little to convince farmers that lime was actually needed, resulting in most liming materials being used in subsidized conservation programs. Liming row crops remained generally neglected throughout much of the region until the late 1960s, when soils in many areas had become sufficiently acid to cause obvious and near-total crop failures in neglected fields. A major factor in precipitating this need was the widespread use of high rates of acid-forming N fertilizer, beginning in the 1950s.

The beneficial effect of liming acid soils was being researched by state agricultural experiment stations and the USDA at the beginning of the 20th century. For example, the Alabama station conducted numerous experiments between 1906 and 1910 on major crops of the region (Duggar & Funchess, 1911). They concluded that "practically all farm crops are less thrifty in . . . an acid soil . . ." It was to be many years, however, before liming recommendations were to become soundly based on research.

Methods for identifying soils needing lime were generally ineffective until the pH concept was adopted. This concept permitted an easy and quick assessment of soil acidity, and pH is probably the most commonly measured soil chemical property today. It is also one of the most useful and informative measures because of its relationship to many aspects of soil fertility and plant growth. An essential corollary to soil pH was the advent of lime requirement methods, the procedure by which liming rates are determined.

Although the separate factors of acid soil infertility have been clearly delineated in many instances, the procedures for measuring soil acidity, determining lime requirements, and predicting yield responses remain essentially empirical. The particular factor causing poor plant growth at low


*Soil Acidity and Liming*—Agronomy Monograph no. 12 (2nd Edition).

soil pH varies among soils and among plant species, but an adequate liming program corrects any and all factors that render acid soils infertile.

## I. SOILS

Soil orders in the southern United States are generally divided along physiographic boundaries. Each physiographic province, occupied partially or wholly by acid soils, is briefly described below. Greater detail can be found in Daniels et al. (1973).

### A. Physiographic Provinces

The Coastal Plain Province extends as a broad belt along the Atlantic seaboard from Virginia to Florida and along the Gulf of Mexico from Florida to Texas. It is about 240 km wide along the Atlantic and up to 640 km wide along the Gulf. The terrain varies from coastal lowlands to plains to hilly uplands. Surface and internal drainage range from poor to good. Parent materials are generally unconsolidated beds of sand, sandy clays, and clays.

The Piedmont Province extends from northern Virginia through central Virginia and North Carolina, western South Carolina, and northern Georgia into eastern Alabama. It is a product of erosion, with gentle slopes and rounded summits. It lies between the mountainous Appalachian Highlands and the Coastal Plains.

The Blue Ridge Province is a mountainous, narrow belt immediately west of the Piedmont, extending across Virginia, North Carolina, and Tennessee into northern Georgia. It varies in width from about 20 to 110 km.

The Ridge and Valley Province is west of and parallel to the Blue Ridge Mountains. It is a narrow belt of valley floors separated by narrow, even-topped mountain ridges. It extends across western Virginia, eastern Tennessee, and northwestern Georgia into northeastern Alabama.

The Appalachian Plateau Province lies just west of the Ridge and Valley Province across eastern Kentucky and Tennessee and extending into northeastern Alabama. The area is underlain by nearly horizontal sandstone beds, and its topography is characterized by narrow-crested ridges and deep, narrow valleys.

The Interior Low Plateau Province lies west of the Appalachian Plateau Province in western Kentucky, central Tennessee, and northern Alabama. It is known as the Pennroyal area in Kentucky and as the Highland Rim in Tennessee and Alabama. The topography is uneven to rolling, and its soils are known for their strong fragipans.

The Mississippi Alluvial Valley Province extends the north-south length of eastern Arkansas and Louisiana and into northwestern Mississippi. It is occupied primarily by recent alluvium, but there are fairly large areas of older alluvium and upland tracts. Crowleys Ridge in Arkansas and Macon Ridge in Arkansas and Louisiana are examples of upland tracts.

The Ozark Plateau Province occupies northwestern Arkansas. Topography varies from very rough to near-level mountain tops to plateaus to steep-sided valleys. It is generally underlain by sandstone and cherty limestones.

The Ouachita Province consists of the Arkansas Valley and the Ouachita Mountains and is located in western Arkansas and eastern Oklahoma. It is bordered on the north by the Ozark Plateau and on the south by the Coastal Plain.

## B. Soil Orders

Acid soils of the region include Alfisols, Entisols, Histosols, Inceptisols, Spodosols, Ultisols, and Vertisols. Land areas occupied by these soil orders vary from small, localized areas to expansive land masses (Fig. 1). A more comprehensive discussion is available in a joint publication issued by the agricultural experiment stations and USDA-SCS (Buol, 1973).

### 1. ULTISOLS

Ultisols are the most extensive soils of the region, dominating approximately 880 000 km$^2$ (Perkins et al., 1973). They occupy much of the Coastal Plain, Piedmont, Appalachian Plateau, Interior Low Plateau, Ozark Plateau, and Ouachita provinces. Ultisols have low base saturated, acid, argillic B horizons and generally have sandy or loamy surface horizons where erosion has not removed them.

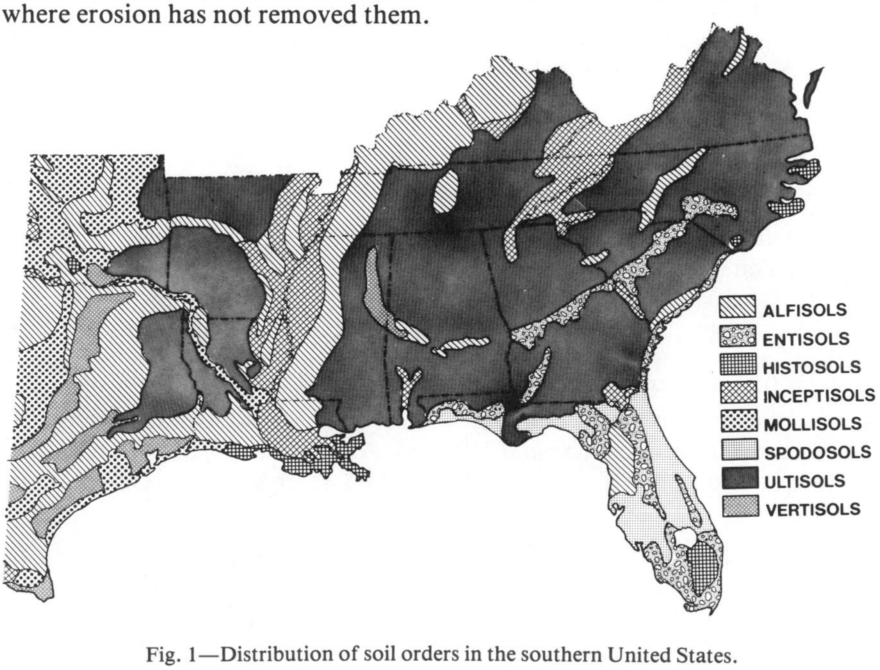


Fig. 1—Distribution of soil orders in the southern United States.

**a. Paleudults**—Paleudults are the dominant Ultisols in the Coastal Plains and parts of north central Alabama. They have thick argillic horizons, many containing plinthite. Subsurface horizons range from yellow to red to brown, often with mottles. Kaolinite is the dominant clay mineral, most often in association with hydroxyinterlayer vermiculite, gibbsite, and goethite.

**b. Hapludults**—Hapludults occupy extensive areas that are underlain by consolidated rocks. They dominate the soils of the Piedmont, Appalachian Plateau, Ozark Plateau, and Ouachita provinces. They are not as thick or as weathered as the Paleudults but are well drained. Subsurface horizons range from yellow to brown to red with some mottles. Kaolinite is generally dominant in the clay fraction, but significant amounts of 2:1 minerals are present, usually interstratified.

**c. Rhodudults**—Rhodudults are widely distributed in the Piedmont Province but are the dominant group in only a small area of South Carolina. These are well-drained soils with dark color values in all horizons. They are moderately acid, containing more basic cations and more Fe and Mn oxides than their associated soils. Surface horizons are loamy to clayey with little evidence of eluviation. Their darker-colored A and B horizons distinguish them from other Ultisols. Kaolinite is the major clay mineral, but it occurs in association with gibbsite and goethite.

**d. Ochraquults**—Ochraquults are poorly drained soils on the relatively level landscapes of the Atlantic and Gulf Coast flatwoods. They are usually water saturated in winter and spring, becoming relatively dry in autumn of most years. They occur primarily in Virginia, North Carolina, and Florida. Organic matter accumulation in the surface horizon gives it a thin, gray to black A horizon. It has a bleached E horizon overlying a gray argillic horizon.

## 2. ALFISOLS

Alfisols occupy extensive areas in the central and western portion of the Coastal Plains. The alluvial plains of the Mississippi and Ohio rivers are bordered on the east and south by a band of Alfisols extending from southeastern Louisiana into north central Kentucky. Smaller areas border the Mississippi alluvium in Arkansas and Louisiana. Relatively large areas of Alfisols also occur along the Arkansas River, in southwestern Louisiana and southeastern Texas, in eastern Oklahoma, and in central Tennessee. Less extensive but important areas occur in central Alabama, Florida, North and South Carolina, and Virginia. Alfisols have light-colored, acid surface horizons and B horizons that are less weathered and less acid than Ultisols.

**a. Fragiudalfs**—Fragiudalfs have subsoil fragipans that restrict root penetration and water movement. They occupy a large portion of the Alfisols in western Kentucky, Tennessee, and Mississippi. They also dominate Crowley's Ridge in Arkansas and Macon Ridge in Louisiana.

**b. Paleudalfs**—Paleudalfs are thick, reddish soils with relatively low base saturation for Alfisols. They occupy rather extensive areas in eastern Texas.

**c. Hapludalfs**—Hapludalfs are more weathered than Paleudalfs and are usually not as thick. They are brownish to reddish soils that occur in Arkansas along the Arkansas River. They also occur in central Alabama, central Tennessee, northern Kentucky, west central Florida, and as small areas in the mountains of Virginia, North Carolina, and South Carolina.

**d. Albaqualfs**—Albaqualfs are poorly drained soils with seasonably perched water tables. Surface horizons are light-colored loams overlying a clayey subsoil with little or no transitional layer. They occur in southern Louisiana, along the South Carolina coast, and in central Arkansas.

**e. Glossaqualfs**—Glossaqualfs are wet soils with loamy surface horizons that extend as vertical tongues into clayey subsoils. They occupy areas in southwestern, central, and northwestern Louisiana and a small area in southwestern Arkansas.

### 3. SPODOSOLS

Spodosols in the southern United States are confined primarily to the Coastal Plain in Florida. A few minor areas occur near the coast in North Carolina, South Carolina, Georgia, Alabama, and Mississippi. They occupy nearly level to gently sloping landscapes with fluctuating water tables at or near the surface during the wet season. Their elevations are almost always $< 40$ m above sea level. The spodic horizon of these soils is characterized by an accumulation of organic matter with traces to moderate amounts of amorphous Al (and sometimes Fe). This horizon is composed of material that gives it a high cation exchange capacity (CEC), large surface area, and high water-holding capacity; it is low in basic cations but high in exchangeable acidity. It usually underlies an E horizon at $< 75$ cm below the surface. Most Spodosols of the region formed in quartz-rich sands and have a sandy texture; a few have loamy material below the spodic horizon.

**a. Haplaquods**—Haplaquods in association with Psammaquents and Humaquepts are dominant in Florida, where Spodosols are extensive. These are wet soils with fluctuating water tables. They commonly have a white, light gray, or gray horizon overlying a spodic horizon of organic matter and Al (Fe is virtually absent).

### 4. INCEPTISOLS

Inceptisols of major agricultural importance in the region are located in the alluvium of the Mississippi River and westward along some of its tributaries into east central Louisiana and Arkansas. Other Inceptisols occupy significant areas in the Appalachian, Blue Ridge, and Cumberland mountains. These are young soils, lacking argillic horizons, with weak to moderate profile horizonation.

**a. Haplaquepts**—Haplaquepts occupy much of the alluvium in the Mississippi Valley. They occur on broad, nearly level floodplains and in swampy depressions. They have light-colored or thin, dark-colored surface horizons overlying dull-colored, mottled subsoils. Many have high clay contents, dominated by montmorillonite.

**b. Humaquepts**—Humaquepts dominate only in isolated areas in Georgia and Florida, where they developed from sandy and loamy marine or fluvial sediments in shallow depressions. They have thick, dark surface horizons and strongly acid subsoils.

**c. Dystrochrepts**—Dystrochrepts occupy the more mountainous areas in Kentucky, Virginia, Tennessee, and North Carolina. They are moderately acid and are of little agricultural importance.

## 5. ENTISOLS

Entisols cover the sandhills of North Carolina, South Carolina, and Georgia; a narrow strip along the South Carolina and Georgia coastline; and large areas of peninsular and panhandle Florida. These soils show little or no evidence of horizon development; sometimes thin A horizons have accumulated organic matter.

**a. Quartzipsamments**—Quartzipsamments are coarse sands or loamy sand soils. They contain at least 95% quartz, are loose, incoherent, and droughty. They occupy the sandhills of the Carolinas and Georgia, large areas in the northern half of Florida, and small areas along the coast of South Carolina.

**b. Haplaquents**—Haplaquents occupy level terrain that is water saturated much of the year. They are primarily in southernmost Florida.

**c. Psammaquents** are wet, sandy soils occupying relatively large areas in southern Florida.

## 6. HISTOSOLS

Histosols are organic soils that occur in the Florida Everglades, along the Louisiana coast, and along the North Carolina coast. Most of these areas serve as wildlife habitats. Where agriculture has been introduced by drainage schemes, subsidence and oxidation pose continuing problems.

## 7. VERTISOLS

Vertisols are dominantly neutral to alkaline, but some are quite acid ($<$ pH 5) in southeastern Texas, northeastern Mississippi, and west central Alabama. These soils are clayey in all horizons and are characterized by wide cracks during droughts because of their high montmorillonite contents. The more acid soils are high in exchangeable $Al^{3+}$.

## II. CROPS

An array of agronomic and horticultural crops are grown in the southern United States. They vary from those that are widely adapted and grown in all states to those that are restricted to specific localities because of climate, soil, or market limitations (Table 1). The most extensively grown crop is soybean. Its range in climatic adaptation permits it to be a major cash crop throughout the region. In contrast, sugarcane and citrus are important crops only in the semitropical climates of Florida, Louisiana, and Texas.

Plant species differ widely in their tolerances of soil acidity. Even cultivars of the same species may differ widely in tolerances, and this broad genetic base in acidity tolerances has stimulated much interest among plant breeders in developing acid-tolerant cultivars for the region.

The soybean is generally adapted to the pH range between 5 and 8. This has allowed it to become the major crop of the acid Coastal Plain soils and the near-neutral, Mississippi alluvium soils. Its planted area is exceeded by another crop only in Florida, Oklahoma, Texas, and Virginia.

Hay and pasture crops occupy large areas of land throughout the region. Most of these crops are used directly on the farm, with probably $< 10\%$ of harvested hay being sold off the farm. A wide variety of hay crops is grown, depending on soil and climate. Mixtures that include alfalfa or lespedeza are grown primarily in Kentucky, Virginia, Tennessee, and North Carolina and do not tolerate low soil pH very well. Alfalfa, in particular, is intolerant of low pH and thrives only on well-limed soils or neutral to calcareous soils. Clover-grass pastures are grown throughout the region, with white clover being the dominant clover. Grass hay and pasture crops, with and without clovers, are dominated by tall fescue, bluegrass, and orchardgrass in the northern tier of states and by bermudagrass, bahiagrass, and dallisgrass in the remaining states. Bermudagrass and bahiagrass are particularly tolerant of low soil pH, whereas tall fescue and dallisgrass have intermediate tolerances.

Wheat is the major crop of the Texas and Oklahoma plains, where soils are neutral to alkaline. However, about 1.5 million ha of wheat are planted annually on the more acid soils of the region. Cultivar tolerances to low soil pH are wide so that choosing the more adapted cultivars is important.

The area planted to cotton has declined markedly since 1930. Most cotton is now grown on the nonacid soils of the Texas and Oklahoma plains and the Mississippi alluvium in Arkansas, Louisiana, and Mississippi. Cotton is one of the less tolerant species to soil acidity in the region, and it has disappeared from many communities in the acid Coastal Plain and Piedmont provinces, where it once reigned as king.

Corn plantings in the southern United States have also declined considerably in recent decades. Without irrigation, the more droughty, acid soils of the Coastal Plain and Piedmont provinces have been big losers, and very little corn is planted in the fertile Mississippi alluvium. Corn is moder-

Table 1—Approximate 1980 harvested area of major crops in the southern United States (USDA, 1981).†

| State | Soybean | Hay | Wheat | Cotton | Corn | Grain sorghum | Vege-tables | Rice | Peanut | Tobacco | Sugar-cane |
|---|---|---|---|---|---|---|---|---|---|---|---|
| | ha × $10^3$ | | | | | | | | | | |
| Alabama | 851 | 253 | 95 | 130 | 164 | 14 | 28 | 0 | 81 | 0.2 | -- |
| Arkansas | 1762 | 373 | 332 | 261 | 15 | 82 | 11 | 518 | -- | 0.0 | -- |
| Florida | 186 | 97 | -- | 2 | 133 | -- | 895 | 0 | 22 | 4.2 | 136 |
| Georgia | 867 | 186 | 243 | 65 | 526 | 33 | 72 | 0 | 208 | 22.2 | -- |
| Kentucky | 648 | 632 | 142 | 0 | 599 | 11 | 6 | 0 | -- | 81.2 | -- |
| Louisiana | 1357 | 144 | 27 | 227 | 12 | 6 | 2 | 237 | -- | 0.1 | 103 |
| Mississippi | 1559 | 263 | 101 | 456 | 36 | 15 | 16 | 97 | 3 | 0.0 | -- |
| North Carolina | 782 | 155 | 121 | 26 | 701 | 25 | 100 | 0 | 67 | 153.0 | -- |
| Oklahoma | 121 | 636 | 2633 | 229 | 30 | 194 | 12 | 0 | 43 | 0.0 | -- |
| South Carolina | 648 | 84 | 78 | 49 | 209 | 6 | 66 | 0 | 5 | 26.3 | -- |
| Tennessee | 1033 | 480 | 142 | 111 | 259 | 14 | 20 | 0 | -- | 26.2 | -- |
| Texas | 255 | 1081 | 2106 | 2774 | 526 | 1600 | 469 | 237 | 93 | 0.0 | 14 |
| Virginia | 247 | 392 | 116 | 1 | 241 | 4 | 39 | 0 | 41 | 26.4 | -- |
| Total | 10 315 | 6512 | 6136 | 4330 | 3452 | 2004 | 1736 | 1090 | 564 | 339.6 | 253 |

† Citrus area = 369 000 ha in Florida and 35 500 ha in Texas (USDC, 1978).

ately tolerant of low soil pH, and it is widely adapted from that standpoint. However, it declined in popularity among growers because of its intolerance to periods of drought during flowering and grain filling.

Grain sorghum is grown primarily on the nonacid soils of Texas and Oklahoma. Even though it is moderately sensitive to acid soils and is more tolerant of drought than corn, it has found little favor among growers in the more humid parts of the region, probably because of marketing problems.

Vegetables are grown throughout the region, frequently in small, isolated areas for local markets. However, winter vegetables are grown extensively in peninsular Florida, particularly the Everglades area, and in the Rio Grande Valley of Texas. A liming program is normally required where these crops are grown on acid soils.

Rice plantings are limited to special areas of wet soils in Louisiana, Arkansas, Texas, and Mississippi. Liming is not a needed practice for rice culture on these soils.

The area planted to peanut has been limited for many years by production control programs of the federal government. It is a major cash crop in seven states, where it is grown primarily on acid, sandy soils of the Coastal Plain. Marketing programs have resulted in large-seeded cultivars in Virginia and North Carolina; medium-seeded cultivars being grown primarily in Georgia, Florida, and Alabama; and small-seeded cultivars in Texas and Oklahoma. Peanuts are very tolerant to most aspects of low soil pH except low available Ca. This makes peanut unique among the major agricultural crops of the region.

The area planted to tobacco is relatively small, because it, too, is rigidly determined by production controls as mandated by the federal government. It is limited to specific states where it is, indeed, a major cash crop. Tobacco is moderately sensitive to low soil pH, tending to suffer from Mn toxicity and Ca deficiency.

Sugarcane is grown for sugar production in Florida, Louisiana, and Texas. It is not very tolerant of low soil pH, and large yield increases result from liming acid mucks on which it is sometimes grown.

Citrus is a major crop on acid soils only in Florida, where it grows on the well-drained sands in the central peninsular region and in the acid, flatwood areas. Other major citrus plantings are in southern Texas.

## III. LIME USE

Although estimates of lime needs continue to outstrip the amount of lime used, more than 10 million t of liming materials are spread annually on acid soils of the southern United States (Table 2). Lime use since 1960 has increased only about one third, but a major shift has occurred in the crops for which soil is being limed. About two thirds of all agricultural lime used in 1960 was applied to government-sponsored conservation programs (primarily pasture) in which the federal government shared the cost. In 1980,

Table 2—Estimated amounts of agricultural liming materials applied to farmland in the southern United States for 1977 to 1979.†

| State | Amount of liming materials | | |
|---|---|---|---|
| | 1977 | 1978 | 1979 |
| | t × $10^3$ | | |
| Alabama | 1 021 | 1 169 | 955 |
| Arkansas | 546 | 415 | 459 |
| Florida | 851 | 891 | 925 |
| Georgia | 1 184 | 1 504 | 1 320 |
| Kentucky | 1 780 | 1 739 | 1 551 |
| Louisiana | 374 | 422 | 286 |
| Mississippi | 728 | 845 | 880 |
| North Carolina | 1 240 | 931 | 1 215 |
| Oklahoma | 206 | 176 | 153 |
| South Carolina | 532 | 639 | 558 |
| Tennessee | 1 083 | 1 087 | 992 |
| Texas | 218 | 227 | 243 |
| Virginia | 756 | 656 | 662 |
| Total | 10 519 | 10 701 | 10 199 |

† National Limestone Institute, unpublished data, Fairfax, Va.

more than 80% of all liming materials was purchased without government cost sharing, according to data compiled by the National Limestone Institute. This is a major shift from liming pasturelands, where cost sharing is available, to such row crops as soybean, corn, cotton, and peanut, where farmers must bear the total cost of liming.

Lime is generally spread on the land by the most practicable means available to the farmer. Most fields are limed by lime vendors using truck spreaders. Although equipment is available for obtaining a reasonably uniform spread of material, it is seldom used. In a study of lime vending and spreading in the Tennessee Valley area of western Tennessee and northern Alabama, Stewart (1977) reached the following conclusions:

1) Most lime vendors purchased used equipment, frequently worn-out fertilizer trucks.
2) Poor spread patterns resulted from poor or worn-out equipment, tunneling because lime was wet, interruption of spreading caused by large pieces of limestone being caught in the gate, poor driving patterns in the field, and inadequate adjustment of lime-spreader gate.
3) Drivers failed to adjust spreader gates.
4) Half of the trucks had no speedometers.
5) Drivers were generally ignorant of agronomic and economic benefits of liming, hence the value of precise and uniform spreading.
6) Many trucks were missing spinner blades, and their gates were frozen with rust.
7) Overlapping of spread and missed areas were common.

These findings are probably representative of much of the selling and spreading of agricultural liming materials throughout the southern United States.

Lime can be spread anytime that it fits conveniently into the other farming operations. The past recommendation that lime be applied and mixed with the soil 3 to 6 months prior to seeding is no longer valid. That particular recommendation originated during a time when available farm implements were incapable of thorough soil-lime mixing in a single operation. Since farm machinery now has that capability, seeding may follow liming immediately. The reaction rate between lime and acid soil is rapid, provided the lime particle and the soil particle are brought into intimate contact with one another.

Lime rates usually range between about 2.2 and 6.7 t $ha^{-1}$. It is impractical to spread $< 2.2$ t $ha^{-1}$; lime in excess of 6.7 t $ha^{-1}$ is usually wasteful and inefficient. Lime rates need to be controlled carefully only with certain crops. For example, potatoes are subject to scab disease in certain soils if pH is much above 5.5. Most crops are tolerant of a relatively wide soil pH range, and more lime is recommended than is needed for the immediate crop. The excess limestone serves as a reserve, which reacts only slowly with the soil, and allows lime applications to be spaced at intervals of 3 yr or more. Lime rates for cultivated crops should match that required to raise pH of the entire plow layer to the desired level, usually pH 6.0 to 6.5. Lime rates for undisturbed sod crops are less well defined and need not exceed that required to raise pH of the upper 10 to 15 cm to the desired pH level.

Liming no-till crops may need to be patterned more after sod crop liming than after cultivated crop liming, particularly where $NH_4^+$ fertilizers are used as N sources. No-till experiments with corn in Virginia (Moschler et al., 1973) and in Kentucky (Blevins et al., 1978) showed the development of intense acidity from nitrification in the upper 5 cm or so of soil. Because root density is greatest in the upper few centimeters of soil, low soil pH in this zone was highly detrimental to corn yields. Blevins et al. (1978) recommended that lime be surface applied with shallow or no incorporation if no-till corn is to be continued for more than 1 or 2 yr.

### A. Lime Requirement

Soil pH will indicate the need for lime, but it gives no hint of the amount needed. The amount needed is the lime requirement and may be determined by one of several methods. No single method for estimating the amount of lime needed on a particular soil has been accepted as the official method by all state soil testing laboratories of the region (Table 3). Each state laboratory adopted a lime requirement method at its inception, and changes have come only slowly. The early method was based almost entirely on soil pH, soil texture, type of clay mineral, and organic matter content. As titration methods were developed, they have gradually replaced the original method in most state laboratories (Sabbe & Breland, 1974).

Whereas Arkansas, Texas, and Virginia still use the original method of estimating lime requirement, all other state laboratories have adopted some chemical procedure. The chemical methods need not be highly accurate to be satisfactory because most plants are insensitive to a wide range in soil

Table 3—Lime requirement methods used by the state soil testing laboratories.

| State | Method |
|---|---|
| Alabama | Adams-Evans buffer |
| Arkansas | Soil pH, exchangeable $Ca^{2+}$, soil texture, crop |
| Florida | Adams-Evans buffer |
| Georgia | Adams-Evans buffer |
| Kentucky | SMP buffer |
| Louisiana | $Ca(OH)_2$ titration |
| Mississippi | Woodruff buffer, modified |
| North Carolina | Mehlich buffer |
| Oklahoma | SMP buffer, modified |
| South Carolina | Adams-Evans buffer |
| Tennessee | Adams-Evans buffer |
| Texas | Soil pH, soil texture |
| Virginia | Soil pH, soil texture |

pH. Lime need be added only in an amount that raises pH above a minimum, "critical" value but short of an excess that is wasteful or likely to cause an unnecessary nutrient deficiency. Properly applied lime is quite effective in raising soil pH to about 6.0; however, lime becomes increasingly inefficient as soil pH is raised above this value. Where lime recommendations are needed for large numbers of soil samples, the laboratory should choose a method that is rapid and reasonably accurate. A high degree of precision in methods is usually wasted because of problems encountered in the field application of lime. These problems include (i) soil variation, (ii) lime-spreading irregularities, (iii) lime quality, and (iv) mixing of lime with soil.

The most accurate method is probably the $Ca(OH)_2$ titration procedure (Abruna & Vincente, 1955). It is used by the Louisiana laboratory, but it is not well suited for large numbers of samples. All quick methods rely on the titration of soil acidity by a calibrated buffered solution. The most widely used method is the Adams-Evans method (Adams & Evans, 1962). This method is based on the empirical "soil pH–base saturation" relationship of similar soils (Fig. 2) and the amount of acidity titrated by a buffered solution of *p*-nitrophenol and $H_3BO_3$ in 1*M* KCl. The SMP method (Shoemaker et al., 1961) is used by Kentucky and Oklahoma laboratories. It is based on the pH response of selected reference soils to $CaCO_3$ rates and the amount of acidity titrated by a buffered solution. In general, it is not well suited to most soils of the southern United States. Other buffer methods in use include a modification of Woodruff's method (Woodruff, 1948) in Mississippi and Mehlich's method (Mehlich, 1976) in North Carolina. Yuan (1976) developed a lime requirement procedure based on pH changes in two buffered solutions. Although the method is based on sound chemical principles, it is not used by any state laboratory.

Soil pH is used in all states as the criterion for determining that lime is needed. The electrometric pH determination with the glass electrode and calomel electrode assembly is the laboratory method used in all states (Sabbe & Breland, 1974). However, field-measured pH by color kits is still used and gives satisfactory measurements. In measuring soil pH electro-

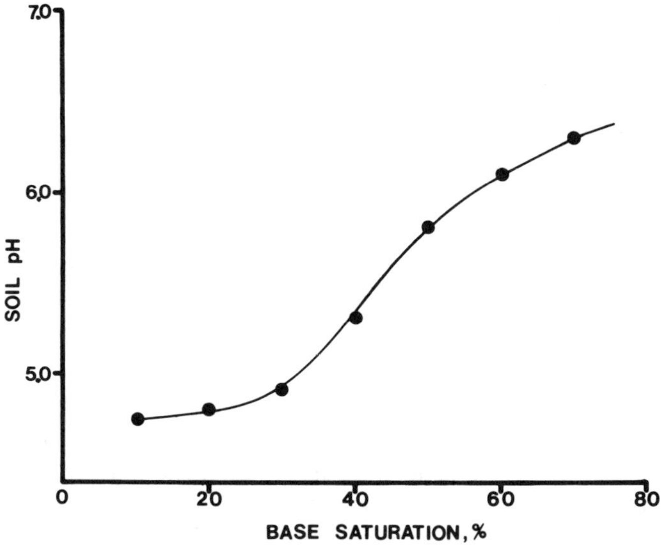


Fig. 2—Average relationship between soil pH and base saturation (at pH 7.0) of Ultisols in Alabama (Adams & Evans, 1962).

metrically, most laboratories use soil/water slurries of 1:1 or 1:3 (vol/vol) and allow them to stand for 1 hr or less before reading.

For soil pH to be a suitable criterion for triggering a lime recommendation, it should be consistent and reproducible. Although laboratory methods can be standardized rather easily, it is less easy to standardize samples of soil for constant behavior. Soluble fertilizer salts, for example, will often lower the pH of low CEC, acid soils by more than 0.5 pH unit (Fig. 3). Thus, measured soil pH is lowered by fertilizer without a corre-

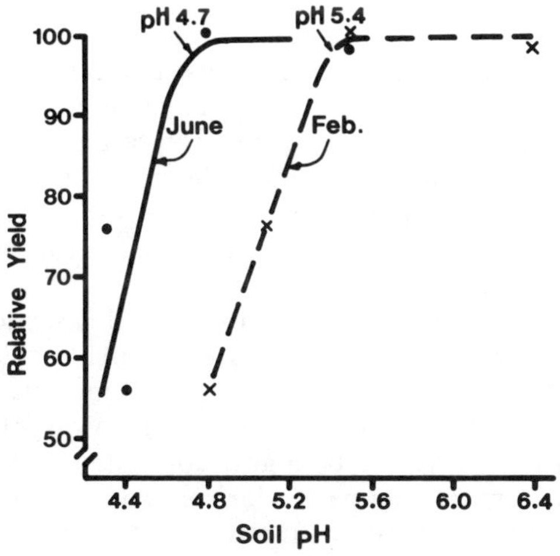


Fig. 3—Effect of fertilizer application and sampling date on pH of an Ultisol; February sample preceded fertilizer application to potato plants (C. E. Evans, 1974, unpublished data, Auburn Univ.).

sponding increase in total soil acidity; there is a minor shift of $H^+$ and possibly $Al^{3+}$ from exchange sites to the solution, where it is manifested as a lower pH. If a soil is sampled after a heavy fertilizer application instead of before, soil pH may indicate a need for lime where there is none. This problem is handled by some laboratories by leaching suspected samples prior to measuring soil pH. If the effect of fertilizer on pH is ignored, too much lime will be recommended, and this can cause serious problems with such crops as potatoes.

All lime requirement methods should account for the fact that only about two thirds of most agricultural limestones will react effectively with soil acidity. Even a lower percentage reacts if the material is unduly coarse, is poorly mixed with the soil, or if soil pH is being raised much above 6.0.

The crop to be grown is a major criterion that is used to determine when and if lime is to be used in the rotation. However, crops are not rigidly categorized according to soil pH requirements in most states. In several states, most crops are considered to have about the same minimum ("critical") soil pH value for lime recommendation purposes. Yet, there is clear evidence of great differences in tolerances to low soil pH. For example, alfalfa requires a much higher soil pH than bermudagrass for optimum forage production.

## B. Soil pH and Crop Response

The advent of a rapid and precise method for measuring soil pH led to the attractive concept of a critical soil pH value, i.e., the maximum pH at which liming increases crop yield. Neither complete success nor complete failure can be claimed for research directed toward this end.

Early failures were not uncommon, as illustrated by Pierre (1931) at the Alabama Agricultural Experiment Station when he concluded that soil pH "cannot be considered the direct cause of poor plant growth nor the main factor governing plant distribution or response to liming."

Partial success in defining critical soil pH values in the southern United States is attested to by the fact that all state soil testing laboratories use soil pH as the sole criterion that determines whether or not a field should be limed. Unfortunately, laboratories often use the same critical pH value for most crops on most soils, even though research has clearly shown great differences in both (Fig. 4).

Crop response to liming is not discussed in this section with the presumption that yields are increased because soil pH is raised above some magic value. It is fully recognized that some yield-limiting factor, e.g., Al, Mn, or Ca, is corrected by liming, and that is the reason for increased yield. However, soil pH is expected to be highly correlated with one or more of these yield-limiting factors on a particular soil, so that soil pH actually becomes a measure of the intensity of the factor, such as exchangeable $Al^{3+}$. Furthermore, soil testing laboratories are much better equipped and staffed to measure and interpret soil pHs than Al and Ca saturations of soils. Thus, identifying critical pH for a particular crop is needed less for research into

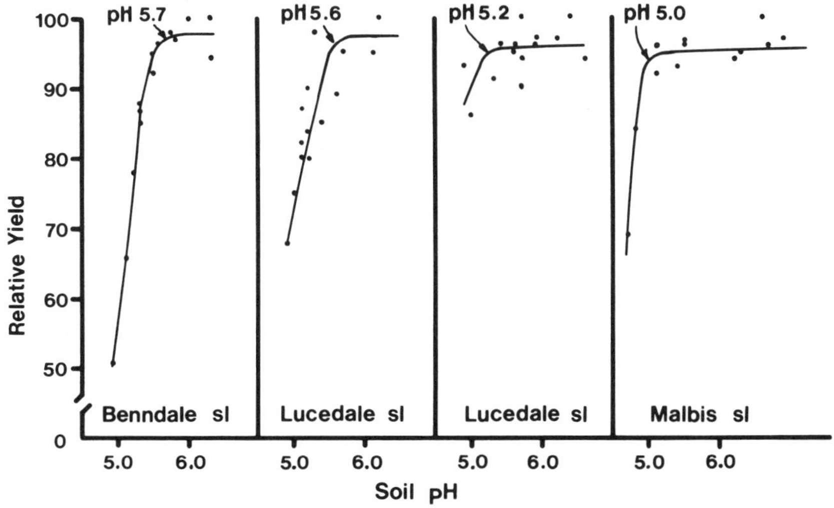


Fig. 4—Effect of soil pH on soybean yields on four Ultisols in Alabama (Rogers et al., 1973). Critical pH is indicated for each.

causes of acid soil infertility than for advising growers on the need for liming.

Greenhouse experiments are not particularly useful in helping to identify critical soil pHs, because they usually test only the A horizon under a demanding growth regime. Field experiments, on the other hand, test the combined effects of surface and subsurface soil horizons on crop yield. Consequently, only field experiments are considered in the following discussion of critical soil pH.

Although the genetic character of a cultivar greatly influences its response to liming, this factor has not been sufficiently delineated and cataloged for use in a general discussion on critical soil pHs. It is tacitly assumed that the genetic base for acid tolerance in a particular crop is about the same for most cultivars grown in the region.

## 1. COTTON

Most liming experiments with cotton have been conducted on the acid soils of the Coastal Plains, whereas cotton is now largely concentrated on the nonacid soils of the Texas and Oklahoma plains and the Mississippi Alluvial Valley. However, cotton remains a major crop on acid soils in localized areas of the Coastal Plains and the Interior Low Plateau provinces in northern Alabama. Examples of yield responses to liming on these soils at low pHs are given in Table 4.

From the data of Adams (1956; 1968a) and Walker and White (1963), the critical soil pH for cotton on Coastal Plain soils ranges between 5.0 and 5.7, with most being about 5.5. Critical pH of soils on the Interior Low Plateau Province ranges between 5.0 and 5.3, with reddish and brownish soils having lower critical pHs than the grayish soils (Burmester et al., 1981; Mitchell et al., 1977).

Table 4—Examples of effects of liming on yield of cotton in field experiments in Alabama.

| Soil | Years | pH of unlimed soil | Seed-cotton yield Unlimed | Limed | Reference |
|---|---|---|---|---|---|
| | | | —— kg ha$^{-1}$ —— | | |
| Benndale sl (Typic Paleudults)† | 1959–1961 | 5.0 | 990 | 1920 | Adams (1968a) |
| Lucedale fsl (Rhodic Paleudults)† | 1959–1961 | 5.0 | 990 | 2000 | Adams (1968a) |
| Lucedale scl† | 1962–1964 | 5.1 | 2150 | 2320 | Adams (1968a) |
| Dothan sl (Plinthic Paleudults)† | 1960–1966 | 5.2 | 2590 | 3080 | Adams (1968a) |
| Hartsells fsl (Typic Hapludults)‡ | 1960–1966 | 5.2 | 2430 | 2800 | Adams (1968a) |
| Decatur cl (Rhodic Paleudults)§ | 1960–1966 | 5.1 | 2520 | 2590 | Adams (1968a) |
| Decatur sil§ | 1977–1980 | 4.9 | 1710 | 2040 | Burmester et al. (1981) |
| Bewleyville sil (Typic Paleudults)§ | 1977–1980 | 5.3 | 2020 | 2390 | Burmester et al. (1981) |
| Locust sil (Glossic Fragiudults)§ | 1977–1980 | 5.4 | 1880 | 2150 | Burmester et al. (1981) |
| Mountview sil (Typic Paleudults)§ | 1975–1976 | 4.9 | 1550 | 1870 | Mitchell et al. (1977) |

† Coastal Plain Province.
‡ Appalachian Plateau Province.
§ Interior Low Plateau Province.

Table 5—Examples of effects of liming on corn yield in field experiments in the southern United States.

| State | Soil | Years | pH of unlimed soil | Grain yield Unlimed | Limed | Reference |
|---|---|---|---|---|---|---|
| | | | | —— kg ha$^{-1}$ —— | | |
| Alabama | Benndale sl | 1962–1966 | 5.2 | 2970 | 3920 | Adams (1969) |
| Alabama | Lucedale fsl | 1962–1966 | 5.1 | 3020 | 3700 | Adams (1969) |
| Alabama | Lucedale scl | 1967–1968 | 4.9 | 4480 | 5770 | Adams (1969) |
| Alabama | Malbis fsl (Plinthic Paleudults) | 1966–1968 | 4.7 | 2740 | 3700 | Adams (1969) |
| Alabama | Harsells fsl | 1960–1966 | 5.2 | 4700 | 5100 | Adams (1969) |
| Alabama | Decatur cl | 1960–1967 | 5.1 | 3530 | 3360 | Adams (1969) |
| Georgia | Tifton ls (Plinthic Paleudults) | 1957–1961 | 5.1 | 4700 | 5450 | Walker & White (1963) |
| Florida | Red Bay fsl (Rhodic Paleudults) | 1950–1955 | 5.1 | 3210 | 3250 | Robertson et al. (1957) |
| Florida | Ruston fsl (Typic Paleudults) | 1951–1957 | 5.2 | 3710 | 3690 | Robertson et al. (1957) |
| Texas | Katy fsl (Aquic Paleudalfs) | 1977 | 5.2 | 4450 | 4320 | Haby et al. (1979) |
| Virginia | Goldsboro sl (Aquic Paleudults) | 1980 | 4.1† | 1440 | 8150 | Alley (1981) |
| Virginia | Frederick sil (Typic Paleudults) | 1965–1972 | 5.1 | 6130 | 6960 | Moschler et al. (1973) |

† Soil pH is unusually low, probably because soil was sampled after high fertilizer rate was applied.

## 2. CORN

Liming experiments with corn seem to have been concentrated on soils of the Coastal Plain Province, even though corn is equally important as a crop in areas of the Piedmont and Interior Low Plateau provinces. Examples of yield responses to liming are given in Table 5.

From the data of Alley (1981), Adams (1956; 1969), and Walker and White (1963), critical pH on soils of the Paleudult great soil group ranged between 5.0 and 5.5. The reddish and brownish soils tended to have lower critical pHs. Critical pH of soils on the Appalachian Plateau appears to be about 5.2; it is about 5.0 on soils of the Interior Low Plateau Province.

## 3. SOYBEAN

Since soybean is a major agronomic crop in all states of the South, there is much interest in its performance on acid soils throughout the region. Examples of yield responses to liming on a range of soils are listed in Table 6.

Based primarily on results from field experiments in Alabama (Rogers et al., 1973), critical soil pH on Coastal Plain soils ranged between 5.0 and 5.7. Soils in other physiographical provinces had critical soil pHs in the general range of 5.2 to 5.4. Where soils are extremely deficient in Mo and it is not applied as a nutrient, critical pH can exceed 6.0 (Parker & Harris, 1962).

## 4. WHEAT

Although extensive areas are planted to wheat, published data from liming experiments in the region are lacking. In seven experiments in Alabama in which wheat was double-cropped with soybean, liming increased grain yields by 130 to 1310 kg $ha^{-1}$ (Table 7). The data suggest that critical soil pH for wheat is about 5.5 for most of these soils.

## 5. PEANUT

In a comprehensive review of field research with peanut, Cox et al. (1982) concluded that critical soil pH per se was not an identifiable criterion for liming. Adams and Hartzog (1980) summarized results from 78 field experiments in Alabama from 1972 to 1979 and found no correlation between soil pH and yield response of peanut to liming. Instead, yield response was highly correlated with exchangeable $Ca^{2+}$. Thus, critical soil pH is generally meaningless for peanut, except as it defines available soil Ca.

## 6. FORAGE CROPS

There is a diversity of forage crops grown on acid soils in the South, ranging from highly sensitive alfalfa to relatively insensitive bermudagrass. Critical soil pH for alfalfa is about 6.0 (Table 7), whereas that for 'Coastal' bermudagrass is about 5.0. Among winter annuals, ryegrass is more acid

Table 6—Examples of effects of liming on soybean grain yield in field experiments in the southern United States.

| State | Soil | Years | pH of unlimed soil | Grain yield | | Reference |
|---|---|---|---|---|---|---|
| | | | | Unlimed | Limed | |
| | | | | kg ha$^{-1}$ | | |
| Alabama | Benndale sl | 1968–1971 | 5.0 | 750 | 2030 | Rogers et al. (1973) |
| Alabama | Malbis fsl | 1966–1972 | 5.1 | 2000 | 2330 | Rogers et al. (1973) |
| Alabama | Lucedale sl | 1968–1971 | 4.9 | 1650 | 2360 | Rogers et al. (1973) |
| Alabama | Hartsells fsl | 1968–1972 | 5.1 | 1210 | 2550 | Rogers et al. (1973) |
| Alabama | Mountview sil | 1975–1976 | 4.9 | 1830 | 2360 | Mitchell et al. (1977) |
| Alabama | Dickson sil (Glossic Fragiudults) | 1977–1980 | 5.3 | 2840 | 3100 | Burmester et al. (1981) |
| Alabama | Tupelo sil (Aquic Hapludalfs) | 1977–1980 | 5.0 | 700 | 1980 | Burmester et al. (1981) |
| Alabama | Etowah l (Typic Paleudults) | 1977–1980 | 5.1 | 4420 | 4600 | Burmester et al. (1981) |
| Alabama | Wynnville sl (Glossic Fragiudults) | 1977–1980 | 5.1 | 1950 | 2740 | Burmester et al. (1981) |
| Alabama | Eutaw c (Vertic Ochraqualfs) | 1967–1972 | 4.9 | 1160 | 1920 | Rogers et al. (1973) |
| Alabama | Vaiden sicl (Vertic Hapludalfs) | 1972 | 5.1 | 980 | 2510 | Rogers et al. (1973) |
| Florida | Red Bay sl | 1975–1976 | 5.4 | 1880 | 2510 | Yuan et al. (1979) |
| Georgia | Tusquitel l (Humic Hapludults) | 1960 | 5.6 | 2040 | 3210 | Parker & Harris (1962) |
| Georgia | State l (Typic Hapludults) | 1961 | 5.5 | 2130 | 2740 | Parker & Harris (1962) |
| Louisiana | Beauregard sil (Plinthaquic Paleudults) | 1964 | 5.4 | 1790 | 2050 | Sedberry et al. (1973) |
| Louisiana | Calhoun sil (Typic Glossaqualfs) | 1967 | 5.1 | 1380 | 1850 | Sedberry et al. (1973) |
| Louisiana | Wrightsville sil (Typic Glossaqualfs) | 1971 | 5.1 | 2620 | 3060 | Sedberry et al. (1973) |
| Mississippi | Bude sil (Glossaquic Fragiudalfs) | 1974–1976 | 5.0 | 1870 | 2450 | Jones & Nelson (1978) |
| Mississippi | Prentiss fsl (Glossic Fragiudults) | 1975–1976 | 4.7 | 1970 | 2730 | Jones & Nelson (1978) |
| Tennessee | Loring sil (Typic Fragiudalfs) | 1969–1970 | 5.3 | 1340 | 2350 | Smith & Connell (1973) |
| Tennessee | Calloway sil (Glossaquic Fragiudalfs) | 1966–1969 | 5.2 | 1790 | 2240 | Howard (1970) |
| Tennessee | Memphis sil (Typic Hapludalfs) | 1963–1964 | 5.2 | 1400 | 2330 | Long et al. (1965) |
| Virginia | Craven fsl (Aquic Hapludults) | 1959 | 5.1 | 2080 | 2280 | Hallock (1964) |
| Virginia | Bladen fsl (Typic Albaquults) | 1960 | 5.0 | 1950 | 2280 | Hallock (1964) |

Table 7—Examples of effects of liming on grain yield of wheat, double-cropped with soybean, in field experiments in Alabama (unpublished data).

| Soil | Years | pH of unlimed soil | Grain yield Unlimed | Grain yield Limed |
|---|---|---|---|---|
| | | | kg ha$^{-1}$ | |
| Benndale sl† | 1968–1978 | 5.0 | 1140 | 1610 |
| Lucedale sl† | 1968–1978 | 4.8 | 1560 | 1720 |
| Lucedale fsl† | 1968–1978 | 4.8 | 2050 | 2610 |
| Dothan sl† | 1968–1978 | 5.2 | 2590 | 2720 |
| Hartsells fsl‡ | 1968–1978 | 4.9 | 1710 | 3020 |
| Decatur sil§ | 1968–1978 | 5.5 | 2860 | 3020 |
| Allen l (Typic Paleudults)§ | 1981 | 5.2 | 1830 | 2460 |

† Coastal Plain Province.
‡ Appalachian Plateau Province.
§ Interior Low Plateau Province.

tolerant than crimson clover; among summer annuals, browntop millet is more acid tolerant than sorghum-sudangrass hybrids (Table 8). The majority of forage crops have critical soil pHs somewhere between that for bermudagrass and alfalfa.

## 7. GRAIN SORGHUM

Grain sorghum is recommended as an alternate grain crop for corn in much of the drought-prone Southeast, where corn often fails. Although problems with birds, feeding quality, and markets have limited plantings to small areas outside Texas and Oklahoma, research on soil fertility requirements continues. In four widely scattered experiments (Table 9), critical soil pH appears to be about 5.3 to 5.5.

## 8. MISCELLANEOUS CROPS

Special attention is paid to soil pH adjustment on certain crops. Potatoes are subject to scab disease, which makes them unmarketable, if soil pH is much above 5.5; yet, yields are sharply reduced below pH 5.0 to 5.2 (C. E. Evans, 1974, unpublished data, Auburn Univ.). Sweet potatoes are equally afflicted above pH 5.5 by soil rot lesions, which can greatly reduce the marketable portion of the crop (Jones et al., 1977). Burley tobacco is subject to black root rot (*Thielaviopsis basicola*) if soil pH exceeds about 5.5 (Link, 1979).

Sugarcane production in the southern United States is limited primarily to the southernmost areas of Florida and Louisiana. The alluvial soils of southern Louisiana are mostly near neutral and do not need to be limed for sugarcane. The Histosols of southern Florida, however, may be quite acid, and liming can greatly increase yield. In an experiment on Pahokee muck (Lithic Medisaprists) at pH 4.8, cane yield and sugar yield were more than doubled by liming with a silicate slag or cement at 15 t ha$^{-1}$ (Elawad et al., 1982). The authors attributed this yield response to the silica content of the

Table 8—Examples of effects of liming on forage production in field experiments in the southern United States.

| State | Soil | Years | pH of unlimed soil | Forage yield: Unlimed | Forage yield: Limed | Reference |
|---|---|---|---|---|---|---|
| | | | | —— kg ha$^{-1}$ —— | | |
| | | **Alfalfa** | | | | |
| Alabama | Hartsells fsl | 1947–1949 | 5.1 | 450 | 7 390 | Adams (1956) |
| Georgia | Cecil-Lloyd sl | 1955–1959 | 5.5 | 670 | 6 870 | Giddens & Perkins (1960) |
| Mississippi | Grenada sil (Glossic Fragiudalfs) | 1959–1962 | 5.3 | 3 470 | 8 520 | Sanford et al. (1968) |
| Tennessee | Mountview sil | 1958–1959 | 5.2 | 1 680 | 6 270 | Long & Safley (1962) |
| Virginia | Tatum sil (Typic Hapludults) | 1957–1959 | 4.9 | 200 | 5 840 | Moschler et al. (1960) |
| | | **White clover** | | | | |
| Florida† | Myakka fs (Aeric Haplaquods) | 1973–1977 | 4.3 | 3 720 | 13 320 | Blue (1979) |
| Mississippi | Falaya sil (Aeric Fluvaquents) | 1959–1962 | 5.2 | 2 640 | 3 790 | Sanford et al. (1968) |
| | | **Bermudagrass‡** | | | | |
| Georgia | Cecil sl | 1959–1963 | 4.4 | 18 070 | 19 410 | W. E. Adams et al. (1967a) |
| Mississippi | Falaya sil | 1959–1962 | 5.4 | 8 390 | 8 680 | Sanford et al. (1968) |
| North Carolina | Eustis s (Psammentic Paleudults) | 1961–1965 | 4.6 | 5 900 | 11 500 | Woodhouse (1969) |
| Oklahoma | Chickasha fsl (Udic Argiustolls) | 1972–1975 | 5.0 | 6 360 | 6 950 | Altom & Rogers (1977) |
| | | **Crimson clover** | | | | |
| Alabama | Benndale sl | 1962–1964 | 5.1 | 1 230 | 1 900 | Adams (1968b) |
| Georgia | Lynchburg ls (Aeric Paleaquults) | 1955–1956 | 4.9 | 2 560 | 3 180 | Perkins et al. (1957) |
| South Carolina | Cecil sl | 1950 | 5.5 | 3 160 | 3 660 | Stewart & Pearson (1952) |
| | | **Millet** | | | | |
| Alabama | Benndale sl | 1962–1964 | 5.1 | 2 900 | 4 370 | Adams (1968b) |
| Georgia | Greenville scl (Rhodic Paleudults) | 1968–1970 | 5.2 | 8 520 | 8 520 | Walker et al. (1975) |
| | | **Sorghum** | | | | |
| Alabama§ | Benndale sl | 1962–1964 | 5.1 | 3 580 | 9 520 | Adams (1968b) |
| Georgia | Greenville scl | 1968–1970 | 5.2 | 6 240 | 9 430 | Walker et al. (1975) |
| | | **Ryegrass, annual** | | | | |
| Alabama | Benndale sl | 1962–1964 | 5.1 | 2 800 | 3 020 | Adams (1968b) |

† Sward included 'Pensacola' bahiagrass.
‡ 'Coastal' bermudagrass except for the experiment in Oklahoma where 'Midland' variety was grown.
§ The crop was a sorghum-sudangrass hybrid.

Table 9—Examples of effects of liming on yield of grain sorghum.

| State | Soil | Years | pH of unlimed soil | Grain yield Unlimed | Limed | Reference |
|---|---|---|---|---|---|---|
| | | | | —— kg ha$^{-1}$ —— | | |
| Alabama | Benndale sl | 1955–1960 | 5.1 | 1190 | 1820 | Adams (1968b) |
| Florida | Orangeburg fsl (Typic Paleudults) | 1972–1976 | 5.3 | 4630 | 4250 | Lutrick & Martin (1977) |
| Georgia | Congaree sl (Typic Udifluvents) | 1977–1979 | 5.1 | 3560 | 4280 | Duncan et al. (1980) |
| Louisiana | Olivier sil (Aquic Fragiudalfs) | 1967–1977 | 4.8 | 3490 | 4530 | Peevy et al. (1978) |

liming material, but since there was no $CaCO_3$ treatment to serve as a silica check, the yield response was caused more likely by the pH change than by the silica amendment.

Peach tree mortality is a major problem on sites that have been devoted to peach production for many years. The problem is referred to as "peach tree short life," and some research has suggested that it is associated with increased acidity in the soil profile (Perkins et al., 1979). However, in an experiment involving several practices as variables, Wehunt et al. (1980) concluded that liming increased yield but did not affect tree growth or survival.

## C. Reaction Rate of Lime

The efficiency or rate at which lime reacts with soil is largely determined by soil pH, limestone particle size, and extent of lime-soil intermixing. Soil pH greatly affects the dissolution rate of limestone. The reaction rate between lime and soil is greater at lower pH because of the pH-dependent solubility of liming materials (Pierre, 1930).

A thorough distribution of lime particles within the Ap horizon is realized only after considerable mechanical mixing of lime with soil. This is done by plowing, disking, harrowing, and cultivating, none of which will achieve complete and uniform mixing with the plow layer in one operation. However, complete and uniform mixing is not needed for lime to be effective immediately in neutralizing soil acidity. If the recommended rate of agricultural-grade limestone is properly mixed with the soil, planting may follow without delay because enough fines will be present to raise soil pH immediately above toxic Al and Mn levels and to correct a Ca deficiency. This has been demonstrated by the work of Alley (1981) and by the many on-farm experiments of Hartzog and Adams (1973), Adams and Hartzog (1980), and Burmester et al. (1981), in which planting immediately followed the incorporation of lime. In earlier days, when neither the implements nor the power was available for thorough lime-soil mixing, it was recommended throughout the region that lime be applied at least 2 to 3 months prior to planting. The basis for that recommendation no longer exists; liming may be done anytime between the harvesting of one crop and the planting of the next.

Only the availability of lime and the convenience of getting it spread should govern when a field is limed, because the neutralizing reaction is essentially a reaction between two surfaces: (i) the soil's surface occupied by exchangeable $H^+$ and $Al^{3+}$ and (ii) the lime particle's surface, which is the source of the neutralizing $CO_3^{2-}$ ion. The pseudosurface reaction between soil and lime is also evident by the effect that limestone particle size has on the reaction rate; the smaller the lime particles (greater surface area per unit weight), the faster the reaction rate (Pierre, 1930).

In an attempt to capitalize on the rapid reaction rate between small limestone particles and the soil, some lime vendors have promoted the use of a lime slurry. It is a suspension consisting of about 50% fine limestone, 48% water, and 2% attapulgite clay. By the use of fluid fertilizer equipment, it can be spread more uniformly than dry material, which is another advantage. In an evaluation of lime slurries in the field, Adams and Hartzog (1979) conducted four experiments with peanut on sandy soils and concluded that the slurry and dry material at equal $CaCO_3$ rates produced equal results. However, the slurry rate recommended by the vendor was too low to correct completely the effects of low Ca and low pH on peanut yields.

Heterogeneity of the soil-lime mixture typically follows liming for a considerable time, even where uniform spreading and thorough mixing are attempted. Coarse lime particles expose too little surface to react completely, whereas fine particles clump together, resulting in the exposed surface being much less than the total surface. Cycles of wetting and drying, along with additional mechanical mixing, appear to hasten the reaction rate because of bringing fresh soil particles into physical contact with partially reacted lime particles. This is true even for the most reactive of liming materials. For example, soil pH data by Hutcheson and Freeman (1965) suggested a large amount of unreacted $Ca(OH)_2$ 6 weeks after "thorough" mixing with a silty clay loam. Soil pH values measured during a period when fine particles of lime remain unreacted in the soil because of clumping are not representative of the actual soil pH because of the dissolution of lime particles that occurs during the procedure for measuring pH.

Maximum soil pH from liming is usually not attained in cultivated fields until 2 to 3 yr after liming, probably because of the difficulty of complete mechanical mixing of lime particles with soil to give a complete and homogeneous mixture and because of the slow dissolution rate of the larger particles. The reaction rate is influenced significantly by the fertilizer salt level in the soil: first, higher salt levels result in greater displacement of exchangeable $Al^{3+}$ and $H^+$ into soil solution; second, higher salt levels result in a greater rate of ion diffusion between microregions of soil acidity and microregions of adjacent limestone alkalinity.

The dissolution rate of limestone in untilled soil has not been adequately characterized, but it is certainly less than in cultivated fields (Moschler et al., 1962) since the reaction rate is greatly affected by the degree of physical mixing. Reaction rate is expected to be minimum where untilled soil receives surface-applied lime and no incorporation is attempted. Under such conditions, soil pH values should be considered by profile layers, with the surface sample extending to a depth of only 2 to 5 cm.

One of the anomalies of soil pH that complicates interpretations of lime efficiency, as well as critical pH values, is the fact that the pH of unlimed soil changes from year to year as well as during the year. Effects of seasonal changes in soil pH can be avoided by sampling soil following winter rains (or summer rains in the case of winter vegetable crops) when salt effect from fertilizers is at a minimum (Volk & Bell, 1944; Adams, 1978), as illustrated in Figure 1. However, the differences in soil pH that occur on unlimed soil from year to year do not appear to be related to salt level since the difference in pH will also be evident when pH values are measured in a dilute salt solution (Moschler et al., 1962) or when soil samples are leached prior to the pH measurement. In lime rate experiments, there appears to be a definite tendency for the pH values of all treatments in a particular experiment to increase or decrease in concert from year to year.

## IV. NITROGEN AND SOIL ACIDITY

The highly leached, weakly buffered soils of the southern United States are especially subject to pH changes caused by ammoniacal fertilizers. Pierre (1928), working in Alabama during the 1920s, was highly instrumental in creating the great concern for the effect of "acid-forming" N fertilizers on soil pH and crop yields. He demonstrated that nitrification of an $NH_4^+$ fertilizer produced soil acidity, but it was always less than that expected from the oxidation reaction. See Table 10 for a summary of reactions and resulting acidities. For example, $NH_4NO_3$ oxidizes according to

$$NH_4NO_3 + 2O_2 \rightarrow 2NO_3^- + 2H^+ + H_2O. \qquad [1]$$

Reaction [1] shows that each 28 g of N from $NH_4NO_3$ produce acidity equivalent to 100 g of $CaCO_3$, or 3.57 kg of $CaCO_3$ per kg of N. The same $CaCO_3$/N ratio applies to the nitrification of urea,

$$(NH_2)_2CO + 4O_2 \rightarrow 2NO_3^- + 2H^+ + CO_2 + H_2O, \qquad [2]$$

and to the nitrification of anhydrous ammonia,

$$NH_3 + 2O_2 \rightarrow NO_3^- + H^+ + H_2O. \qquad [3]$$

However, in each of the above examples, only about one half of the predicted acidity will actually develop, or 1.8 kg of $CaCO_3$ per kg of N (Pierre, 1928).

In the case of N fertilizers containing acidic anions other than $NO_3^-$, more than half of the predicted acidity is produced. For example, ammonium sulfate oxidation,

Table 10—Soil acidity produced by major N fertilizers.

| N source | Nitrification reaction | Residual soil acidity: Maximum, Acid residue | Maximum, $CaCO_3$ equiv. | Minimum, Acid residue | Minimum, $CaCO_3$ equiv. | Official value† |
|---|---|---|---|---|---|---|
| | | | kg $CaCO_3$ $kg^{-1}$ of N | | —— kg $CaCO_3$ $kg^{-1}$ of N —— | |
| Anhydrous ammonia | $NH_3(g) + 2O_2 \rightarrow H^+ + NO_3^- + H_2O$ | $H^+$ $NO_3^-$ | 50/14 = 3.6 | None | 0 | 1.8 |
| Urea | $(NH_2)_2CO + 4O_2 \rightarrow 2H^+ + 2NO_3^- + CO_2 + H_2O$ | $2H^+$ $2NO_3^-$ | 100/28 = 3.6 | None | 0 | 1.8 |
| Ammonium nitrate | $NH_4NO_3 + 2O_2 \rightarrow 2H^+ + 2NO_3^- + H_2O$ | $2H^+$ $2NO_3^-$ | 100/28 = 3.6 | None | 0 | 1.8 |
| Ammonium sulfate | $(NH_4)_2SO_4 + 4O_2 \rightarrow 4H^+ + 2NO_3^- + SO_4^{2-} + 2H_2O$ | $4H^+$ $2NO_3^-$ $SO_4^{2-}$ | 200/28 = 7.2 | $2H^+$ $SO_4^{2-}$ | 100/28 = 3.6 | 5.4 |
| Monoammonium phosphate | $NH_4H_2PO_4 + O_2 \rightarrow 2H^+ + NO_3^- + H_2PO_4^- + H_2O$ | $2H^+$ $NO_3^-$ $H_2PO_4^-$ | 100/14 = 7.2 | $H^+$ $H_2PO_4^-$ | 50/14 = 3.6 | 5.4 |
| Diammonium phosphate | $(NH_4)_2HPO_4 + O_2 \rightarrow 3H^+ + 2NO_3^- + H_2PO_4^- + H_2O$ | $3H^+$ $2NO_3^-$ $H_2PO_4^-$ | 150/28 = 5.4 | $H^+$ $H_2PO_4^-$ | 50/28 = 1.8 | 3.6 |

† Value adopted by Association of Official Analytical Chemists (Pierre, 1934).

$$(NH_4)_2SO_4 + 4O_2 \rightarrow 2NO_3^- + SO_4^{2-} + 4H^+ + 2H_2O, \qquad [4]$$

is estimated to be three fourths of the predicted value, i.e., 5.36 kg of $CaCO_3$ per kg of N instead of 7.14; and diammonium phosphate oxidation,

$$(NH_4)_2HPO_4 + 4O_2 \rightarrow 2NO_3^- + H_2PO_4^- + 3H^+ + 2H_2O, \qquad [5]$$

is estimated to give two thirds of the predicted value, i.e., 3.57 kg of $CaCO_3$ per kg of N instead of 5.36.

To explain his results, Pierre (1928) invoked the concept of physiological acidity and basicity, which results when plants take up unequal equivalents of anions and cations. Thus, the uptake of $NO_3^-$ in excess of cations neutralizes some of the acidity produced by nitrification. In the case of $SO_4^{2-}$ and $H_2PO_4^-$ portions of $NH_4^+$ fertilizers, neither ion is taken up in amounts sufficient to affect soil acidity and, consequently, each can be expected to express its full potential acidity (2 $H^+$ for each $SO_4^{2-}$ in reaction [4] and 1 $H^+$ for each $H_2PO_4^-$ in reaction [5]).

In addition to the physiological basicity resulting from excess uptake of $NO_3^-$, $NH_4^+$ fertilizers may not realize their potential acidity because of volatilization of $NH_3$(g) at high pH or because of denitrification at any pH. Hiltbold and Adams (1960) demonstrated that denitrification in soil results in an equivalent decrease in $H^+$, probably according to a reaction such as

$$NO_3^- + (CH_2O) \rightarrow \tfrac{1}{2}N_2O + CO_2 + \tfrac{1}{2}H_2O + OH^-. \qquad [6]$$

The extreme possibility from the combined effects of excess anion uptake, $NH_3$ volatilization, and denitrification is the total negation of acidity from the N portion of the fertilizer. For example, it is possible for the net acidity from $NH_4NO_3$, urea, or $NH_3$ to be zero. Obviously, this is highly unlikely. It is equally unlikely that the full acidity of reactions [1], [2], and [3] will develop. The actual soil acidity that develops from these N fertilizers will lie somewhere between the above limits. The midpoint was first chosen by Pierre (1928), and his method was subsequently adopted as the official method by the Association of Official Analytical Chemists (Pierre, 1934).

The least possible acidity from ammoniacal fertilizers containing $SO_4^{2-}$, $H_2PO_4^-$, or $HPO_4^{2-}$ is not zero. Regardless of what happens to the N portion, the $SO_4^{2-}$, $H_2PO_4^-$, and $HPO_4^{2-}$ portions will contribute residual acidity. In the case of $(NH_4)_2SO_4$, the least possible acidity is that caused by $2H^+ + SO_4^{2-}$, i.e., 2 mol of $H^+$ per mol of $SO_4^{2-}$ or per 2 mol of $NH_4^+$. In $CaCO_3$ equivalence, it is 3.6 kg of $CaCO_3$ per kg of N. In the case of $NH_4H_2PO_4$, the least possible acidity is that caused by $H^+ + H_2PO_4^-$, i.e., 1 mol of $H^+$ per 1 mol of $H_2PO_4^-$ or per 1 mol of $NH_4^+$. It, too, is equivalent to 3.6 kg of $CaCO_3$ per kg of N. For $(NH_4)_2HPO_4$, the least possible acidity is that caused by $H^+ + H_2PO_4^-$, i.e., 1 mol of $H^+$ per 1 mol of $H_2PO_4^-$ or per 2 mol of $NH_4^+$. It is equivalent to 1.8 kg of $CaCO_3$ per kg of N. Therefore, the acidity to expect from $(NH_4)_2SO_4$ and $NH_4H_2PO_4$ is between 3.6 and 7.2 kg of $CaCO_3$ per kg of N (midpoint is 5.4 kg) and between 1.8 and 5.4 kg of $CaCO_3$ per kg of N (midpoint is 3.6 kg) for $(NH_4)_2HPO_4$.

Table 11—Effect of surface-applied N on soil pH of Lakeland fine sand (Typic Quartzipsamments) under grass sods in lysimeters (Volk, 1956).†

| Soil depth, cm | Initial soil pH | Final soil pH after N applications‡ | | |
|---|---|---|---|---|
| | | 940 kg ha$^{-1}$ of N | | 470 kg ha$^{-1}$ of N as $NH_4NO_3$ |
| | | Urea | $NH_4NO_3$ | |
| 0–5 | 6.2 | 5.4 | 4.9 | 5.2 |
| 5–25 | 6.2 | 5.6 | 5.4 | 5.7 |
| 25–45 | 5.5 | 6.0 | 5.8 | 5.7 |

† Fertilizer applied in 1952 and 1953; soil pH measured in 1954.
‡ Average soil pH values under sods of bahiagrass, pangolagrass, and bermudagrass.

For acidity produced by nitrification to affect a soil permanently, an equivalence of basic cations must be leached. Leaching of cations from a soil depends on the supply of anions in soil solution. In neutral and alkaline soils, the $HCO_3^-$ ion is of major importance. In acid soils (< pH 6), however, its importance diminishes to insignificance because $H_2CO_3$ is only weakly ionized at low pH (p$K$ = 6.35). Thus, the leaching of bases from acid soils is activated primarily by $SO_4^{2-}$, $NO_3^-$, and $Cl^-$ ions. These anions serve not only as the counterpart for bases in the leachate but also as the means for the downward movement of $H^+$ and $Al^{3+}$ cations from strongly acid soil horizons (Volk, 1956).

The acidity that develops from an $NH_4^+$ fertilizer is not a constant value, but it is an integrated value reflecting soil characteristics, cropping systems, environmental variables, and chemical composition of the fertilizer. Data in Tables 10 and 11 show that soil acidity from a fertilizer is not constant. If acidity were based solely on N content, the acidity per kilogram of N from urea and $NH_4NO_3$ would be equal (see reactions [1] and [2]). The data in Table 11 show, however, that their predicted equalities were greatly altered when each was surface applied to grass sods (Volk, 1956). In this particular experiment, more acidity developed from 470 kg of N as $NH_4NO_3$ than from 940 kg of N as urea. This discrepancy in acidity resulted primarily because surface-applied urea is more susceptible to loss by volatilization of $NH_3$ at the high pH resulting from urea hydrolysis.

The amount of acidity that develops from each kilogram of N will also be affected by the rate of N, as shown by the data in Table 12. The acidity

Table 12—Comparison of potential and actual acidity that developed after applying $NH_4NO_3$ for 30 months to a Coastal bermudagrass sod on a Cecil sandy loam (Abruna et al., 1958).

| Total N added | Potential acidity† | Actual acidity‡ | |
|---|---|---|---|
| | | Total | per kg of N |
| kg ha$^{-1}$ | | | |
| 266 | 479 | 67 | 0.25 |
| 532 | 956 | 258 | 0.48 |
| 1064 | 1915 | 974 | 0.92 |

† $CaCO_3$ equivalence based on $CaCO_3$/N ratio of 1.8 for $NH_4NO_3$.
‡ $CaCO_3$ equivalence based on loss of bases from soil.

per kilogram of N from $NH_4NO_3$ applied to a Coastal bermudagrass sod was equivalent to 0.92 kg of $CaCO_3$ at the highest N rate but only 0.25 kg of $CaCO_3$ at the lowest N rate (Abruna et al., 1958). These acidity values were based on loss of basic cations from soils, which may have seriously underestimated actual acidity that developed from the fertilizer. It is probable that the increased acidity from nitrification accelerated the weathering rate of soil minerals, thereby releasing additional basic cations to the soil, which would give the appearance of less acidity than was actually produced. This is particularly true for soils of the Piedmont Province, where this experiment was conducted.

## V. SUBSOIL ACIDITY

The effect of soil acidity from $NH_4^+$ fertilizers is not confined to the surface soil but may extend well below the surface horizon. Undoubtedly, however, the greatest effects of acidity do occur in the surface soil because of the greater activity of microorganisms and plant roots. The increased acidity in subsoil horizons depends on leaching of $H^+$ and $Al^{3+}$ ions from the surface horizon as counterions of anions. The mobilized $H^+$ and $Al^{3+}$ ions from the surface horizon exchange for adsorbed basic cations of the lower horizon, which are subsequently leached along with mobile anions.

An example of profile acidity resulting from high rates of $NH_4^+$ fertilizers on bermudagrass sod is shown by the data in Fig. 5. After 3 yr and either 1680 or 3360 kg ha$^{-1}$ of N as $NH_4NO_3$ or $(NH_4)_2SO_4$, soil pH had dropped significantly. Ammonium sulfate was particularly effective in lowering pH; the overall effect of 1680 kg ha$^{-1}$ of N as $(NH_4)_2SO_4$ was about the same as that from 3360 kg ha$^{-1}$ of N from $NH_4NO_3$, even though their

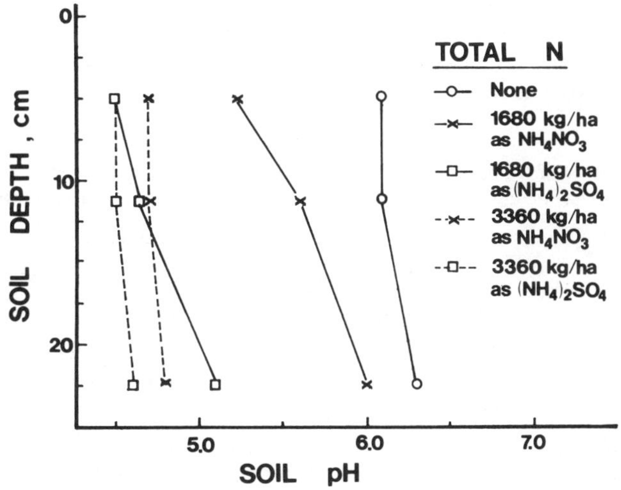


Fig. 5—Effect of surface-applied $NH_4NO_3$ and $(NH_4)_2SO_4$ on subsoil acidity under a Coastal bermudagrass sod on a Pacolet (formerly classified as Lloyd) clay loam (Typic Hapludults) after 3 yr (Adams & Pearson, 1969).

effects differed in the uppermost and lowest horizons. The latter difference was probably the result of differences in amounts of mobile anions available for cation leaching.

## A. Prevention

The inaccessibility of subsoil horizons for direct and effective liming treatments makes the prevention of increased subsoil acidity a highly desirable objective. The essence of such prevention is to maintain a favorable surface-soil pH. Where $NH_4^+$ fertilizers are used regularly, surface-soil liming should be practiced regularly.

In the case of sod crops and high N rates, the data of W. E. Adams et al. (1967b) suggest that maintaining pH of both surface and subsoil horizons might be difficult. With an annual N rate of 900 kg $ha^{-1}$ from $NH_4NO_3$, they found after 4 yr that a single initial lime application of 6.7 t $ha^{-1}$ had allowed surface soil (0–15 cm) pH to drop from 5.8 to 4.6, whereas the pH of a subsurface horizon (45–60 cm) had increased from 5.2 to 5.7. With four annual 6.7 t $ha^{-1}$ applications of limestone, surface soil pH was maintained at 5.8, whereas the pH of the 45- to 60-cm horizon had increased from 5.2 to 6.0. The data suggest that a lower, annual application rate was needed to maintain both surface and subsoil pH.

In a long-term experiment (32 yr) in Alabama with row crops at relatively low N rates and with annual liming treatments, subsoil pH to a depth of 60 cm was found to be greatly affected by surface-applied treatments (Table 13). Both surface-soil and subsoil pH remained essentially unchanged when the N source was $NaNO_3$. In contrast, $(NH_4)_2SO_4$ without lime resulted in a strongly acid profile to at least 60 cm, whereas $(NH_4)_2SO_4$ with lime had maintained soil pH mostly unchanged. When lime was used in conjunction with $NaNO_3$, subsoil pH increased markedly over the years.

In two field experiments with bermudagrass, Adams and Pearson (1969) found that surface-applied lime was more effective in preventing the accumulation of subsoil acidity from $NH_4NO_3$ than from $(NH_4)_2SO_4$ (less

Table 13—Effect of N sources and lime on soil profile acidity in two 2-yr cotton-corn rotation experiments in Alabama after 32 yr, 1930–1962 (Adams, 1981).

| | Soil pH by soil depth after 32 yr† | | | | | | | |
|---|---|---|---|---|---|---|---|---|
| | Dothan ls | | | | Lucedale sl | | | |
| N and lime treatment‡ | 0–15 cm | 15–30 cm | 30–45 cm | 45–60 cm | 0–15 cm | 15–30 cm | 30–45 cm | 45–60 cm |
| $(NH_4)_2SO_4$ | 4.8 | 4.8 | 4.3 | 4.3 | 4.8 | 4.8 | 4.9 | 4.9 |
| $(NH_4)_2SO_4$ + lime | 6.2 | 5.4 | 4.7 | 4.6 | 6.0 | 5.5 | 5.3 | 5.1 |
| $NaNO_3$ | 5.7 | 5.5 | 5.1 | 5.0 | 5.8 | 5.5 | 5.3 | 5.1 |
| $NaNO_3$ + lime | 6.4 | 6.4 | 6.3 | 6.3 | 6.8 | 6.6 | 5.8 | 5.2 |

† Initial surface soil pH was 6.0 at both sites.

‡ Nitrogen rate was 40 kg $ha^{-1}$ during 1930 to 1945 and 53 kg $ha^{-1}$ during 1946 to 1962. Basic slag was lime source; it was applied annually at 500 kg $ha^{-1}$ during 1930 to 1945 and at 800 kg $ha^{-1}$ during 1946 to 1962.

Table 14—Effect of $NH_4NO_3$ at annual N rate of 670 kg ha$^{-1}$ applied to Coastal bermudagrass sod for 5 yr, 1959–1964.

| State | Soil | Lime rate in 1959 | Soil pH in 1960 0–15 cm | Final soil pH by depth 0–15 cm | 15–30 cm | 30–45 cm | 45–60 cm |
|---|---|---|---|---|---|---|---|
| | | t ha$^{-1}$ | | | | | |
| Arkansas | Johnsburg sil | 0 | 5.2 | 4.5 | 4.8 | 4.9 | 4.7 |
| | (Aquic Fragiudults) | 4.5 | 5.5 | 5.1 | 5.0 | 4.8 | 4.7 |
| | | 18 | 6.5 | 6.3 | 5.5 | 4.9 | 4.9 |
| Georgia | Bladen cl | 0 | 4.5 | 4.3 | 4.2 | 4.1 | 4.1 |
| | | 9 | 5.5 | 4.8 | 4.3 | 4.1 | 4.1 |
| | | 36 | 5.7 | 5.9 | 4.2 | 4.1 | 4.1 |
| Mississippi | Grenada sil | 0 | 5.9 | 5.4 | 5.1 | 5.0 | 5.0 |
| | | 2.2 | 6.7 | 6.2 | 5.2 | 5.0 | 5.0 |
| | | 6.7 | 6.9 | 6.4 | 5.3 | 5.1 | 5.0 |
| S. Carolina | Goldsboro sl | 0 | 5.5 | 4.6 | 5.0 | 5.3 | 5.3 |
| | | 2 | 6.0 | 4.9 | 5.5 | 5.4 | 5.2 |
| | | 5 | 6.5 | 5.3 | 5.6 | 5.7 | 5.4 |

anions available for leaching); lime was more effective on Lakeland loamy sand than on Pacolet clay loam, probably because of more effective leaching of the loamy sand; lime was also more effective at lower N rates (ratio of $CaCO_3/N$ was constant).

The effects of surface-applied $NH_4NO_3$ and different liming rates on the acidity of representative soil profiles of the southern United States are given in Table 14. Although the data are inadequate for a quantitative evaluation of acidity accumulation, they do show the unequal development of acidity to various soil depths and the ameliorating influence of surface-applied lime on acidity development in the lower horizons.

### B. Neutralization

If subsoil horizons within a crop's rooting zone are too acid for normal root growth, partial neutralization of these horizons is highly desirable. Correction of shallow subsoil acidity (< 30 cm) by direct incorporation of lime is feasible. Doss et al. (1979) attempted to incorporate lime to a depth of 45 cm but barely got it beyond 30 cm. Direct incorporation of lime below 30 cm is apparently fraught with too many insurmountable hurdles, including excessive power requirements and mixing too much undesirable subsoil material with the surface soil.

Since thorough mixing of lime with subsoil is simply impractical, Pearson et al. (1973) made an effort to determine just how little mixing would be adequate for near-normal root proliferation into acid subsoil material. In a growth chamber experiment, they added strips of lime vertically and horizontally and observed root growth through a glass front. They concluded that about two thirds of the subsoil needed to be limed, with strips of lime being no more than 2.2 cm apart if adequate root growth was to be realized.

Table 15—Effect of high rates of $Ca(NO_3)_2$ on soil profile pH of a Lucedale sandy loam after 4 yr (Adams & Pearson, 1969).

| | | Final soil pH at annual N rates | | |
|---|---|---|---|---|
| Soil depth, cm | Initial soil pH | 450 kg $ha^{-1}$ | 900 kg $ha^{-1}$ | 1350 kg $ha^{-1}$ |
| 0–15 | 5.7 | 6.1 | 6.1 | 5.9 |
| 15–30 | 5.4 | 6.3 | 6.1 | 6.1 |
| 30–45 | 5.2 | 6.0 | 6.1 | 5.9 |
| 45–60 | 5.2 | 5.7 | 5.8 | 5.7 |
| 60–75 | 5.2 | 5.5 | 5.5 | 5.5 |

Since the direct incorporation of lime into subsoils is essentially impractical, judicious surface soil amendments may offer a feasible alternative. The data of W. E. Adams et al. (1967b) showed that high lime rates combined with high rates of $NH_4NO_3$ would significantly increase subsoil pH to a depth of at least 75 cm under a Coastal bermudagrass sod. Moschler et al. (1950) reported similar results in a 6-yr rotation where legumes supplied the N. The data in Table 13 show that combining a low annual rate of lime with a low $NaNO_3$ rate over 32 yr increased subsoil pH to a depth of more than 60 cm. In experiments with Coastal bermudagrass, Adams and Pearson (1969) showed that a high rate of $NaNO_3$ (448 kg $ha^{-1}$ of N annually) for 2 or 3 yr significantly increased subsoil pH to a depth of at least 45 cm in both a loamy sand and a clay loam soil; if lime were also surface applied, subsoil pH was increased even more. They also found that high rates of $Ca(NO_3)_2$ over a 3-yr period were quite effective in raising soil pH to a depth of at least 75 cm (Table 15). Although N rates ranged from 450 to 1350 kg $ha^{-1}$ and pH was increased between 0.3 pH unit (at 60–75 cm) and 0.8 pH unit (at 15–30 cm), there was no difference in final soil pH among N rates at any soil depth. Exchangeable $Ca^{2+}$ was also increased significantly in the subsoil layers, with the highest N rate resulting in the highest Ca level.

The neutralizing action in subsoil horizons of surface-applied lime plus high rates of $NH_4NO_3$ and high rates of surface-applied $NaNO_3$ and $Ca(NO_3)_2$ is a result of the physiological basicity produced by excess anion uptake by plants. Because Coastal bermudagrass is a particularly voracious absorber of nitrates, it proved to be a very effective crop for demonstrating the swiftness with which a root system can alter the pH of its own environment. The fact that maximum increases in pH were obtained with $Ca(NO_3)_2$ at a N rate of 450 kg $ha^{-1}$ (Table 15) suggests that the crop was unable to absorb and use N above this rate.

### C. Effects on Crop Yields

The yield-limiting effect of subsoil acidity in the southern United States became a focal point of research at Auburn University during the early 1960s. Despite considerable progress in delineating root growth responses to subsoil acidity factors, few data have been collected that help to define the significance of the problem to growers on a field basis.

Ultisols, by definition, have low base saturated subsoil horizons, and they constitute the major soil order of the southern United States. Thus, the potential is high for acid subsoils to limit crop yields over a wide geographical area. Research on subsoil acidity at Auburn University began with the objective of identifying specific chemical factors responsible for poor plant performance. Pearson (1966) summarized results from these early experiments, which had shown that too little Ca and too much Al were the causal factors. Both directly inhibited or prevented root proliferation. In contrast, high Mn levels (another factor of acid soil infertility) have no direct effect on root growth. Finally, soil pH seems to never be low enough for $H^+$ to be a growth-limiting factor in mineral soils; it can be in organic soils.

The effect of denying subsoil horizons to root systems is to subject the crop to a restricted rooting zone. If the affected horizon is just below the Ap, the crop will have a very shallow root system and most soil water below the Ap horizon will be denied the crop, resulting in frequent drought stresses during the summer. If adequately limed surface soils overlie excessively acid subsoils, plants tend to show no deficiency or phytotoxic symptoms. The crop will appear normal, except for less vigorous growth and early symptoms of water stress after rains.

Although Howard and Adams (1965) and Lyle and Adams (1971) identified the critical soil level for Ca deficiency to occur in acid subsoils, it is generally believed that Ca deficiency in subsoils is unlikely in the field. The concurrent findings of widespread excess Al in acid subsoils and its inhibitory effect on root growth (Adams & Lund, 1966; F. Adams et al., 1967; Adams, 1981; Soileau et al., 1969) were responsible for the view that Ca deficiency in subsoils would probably be rare in the Southeast. Expected exceptions would be the acid, fine sands of Florida (Gammon, 1957).

In a more recent study of acid subsoils, Adams and Moore (1983) carefully sampled the uppermost three subsoil horizons of six Coastal Plain soils and found that Ca deficiency was, indeed, a growth-limiting factor for root proliferation. Of 13 horizons that inhibited root growth because of low pH, Ca deficiency was the sole factor in four and was a cofactor (with excess Al) in three; excess Al was the sole factor in five horizons. Excess Al was a factor only in argillic horizons; Ca deficiency occurred in some argillic horizons, but it was the only inhibitory factor in E, EB, and BE horizons. Thus, Ca deficiency in subsoil horizons of Coastal Plain soils may be considerably more widespread in the Coastal Plain than previously believed.

Field evaluations of subsoil acidity effects on crop yields have been skimpy, indeed. A major reason is the obvious difficulty of establishing uniform subsoil acidity levels. Using plots from long-term fertilizer experiments and plots from an area where the soil profile had been rapidly acidified by high rates of $(NH_4)_2SO_4$, F. Adams et al. (1967) found that cotton yields were reduced by about one third when roots were restricted primarily to the Ap horizon because of strongly acid horizons immediately below it (Table 16). In a subsequent experiment on one of the same field sites, Adams and Pearson (1970) found peanut yields and peanut roots to be insensitive to the same subsoil acidity level that had limited cotton yield and restricted cotton root growth to the Ap horizon.

Table 16—Effect of subsoil pH on cotton yield in field experiments and on cotton root growth in growth chamber experiments (F. Adams et al., 1967).

| Soil | Subsoil pH at 15–30 cm† | Seed-cotton yield* | Rel. root growth after 48 hr* |
|---|---|---|---|
| | | kg ha$^{-1}$ | |
| Dothan sl | 5.0 | 1820b | 0.34c |
| | 5.5 | 2680a | 0.65b |
| | 6.4 | 2620a | 1.00a |
| Lucedale sl | 4.9 | 1910b | 0.26c |
| | 5.4 | 2280a | 0.51b |
| | 6.0 | 2030b | 1.00a |
| Lucedale fsl | 4.2 | 1550c | 0.16d |
| | 4.4 | 1980b | 0.40c |
| | 5.2 | 2430a | 0.69b |
| | 5.9 | 2400a | 1.00a |

* Yields and root growths followed by the same letter within a soil type are not different at 0.01 probability level.
† Surface-soil pH was 6.0 to 6.5.

In a follow-up experiment on another of the above sites, Doss and Lund (1975) found that frequent irrigations reversed the effect of low subsoil pH on cotton yield. Although roots were scanty or absent in low pH horizons and abundant in moderate pH horizons, average cotton yields were actually higher on the lowest subsoil pH plots. This apparent negative yield response to increasing subsoil pH, however, is peculiar to problems associated with controlling boll rot and insects on cotton. Higher subsoil pHs produced larger plants with more foliage, thereby enhancing boll rot and making insect control measures less effective. The combination of larger plants and greater disease and insect damage resulted in lower lint yields.

Doss et al. (1979) incorporated lime to depths of 15, 30, and 45 cm on three different soils of the Coastal Plain. Both corn and cotton yields were increased by incorporating lime to a depth of 30 cm. Increasing the depth of liming beyond 30 cm was not beneficial.

In a field experiment with two tomato cultivars, Doss et al. (1977) found that both height of plants and fruit yield were increased by increasing subsoil pH from pH 4.4 to 5.5 (pH of Ap horizon was 6.0). With subsoil pH above about 5.5, plant heights were greater but yields were not.

## VI. FACTORS OF ACID SOIL INFERTILITY

The relationship between soil acidity and plant growth has been the subject of considerable research in the region, and it was recognized early that soil acidity affected plant growth via multiple mechanisms. Positive delineation and identification of the different factors, however, were made difficult by the interdependence of soil pH, base saturation, exchangeable $Al^{3+}$, water-soluble Mn, and availability of inorganic nutrients.

Many early field experiments, as late as the 1940s, were conducted with little regard to even the most elementary soil characteristics. Consequently, those experiments were not helpful in elucidating the causes of acid soil infertility. At the same time, however, the laboratory and greenhouse experiments at the Alabama Agricultural Experiment Station had made significant contributions by the early 1930s to an understanding of soil acidity and plant growth (Pierre, 1928, 1930, 1931; Pierre & Scarseth, 1931).

The contradictory results obtained by various researchers during the 1920s and early 1930s in defining maximum pH values at which response to lime occurred led some workers to conclude that soil pH was not a direct factor of acid soil infertility. Instead, it was considered to be a symptom of another soil property that actually caused poor plant growth. It was during this early period that Al toxicity became widely recognized as a major factor of infertility.

The most notable early work in the region was that done by Pierre (1931) in Alabama. Corn, sorghum, and barley were grown in the greenhouse on each of 13 soils. Pierre attempted to relate vegetative growth of these plants to either soil pH, base saturation, or soil-solution composition. The following conclusions were reached by Pierre and serve to illustrate the unresolved and conflicting theories on the nature of acid soil infertility that existed in 1930:

1) Soil pH is not "the direct cause of poor plant growth nor the main factor governing plant response to liming."
2) "Soluble Al is not the primary factor affecting plant growth on acid soils."
3) Ca concentration "in the soil solution bore no relationship to plant injury" in acid soils.
4) There is no relationship between "Mn in solution and plant injury from soil acidity."
5) "Percentage base saturation of soils is. . .one of the most important factors in determining the growth of plants on acid soils."

Current concepts on acid soil infertility embrace essentially the same factors as Pierre considered. More than one may simultaneously inhibit plant growth in some cases, whereas a single factor may be responsible for poor growth in others. The factors are H, Al, and Mn toxicities and Ca, Mg, and Mo deficiencies.

### A. Hydrogen Ion Toxicity

It is a widely held view that pH per se is not a major factor of poor plant growth in acid soils in the southern United States. This is based on results from nutrient-solution experiments in which various pH levels were used to evaluate the effect of $H^+$ activity on plant growth. Available data suggest that pH must be below 5.0 for $H^+$ to be toxic to plants. The early work of Arnon and Johnson (1942) showed that the pH required for toxicity was genetically dependent. For example, they showed $H^+$ was toxic at pH 4.0 to tomato but not to bermudagrass. Howard and Adams (1965) found a drastic reduction in the growth rate of cotton taproots below pH 4.2. Lund (1970) found that $H^+$ was toxic to soybean taproots at pH 4.75 and below, provided solution Ca was very low. With increasing solution Ca levels, however, $H^+$ toxicity was increasingly ameliorated.

Although $H^+$ activity has not been identified in the field as the factor responsible for poor plant growth, it can be expected under certain conditions. A prime requisite for such a condition would be a lack of toxic levels of soluble Al and Mn, as might occur in the sands or organic soils of Florida and North Carolina. Evans and Kamprath (1970) in a greenhouse experiment reported sharply reduced corn growth by suspected $H^+$ toxicity in an organic soil where soil pH was $< 4.0$.

### B. Aluminum Toxicity

The most common factor responsible for acid soil infertility in the southern United States is believed to be excess Al. After Miyake (1916) demonstrated the toxic effect of Al on rice plants, Pierre (1931) was probably the first in the region to make an intensive effort to ascribe Al toxicity as the major factor causing poor yields on acid soils. Unfortunately, his data led him to conclude that "soluble Al is not the primary factor in influencing the growth of acid-sensitive crops on acid soils." Although procedures available at that time for the determination of very low Al concentrations lacked desirable levels of accuracy and sensitiveness, his data showed a wide variation in soil-solution Al levels at pH 4.8 for different soils. Pierre et al. (1932) subsequently substantiated and elaborated on the soil pH-Al relationship with additional soils at other pH levels. For any particular soil, solution Al increased with decreasing pH, but the Al concentration at the same pH differed widely among soils. Adams and Lund (1966), Evans and Kamprath (1970), and Richburg and Adams (1970) reported similar findings 40 yr later.

Hester (1935) found that the appearance of Al in leaching waters of three Atlantic Coastal Plain soils was closely related to plant growth but not to soil pH. Plant growth was markedly reduced at the pH where Al appeared in the leachate. Aluminum appeared in the leachate at different pH levels, namely, at pH 4.8 in Portsmouth soil (Typic Umbraquults), at pH 4.9 in Bladen soil (Typic Albaquults), and at pH 5.5 in Norfolk soil (Typic Paleudults).

Since the toxicity of Al is manifested primarily in retarded root growth, roots are excellent material for studying the effect of soil Al on plant growth. By placement of part of the root system in a complete nutrient, nontoxic medium and the remaining part in a medium containing Al, some of the complicating nutritional factors involving Al and plant growth are avoided. This approach was used by Ragland and Coleman (1959) and by several researchers at Auburn University (Adams & Lund, 1966; Adams & Moore, 1983; Richburg & Adams, 1970; Rios & Pearson, 1964; Lance & Pearson, 1969; Lund, 1970; F. Adams et al., 1967; Adams & Pearson, 1970). Although weight of plant tops is a reliable measure of the intensity of

Al toxicity, root weights are not (Rios & Pearson, 1964; Sartain & Kamprath, 1975). Root systems change from sparse branching of large-diameter roots at low pH and high Al to thick branching of small-diameter roots at high pH and low Al. Although surface area of a root system may increase considerably when root medium changes from low to high pH, its weight may change only slightly (except at very low pH).

Predicting the incidence and intensity of Al toxicity would be greatly simplified if Al toxicity were related directly to exchangeable $Al^{3+}$ or Al saturation of the cation exchange sites. This, in fact, is the case for soil from a particular site when it is subdivided into samples and treated to give different pH levels. Unfortunately, the results for one soil site are not quantitatively applicable to other sites; they are not applicable even between horizons of the same site.

By defining "critical" Al saturation as the maximum Al percentage of cation exchange sites (sum of exchangeable $Ca^{2+}$, $Mg^{2+}$, $Mn^{2+}$, $K^{+}$, $Al^{3+}$, and sometimes $H^{+}$) that fails to inhibit plant growth, results from various researchers can be compared. For several Ultisols in North Carolina, critical Al saturation for soybean appeared to be about 20 to 25% (Evans & Kamprath, 1970; Kamprath, 1970; Sartain & Kamprath, 1975); it was about 10% for Histosols (Mengel & Kamprath, 1978). In other experiments, critical Al saturation for soybean was about 20% for Ultisols in Alabama (Pearson et al., 1977), 6% for an Ultisol in South Carolina (Pearson et al., 1977), and 5% for a Spodosol in Florida (Zakaria et al., 1977).

Other examples of critical Al saturation values include 4% for alfalfa (Moschler et al., 1960) and 5% for white clover, tall fescue, and sericea lespedeza (Shoop et al., 1961) on some soils in Virginia. Data for corn range from 40 to 50% for three Ultisols in North Carolina (Kamprath, 1970) to 1 to 8% for six Ultisols in Georgia (Farina et al., 1980a).

There are major problems in comparing Al toxicity data obtained by different researchers: (i) genetic differences in test crops, (ii) procedural differences in extracting Al from soil samples, (iii) differences in cation exchange coefficients of soil material, and (iv) failure to eliminate all growth-inhibiting factors except Al. A resolution of conflicting critical Al saturation values apparently awaits the adoption of methods that are more standardized than those currently in use.

Even where procedures and genetic material are constant, critical Al saturation values are not. In a summary of root growth experiments with cotton over several years, critical Al saturation ranged between 5 and 25% (Adams, 1981). Differences in critical Al levels were just as great among different sites of the same soil series as among different soils. With Al saturation at about 50%, relative root growth in the different soils ranged from 34 to 77%, a twofold difference. In a study of six soil profiles of the Coastal Plain in Alabama, Adams and Moore (1983) found similar results: Al toxicity occurred in the Bt2 horizon of one soil with an Al saturation of only 2% but not in the Bt1 horizon of another soil with an Al saturation of 56%. Obviously, a more definitive method for identifying the toxic fraction of soil Al is needed.

It is not exchangeable $Al^{3+}$ per se that is toxic to plants; it is solution Al. An early effort to relate plant growth at low soil pH to Al toxicity and soil-solution Al concentration was made by Pierre (1931) at the Alabama Agricultural Experiment Station. He found that the threshold Al concentration for toxicity varied considerably among the soils he evaluated. Others have reported similar results for the region (Adams & Lund, 1966; Adams & Moore, 1983; Adams & Pearson, 1970; F. Adams et al., 1967; Ragland & Coleman, 1959; Richburg & Adams, 1970; Evans & Kamprath, 1970). In an effort to define threshold toxic levels of solution Al that will be applicable to all soil solutions, Adams and Lund (1966) introduced the ion activity concept. Although this approach greatly improved the correlation between solution Al and its phytotoxic effect, it is not the sole criterion. Other cations, particularly $Ca^{2+}$, have a decided ameliorating effect on Al toxicity (Lund, 1970; Lance & Pearson, 1969). It is also possible for $H^+$ activity to influence Al toxicity because of its effect on cell membrane permeability and its effect on speciation of Al ions in solution.

Aluminum is probably present in soil solution as organic compounds or chelates as well as inorganic ionic species. It is quite likely that chelated Al is nontoxic to plants, or at least, much less than inorganic $Al^{3+}$ ions (Bartlett & Riego, 1972). Adams and Moore (1983) speculated that chelation was the reason that relatively high soil-solution Al levels were nontoxic to cotton roots in several E horizons of Coastal Plain soils in Alabama.

It seems clear that more definitive data are needed for either Al saturation of cation exchange sites or soil-solution Al activity to be reliable predictors of Al toxicity under a range of soil conditions. The chemistry of solution, exchangeable, and solid-phase Al in soils still needs further elucidation.

### C. Manganese Toxicity

Manganese toxicity was first proposed as the agent responsible for poor cotton growth on some very acid soils in Alabama in 1918 by Funchess (1918). Although he subsequently realized that his conclusion was erroneous (Funchess, 1919), Mn toxicity has since been identified as a major cause of acid soil infertility in the southern United States.

The potential for Mn toxicity lies in the amount of easily reducible Mn present in the soil. As shown by data in Table 17, the Mn content of representative soils of the region varies from a negligible amount to $>$ 3000 mg $kg^{-1}$. Although the amount of reducible soil Mn needed to be potentially toxic is not clearly defined, it appears to be on the order of 50 to 100 mg $kg^{-1}$ Mn. Soils of the Coastal Plain along the Atlantic Seaboard are generally low in Mn, much too low for potentially toxic levels of Mn. Soils of the Gulf Coastal Plain are much higher in reducible Mn, with many having potentially toxic levels. An early report by Carlyle (1931) showed that the Mn contents of Coastal Plain soils in eastern Texas were similar to those listed for the Gulf Coastal Plain in Table 17. The greatest potential for Mn toxicity is with soils derived from limestone residuum on the Interior Low Plateau of Alabama, Tennessee, and Kentucky.

Table 17—Easily reducible Mn contents of representative, acid, surface soils of the southern United States (F. Adams, 1960, unpublished data).

| Soil series | State | Soil Mn, mg $kg^{-1}$ |
|---|---|---|
| **Well-drained Coastal Plains** | | |
| Bowie (Plinthic Paleudult) | Louisiana | 103 |
| Kalmia | Alabama | 176 |
| Lakeland | Georgia | 3 |
| Lakeland | S. Carolina | 28 |
| Lucedale | Alabama | 1540 |
| Lucedale | Georgia | 108 |
| Marlboro (Typic Paleudults) | Georgia | 36 |
| Marlboro | S. Carolina | 8 |
| Norfolk | Alabama | 76 |
| Norfolk | Georgia | 21 |
| Norfolk | N. Carolina | 4 |
| Norfolk | S. Carolina | 5 |
| Orangeburg | Alabama | 314 |
| Rumford (Typic Hapludults) | Virginia | 4 |
| Ruston (Typic Paleudults) | Georgia | 40 |
| Ruston | Lousiana | 84 |
| Sassafras (Typic Hapludults) | Virginia | 104 |
| Shubuta (Typic Paleudults) | Louisiana | 20 |
| Tifton | Georgia | 26 |
| **Poorly drained Coastal Plains** | | |
| Beauregard | Louisiana | 116 |
| Bladen | N. Carolina | 3 |
| Charleston (Aquultic Hapludalfs) | S. Carolina | 13 |
| Coxville (Typic Paleudults) | S. Carolina | 6 |
| Crowley (Typic Albaqualfs) | Arkansas | 11 |
| Crowley | Louisiana | 222 |
| Dunbar (Aeric Paleaquults) | S. Carolina | 7 |
| Duplin (Aquic Paleudults) | N. Carolina | 6 |
| Eulonia (Aquic Hapludults) | S. Carolina | 17 |
| Portsmouth | N. Carolina | 4 |
| Rains (Typic Paleaquults) | N. Carolina | 5 |
| **Appalachian Plateau** | | |
| Fannin (Typic Hapludults) | Georgia | 18 |
| Hartsells | Alabama | 114 |
| **Appalachian Plateau, continued** | | |
| Hartsells | Tennessee | 56 |
| Holston | Tennessee | 45 |
| **Interior Low Plateau** | | |
| Clarksville (Typic Paleudults) | Tennessee | 357 |
| Decatur | Alabama | 2840 |
| Decatur | Alabama | 333 |
| Decatur | Tennessee | 3460 |
| Dewey (Typic Paleudults) | Tennessee | 212 |
| Dewey | Tennessee | 1050 |
| Loradale (Typic Argiudolls) | Kentucky | 1130 |
| Maury (Typic Paleudalfs) | Tennessee | 715 |
| Pembroke (Mollic Paleudalfs) | Kentucky | 1230 |
| Sango (Glossic Fragiudults) | Kentucky | 170 |
| Sango | Kentucky | 130 |
| Shelbyville (Mollic Hapludalfs) | Kentucky | 416 |
| **Piedmont** | | |
| Appling (Typic Hapludults) | Georgia | 68 |
| Cecil | Alabama | 29 |
| Cecil | Georgia | 205 |
| Davidson (Rhodic Paleudults) | Georgia | 465 |
| Davidson | S. Carolina | 107 |
| Georgeville (Typic Hapludults) | S. Carolina | 260 |
| **Alluvium** | | |
| Commerce | Louisiana | 86 |
| Dubbs | Arkansas | 143 |
| Sharkey (Vertic Haplaquepts) | Arkansas | 700 |
| Sharkey | Louisiana | 10 |
| **Loess** | | |
| Grenada | Tennessee | 320 |
| Olivier | Louisiana | 760 |
| **Black Belt** | | |
| Vaiden (Vertic Hapludalfs) | Alabama | 182 |

Absorption of Mn by plants depends primarily on the amount of divalent Mn in the ambient solution surrounding the root system. However, less Mn is absorbed as the concentration of other cations is increased. Under most conditions, only a few milligrams per liter of solution Mn is required to produce symptoms of Mn toxicity. Nutrient solution experiments have shown toxicity symptoms to occur in tobacco with solution Mn at 15 to 20 mg $L^{-1}$ (Bortner, 1935; Hiatt & Ragland, 1963) and in cotton at 10 mg $L^{-1}$ (Adams & Wear, 1957). In soil experiments, Morris (1949) found 13 mg $L^{-1}$ Mn in soil solution of a Cecil sandy loam (Typic Hapludults) to be toxic to lespedeza and sweetclover, and Adams and Wear (1957) found 11 mg $L^{-1}$ Mn in soil solution of a Kalmia fine sandy loam (Typic Hapludults) to be toxic to cotton. Thus, it appears that toxic levels of Mn in soil solution correspond closely to that in nutrient solutions.

In alkaline and near-neutral soils, only a small fraction of soil Mn is in the available, divalent form. As soils become more acid, however, this fraction increases. For example, a Kalmia fine sandy loam at pH 4.5 contained almost half of its reducible Mn in the soluble, divalent form (Adams & Wear, 1957). The maximum pH at which soluble Mn can be high enough to be toxic appears to be about 5.5.

Because of striking visual symptoms, Mn toxicity is frequently observed on fields of cotton, soybean, and tobacco. Bortner (1935) reported Mn toxicity on tobacco in Kentucky with soil pHs from 4.6 to 5.5. Probably because of fertilizer effects on measured pH, Link (1979) found Mn toxicity on tobacco in Virginia only at pH 5.0 or less. Lancaster and Gholston (1953) reported Mn toxicity on cotton, a condition known as *crinkle leaf,* to occur on a number of soils in Mississippi that ranged in pH from 4.2 to 5.4. Additional incidences of crinkle leaf on cotton have been reported on soils within the same pH range from Louisiana, Arkansas, and Alabama (Neal & Lovett, 1938; Smith et al., 1955; Young & Beacher, 1952). The writer observed Mn toxicity symptoms on many of the liming experiments on soybean and cotton in northern Alabama as reported by Mitchell et al. (1977) and Burmester et al. (1981). The field-observed toxicities have been on soils that are typically high in easily reducible Mn.

The toxic action of Mn is evidenced on the aerial portion of the plants; roots are insensitive to excess Mn, except as they are affected by reduced top growth. Symptoms of Mn toxicity appear quite differently on different species; e.g., tobacco leaves are stunted and chlorotic with necrotic spots (Bortner, 1935); alfalfa leaves are afflicted with tip and marginal chlorosis (Foy, 1964); cotton plants are stunted, leaves are puckered and distorted, and normal branches frequently develop from the basal nodes of main stems (Neal, 1937).

The threshold concentration of leaf Mn for toxicity symptoms varies widely among species. Hiatt and Ragland (1963) found that burley tobacco leaves contained about 3000 mg $kg^{-1}$ Mn at first appearance of symptoms in nutrient solution experiments. In field experiments, however, Link (1979) found that the lowest leaf-Mn concentration associated with reduced yield was 1710 mg $kg^{-1}$; toxicity symptoms appeared with leaf Mn at only 500 mg

$kg^{-1}$ during one year, but they failed to appear the next year with leaf Mn as high as 1800 mg $kg^{-1}$. No explanation was offered for this apparent anomaly. Leaf-Mn content of cotton leaves associated with incipient toxicity symptoms is about 3000 mg $kg^{-1}$ (Adams & Wear, 1957). Toxicity symptoms appear in soybean with leaf Mn at 300 to 500 mg $kg^{-1}$.

The potential toxicity at any one level of easily reducible Mn and soil pH is less at higher levels of soil organic matter. Apparently, organic matter has a great capacity for complexing $Mn^{2+}$ ions and, thus, removing them from soil solution. Potential toxicity is also less in soils of high CEC because more $Mn^{2+}$ occupies exchange sites, leaving less in solution for uptake by plants.

The addition of a neutral salt, such as KCl, increases soil-solution Mn because of cation exchange reactions and intensifies Mn toxicity. Poor soil aeration also increases the potential for Mn toxicity. The valency of soil Mn is quite responsive to changes in the redox potential of soils as well as to pH changes. However, this is probably of significance only in soils that are subject to periodic flooding.

Although soil-solution Mn is a function of pH, soil pH is not a reliable predictor of Mn toxicity. That is because of the sensitivity of Mn valency to redox potential, microbial activity, soil drying, and heat (the latter two are particularly important in laboratory samples). Heating and drying soil samples greatly increases solution Mn levels, which means that Mn levels in fields are probably much lower than many laboratory-reported values (Nelson, 1977).

Frequently, toxic effects from Al and Mn occur simultaneously on high Mn soils. Since Al toxicity affects root systems and Mn toxicity affects leaves, Mn toxicity symptoms are highly visible and may be given undue credit for inhibited plant growth (Foy, 1964). As pointed out by Hiatt and Ragland (1963), symptoms of Mn toxicity must be rather severe before plant growth is greatly inhibited in the absence of Al toxicity. In many soils of the region, solution Al can be expected to be at toxic levels at about the same pH that incipient Mn toxicity occurs.

### D. Calcium Deficiency

Among the list of benefits popularly claimed for liming acid soils is the correction or prevention of Ca deficiencies. Although this claim seems logical, Ca deficiency per se is unusual in the southern United States except for special crops. Under many acid soil conditions, the addition of soluble Ca salt actually aggravates the infertility of acid soils (Adams & Wear, 1957; Morris, 1949; Ragland & Coleman, 1959).

The capacity of a soil to supply Ca for plants depends on the CEC, Ca saturation of the CEC, and cation exchange coefficients. Acid soils of very low CEC, e.g., fine sands with little organic matter, may inherently contain inadequate Ca for many crops (Gammon, 1957). In very dilute soil solutions, with pHs near 5.0, Adams and Moore (1983) found that cotton taproots suffered Ca deficiency if solution Ca was $< 0.3$ m$M$. In dilute nu-

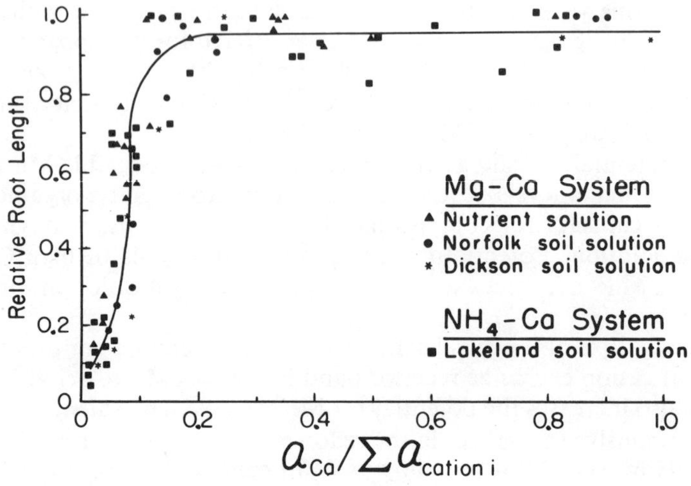


Fig. 6—Effect of the activity ratio of Ca/total cations in nutrient solutions and soil solutions in situ on growth rate of cotton taproots (Howard & Adams, 1965; Adams, 1966).

trient solutions, however, Lund (1970) reported that as little as 60 $\mu M$ of Ca at pH 4.5 was enough to prevent Ca deficiency in soybean taproots; even less Ca was needed at higher pHs.

If such cations as $Mg^{2+}$, $K^+$, $Na^+$, and $NH_4^+$ are added in large amounts to a soil already low in Ca (but not deficient), the result may well be Ca deficiency problems with that crop. This antagonistic effect of cations on Ca availability has been demonstrated in the region with sweetclover roots (Gammon, 1957), cotton roots (Howard & Adams, 1965; Adams, 1966; Bennett & Adams, 1970), sudangrass (Bennett & Adams, 1970), soybean roots (Lund, 1970), and loblolly pine roots (Lyle & Adams, 1971). These data have shown clearly that Ca availability is a function of the ratio of Ca to other cations (Fig. 6), with divalent cations being more antagonistic than monovalent cations on Ca availability.

The ratio concept of Ca availability, as validated in solution experiments, is equally applicable to exchangeable cations. The importance of exchangeable $Ca^{2+}$ for plant growth in acid soils was pointed out by Pierre (1931) and later by Mehlich and co-workers (Mehlich, 1942; Mehlich & Colwell, 1944; Mehlich & Reed, 1947), even though they credited Ca with benefits that actually belonged to decreasing H, Al, and Mn toxicities. It is only in the absence of H, Al, and Mn toxicities that Ca saturation becomes a useful measure of Ca availability.

Calcium deficiencies in the field are usually relegated to specific crops, which have specific internal problems with delivering Ca to certain tissue. Such crops suffer Ca deficiencies on the same soil that supplies adequate Ca for other crops. The most notable examples of such crops in the region are peanut (Cox et al., 1982), tomato (Geraldson, 1957), tobacco (Peedin & McCants, 1977), and celery (Geraldson, 1954).

The unusual fruiting habit of the peanut plant makes it one of the most susceptible crops to Ca deficiency for fruit development. The Ca requirement for vegetative and flower production is not unlike that of most other plants (Wolt & Adams, 1979), but it is appreciably higher for nut production. Field-observed Ca deficiency is manifested by unfilled ovarian cavities, known as *pops.* Bledsoe et al. (1949) used $^{45}Ca$ to show that Ca translocation from roots to gynophore and ovary was adequate until the peg (fertilized ovary) penetrated the soil. Once the developing gynophore penetrates the soil, it ceases to transpire root-absorbed water and, consequently, loses access to root-absorbed Ca (Ca does not move in phloem tissue). Consequently, the peanut fruit must absorb Ca directly from its ambient soil solution. Because the fruit is an inefficient absorber of Ca and each fruit is surrounded by a very small volume of soil solution, the Ca concentration of the ambient solution needs to be relatively high.

Calcium deficiency was recognized as a major factor of peanut production in the southern United States during the 1940s. A series of research papers by the North Carolina Agricultural Experiment Station was highly instrumental in defining the effect of Ca deficiency on peanut and relating the incidence of Ca deficiency to soil properties (Brady & Colwell, 1945; Burkhart & Collins, 1942; Colwell & Brady, 1945; Colwell et al., 1945; Middleton et al., 1945). Rogers (1948) found similar results in Alabama.

The Ca level required to prevent pops varies among cultivars. In general, large-seeded cultivars require a higher level than small-seeded ones. However, it seems unlikely that seed size per se is the controlling factor; more likely, there is a gene linkage between seed size and Ca transfer from shell to kernel (Beringer & Taha, 1976). There is evidence for this in the large-seeded cv. Florigiant, which has a Ca requirement closer to that of small-seeded cultivars than to other large-seeded ones (Hartzog & Adams, 1973; Walker et al., 1976).

Since peanuts are generally grown on soils of low CEC and low Ca, it is frequently essential to supplement soil Ca with lime, gypsum, or both. The Ca need of small-seeded cultivars is readily met by a good liming program (Adams & Hartzog, 1980). Some large-seeded cultivars may need supplemental gypsum applications. The objective of fertilizing and liming peanut is to add Ca to the pegging zone; thus, care must be taken to ensure that the added Ca is in that zone during fruit development.

Blossom-end rot of tomato is a physiological disorder caused by a deficiency of Ca at the flower end of the fruit. Although the genetically controlled mechanism that prevents translocation of Ca to the fruit at a rate sufficient for normal fruit development is still obscure (Greenleaf & Adams, 1969), effective treatment can be made in the field to control its occurrence. Geraldson (1957) effectively controlled blossom-end rot under a variety of conditions by spraying a solution of $CaCl_2$ onto the fruit. He also obtained satisfactory control in the field by soil applications of Ca salts. He further showed that blossom-end rot was associated with neither low soil pH per se nor Ca concentration per se; it was associated with a low ratio of Ca to other cations.

Heart rot of celery, commonly called *black heart,* is also a physiological disorder of unknown cause that is evidenced by a Ca deficiency in the celery heart. This disorder is apparently common to all celery-producing areas and is of special importance in Florida. Geraldson (1954) reported that heart rot was effectively controlled by a foliar Ca spray. He also showed that more Mg, K, or Na in the solution, which lowered the ratio of Ca to other cations, increased the incidence of heart rot.

Flue-cured tobacco is grown in the southern United States on acid, sandy soils. Because of fear of black root rot, soil pH is generally maintained at about 5.5 or less, which leaves many coarse-textured soils with low levels of exchangeable $Ca^{2+}$. With the use of Ca-free fertilizers for tobacco production, Ca deficiency is often observed in the field (Peedin & McCants, 1977). Leaf-Ca content was increased significantly by low rates of lime but less so by gypsum in an experiment in Florida (Breland et al., 1965). In a North Carolina experiment (Peedin & McCants, 1977), neither gypsum nor dolomitic limestone effectively corrected Ca deficiency in the field. The researchers concluded that Ca deficiency in tobacco is more related to rapid growth conditions than to soil Ca level. It appears that additional research is needed in this area.

### E. Magnesium Deficiency

Magnesium deficiency is not a major factor of infertility on most acid soils. However, it does occur in the southern United States not only on acid soils but also on some calcareous soils in Florida (Camp, 1947), Alabama (Adams & Henderson, 1962), and Mississippi (Lancaster, 1958).

Conditions conducive to Mg deficiency on acid soils are (i) low CEC, (ii) liming with material that is almost Mg free, (iii) high rate of $NH_4^+$ fertilizer, (iv) high rate of K fertilizer, (v) excessive leaching of soil profile, and (vi) growing crops that require high levels of Mg. The crops on which Mg deficiency has most commonly occurred are tobacco, citrus, potato, cotton, and soybean.

Tobacco was one of the first crops in the southern United States on which Mg deficiency was recognized as a serious problem. *Sand drown* was the term used to describe the Mg-deficient symptoms before the cause was known, because it occurred on deep, sandy soils. The problem is corrected or avoided on tobacco by low rates of dolomitic limestone (Breland et al., 1965). Magnesium deficiency in tobacco is not so much the result of high soil acidity as it is the combined effects of a crop with high Mg requirements grown on a low Mg soil fertilized with a high rate of K. Magnesium deficiency probably affects tobacco quality more than it does yield (McMurtrey, 1947).

Citrus is grown in Florida on acid, coarse-textured soils and on calcareous soils, both of which are inherently low in Mg. The Mg requirement of citrus is relatively high, and Mg deficiency in citrus groves of Florida became prevalent during the 1930s (Camp, 1947; Fudge, 1939). Yields were frequently doubled by Mg applications. Magnesium deficiency on the calcareous soils of Florida results from the combination of low Mg and high Ca contents of the limestone. Magnesium deficiency on acid soils is undoubtedly the combined effects of high Mg demands by the crop, low soil Mg, and high rates of K and $NH_4$ fertilizers.

Potato has sometimes responded to Mg fertilizer on acid, sandy soils, even though the magnitude of response is usually small (Carolus & Brown, 1935; Johnson et al., 1956). Magnesium deficiency in potato is probably the combined effects of low soil Mg, high rates of $NH_4^+$ and K fertilizer, and excess solution Al. Potato is generally grown on highly fertilized soils at a pH near 5.0 because of the fear of potato scab, which occurs above pH 5.5. Potato soils frequently become strongly acid from the use of $NH_4^+$ fertilizer without compensating lime rates. Thus, available Mg is subjected to conditions that favor its leaching from the Ap horizon. Concurrently, increases in soluble Al will further inhibit a plant's ability to absorb and use Mg. In addition, Mg availability is further antagonized by the high rates of K fertilizer commonly applied to potato.

An extensive field evaluation of soil Mg availability for the production of cotton, soybean, corn, and peanut was conducted in Alabama from 1957 to 1973. These experiments used $MgSO_4$ and dolomitic limestone as Mg sources. In 45 experiments with cotton, Mg additions increased seed-cotton yields about 200 to 400 kg $ha^{-1}$ on most sandy soils when available Mg (dilute, double-acid extractable) was less than 15 mg $kg^{-1}$ (Adams, 1975). The incidence of Mg deficiency was increased in several experiments by liming an acid soil with a Mg-free (almost) material and fertilizing with KCl at an excessive rate. In spite of this highly favorably condition for Mg deficiency, the deficiency occurred only when exchangeable $Mg^{2+}$ was about 0.1 cmol ($1/2Mg^{2+}$) $kg^{-1}$ or less in the Ap horizon. It is quite likely that Mg deficiency was absent in some soils because of an adequate Mg supply in subsoil horizons.

In other field experiments in Alabama, soybean yields were increased by 130 to 400 kg $ha^{-1}$ by fertilizing with $MgSO_4$ or liming with dolomite on some soils with $< 15$ mg $kg^{-1}$ of extractable Mg (Adams, 1975; Rogers et al., 1973). In 13 experiments with corn in rotation with cotton or soybean, added Mg had no effect on corn yield, even on the same plots where cotton or soybean yields were increased. Even when foliage symptoms of Mg deficiency occurred, there was no yield effect. In 10 on-farm experiments with peanut, extractable soil Mg ranged from 3 to 14 mg $kg^{-1}$, and only one site produced a yield increase to applied Mg (Hartzog & Adams, 1973; Adams & Hartzog, 1980). The yield increase was 560 kg $ha^{-1}$; the soil was a Troup loamy sand (Grossarenic Paleudults) with 3 mg $kg^{-1}$ extractable Mg in the Ap horizon.

Other reported Mg deficiencies include sweetclover and pearl millet on Jonesville fine sand (Arenic Hapludalts) in Florida (Blue & Eno, 1956) and sweet potato on a deep-phase Norfolk loamy sand in North Carolina (Jackson & Thomas, 1960).

Except for the high pH, chalk-derived soils of Alabama, Florida, and Mississippi (Adams & Henderson, 1962; Camp, 1947; Lancaster, 1958), Mg deficiency has not been reported on soils in the southern United States that have a significant clay content. Numerous liming experiments have shown no advantage for dolomitic over calcitic limestone for various crops. A major reason for the lack of yield response to added Mg is because Mg is a structural component of vermiculite and montmorillonite. Since clay structures are less stable under acid conditions, structural Mg is released to the soil solution or exchange sites as the clay becomes more acid.

The level of exchangeable $Mg^{2+}$ must be quite low in the surface soil for Mg to be deficient. The role that subsoil horizons play in providing Mg to crops has not been explored, but it may be significant in acid, clayey horizons where exchangeable $Mg^{2+}$ sometimes exceeds Ca. In general, Mg deficiency is probable when Mg saturation is no more than about 5% of the $NH_4OAc$-determined CEC (Adams, 1975; Adams & Henderson, 1962; Lancaster, 1958).

### F. Molybdenum Deficiency

The most recently recognized factor of acid soil infertility in the southern United States is Mo deficiency. It was first recognized as a field problem in the region in the early 1950s on citrus and cauliflower in Florida. Subsequently, Gammon et al. (1954) showed that liming a fine sand complex from pH 4.5 to 5.5 greatly reduced Mo deficiency in cauliflower but that added Mo eliminated it.

Other crops in the region responding to increased Mo availability by lime are primarily legumes. Alfalfa readily responds to Mo and liming. In an experiment on a Cecil-Lloyd soil at pH 5.5, Giddens and Perkins (1960) found that liming to pH 6.3 increased alfalfa forage by 3500 kg $ha^{-1}$; Mo alone increased yield by 2840 kg $ha^{-1}$; liming and Mo together increased yield by 6200 kg $ha^{-1}$. Mortvedt (1981) measured yield response of several legumes to Mo and lime on a Decatur silt loam in a greenhouse experiment. Liming produced the same forage as a Mo application with soybean, alfalfa, and subterranean clover; liming increased yield of arrowleaf, crimson, and white clovers but Mo did not.

The availability of Mo has attracted most attention with soybean. Significant yield increases have been reported for liming and for Mo applications in Georgia (Parker & Harris, 1978), Louisiana (Sedberry et al., 1973), Mississippi (Anthony, 1967), and Tennessee (Long et al., 1965; Smith & Connell, 1973). In most experiments, yield responses to Mo application vanish when soil is limed to pH 5.5 to 6.0. In others, however, yield response to Mo occurred above pH 6.0.

The availability of soil Mo is apparently a function of Mo content, pH, and Fe oxide content. Molybdenum deficiency is more prevalent in some areas than others, which means that it should be associated with soil series. Most deficiencies in Louisiana, Mississippi, and Tennessee have been reported on Alfisols, although not exclusively; in Georgia they have been on Ultisols.

## VII. SOIL pH AND NUTRIENT AVAILABILITY

Liming acid soils tends to decrease the availability of several nutrients. This has sometimes resulted in overliming injury because the availability of some nutrient was decreased to a deficient level. Deficiencies of this kind are called *lime induced* and can be expected on soils with marginally available levels of that particular nutrient. The highly weathered, leached, weakly buffered, acid soils of the southern United States are especially subject to changes in the availability of some nutrients as a result of liming. At the same time, the availability of other nutrients may be increased.

### A. Nitrogen

Soil pH has long been recognized as influencing soil microbial activities related to N, e.g., ammonification, nitrification, denitrification, and $N_2$ fixation. However, since most microbial activities in soils proceed more or less uninhibited over a rather wide soil pH range, significant inhibition because of soil acidity probably occurs only in strongly acid soils ($<$ pH 5). Furthermore, the agronomic significance of such effects has been difficult to assess.

In a review of the literature on soil pH and N availability, Adams and Martin (1984) summarized current concepts as follows:

1) Mineralization of organic N occurs over the entire soil pH range, but the rate progressively decreases below pH 6.0 to 6.5.
2) Nitrification rate is optimum at pH 6.6 to 8.0; it progressively decreases with decreasing pH, becoming negligible below pH 4.5.
3) Optimum pH for denitrification by soil microbes is 7.0 to 7.5; the rate is much slower below pH 5.
4) Dinitrogen fixation by rhizobia is highly pH dependent for some legume-*Rhizobium* associations. A soil pH of 6.0 or more is required by some, whereas a pH as low as 5.0 is still optimum for others.
5) Ammonia losses from soil by volatilization is highly pH dependent, becoming increasingly significant above pH 7.0.
6) Plants generally preferentially absorb $NH_4^+$ at high pH and $NO_3^-$ at low pH.

Liming crops in which the soil is not mechanically disturbed, e.g., pasture sods and no-till crops, may require special consideration in the use of N fertilizers and lime. It is quite likely that surface-applied lime on undisturbed sod or soil will result in a volatilization loss of N from $NH_4^+$ and urea fertilizers. Consequently, yields may be reduced by liming because they fail to receive full benefit from the N fertilizer. Volk (1956) found that liming reduced the yield of grasses receiving $NH_4$ or urea fertilizer but did not reduce the yield of those receiving N as $NO_3^-$. Liming a sod crop would be expected to result in a thin layer of alkaline soil at the surface. The high pH of this layer should cause $NH_3$ volatilization when $NH_4^+$ fertilizer comes in contact with it.

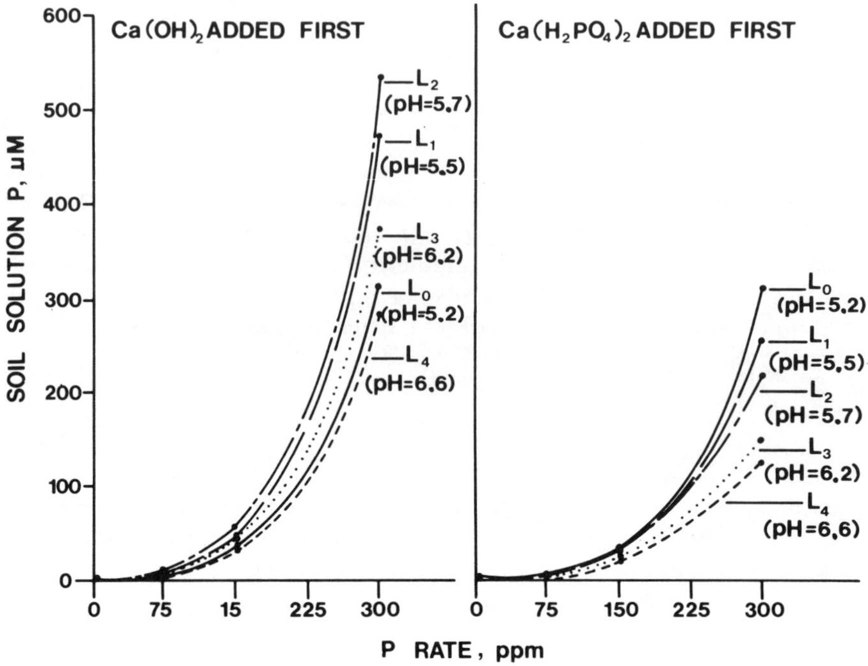


Fig. 7—Effect of $Ca(OH)_2$ and $Ca(H_2PO_4)_2$ rates on soil-solution P concentration in a Lucedale sandy loam (Soltanpour et al., 1974).

## B. Phosphorus

Liming an acid soil is generally credited with rendering P more available. However, much ambiguity surrounds this concept. Whether or not liming increases P availability depends on a combination of pH and P factors. If soil pH is low enough for exchangeable $Al^{3+}$ to be appreciable, liming to pH 6.0 or 6.5 will increase the availability of subsequently added fertilizer P; liming under the same conditions after fertilizer P has reacted with the soil will decrease P availability (Singh & Seatz, 1961; Soltanpour et al., 1974). This is illustrated by the data in Figure 7. Coleman et al. (1960) showed that adding fertilizer P to an acid soil resulted in a precipitation reaction between exchangeable $Al^{3+}$ and added P and the formation of a highly insoluble Al-P compound. Subsequent liming does not affect the availability of P in this compound even though thermodynamics show that it should. The Al-P compound may become surface-coated with $Al(OH)_3$ upon liming, which effectively occludes the P compound from the dissolving action of soil solution. Robertson et al. (1954) found that liming Florida soils, high in Fe and Al oxides, increased the availability of subsequently added P fertilizer, whereas liming soils low in these oxides did not. Liming in the absence of exchangeable $Al^{3+}$ will reduce available P because P will be precipitated as apatite, and the solubility of apatite decreases with increasing pH.

Under certain conditions, liming a soil without exchangeable $Al^{3+}$ will increase available P. Under such conditions, however, the increased P is not coming from mineral-P compounds. It is coming from organic matter because of an accelerated decomposition rate at the higher pH.

### C. Potassium

Liming acid soils results in decreased availability of the exchangeable $K^+$ that is present at time of liming in most soils of the southern United States. This is caused by the relative high proportion of pH-dependent CEC, which results in an important shift of solution K to the exchangeable phase as pH increases. This same property allows limed soils to retain more fertilizer K against leaching losses, thereby increasing fertilizer-K efficiency. In earlier days, when 20 kg $ha^{-1}$ of K was considered a good rate of fertilizer, lime-induced K deficiency was rather striking at times (Pearson, 1958). However, current rates of K fertilizers essentially preclude the occurrence of such deficiencies in the region.

### D. Calcium and Magnesium

Since all liming materials contain Ca, liming should always result in increased Ca availability. Similarly, if liming material contains appreciable Mg, liming should increase Mg availability. If, however, a soil is limed with a Mg-free material, Mg availability is decreased. In a number of field experiments in Alabama, liming with a high-grade calcitic material caused a Mg deficiency where none existed at the lower pH (Adams, 1975). This may have resulted from decreased solution Mg because it moved to exchange sites, or it may have resulted from an undefined Mg fixation mechanism (Farina et al., 1980b; Sumner et al., 1978).

### E. Sulfur

Liming an acid soil in the southern United States can be expected to increase available S in most cases. For one thing, it will accelerate organic matter decomposition, thereby increasing the rate at which organic S is mineralized. For another, the capacity of a soil to retain $SO_4^{2-}$ in a solid-phase mineral form is highly pH dependent, particularly in soils high in Al and Fe oxides. As soil pH progressively decreases below pH 6 in such soils, the soil's capacity to remove $SO_4^{2-}$ from solution by either surface adsorption (Chang & Thomas, 1963) or precipitation (Adams & Rawajfih, 1977) increases sharply. Liming such soils has been shown to increase extractable $SO_4^{2-}$ appreciably (Kamprath et al., 1956; Ensminger, 1954; Nelson, 1964).

Data by Elkins and Ensminger (1971), in which an acid Dothan subsoil material (Plinthic Paleudults) was limed to five pH levels, present an excellent example of the effect that liming has on $SO_4^{2-}$ availability. They found that both S uptake by plants and $SO_4^{2-}$ concentrations in soil solutions were increased by each increment of lime. The effect noted by Elkins and Ensminger (1971) will be much more pronounced in argillic subsoil horizons than in surface horizons where there is less clay but more organic matter.

### F. Micronutrients

Although foliage symptoms of Fe deficiency are observed on Fe-sensitive crops throughout the southern United States, there has been no definitive work on its pH dependency. Availability of soil Fe is dependent on the genetic character of a plant as well as soil pH. Consequently, when experience shows that certain crops are subject to Fe deficiency on neutral and alkaline soils, the grower switches to crops or cultivars that are not. With few exceptions, e.g., blueberry, liming an acid soil at recommended rates will not affect Fe availability enough to be a practical problem.

Zinc deficiency is widespread in the southern United States, particularly in corn, pecan, peach, and citrus. In general, Zn deficiency is not evident at low soil pH (< 6.0) and is only mild at higher pHs (Cox & Wear, 1977). The work of Wear (1956) and Adams et al. (1982) clearly demonstrated that Zn availability progressively diminishes with increasing lime rates. It becomes a practical problem for Zn-sensitive crops on low Zn soils when they are limed to pH 6 or more.

Manganese availability is highly sensitive to soil pH changes. At low pH, Mn availability may be high enough to be toxic to many crops (section VIC). Some soils of the Southeast, however, contain too little Mn to be toxic at any pH level. Such soils are rather widespread in the lower Coastal Plain Province along the Atlantic Coast. Liming these soils to pH 6.5 or more has induced striking Mn deficiencies on many fields in that area (Cox, 1968). Only minor, isolated instances of Mn deficiency have been lime induced in other areas of the southern United States.

Lime-induced Cu deficiencies have been rare, if they occur at all, in the southern United States. Copper deficiency has been reported only for organic soils with little regard for soil pH. However, liming to about pH 7 has played a special role in reducing the availability of toxic levels of Cu in some Florida citrus groves (Reuther & Smith, 1952; Spencer, 1966).

Boron deficiency began to be recognized in the field in the southern United States during the late 1930s and was frequently associated with liming (Page & Paden, 1954; Wear, 1957). Since its early recognition, B deficiency has frequently occurred on alfalfa, apple, clover, citrus, turnip, beet, cauliflower, celery, peanut, and sweet potato. Soils of the region that are coarse textured and low in organic matter tend to be inherently low in available B. Since increasing soil pH resulted in a significant reduction in available B, deficiencies are especially likely on B-sensitive crops after liming very sandy soils (Wear & Patterson, 1962). However, lime-induced B deficiency per se probably occurs only rarely below pH 6.5.

Molybdenum availability is highly pH dependent. Deficiencies have been reported above pH 6, but most will occur below pH 5.5. In a review of research with Mo and soybean, Boswell (1980) noted that Mo deficiency is likely on soils that are highly weathered, have low pH, are low in organic matter, and are high in Fe oxides. Liming to about pH 6.2 or more will usually increase Mo availability enough to correct the deficiency.

## REFERENCES

Abruna, F., R. W. Pearson, and C. B. Elkins. 1958. Quantitative evaluation of soil reaction and base status changes resulting from field application of residually acid-forming nitrogen fertilizers. Soil Sci. Soc. Am. Proc. 22:539–542.

Abruna, F., and J. Vicente. 1955. Refinement of a quantitative method for determining the lime requirements of soils. J. Agric. Univ. P.R. 39:41–45.

Adams, F. 1956. Response of crops to lime in Alabama. Alabama Agric. Exp. Stn. Bull. 301.

Adams, F. 1966. Calcium deficiency as a causal agent of ammonium phosphate injury to cotton seedlings. Soil Sci. Soc. Am. Proc. 30:485–488.

Adams, F. 1968a. Response of cotton to lime in field experiments. Alabama Agric. Exp. Stn. Bull. 376.

Adams, F. 1968b. Relative responses of grain and annual forage crops to lime, phosphorus, and potassium. Alabama Agric. Exp. Stn. Circ. 161.

Adams, F. 1969. Response of corn to lime in field experiments on Alabama soils. Alabama Agric. Exp. Stn. Bull. 391.

Adams, F. 1975. Field experiments with magnesium in Alabama—cotton, corn, soybeans, peanuts. Alabama Agric. Exp. Stn. Bull. 472.

Adams, F. 1978. Liming and fertilization of Ultisols and Oxisols. p. 377–394. *In* C. S. Andrew and E. J. Kamprath (ed.) Mineral nutrition of legumes in tropical and subtropical soils. CSIRO, Melbourne, Australia.

Adams, F. 1981. Alleviating chemical toxicities: Liming acid soils. p. 267–301. *In* G. F. Arkin and H. M. Taylor (ed.) Modifying the root environment to reduce crop stress. ASAE Monogr. 4. Am. Soc. of Agric. Engineers, St. Joseph, Mich.

Adams, F., and C. E. Evans. 1962. A rapid method for measuring lime requirement of Red-Yellow Podzolic soils. Soil Sci. Soc. Am. Proc. 26:355–357.

Adams, F., and D. D. Hartzog. 1979. Effects of a lime slurry on soil pH, exchangeable calcium, and peanut yields. Peanut Sci. 6:73–76.

Adams, F., and D. Hartzog. 1980. The nature of yield response of Florunner peanuts to lime. Peanut Sci. 7:120–123.

Adams, F., and J. B. Henderson. 1962. Magnesium availability as affected by deficient and adequate levels of potassium and lime. Soil Sci. Soc. Am. Proc. 26:65–68.

Adams, F., and Z. F. Lund. 1966. Effect of chemical activity of soil solution aluminum on cotton root penetration of acid subsoils. Soil Sci. 101:193–198.

Adams, F., and J. B. Martin. 1984. Liming effects on nitrogen use and efficiency. *In* R. D. Hauck (ed.) Nitrogen in crop production. Am. Soc. of Agron., Madison, Wis. (In press.)

Adams, F., and B. L. Moore. 1983. Chemical factors affecting root growth in subsoil horizons of Coastal Plain soils. Soil Sci. Soc. Am. J. 47:99–102.

Adams, F., and R. W. Pearson. 1969. Neutralizing soil acidity under bermudagrass sod. Soil Sci. Soc. Am. Proc. 33:737–742.

Adams, F., and R. W. Pearson. 1970. Differential response of cotton and peanuts to subsoil acidity. Agron. J. 62:9–12.

Adams, F., R. W. Pearson, and B. D. Doss. 1967. Relative effects of acid subsoils on cotton yields in field experiments and on cotton roots in growth-chamber experiments. Agron. J. 59:453–456.

Adams, F., and Z. Rawajfih. 1977. Basaluminite and alunite: A possible cause of sulfate retention by acid soils. Soil Sci. Soc. Am. J. 41:686–692.

Adams, F., and J. I. Wear. 1957. Manganese toxicity and soil acidity in relation to crinkle leaf of cotton. Soil Sci. Soc. Am. Proc. 21:305–308.

Adams, J. F., F. Adams, and J. W. Odom. 1982. Interaction of phosphorus rates and soil pH on soybean yield and soil solution composition of two phosphorus-sufficient Ultisols. Soil Sci. Soc. Am. J. 46:323–328.

Adams, W. E., R. W. Pearson, W. A. Jackson, and R. A. McCreery. 1967a. Influence of limestone and nitrogen on soil pH and 'Coastal' bermudagrass yield. Agron. J. 59:450–453.

Adams, W. E., A. W. White, and R. N. Dawson. 1967b. Influence of lime sources and rates on 'Coastal' bermudagrass production, soil profile reaction, exchangeable Ca and Mg. Agron. J. 59:147–149.

Alley, M. M. 1981. Short-term soil chemical and crop yield responses to limestone applications. Agron. J. 73:687–689.

Altom, W., and J. Rogers. 1977. The effects of various fertilizers and lime on Midland bermudagrass forage yields and soil test values of a Chickasha fine sandy loam. Agric. Div., Noble Foundation, Ardmore, Okla.

Anthony, J. L. 1967. Fertilizing soybeans in the hill section of Mississippi. Mississippi Agric. Exp. Stn. Bull. 743.

Arnon, D. I., and C. M. Johnson. 1942. Influence of hydrogen ion concentration on growth of higher plants under controlled conditions. Plant Physiol. 17:525–539.

Bartlett, R. J., and D. C. Riego. 1972. Effect of chelation on the toxicity of aluminum. Plant Soil 37:419–423.

Bennett, A. C., and F. Adams. 1970. Calcium deficiency and ammonia toxicity as separate causal factors of $(NH_4)_2HPO_4$-injury to seedlings. Soil Sci. Soc. Am. Proc. 34:255–259.

Beringer, H., and M. A. Taha. 1976. Calcium-45 absorption by two cultivars of groundnut (*Arachis hypogaea*). Exp. Agric. 12:1–7.

Bledsoe, R. W., C. L. Comar, and H. C. Harris. 1949. Absorption of radioactive calcium by the peanut fruit. Science (Washington, D.C.) 109:329–330.

Blevins, R. L., L. W. Murdock, and G. W. Thomas. 1978. Effect of lime application on no-tillage and conventionally tilled corn. Agron. J. 70:322–326.

Blue, W. G. 1979. Forage production and N contents, and soil changes during 25 years of continuous white clover-Pensacola bahiagrass growth on a Florida Spodosol. Agron. J. 71: 795–798.

Blue, W. G., and C. F. Eno. 1956. Magnesium and lime are needed in the Suwannee Valley area. Florida Agric. Exp. Stn. Bull. 577.

Bortner, C. E. 1935. Toxicity of manganese to Turkish tobacco in acid Kentucky soils. Soil Sci. 39:15–33.

Boswell, F. C. 1980. Factors affecting the response of soybeans to molybdenum application. p. 417–432. *In* F. T. Corbin (ed.) World soybean research conference II Proc., Raleigh, N.C. 26–29 Mar. 1979. Westview Press, Boulder, Colo.

Brady, N. C., and W. E. Colwell. 1945. Yield and quality of large-seeded type peanuts as affected by potassium and certain combinations of potassium, magnesium, and calcium. J. Am. Soc. Agron. 37:429–442.

Breland, H. L., W. L. Pritchett, and H. W. Lundy. 1965. Liming soils for flue-cured tobacco in the Suwannee Valley area. Florida Agric. Exp. Stn. Bull. 702.

Buol, S. W. (ed.) 1973. Soils of the southern states and Puerto Rico. Southern Coop. Ser. Bull. 174.

Burkhart, L., and E. R. Collins. 1942. Mineral nutrients in peanut plant growth. Soil Sci. Soc. Am. Proc. 6:272–280.

Burmester, C. H., A. Waggoner, and F. Adams. 1981. On-farm lime and fertilizer experiments with soybeans and cotton in northern Alabama, 1977–1980. Alabama Agric. Exp. Stn. Circ. 254.

Camp, A. F. 1947. Magnesium in citrus fertilization in Florida. Soil Sci. 63:43–52.

Carlyle, E. C. 1931. Manganese in Texas soils and its relation to crops. Texas Agric. Exp. Stn. Bull. 432.

Carolus, R. L., and B. E. Brown. 1935. Truck crop investigations. Virginia Truck Exp. Stn. Bull. 89.

Chang, M. L., and G. W. Thomas. 1963. A suggested mechanism for sulfate adsorption by soils. Soil Sci. Soc. Am. Proc. 27:281–283.

Coleman, N. T., J. T. Torup, and W. A. Jackson. 1960. Phosphate sorption reactions that involve exchangeable aluminum. Soil Sci. 90:1–7.

Colwell, W. E., and N. C. Brady. 1945. The effect of calcium on yield and quality of large-seeded type peanuts. J. Am. Soc. Agron. 37:413–428.

Colwell, W. E., N. C. Brady, and J. R. Piland. 1945. Composition of peanut shells of filled and unfilled fruits as affected by treatments. J. Am. Soc. Agron. 37:792–805.

Cox, F. R. 1968. Development of a yield response prediction and manganese soil test interpretation for soybeans. Agron. J. 60:521–524.

Cox, F. R., F. Adams, and B. B. Tucker. 1982. Liming, fertilization, and mineral nutrition. p. 139–163. *In* H. E. Pattee and C. T. Young (ed.) Peanut science and technology. North Carolina State Univ., Raleigh.

Cox, F. R., and J. I. Wear. 1977. Diagnosis and correction of zinc problems in corn and rice production. Southern Coop. Ser. Bull. 222. North Carolina State Univ., Raleigh.

Daniels, R. B., B. L. Allen, H. H. Bailey, and F. H. Beinroth. 1973. Physiography. p. 3–16. *In* S. W. Buol (ed.) Soils of the southern states and Puerto Rico. Southern Coop. Ser. Bull. 174.

Doss, B. D., W. T. Dumas, and Z. F. Lund. 1979. Depth of lime incorporation for correction of subsoil acidity. Agron. J. 71:541–544.

Doss, B. D., C. E. Evans, and J. L. Turner. 1977. Influence of subsoil acidity on tomato yield and fruit size. J. Am. Soc. Hortic. Sci. 102:643–645.

Doss, B. D., and Z. F. Lund. 1975. Subsoil pH effects on growth and yield of cotton. Agron. J. 67:193–196.

Duggar, J. F., and M. J. Funchess. 1911. Lime for Alabama soils. Alabama Agric. Exp. Stn. Bull. 161.

Duncan, R. R., J. W. Dobson, and C. D. Fisher. 1980. Leaf elemental concentrations and grain yield of sorghum grown on an acid soil. Commun. Soil Sci. Plant Anal. 11:699–707.

Elawad, S. H., G. J. Gascho, and J. J. Street. 1982. Response of sugarcane to silicate source and rate: I. Growth and yield. Agron. J. 74:481–484.

Elkins, D. M., and L. E. Ensminger. 1971. Effect of soil pH on the availability of adsorbed sulfate. Soil Sci. Soc. Am. Proc. 35:931–934.

Ensminger, L. E. 1954. Some factors affecting the absorption of sulfate by Alabama soils. Soil Sci. Soc. Am. Proc. 18:259–264.

Evans, C. E., and E. J. Kamprath. 1970. Lime response as related to percent Al saturation, solution Al, and organic matter content. Soil Sci. Soc. Am. Proc. 34:893–896.

Farina, M. P. W., M. E. Sumner, C. O. Plank, and W. S. Letzsch. 1980a. Exchangeable aluminum and pH as indicators of lime requirement for corn. Soil Sci. Soc. Am. J. 44: 1036–1041.

Farina, M. P. W., M. E. Sumner, C. O. Plank, and W. S. Letzsch. 1980b. Effect of pH on soil magnesium and its absorption by corn. Commun. Soil Sci. Plant Anal. 11:981–992.

Foy, C. D. 1964. Toxic factors in acid soils of the southeastern United States as related to the response of alfalfa to lime. USDA Production Res. Rep. 80.

Fudge, B. R. 1939. Relation of magnesium deficiency in grapefruit leaves to yield and chemical composition of fruit. Florida Agric. Exp. Stn. Bull. 331.

Funchess, M. J. 1918. The development of soluble manganese in acid soils as influenced by certain nitrogenous fertilizers. Alabama Agric. Exp. Stn. Bull. 201.

Funchess, M. J. 1919. Acid soils and the toxicity of manganese. Soil Sci. 8:69.

Gammon, N., Jr. 1957. Root growth responses to soil pH adjustments made with carbonates of calcium, sodium, or potassium. Proc. Soil Crop Sci. Soc. Fla. 17:249–254.

Gammon, N., Jr., G. M. Volk, E. N. McCubbin, and A. H. Eddins. 1954. Soil factors affecting molybdenum uptake by cauliflower. Soil Sci. Soc. Am. Proc. 18:302–305.

Geraldson, C. M. 1954. The control of blackheart of celery. Proc. Am. Soc. Hortic. Sci. 63: 353–358.

Geraldson, C. M. 1957. Control of blossom-end rot of tomatoes. Proc. Am. Soc. Hortic. Sci. 69:309–317.

Giddens, J., and H. F. Perkins. 1960. Influence of molybdenum on growth and composition of alfalfa and distribution of molybdenum in a Cecil-Lloyd soil. Soil Sci. Soc. Am. Proc. 24:496–497.

Greenleaf, W. H., and F. Adams. 1969. Genetic control of blossom-end rot disease in tomatoes through calcium metabolism. J. Am. Soc. Hortic. Sci. 94:248–250.

Haby, V. A., W. B. Anderson, and C. D. Welch. 1979. Effect of limestone variables on amendment of acid soils and production of corn and Coastal bermudagrass. Soil Sci. Soc. Am. J. 43:343–347.

Hallock, D. L. 1964. Response of soybeans to gypsum, lime, and fertilizer on three soils in southeastern Virginia. Virginia Agric. Exp. Stn. Res. Rep. 89.

Hartzog, D., and F. Adams. 1973. Fertilizer, gypsum, and lime experiments with peanuts in Alabama. Alabama Agric. Exp. Stn. Bull. 448.

Hester, J. B. 1935. The amphoteric nature of 3 Coastal Plains Soils. I. In relation to plant growth. Soil Sci. 39:237–245.

Hiatt, A. J., and J. L. Ragland. 1963. Manganese toxicity of burley tobacco. Agron. J. 55: 47–49.

Hiltbold, A. E., and F. Adams. 1960. Effect of nitrogen volatilization on soil acidity changes due to applied nitrogen. Soil Sci. Soc. Am. Proc. 24:45–47.

Howard, D. D. 1970. Adjust soil pH for profit. Tennessee Agric. Ext. Ser. Pub. 599.

Howard, D. D., and F. Adams. 1965. Calcium requirement for penetration of subsoils by primary cotton roots. Soil Sci. Soc. Am. Proc. 29:558–562.

Hutcheson, T. B., Jr., and J. F. Freeman. 1965. Longevity of soil reaction effects following lime and alum additions. Agron. J. 57:89–90.

Jackson, W. A., and G. W. Thomas. 1960. Effects of KCl and dolomitic limestone on growth and ion uptake of the sweet potato. Soil Sci. 89:347–352.

Johnson, W. A., J. I. Wear, and F. Garrett. 1956. Effects of fertilizer and use of magnesium and minor elements on yields and storage quality of potatoes in Baldwin County, Alabama. Am. Potato J. 33:103–112.

Jones, L. G., R. J. Constantin, J. M. Cannon, W. J. Martin, and T. P. Hernandez. 1977. Effects of soil amendment and fertilizer applications on sweet potato growth, production, and quality. Louisiana Agric. Exp. Stn. Bull. 704.

Jones, W. F., and L. E. Nelson. 1978. Response of field-grown soybeans to lime. Commun. Soil Sci. Plant Anal. 9:607–614.

Kamprath, E. J. 1970. Exchangeable aluminum as a criterion for liming leached mineral soils. Soil Sci. Soc. Am. Proc. 34:252–254.

Kamprath, E. J., W. L. Nelson, and J. W. Fitts. 1956. The effect of pH, sulfate, and phosphate concentrations on the adsorption of sulfate by soils. Soil Sci. Soc. Am. Proc. 20: 463–466.

Lancaster, J. D. 1958. Magnesium status of blackland soils of northeast Mississippi for cotton production. Mississippi Agric. Exp. Stn. Bull. 560.

Lancaster, J. D., and L. E. Gholston. 1953. Crinkle leaf of cotton corrected by use of lime. Miss. Farm Res. 16(12):7.

Lance, J. C., and R. W. Pearson. 1969. Effect of low concentrations of aluminum on growth and water and nutrient uptake by cotton roots. Soil Sci. Soc. Am. Proc. 33:95–98.

Link, L. A. 1979. Critical pH for the expression of manganese toxicity on burley tobacco and the effect of liming on growth. Tob. Int. 23:99–101.

Long, O. H., J. R. Overton, E. W. Counce, and T. McCutchen. 1965. Lime and fertilizer on soybeans. Tennessee Agric. Exp. Stn. Bull. 391.

Long, O. H., and L. M. Safley. 1962. Alfalfa production on Dickson and Mountview soils. Tennessee Agric. Exp. Stn. Bull. 350.

Lund, Z. F. 1970. The effect of calcium and its relation to several cations in soybean root growth. Soil Sci. Soc. Am. Proc. 34:456–459.

Lutrick, M. C., and F. G. Martin. 1977. Grain sorghum response to lime, phosphorus, and potassium. Proc. Soil Crop Sci. Soc. Florida 36:55–57.

Lyle, E. S., and F. Adams. 1971. Effect of available soil calcium on taproot elongation of loblolly pine (*Pinus taeda* L.) seedlings. Soil Sci. Soc. Am. Proc. 35:800–805.

McMurtrey, J. E., Jr. 1947. Effect of magnesium on growth and composition of tobacco. Soil Sci. 63:59–67.

Mehlich, A. 1942. Base unsaturation and pH in relation to soil type. Soil Sci. Soc. Am. Proc. 6:150–156.

Mehlich, A. 1976. New buffer pH method for rapid estimation of exchangeable acidity and lime requirement of soils. Commun. Soil Sci. Plant Anal. 7:637–652.

Mehlich, A., and W.E. Colwell. 1944. Influence of the nature of soil colloids and degree of base saturation on growth and nutrient uptake by cotton and soybeans. Soil Sci. Soc. Am. Proc. 8:179–184.

Mehlich, A., and J. F. Reed. 1947. The influence of type of colloid and degree of calcium saturation on fruit characteristics of peanuts. Soil Sci. Soc. Am. Proc. 11:201–205.

Mengel, D. B., and E. J. Kamprath. 1978. Effect of soil pH and liming on growth and nodulation of soybeans in Histosols. Agron. J. 70:959–963.

Middleton, G. K., W. E. Colwell, W. C. Brady, and E. F. Schultz. 1945. The behavior of four varieties of peanuts as affected by calcium and potassium variables. J. Am. Soc. Agron. 37:443–457.

Mitchell, C. C., S. M. Eich, and F. Adams. 1977. On-farm lime and fertilizer experiments with soybeans and cotton in northern Alabama, 1975–1976. Alabama Agric. Exp. Stn. Prog. Rep. Ser. 110.

Miyake, K. 1916. The toxic action of soluble Al salts upon the growth of the rice plant. J. Biol. Chem. 25:23–28.

Morris, H. D. 1949. The soluble manganese content of acid soils and its relation to the growth and manganese content of sweet clover and lespedeza. Soil Sci. Soc. Am. Proc. 13:362–371.

Mortvedt, J. J. 1981. Nitrogen and molybdenum uptake and dry matter relationships of soybeans and forage legumes in response to applied molybdenum on acid soil. J. Plant Nutr. 3:245–256.

Moschler, W. W., G. D. Jones, and G. W. Thomas. 1960. Lime and soil acidity effects on alfalfa growth in a Red-Yellow Podzolic soil. Soil Sci. Soc. Am. Proc. 24:507–509.

Moschler, W. W., D. C. Martens, C. I. Rich, and G. M. Shear. 1973. Comparative lime effects on continuous no-tillage and conventionally tilled corn. Agron. J. 65:781–783.

Moschler, W. W., S. S. Obenshain, R. P. Cocke, and H. M. Camper. 1950. Effect of varying amounts of ground limestone on the pH and base exchange properties of Sassafras fine sandy loam. Soil Sci. Soc. Am. Proc. 14:123–125.

Moschler, W. W., R. K. Stivers, D. L. Hallock, R. D. Sears, H. M. Camper, M. J. Rogers, G. D. Jones, M. T. Carter, and F. S. McClaugherty. 1962. Lime effects on soil reaction and base content of eleven soil types in Virginia. Virginia Agric. Exp. Stn. Tech. Bull. 159.

Neal, D. C. 1937. Crinkle leaf, a new disease of cotton in Louisiana. Phytopathology 27:1171–1175.

Neal, D. C., and H. C. Lovett. 1938. Further studies of crinkle leaf, a disorder of cotton plants prevalent in Lintonia and Olivier silt loam soils of Louisiana. Phytopathology 28:582–587.

Nelson, L. E. 1964. The effect of pH on the acetate-soluble sulfur content of a Mayhew soil in Mississippi before and after incubation. Soil Sci. Soc. Am. Proc. 28:290–291.

Nelson, L. E. 1977. Changes in water-soluble Mn due to soil sample preparation and storage. Commun. Soil Sci. Plant Anal. 8:479–487.

Page, N. R., and W. R. Paden. 1954. Boron-supplying power of several South Carolina soils. Soil Sci. 77:427–434.

Parker, M. B., and H. B. Harris. 1962. Soybean response to molybdenum and lime and the relationship between yield and chemical composition. Agron. J. 54:480–483.

Parker, M. B., and H. B. Harris. 1978. Molybdenum studies on soybeans. Georgia Agric. Exp. Stn. Res. Bull. 215.

Pearson, R. W. 1958. Liming and fertilizer efficiency. Agron. J. 50:356–362.

Pearson, R.W. 1966. Soil environment and root development. p. 95–126. *In* W.H. Pierre et al. (ed.) Plant environment and efficient water use. Am. Soc. of Agron., Madison, Wis.

Pearson, R. W., J. Childs, and Z. F. Lund. 1973. Uniformity of limestone mixing in acid subsoil as a factor in cotton root penetration. Soil Sci. Soc. Am. Proc. 37:727–732.

Pearson, R. W., R. Perez-Escolar, F. Abruna, Z. F. Lund, and E. J. Brenes. 1977. Comparative responses of three crop species to liming several soils of the southeastern United States and of Puerto Rico. J. Agric. Univ. P.R. 61:361–382.

Peedin, G. F., and C. B. McCants. 1977. Influence of variety and soil applications of Ca on development of Ca deficiency in tobacco. Agron. J. 69:71–76.

Peevy, W. J., K. W. Tipton, J. E. Sedbury, H. Viator, and R. H. Brupbacher. 1978. Effects of soil reaction and residual and fertilizer phosphorus on the growth and yield of grain sorghum on Olivier silt loam soil. Louisiana Agric. Exp. Stn. Bull. 711.

Perkins, H. F., H. J. Byrd, and F. T. Ritchie, Jr. 1973. Ultisols—light-colored soils of the warm temperate forest lands. *In* S. W. Buol (ed.) Soils of the southern states and Puerto Rico. Southern Coop. Ser. Bull. 174.

Perkins, H. F., J. E. Giddens, and J. W. Daniell. 1979. Effect of lime and fertilizer and peach root residues on peach trees in a decline area. Georgia Agric. Exp. Stn. Res. Bull. 246.

Perkins, H. F., R. L. Wehunt, and H. D. Morris. 1957. Lime for Georgia soils. Georgia Agric. Exp. Stn. Bull. N.S. 45.

Pierre, W. H. 1928. Nitrogenous fertilizers and soil acidity: I. Effect of various nitrogenous fertilizers on soil reaction. J. Am. Soc. Agron. 20:2–16.

Pierre, W. H. 1930. Neutralizing values and rates of reaction with acid soils of different grades and kinds of liming materials. Soil Sci. 29:137–158.

Pierre, W. H. 1931. Hydrogen-ion concentration, aluminum concentration in the soil solution, and percentage base saturation as factors affecting plant growth on acid soils. Soil Sci. 31:183–207.

Pierre, W. H. 1934. The equivalent acidity and basicity of fertilizers as determined by a newly proposed method. Assoc. Off. Agric. Chem. J. 17:101–107.

Pierre, W. H., G. G. Pohlman, and T. C. McIlvaine. 1932. Soluble aluminum studies: I. The concentration of aluminum in the displaced soil solution of naturally acid soils. Soil Sci. 34:145–160.

Pierre, W. H., and G. D. Scarseth. 1931. Determination of the percentage base saturation of soils and its value in different soils at definite pH values. Soil Sci. 31:99–114.

Ragland, J. L., and N. T. Coleman. 1959. The effect of soil solution aluminum and calcium on root growth. Soil Sci. Soc. Am. Proc. 23:355–357.

Reuther, W., and P. F. Smith. 1952. Iron chlorosis in Florida citrus groves in relation to certain soil constitutents. Proc. Florida State Hortic. Soc. 65:62–69.

Richburg, J. S., and F. Adams. 1970. Solubility and hydrolysis of aluminum in soil solutions and saturated-paste extracts. Soil Sci. Soc. Am. Proc. 34:728–734.

Rios, M. A., and R. W. Pearson. 1964. The effect of some chemical environmental factors on cotton root behavior. Soil Sci. Soc. Am. Proc. 28:232–235.

Robertson, W. K., C. E. Hutton, H. W. Lundy, L. G. Thompson, and R. W. Lipscomb. 1957. Effect of lime on some north Florida soils. Proc. Soil Crop Sci. Soc. Florida 17:72–85.

Robertson, W. K., J. R. Neller, and F. D. Bartlett. 1954. Effect of lime on the availability of phosphorus in soils of high to low sesquioxide content. Soil Sci. Soc. Am. Proc. 18:184–187.

Rogers, H. T. 1948. Liming for peanuts in relation to exchangeable soil calcium and effect on yield, quality, and uptake of calcium and potassium. J. Am. Soc. Agron. 40:15–31.

Rogers, H. T., F. Adams, and D. L. Thurlow. 1973. Lime needs of soybeans on Alabama soils. Alabama Agric. Exp. Stn. Bull., 452.

Sabbe, W. E., and H. L. Breland (ed.) 1974. Procedures used by state soil testing laboratories in the southern region of the United States. Southern Coop. Ser. Bull. 190. Alabama Agric. Exp. Stn., Auburn.

Sanford, J. O., D. L. Myhre, B. L. Arnold, and R. E. Coats. 1968. Crop response to lime on loess soils. Mississippi Agric. Exp. Stn. Bull. 764.

Sartain, J. B., and E. J. Kamprath. 1975. Effect of liming a highly Al-saturated soil on the top and root growth and soybean nodulation. Agron. J. 67:507–510.

Sedberry, J. E., T. S. Dharmaputra, R. H. Brupbacher, S. A. Phillips, J. G. Marshall, L. W. Sloane, D. R. Melville, J. L. Rabb, and J. H. Davis. 1973. Molybdenum investigations with soybeans in Louisiana. Louisiana Agric. Exp. Stn. Bull. 670.

Shoemaker, H. E., E. O. McLean, and P. F. Pratt. 1961. Buffer methods for determining lime requirement of soils with appreciable amounts of extractable aluminum. Soil Sci. Soc. Am. Proc. 25:274–277.

Shoop, G. J., C. R. Brooks, R. E. Blaser, and G. W. Thomas. 1961. Differential responses of grasses and legumes to liming and phosphorus fertilization. Agron. J. 53:111–115.

Singh, R. N., and L. F. Seatz. 1961. Alfalfa yield and composition after different times and rates of lime and phosphorus application. Soil Sci. Soc. Am. Proc. 25:307–309.

Smith, A. L., O. N. Andrews, and J. I. Wear. 1955. Distribution of crinkle leaf of cotton in Alabama and adjoining states. Plant Dis. Rep. 39:773–775.

Smith, H. C., and J. Connell. 1973. Lime and molybdenum in soybean production. Tenn. Farm Home Sci. Prog. Rep. 85:28–31.

Soileau, J. M., O. P. Engelstad, and J. B. Martin, Jr. 1969. Cotton growth in an acid fragipan subsoil: II. Effects of soluble calcium, magnesium, and aluminum on roots and tops. Soil Sci. Soc. Am. Proc. 33:919–924.

Soltanpour, P. N., F. Adams, and A. C. Bennett. 1974. Soil phosphorus availability as measured by displaced soil solutions, calcium-chloride extracts, dilute-acid extracts, and labile phosphorus. Soil Sci. Soc. Am. Proc. 38:225–228.

Spencer, W. F. 1966. Effect of copper on yield and uptake of phosphorus and iron by citrus seedlings grown at various phosphorus levels. Soil Sci. 102:296–299.

Stewart, E., and R. W. Pearson. 1952. Utilization of phosphorus by Crimson clover as affected by fertilizer placement and rate of liming. Agron. J. 44:501–502.

Stewart, W. S. 1977. An economic analysis of agricultural lime vending—a case study. Ph.D. Diss. Univ. of Mississippi, Oxford.

Sumner, M. E., P. M. W. Farina, and V. J. Hurst. 1978. Magnesium fixation—a possible cause of negative yield responses to lime applications. Commun. Soil Sci. Plant Anal. 9: 995–1007.

U.S. Department of Agriculture. 1981. Agricultural statistics. USDA, U.S. Government Printing Office, Washington, D.C.

U.S. Department of Commerce. 1978. Agricultural census. U.S. Dep. of Commerce, U.S. Government Printing Office, Washington, D.C.

Volk, G. M. 1956. Efficiency of various nitrogen sources for pasture grasses in large lysimeters of Lakeland fine sand. Soil Sci. Soc. Am. Proc. 20:41–45.

Volk, G. M., and C. E. Bell. 1944. Soil reaction (pH). Some critical factors in its determination, control and significance. Florida Agric. Exp. Stn. Bull. 400.

Walker, M. E., T. C. Keisling, and J. S. Drexler. 1976. Responses of three peanut cultivars to gypsum. Agron. J. 68:527–528.

Walker, M. E., W. H. Marchant, and W. J. Ethredge. 1975. Effects of soil pH on forage yield and chemical composition of sorghum and millet. Agron. J. 67:191–193.

Walker, M. E., and A. W. White. 1963. Effect of liming on crop yields and chemical properties of Tifton and Greenville soils. Georgia Agric. Exp. Stn. Bull. N.S. 108.

Wear, J. I. 1956. Effect of soil pH and calcium on uptake of zinc by plants. Soil Sci. 81:311–315.

Wear, J. I. 1957. Boron requirements of crops in Alabama. Alabama Agric. Exp. Stn. Bull. 305.

Wear, J. I., and R. M. Patterson. 1962. Effect of soil pH and texture on the availability of water-soluble boron in the soil. Soil Sci. Soc. Am. Proc. 26:344–345.

Wehunt, E. J., B. D. Horton, and V. E. Prince. 1980. Effects of nematicides, lime, and herbicides on peach tree short life in Georgia. J. Nematol. 12:183–189.

Wolt, J. D., and F. Adams. 1979. Critical levels of soil- and nutrient-solution calcium for vegetative growth and fruit development of Florunner peanuts. Soil Sci. Soc. Am. J. 43: 1159–1164.

Woodhouse, W. W. 1969. Long-term fertility requirements of Coastal bermuda: II. Nitrogen, phosphorus and lime. Agron. J. 61:251–256.

Woodruff, C. M. 1948. Testing soils for lime requirement by means of a buffered solution and the glass electrode. Soil Sci. 66:53–63.

Young, V. H., and R. L. Beacher. 1952. Crinkle leaf of cotton, a result of manganese toxicity. Arkansas Farm Res., Vol. 1, Issue no. 2.

Yuan, T. L. 1976. Anomaly and modification of pH-acidity relationship in the double buffer method for lime requirement determination. Soil Sci. Soc. Am. J. 40:800–802.

Yuan, T. L., M. S. Lutrick, and W. K. Robertson. 1979. Response of soybeans and oats to lime, phosphorus, and potassium on a Paleudult. Proc. Soil Crop Sci. Soc. Fla. 38:116–121.

Zakaria, Z. Z., V. N. Schroder, and K. J. Boote. 1977. Soybean response to calcium and phosphorus under aluminum saturation. Proc. Soil Crop Sci. Soc. Fla. 36:178–181.

# 6 Crop Response to Lime in the Midwestern United States[1,2]

E. O. MC LEAN

*Ohio State University*
*Columbus, Ohio*

J. R. BROWN

*University of Missouri*
*Columbia, Missouri*

The large expanse of level to gently rolling, deep, productive soils that extends from the foothills of the Appalachians westward to the Great Plains and from the Canadian boundary southward to the confluence of the Ohio and Mississippi rivers comprises the midwestern United States, or Midwest. The region was originally covered with forest and tall grass prairies.

The first agriculture of the region was a mixed type that provided for the various needs of the pioneers. As transportation facilities developed, permitting exchange of products with other areas, the agriculture became more specialized. Soil and climatic conditions were particularly favorable for the production of corn, and the region soon developed into what has become known as the Corn Belt of the United States.

The first half-century of agriculture in the region (1800–1850) was characterized by bounteous harvests of all kinds of crops on the new lands, and little concern was evidenced for the depletion of soil fertility. Total production of crops was maintained in the second half-century of agriculture (1850–1900) by clearing of additional land. During this period, concern about declining fertility of the older cropland led to the establishment of the agricultural experiment stations. However, it was not until early in the 20th century that extensive field investigations of crop responses to added lime were conducted.

The fundamental principles of soil fertility maintenance were established in the third half-century of agriculture (1900–1950), and measures

[1] This chapter in the original monograph of 1967 was written by Dr. C. M. Woodruff, now emeritus professor, University of Missouri. Although most sections have been updated and rather drastically revised, those we consider more than adequate without revision remain as our grateful tribute to "Woody."

[2] Joint contribution from the Ohio Agricultural Research and Development Center, Ohio State University, Journal Article no. 160-82 and Missouri Agricultural Experiment Station Journal Series no. 9178.


*Soil Acidity and Liming*—Agronomy Monograph no. 12 (2nd Edition).

were taken to inform a receptive public. Low crop yields, a series of drought years, and the Great Depression of the 1930s proved the need for conserving and protecting soils. Guided by the results of soil fertility and erosion investigations, farmers grew corn in rotation with small grain and leguminous meadow crops as a means of maintaining corn production. Each farm enterprise became a highly integrated operation. Livestock was the means through which the soil-crop-animal-manure-soil cycle of plant nutrients was completed, returning nutrients to the soil for use by subsequent crops. This system of soil fertility maintenance produced a high degree of uniformity in farm operations over much of the humid portion of the region. Dairying and the production of feeder cattle, both of which required considerable pasture and forage acreage, were concentrated on the rolling lands that were more subject to erosion. The more level lands where grain crops were produced without serious erosion problems were used for cattle-feeding operations and hog production. In the semiarid part of the region, from western Kansas through North Dakota, wheat, the dominant crop, was usually grown continuously. In contrast, in the subhumid and humid portions of the region, rotation of crops was common.

With the introduction of abundant and cheap sources of commercial N fertilizers in the early 1950s, average yields of grain crops doubled within a period of 15 yr. In many states, they have doubled again in the past 15 yr. Also, the use of legume crops as a source of N was virtually abandoned. As a consequence, the amount of lime applied to the land declined rapidly but has increased almost to the point where decline started. The amounts of P and K that were used increased. Concurrently, with the expansion of fertilizer facilities, farm machinery and pesticide chemicals changed markedly. These technological advances caused a major shift from pasture and forage crops to greater acreages of cultivated crops, especially corn and soybean, on the level lands. The shift to row crops on the more rolling lands of the region was initially much less extensive because cultivated crops had to be grown on contoured or terraced land in rotations with meadow crops to control erosion. Thus, as a soil-saving principle, the trend on hill-land farms was toward larger farm units with greater percentages of pasture and forage crops. Unfortunately, with time and the resulting buildup of economic pressures, this principle has been violated to varying degrees.

The initial impact of the changes in cropping systems was a decrease in the quantities of lime used on both the level and the rolling lands of the region. However, this initial decrease in lime applications was merely a transitory phase. Legume crops, essential as fixers of N for pasture and forage crops on the hill lands, require large amounts of lime to function best. High rates of fertilization, especially N, and the removal of large yields of grain hastened the depletion of lime reserves in cultivated soils so that they, too, now require regular applications of lime to maintain production. However, the soil pH required for optimum performance of grain crops fertilized with synthetic N will probably be less than that formerly required for these same crops where legumes were grown in the rotations as the source of N.

# I. SOILS

The kinds of soil, as well as the climatic environment, determine to a great extent the type of agriculture within an area. Furthermore, the integrated effects of soil, climate, and agriculture determine the lime and fertilizer requirements necessary for sustained crop production. Soils may be grouped according to their parent materials and the climate under which they have developed. A general map of the soil orders of the Midwest is shown in Figure 1 (Soil Survey Staff, 1975). The soils of this region are discussed in more detail in section IC.

## A. Parent Materials of Soils

The soils of the Midwest developed principally from loess and glacial till that were spread over the area during the glacial epochs. Residual soils formed from sedimentary rocks occupy a major portion of the land in the southern portion of the states of Kansas, Missouri, Illinois, Indiana and Ohio. There are, of course, major areas of soils formed from lacustrine and alluvial materials throughout the region.

Soils of the glaciated area exist as broad expanses of gently undulating plains, well suited to multirow farm machinery. Their recent origin and a favorable climate make these soils some of the most productive in the region. Sorting by meltwater from the retreating glaciers left extensive areas of very sandy soils along the northern border of the region, principally in northern Wisconsin and Michigan. The finer materials were carried southward by water and then spread by wind as a blanket of loess over large areas

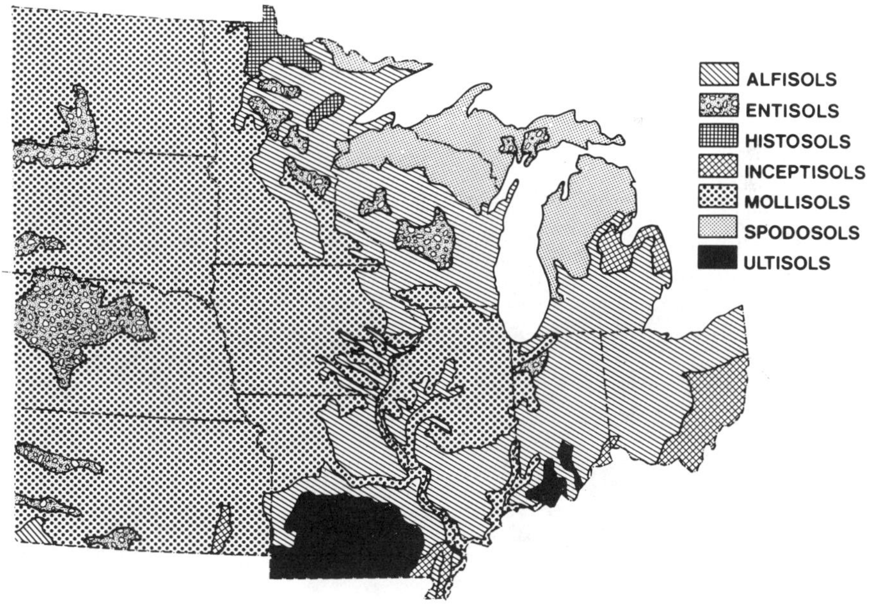


Fig. 1—Major soil orders of the Midwest.

in Iowa, Missouri, Illinois, and Indiana. Adjacent to the major streams, the deposits of loess are deep and highly dissected, giving narrow bands of hilly topography, suited best for the production of pasture and forage crops. Some areas of level lake-laid clays occur mainly in northern Ohio.

The soils found in the nonglaciated area along the western border of the region exist as gently undulating plains. Through the more highly dissected central and eastern portions of the region, soils formed in the unglaciated material occur on rolling to steep topography. These soils, primarily those of limestone valleys and uplands, are used for a mixed type of farming, including cultivated crops, small grain, forage, and pasture crops.

### B. Climatic Zones

The soils of the Midwest and the agriculture associated with them have developed under a wide range of climatic conditions that have greatly influenced the liming needs of the soils. Climate is a complex of temperature, precipitation, and evapotranspiration potentials. North of the 39th parallel (39° N Lat), which passes through the central parts of Kansas, Missouri, and Illinois and the southern parts of Indiana and Ohio, average daily temperatures in winter are below freezing, and biological and chemical processes and leaching are relatively inactive, thus favoring the accumulation of soil organic matter and more efficient use of lime and fertilizer. Since winter temperatures south of this line are low enough to keep vegetation dormant but warm enough for biological and chemical processes to remain active, serious leaching has resulted in very acid soils, creating a great need for lime.

However, the effect of excess precipitation on soil acidity is of far greater significance than are temperature effects. Amounts of surplus pre-

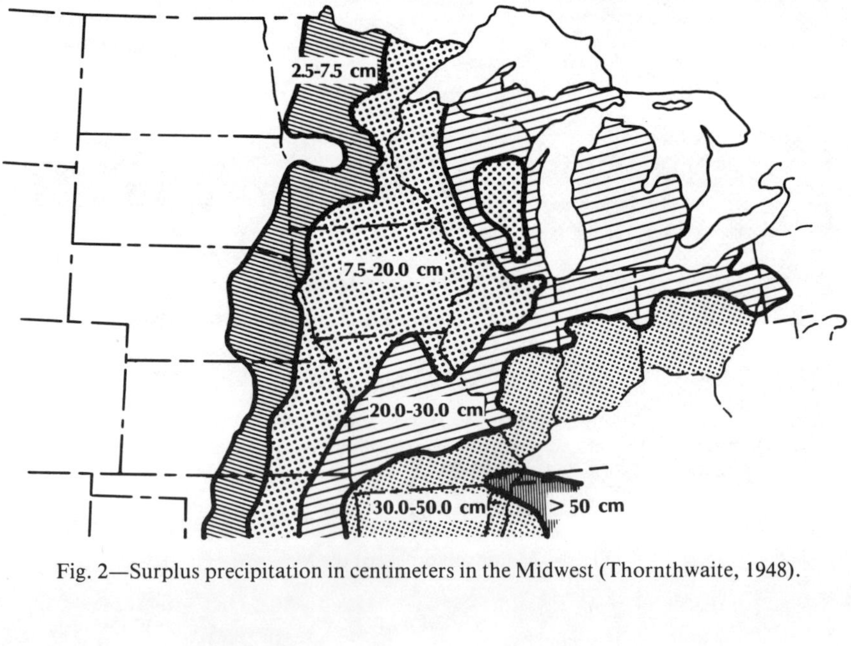


Fig. 2—Surplus precipitation in centimeters in the Midwest (Thornthwaite, 1948).

cipitation for the region, as estimated by Thornthwaite (1948), are shown in Figure 2. The zone of 2.5 to 7.5 cm of surplus annual precipitation occupies the transition from calcareous soils of the Great Plains to moderately acid soils on which most of the corn is grown in the region. The zone of 20 to 30 cm of surplus annual precipitation encompasses large areas of good agricultural land with soils that are acid throughout their profiles. The zone with 30 to 50 cm of surplus precipitation covers an area of light-colored, forested soils that tend to be very acid. Fifty centimeters of surplus precipitation affects only a small area in southeatern Missouri.

## C. Soils of the Midwest

The soils of the Midwest fall into relatively major areas of the orders Mollisols (52%) and Alfisols (26%) with progressively lesser areas of Spodosols (7%), Entisols (7%), Ultisols (4%), Inceptisols (3%), and Histosols (1%) (Fig. 1).

### 1. SOILS OF THE WESTERN MAJOR SUBREGION (MOLLISOLS)

The western major subregion of the Midwest consists of South Dakota, Nebraska, Kansas, western one third of Minnesota, western three fourths of Iowa, northwestern one fourth of Missouri, and northern two thirds of Illinois. The soils in the more central part of this subregion have slight differences depending on whether or not clay has accumulated in the subhorizons. Within this portion of the subregion, the surplus precipitation has been sufficient to leach the carbonates from the upper horizons of the soil profile, leaving them moderately acid. The soils are ideally suited to the production of both corn and soybean because (i) they have developed from deep, moderately weathered, glacial and loessial deposits; (ii) they occupy broad expanses of gently undulating landscape; (iii) the combination of low winter temperatures and moderately long growing seasons favors the production and preservation of soil organic matter; and (iv) there is an adequate supply of available soil moisture without undue excesses. This area is known as the central Corn Belt.

The soils of the western part of this subregion with relatively low rainfall differ according to whether salts or carbonates have accumulated or whether bases are high but clay, $CaCO_3$, or $CaSO_4$ have not accumulated. This subregion contains much of the hard winter wheat, grain sorghum, and cattle ranching areas. The relatively low rainfall makes irrigation essential even in the stream bottoms for most corn, soybean, and alfalfa production. Since the soils are generally high in bases, $CaCO_3$ accumulations, or both, little need for lime exists in this portion of the subregion.

The soils of the northwestern part of this subregion are the cool Mollisols with or without clay accumulations in a subsurface horizon. Others have materials altered but with no accumulation of clay or $CaCO_3$, and still others have $CaCO_3$ accumulated near the surface. This portion of the subregion contains much of the hard spring wheat, sunflower, and cattle ranching areas. Since these soils are also generally high in bases or $CaCO_3$ accumulations (or both) in the subsurface horizons, very little need for lime exists in the area.

## 2. SOILS OF THE EASTERN MAJOR SUBREGION (ALFISOLS)

The eastern major subregion of the Midwest consists of eastern Minnesota, Wisconsin, Michigan, Ohio, Indiana, southern Illinois, and northeast and southern Missouri. The soils in this crescent-shaped area surrounding (on the north, east, and south) the central portion of the Mollisolic subregion are primarily the Alfisols, with a relatively thin or brownish subsurface horizon of clay accumulation. Large-scale inclusions of soils known earlier as the flat claypan and fine-textured lakebed soils occur in the subregion. One of these has a bleached upper horizon with heavy concentration of clay in a lower horizon, whereas the other is very fine textured with very gradual change in texture or color (or both) with depth. This area, which surrounds the central Corn Belt on three sides, is very much influenced by glacial activity, much of the area having rolling moraines on which these well-developed Alfisols occur. They support a general type of farming with emphasis on production of corn, soybean, alfalfa, and small grains. The soils tend to be quite acid and are treated extensively with lime and fertilizers. As in the Corn Belt to the south and west, most of the land of this area is used for the production of agricultural crops.

Under certain conditions, poor surface drainage has intensified leaching of the shale-derived soils on level land near the streams in southeastern Kansas and southwestern Missouri and on the loessial plains of northeastern Missouri, south central Illinois, and southern Indiana and Ohio. The excessive leaching has produced not only soils that are very acid but also soils with high concentrations of acid clay in the subsoils. These are the previously mentioned claypan soils. Because of the level topography and the silty texture of the surface horizons, these areas are well suited to the production of cultivated crops. It is from these soils, which originally were not very productive, that lime and fertilizer treatments have resulted in some of the largest increases in crop yields in the Midwest.

## 3. SOILS OF THE NORTHEASTERN MINOR SUBREGION (SPODOSOLS)

The northeastern minor subregion of the Midwest consists of northern Wisconsin, Michigan, and Minnesota. The soils are primarily Spodosols of the cold region, which have organic matter and Fe and Al accumulated but no fragipan. Because of short, cool summers and cold winters, humid conditions exist in this subregion with only 60 to 75 cm of annual precipitation. The soils of this area have developed under a coniferous type of forest and are highly podzolized. Granitic rocks, swept free of soil during glaciation, and organic soils in poorly drained areas are common in portions of northern Minnesota. Extensive areas of leached, sandy soils occur in northern Wisconsin and Michigan. The soils in these areas require extensive treatments, including considerable lime for the production of agricultural crops.

## 4. SOILS OF THE WESTERN AND NORTH CENTRAL MINOR SUBREGION (ENTISOLS)

The western and north central minor subregion of the Midwest is made up of relatively large numbers of smaller areas in the Dakotas, Nebraska, Kansas, Minnesota, Wisconsin, and northwestern Indiana. These soils (Entisols) are coarse textured and contain easily weathered minerals. The areas in the West include the Sand Hills area of Nebraska, where recent developments in usage of large-scale irrigation systems have intensified the production of grains such as corn and increased the potential future use of lime to offset the acidity equivalent of N fertilizers now being applied. Areas in Wisconsin, Minnesota, and Indiana are largely outwash from glacial activity. Here, under these more humid conditions, the more leachable, coarse-textured soils coupled with intensive cropping generally dictate moderate rates of fertilizers and lime applied rather frequently to maintain high yields of crops.

## 5. SOILS OF THE SOUTH CENTRAL MINOR SUBREGION (ULTISOLS)

The south central minor subregion of the Midwest consists of the Ozark Highlands of south central to southwest Missouri and southern Indiana. The soils are primarily Ultisols either with a thick horizon of clay accumulation without appreciable weatherable minerals or with thin clay accumulation with weatherable minerals. The steeply rolling, limestone valleys and uplands of this subregion comprise an area of highly weathered, shallow soils, which have developed under a forest type of vegetation and have been subjected to high rainfall in both summer and winter. Large portions of the area remain in native forest. Many farms of the area are so located that they make use of alluvial soils along stream valleys for the production of grain crops and grow pasture and forage crops on the sloping uplands. Many of these pastures and meadows have now undergone renovation with brush clearing, liming, fertilizing, and reseeding of improved forage species. Both lime and fertilizers are used extensively on the cropped land. On these soils, the amount of lime used per unit area of cropped land generally equals or exceeds that used anywhere else in the Midwest. However, the proportion of the total land area devoted to crops is relatively low.

## 6. SOILS OF THE EASTERN AND SOUTH CENTRAL MINOR SUBREGION (INCEPTISOLS)

The eastern and south central minor subregion of the Midwest consists of southeastern Ohio and north Michigan. It is part of a larger area consisting of southern New York, Pennsylvania, West Virginia, and Kentucky. The soils are primarily Inceptisols, which are usually moist and low in bases with no $CaCO_3$ in subsurface horizons. These are highly weathered soils that require heavy rates of lime and fertilizers to make them and keep them productive. Only the more level uplands and the alluvial valleys are cultivated intensively, whereas the steeper land is maintained in permanent forages or forest cover.

### 7. SOILS OF THE FAR NORTH MINOR SUBREGION (HISTOSOLS)

The far north minor subregion of the Midwest is located primarily in northern Minnesota, but other areas of Histosols too small to be shown on a soil map of this scale (Fig. 1) occur in northern Illinois, Indiana, Michigan, Wisconsin, and Ohio. These organic soils, valued highly for the production of truck and vegetable crops, vary in acidity from neutral to very acid. However, high acidity in organic soils is not usually objectionable; a pH of 5.0 to 5.5 is generally acceptable. Even at this pH range, these soils may contain much more Ca than mineral soils of corresponding degrees of acidity. Only where the material contains enough clay to yield toxic amounts of $Al^{3+}$, $Fe^{3+}$, and $Mn^{2+}$ is liming of organic soils generally practiced.

### 8. SOILS OF WATER-LAID ORIGIN (LACUSTRINE AND ALLUVIAL)

Where sufficiently large to be shown on the map, the areas with soils of water-laid origin are designated as Mollisols or Alfisols. Substantial lake-laid soils can be found in the northern parts of Ohio, Indiana, and Illinois and in the Red River Valley of Minnesota and North Dakota as well, but these are not shown. The high clay content of these soils, the cool climate, and the lack of excess precipitation do not subject them to rapid leaching or to the development of high acidity. Bands of deep productive alluvial soils occur in the stream valleys throughout the Midwest. These soils tend to be either very sandy and subject to much leaching or very clayey and imperfectly drained. They occur both as first bottoms, subject to frequent flooding, and as second bottoms with highly developed profiles. Regardless of their texture or state of development, the alluvial soils are used intensively for high-income crops. Corn and soybean are produced on most of them. Alfalfa for dehydrating and processing as poultry feed is grown on many of the sandier first bottoms that generally require relatively low rates of lime. Although it is a rather isolated area in the region, the seven-county alluvial section of southeastern Missouri produces high yields of cotton and rice.

Many of the second bottoms where corn is fertilized with high rates of N have developed very high levels of acidity. Even though farmers have been slow to use lime on these soils for corn and soybean, this acidity will have to be dealt with sooner or later. Where irrigation has been practiced, the effect of soil acidity from N fertilizers is somewhat counteracted by the alkaline irrigation water. Thus, the future need of agricultural lime on the alluvial soils of this region will depend to some degree on the development of irrigation and the concentrations of basic ions contained in the water used.

## II. CROPS

The crops best suited to the combination of soil and climatic conditions of a region are responsible for major adjustments to conditions of soil fertility in that region. For example, the potential for producing corn in the

Table 1—Area of major cultivated grain, oil, and small grain crops in the Midwest harvested in 1981 (ha × $10^3$) (Crop Reporting Board, 1982).

| State | Corn | Soy-bean | Sor-ghum | Sun-flower | Wheat | Oat | Barley | Rye | Rice |
|---|---|---|---|---|---|---|---|---|---|
| Illinois | 4560 | 3779 | 36 | 0 | 749 | 83 | 0 | 6 | 0 |
| Indiana | 2430 | 1863 | 5 | 0 | 547 | 34 | 0 | 4 | 0 |
| Iowa | 5548 | 3301 | 8 | 0 | 47 | 389 | 0 | 2 | 0 |
| Kansas | 510 | 612 | 1442 | 0 | 4941 | 69 | 21 | 5 | 0 |
| Michigan | 1154 | 393 | 0 | 0 | 336 | 138 | 11 | 8 | 0 |
| Minnesota | 2742 | 1782 | 0 | 289 | 1462 | 579 | 417 | 38 | 0 |
| Missouri | 786 | 2045 | 381 | 0 | 1114 | 36 | 0 | 2 | 31 |
| Nebraska | 2827 | 859 | 834 | 0 | 1195 | 160 | 9 | 15 | 0 |
| North Dakota | 208 | 93 | 0 | 1064 | 4734 | 389 | 891 | 32 | 0 |
| Ohio | 1519 | 1418 | 0 | 0 | 668 | 109 | 0 | 2 | 0 |
| South Dakota | 1045 | 312 | 184 | 179 | 1547 | 664 | 239 | 47 | 0 |
| Wisconsin | 1418 | 152 | 0 | 0 | 49 | 367 | 13 | 7 | 0 |
| Region total | 24747 | 16609 | 2890 | 1532 | 17389 | 3017 | 1601 | 168 | 31 |

Table 2—Area of forage, silage, tuber, sugar, fiber, and tobacco crops in the Midwest harvested in 1981 (ha × $10^3$) (Crop Reporting Board, 1982).

| | Hay | | Forage | Silage | | | | | |
|---|---|---|---|---|---|---|---|---|---|
| State | Alf.-Alf. mix | All other | Corn + Sorg. | Corn | Sor-ghum | Potato | Sugar beet | Cotton | Tobacco |
| Illinois | 286 | 176 | 6 | 73 | 4 | 0.9 | 0 | 0 | 0 |
| Indiana | 152 | 166 | 7 | 66 | 2 | 1.9 | 0 | 0 | 3.0 |
| Iowa | 709 | 194 | 13 | 202 | 2 | 0.6 | 0 | 0 | 0 |
| Kansas | 405 | 526 | 123 | 69 | 113 | 0 | 5.7 | 0 | 0 |
| Michigan | 405 | 109 | 4 | 130 | 0 | 15.5 | 40.1 | 0 | 0 |
| Minnesota | 810 | 348 | 8 | 328 | 0 | 30.8 | 104.1 | 0 | 0 |
| Missouri | 223 | 1215 | 16 | 45 | 15 | 0 | 0 | 74 | 1.1 |
| Nebraska | 668 | 830 | 30 | 188 | 36 | 3.6 | 31.7 | 0 | 0 |
| North Dakota | 648 | 660 | 6 | 148 | 0 | 47.0 | 58.7 | 0 | 0 |
| Ohio | 186 | 373 | 2 | 115 | 0 | 4.1 | 5.8 | 0 | 5.2 |
| South Dakota | 960 | 786 | 43 | 302 | 24 | 2.2 | 0 | 0 | 0 |
| Wisconsin | 1154 | 304 | 13 | 354 | 0 | 21.7 | 0 | 0 | 5.4 |
| Region totals | 6606 | 5687 | 271 | 2020 | 196 | 128.3 | 246.1 | 74 | 14.7 |

Midwest was realized early because legumes were used as a source of N for the corn. Consequently, the use of lime for legume production on the acid soils of the humid portion of the region became a general practice before synthetic N fertilizers became available.

Many crops grow well over wide ranges in soil acidity when they are supplied with adequate amounts of N. The rapid increase in amounts and rates of fertilizer used on soils in the Midwest after 1945 resulted in a trend away from crop rotations toward one-crop systems. Row crops grown continuously with high rates of fertilization are common on the more level lands, whereas small grain and meadow crops are relegated to the more rolling uplands.

The 1981 data for the various types of crops are given by state in Tables 1 and 2. Recent years have seen decreased areas planted to oat, cotton, barley, and rye and increased areas planted to sunflower, corn, sorghum (grain and forage), wheat, and soybean. The large increase in soybean plantings since 1960 indicates a shift to a corn-soybean rotation in much of the area.

In terms of total area, about 10 major crops are grown in the Midwest. Yet, three grain crops (corn, wheat, and soybean) and the various hay crops account for 85% of the total. Most of the corn and soybean is produced on the more acid soils of the humid portion of the region. Wheat and other small grains and hay are produced extensively on the calcareous soils of the High Plains in the four western states of the region (Table 1 and 2). Corn production in these four states is concentrated along the subhumid eastern borders of each state. Neither corn nor soybean is highly sensitive to soil acidity. However, in most cases, the yields obtained from limed soils are sufficiently greater than those from unlimed, strongly acid soils to justify the use of lime.

Grain sorghum is grown primarily on calcareous soils in the western states of the region where intermittent summer droughts occur.

Cotton is grown on about 72 000 ha in seven counties of southeastern Missouri, with relatively insignificant amounts in southern Illinois. This crop will grow over a wide range in soil acidity but yields best and most consistently on soils that are not strongly acid. A malady termed *crinkle leaf,* which is associated with low soil pH and an excess of available Mn, is prevented by liming.

Wheat occupies extensive areas in all states of the Midwest except Iowa, but the major areas are in Kansas and North Dakota. This crop produces greater yields on soils that are not strongly acid. Therefore, liming for wheat is an accepted practice throughout the humid portion of the region. Oat and, to a lesser extent, rye (neither of which is highly responsive to liming) are grown extensively throughout the Midwest. In the humid states of the region, oat serves as a nurse crop for red clover and alfalfa, both of which grow best on well-limed soils. Hence, most of the oat is produced on soils that have been limed to meet the needs of red clover, alfalfa or other forage legume.

The hay crops of the region consist of alfalfa, alfalfa mixtures, and other mixed hays (Table 2). Both alfalfa and red clover are relatively sensitive to low soil pH; therefore, in the humid portion of the Midwest, most of the soils producing these crops have been limed.

A substantial area of permanent pasture is found in the Midwest. Liming practices on this area are varied and difficult to summarize. All state extension programs recommend liming of permanent pastures. These pastures usually occupy the least valuable sites and hence tend to receive less lime than do rotational pastures. The benefits of interseeding of legumes on animal performance and reduced reliance on synthetic N causes increased interest in liming permanent pastures.

Potato, tobacco, buckwheat (data not given), and millet (data not given) are the most prominent crops least sensitive to low soil pH. Flaxseed (data not given), much of which has been replaced by sunflower, is grown primarily in the region of calcareous soils. In the more humid portion of the region, sweetclover (data not given), sugar beet, and garden pea, and snap bean crops grow best on soils that have received lime.

Temperature and rainfall are the principal factors governing where specialty crops are grown. Vegetable crops tend to be concentrated in the northern states of the humid section. Dried beans and peas are grown principally in Michigan. Potato is grown principally in the northern states from North Dakota eastward to Michigan. Production of sweetclover seed (data not given) is greatest in South Dakota, Nebraska, Kansas, Minnesota, and Ohio. Tobacco occupies small acreages in Wisconsin, Ohio, Indiana, and Missouri. Buckwheat (not shown) is grown in Minnesota, Wisconsin, and Michigan. Sugar beet is grown in half the states. Millet (not shown) is confined primarily to the Dakotas and Nebraska. Sunflower seed is produced principally in North Dakota, Minnesota, and South Dakota. Rice is grown on more than 30 000 ha of alluvial land in the northern extention of the Mississippi Delta in southeastern Missouri.

## III. LIME USE

### A. Early History

Early research on agricultural lime in the Midwest was apparently stimulated by Wheeler et al. (1896) of the Rhode Island Agricultural Experiment Station. His investigations were thorough, well documented, and conducted in a manner that left no doubt that many species of crop plants were sensitive to high soil acidity. Within a short time after the publication of Wheeler's investigations, lime was introduced as a variable in most soil fertility experiments in the Midwest. Significant early experiments on lime were reported by Thorne (1903) in Ohio, Hopkins and Readheimer (1905, 1907) in Illinois, Miller et al. (1915) in Missouri, Whitson and Weir (1913) in Wisconsin, Brown et al. (1914) in Iowa, Wiancko et al. (1918) in Indiana, McCool and Millar (1918) in Michigan, and Call and Throckmorton (1918) in Kansas.

Following the results of these early experiments, the practice of liming acid soils was promoted vigorously by the agricultural experiment stations in the Midwest. The need for liming was based on the beneficial effects that lime had on legume crops, which were then used as the source of N for corn grown in rotation with the legume.

### B. Lime Use by States

Reliable records of lime use prior to 1929 are not available. The National Limestone Institute has such records from 1929 to date. These data show a rapid increase in limestone use in the Midwest up to post–World War II due to the efforts of research and extension workers in the more humid part of the region. The peak use period was in the immediate post–World War II years (Table 3).

Table 3—Estimated total amounts of agricultural lime used by 5-year periods in the states of the Midwest since World War II (t × $10^3$) (Natl. Limestone Inst., Inc., unpublished data, Washington, D.C.).

| State | 1946–1950 | 1951–1955 | 1956–1960 | 1961–1965 | 1966–1970 | 1971–1975 | 1976–1980 | State totals |
|---|---|---|---|---|---|---|---|---|
| Illinois | 22 843 | 14 113 | 14 570 | 18 674 | 18 248 | 14 592 | 17 337 | 120 377 |
| Indiana | 12 035 | 10 610 | 10 076 | 8 608 | 6 659 | 5 487 | 9 995 | 63 470 |
| Iowa | 15 054 | 9 366 | 6 051 | 5 914 | 12 567 | 11 913 | 15 886 | 76 751 |
| Kansas | 4 385 | 2 637 | 1 983 | 1 857 | 2 410 | 2 144 | 2 267 | 17 683 |
| Michigan | 3 783 | 2 361 | 2 077 | 1 926 | 2 059 | 2 049 | 2 513 | 16 768 |
| Minnesota | 1 693 | 2 028 | 2 606 | 2 100 | 1 373 | 789 | 1 583 | 12 172 |
| Missouri | 14 401 | 11 041 | 10 220 | 12 938 | 16 854 | 13 091 | 17 498 | 96 043 |
| Nebraska | 34 | 726 | 591 | 700 | 687 | 872 | 1 061 | 4 671 |
| North Dakota | -- | -- | -- | -- | -- | -- | -- | -- |
| Ohio | 9 774 | 10 005 | 8 676 | 8 461 | 6 986 | 6 496 | 6 934 | 57 332 |
| South Dakota | -- | 1 | 1 | -- | -- | 8 | 15 | 25 |
| Wisconsin | 9 704 | 6 830 | 7 228 | 5 693 | 4 372 | 4 229 | 6 235 | 44 291 |
| Region totals | 93 706 | 69 718 | 64 079 | 66 871 | 72 215 | 61 670 | 81 324 | 509 583 |

The rapid development of the agricultural lime industry before 1950 was a consequence of numerous complementary factors. Soil fertility had become depleted through crop removal, erosion, and leaching. Farmers, many of whom had received training in vocational agriculture, were informed of methods of maintaining high levels of soil fertility. Federal subsidies for applying lime in the interest of soil conservation encouraged the use of lime. The development of paved highways throughout the Midwest resulted in the accumulation at rock quarries of vast quantities of screenings from crushed limestone that was used in making concrete; and with the highways came trucks and equipment for transporting and spreading lime on farmers' fields.

The tonnage of agricultural limestone declined as synthetic N fertilizers replaced legumes as the major source of N. By combining synthetic N with ample supplies of P and K, farmers were able to achieve higher yields of grain crops than had formerly been possible. Consequently, crop rotations were replaced rapidly, beginning about 1950, with a cropping system that consisted of a single crop that was grown continuously. A sharp reduction in lime use followed, probably because (i) the increased fertilizer cost was partially offset by lower expenditures for lime, (ii) the need for lime had become associated with the growing of forage legumes rather than with the growing of grain crops, and (iii) cost-sharing practices of the federal government for lime were beginning to decline. Although the use of lime after 1950 decreased markedly the first 5 yr, and less so the second 5 yr (1956–1960), the trend was reversed the third 5-yr period (1961–1965) and has generally continued to increase through the 1976 to 1980 period. The lime use level in the latter period has now reached 87% of the peak period (1946–1950). This regained rate of usage is undoubtedly a reflection of the increased soil acidity resulting from additions of large quantities of residually acid fertilizers. In this connection, Iowa and Missouri provide interesting contrasts. Both states reached highest levels of lime usage in the most recent 5-yr period (1976–1980). Yet, they probably did so for different reasons, i.e., increased N usage on corn and expanding livestock industry, respectively, as described below in section VC2.

### C. Lime Requirement of Soils

In general, the lime requirement (LR) of a soil is both the degree of deficit in basic cations and the degree of accumulation of acidic cations. More specifically, LR of a soil is the amount of lime required: (i) to increase its pH (or base saturation) from an initial acid condition to a near-neutral pH level, or (ii) to decrease its content of toxic elements to a lower level, where optimum yields of crops are attained. Liming to attain optimum conditions for maximum yields assures multiconditions favorable to plants. Adequate base saturation of the soil CEC is provided; concentrations of elements toxic to plants are decreased to nontoxic levels, solubilities (availabilities) of most other plant-essential elements are enhanced; and the activities of most microorganisms favorable to higher plants are enhanced.

In most cases, the basis for selecting the point to which a soil is to be limed is the pH range most favorable for plant growth. This may include such a criterion as the inactivation of soluble $Al^{3+}$ or $Mn^{2+}$. In the Midwest, there seems to be little economic advantage for liming soils above pH 6.0 or 6.2 for corn and soybean, whereas such forage legumes as alfalfa generally respond well to liming to pH 6.5 or above (see section IVA). In contrast, peat and muck soils do not generally produce higher yields above pH 5.2.

Veitch (1902) observed very early that maximum crop yields were obtained by liming soils to $<$ pH 7.0, an observation that became almost universally accepted in years to follow. It is evident then that the specific causes of poor plant growth in acid soils are generally counteracted by rates of limestone that give soil pH values considerably $<$ pH 7.0. It also appears that maximum crop yields are realized as long as there are enough neutralized zones of soil to meet the crop's nutrient requirements and tolerances of toxic ions. This is evidenced by the fact that the greatest yield responses to added lime are for the first increment. Extreme acidity (i.e., $<$ pH 5) in most soils probably reflects a combination of factors that are unfavorable for plant growth, e.g., Ca deficiency, Al toxicity, and Mn toxicity.

In reviewing the early European methods of testing soils for acidity, Harris (1914) found them to be less than satisfactory. He noted that litmus paper in direct contact with an acid soil gave an acid reaction, whereas litmus in contact with only the soil solution of the same soil gave a neutral reaction unless a salt was added, in which case the reaction was acid. Although others had noted 20 yr earlier that the pH varies with changes in concentrations of salts in soils, only after Schofield and Taylor (1955) proposed the use of $0.01M$ $CaCl_2$ for measuring soil pH was the salt content of soils considered very seriously in the determination of soil pH. Almost 30 yr later, most soil testing laboratories still measure soil pH in a water suspension. Exceptions are the state soil testing laboratories in Missouri, which have measured soil pH in $0.01M$ $CaCl_2$ since 1957 (Graham, 1959). Extensive studies indicate $0.01M$ $CaCl_2$ lowers the pH by 0.4 to 0.6 unit in most soils relative to that in water. However, ΔpH in highly leached, acid sands may be considerably more, Hence, even though pH measurements are made with electronic equipment accurate to a few hundredths of a pH unit, uncertainties with regard to salt contents of soils introduce variations that in some cases may exceed 1 unit of pH.

## 1. SOIL FACTORS

Factors affecting the measured LR are essentially those affecting total soil acidity plus those that are method dependent. They are degree of soil weathering, type of parent material, clay content, organic matter content, forms of acidity, initial and final soil pH, LR method, and equilibration time.

**a. Degree of Weathering**—A soil becomes more acid when cropping, leaching, and erosion remove basic cations more rapidly than dissolution of weatherable minerals replaces them. As basic cations are removed and are only partially replaced by breakdown of basic minerals, acidic cations accumulate in their places. Primary sources of $H^+$ for various weathering reactions are dissociation of carbonic acid and the oxidation of sulfide-containing minerals. Secondary $H^+$ sources are the acidic components of fertilizers, especially N. Increased time of exposure of any soil in a humid climate results in greater accumulation of soil acidity. Weathering of the 2:1 type of clays (micas, smectites, chlorites, and vermiculites) occurs with a corresponding loss of permanent negative charge. Then, pH-dependent acidity predominates over unbuffered salt-extractable acidity.

**b. Parent Material**—The materials from which soils develop vary widely in contents of basic ions. They vary all the way from calcareous glacial till to highly weathered acid Coastal Plain sediments to sulfide-containing minerals. During the weathering process, basic rocks and minerals initially release basic cations that with time are gradually replaced by the acidic cations, $H^+$ and $Al^{3+}$, as the leaching processes continue. In contrast, low base or acid-forming rocks and minerals release silicic acids of various kinds and accumulate soil material of very low base status even before the leaching of bases occurs. In both cases, however, the accumulation of acidic ions increases the soil LR.

**c. Clay Content**—The greater the concentration of clay in the soil, the greater the cation exchange capacity (CEC), and the greater the amounts of acidic cations that can be adsorbed to cation exchange sites. Also, the higher the cation exchange capacity, the greater the accumulation of polymerized hydroxy-Al ions between the clay layers (Jackson, 1960). Since less water may percolate through a fine-textured than a coarse-textured soil, a fine-textured soil usually has a higher base saturation than a similarly exposed coarse-textured soil. However, a very old, fine-textured soil may be so highly weathered that it is almost devoid of basic cations and almost saturated with acidic cations.

**d. Organic Matter Content**—The greater the concentration of organic matter (OM) in a soil, the greater the tendency for acidic cations to accumulate on the pH-dependent exchange sites. Hence, it is not surprising that OM content and LR are highly correlated in soils (Keeney & Corey, 1963). Organic matter not only adsorbs acidic cations in exchangeable form but also complexes or chelates acidic cations in forms not readily displaced by ordinary exchange reactions (Clark & Nichol, 1966). The $Fe^{3+}$, $Al^{3+}$, and hydroxy-Al and -Fe ion complexes involving the carboxyl groups of the OM cause the latter to act as a weaker acid than it otherwise would be (Martin, 1960; Martin & Reeve, 1960; Schnitzer & Wright, 1960).

**e. Forms of Soil Acidity**—Several forms of acidity account for the total acidity and LR of a soil, not just undissociated $H^+$. In fact, even though $H^+$ ions replace basic cations during acid weathering, they do not accumulate without limit (Jackson, 1960). When a certain concentration builds up in the soil, $H^+$ ions react with the clay crystal, becoming adsorbed internally and simultaneously releasing $Al^{3+}$. Removal of $H^+$ ions from solution raises pH, causing partial neutralization of the $Al^{3+}$ ions. As a consequence, hydroxy-Al ions, $Al(OH)^{2+}$ and $Al(OH)_2^+$, are formed. The latter ions not only are subject to complexation by organic matter but also polymerize on clay surfaces as continuous layers or as islands of nonexchangeable acidity (Schwertmann & Jackson, 1964; de Villiers & Jackson, 1967).

The cationic nature (positive charges) of the hydroxy-Al ions neutralizes the cation exchange sites of the clay and OM. The large size of Al polymers and the strong chelation bond of organically bound Al preclude these Al forms from being exchanged for other cations. Hence, a corresponding decrease in the effective CEC of an acid soil is noted with increasing amounts of Al in these forms (de Villiers & Jackson, 1967). Hydroxy-Fe evidently plays the same role as hydroxy-Al in acid soils, except at a lower pH (Coleman & Thomas, 1964). Because the various hydroxy-Al and hydroxy-Fe polymers or complexes are only partially neutralized and thus are acidic cations, they markedly increase the total LR of a soil. Yet their relative inactivity or inaccessibility to the buffers used in LR tests may cause measured LR to fall short of actual requirement. This, however, can be avoided by the proper use of buffer solutions (Shoemaker et al., 1962).

Many types of OH groups exist in OM and clay minerals that ionize H from coordinate covalent bonds as the pH is increased by the addition of a base. Charges from which $H^+$ ions dissociate are called pH dependent, because they are effective for adsorbing other cations only at higher pH values. The pH-dependent charges adsorb $Ca^{2+}$ and $Mg^{2+}$ from lime as $H^+$ is simultaneously neutralized by $OH^-$. A small decrease in pH reverses the process so that $H^+$ ions are readsorbed to the H-preferring, pH-dependent charges, and basic cations are simultaneously released into solution where they may be taken up by plants or leached into lower horizons of the profile (McLean, 1970).

**f. Initial and Final Soil pH**—The predominant form of acidic cation in a soil differs with pH. A very acid soil (e.g., pH 4.0) might have a small amount of $H^+$ in solution with considerably more adsorbed at the exchange sites. But such a soil would likely have a relatively large amount of adsorbed $Al^{3+}$, and a relatively small amount of adsorbed hydroxy-Al ions. In contrast, if soil pH is slightly higher, it may have a predominance of hydroxy-Al adsorbed to exchange sites. If lime is added and mixed uniformly with the pH 4.0 soil in sufficient quantity to increase the pH to 7, the various forms of acidic cations would be neutralized sequentially in the following order as the pH increases: $H^+$, $Al^{3+}$, $Al(OH)^{2+}$, $Al(OH)_2^+$, and other pH-dependent H (Schwertmann & Jackson, 1964; Coleman & Thomas, 1967). Even though a soil with high percentage $H^+$ and $Al^{3+}$ acidity has no more actual LR than one with more hydroxy-Al and pH-dependent acidity, it will have more indicated by the soil buffer test. In addition, if the lime is not uniformly mixed with the soil, microvolumes of soil will likely go through all of the above steps in the neutralization process while other microvolumes remain at the initial pH.

**g. Lime Requirement Method**—Since there are several forms of acidity in an acid soil and each form seems to react sequentially as the soil pH is increased as noted above, it is not surprising that different LR methods would vary in the extent of reaction with these several forms of acidity and hence in the LR indicated. In fact, the Shoemaker, McLean, and Pratt (1961) (SMP) buffer method resulted from efforts to improve the original Woodruff buffer method (Woodruff, 1948a, b), which was unsatisfactory for use on soils of high LR made so by high levels of extractable Al, much of which was organically complexed. Methods of measurement of LR are given in a following section.

The diversity of forms of acidity in a soil and their different degrees of exchangeability and accessibility to the buffer solutions used in quick test methods affect the indicated LR in a given time. If sufficient time is allowed for reaction, the buffer solution will neutralize all acidity to the pH at which soil and buffer reach equilibrium. Similarly, field experiments allow adequate time for the lime to react and thus are useful for adjusting soil test methods and for correlating results obtained with those from a soil test, but they are too time consuming to be useful as a soil test method as such. In contrast, soil test methods seek to obtain a LR value as accurate as possible in a minimum of time.

## 2. NITROGEN FERTILIZER

In addition to the soil factors affecting the LR of soils, the management factor—rate of N fertilizer—has become very significant in the Midwest. It is ironic that the early liming programs, which were developed primarily to assure successful growth of forage legumes and succeeding grain crops through $N_2$ fixation, may now be preventing disastrous acidity buildup from N fertilizers.

The amounts of N and lime used during 1978 to 1980 in the various states are listed in Table 4 along with averages and totals for the 12-state area. In addition, an average lime equivalent of 1.8 units of $CaCO_3$ per unit

Table 4—Nitrogen and limestone used in the Midwest in 1978, 1979, and 1980 (t × $10^3$).

| | Total N used† | | | | | Total limestone used‡ | | | |
|---|---|---|---|---|---|---|---|---|---|
| State | 1978 | 1979 | 1980 | Avg 1978–1980 | Limestone equivalents 1978–1980§ | 1978 | 1979 | 1980 | Avg 1978–1980 |
| Illinois | 719 | 822 | 938 | 826 | 1487 | 3115 | 3180 | 4679 | 3658 |
| Indiana | 457 | 447 | 466 | 457 | 823 | 1601 | 1906 | 2135 | 1881 |
| Iowa | 836 | 911 | 970 | 906 | 1631 | 2092 | 2205 | 2677 | 2325 |
| Kansas | 501 | 548 | 580 | 543 | 977 | 417 | 453 | 498 | 456 |
| Michigan | 196 | 218 | 214 | 209 | 376 | 493 | 547 | 430 | 490 |
| Minnesota | 446 | 555 | 561 | 521 | 938 | 340 | 327 | 363 | 343 |
| Missouri | 280 | 327 | 373 | 327 | 589 | 3266 | 3666 | 3355 | 3429 |
| Nebraska | 666 | 682 | 749 | 699 | 1258 | 191 | 218 | 254 | 221 |
| North Dakota | 230 | 268 | 303 | 267 | 481 | 0 | 0 | 0 | 0 |
| Ohio | 315 | 356 | 386 | 357 | 634 | 1296 | 1092 | 1552 | 1313 |
| South Dakota | 95 | 105 | 121 | 107 | 193 | 3 | 5 | 5 | 4 |
| Wisconsin | 194 | 196 | 207 | 199 | 358 | 1130 | 1095 | 1185 | 1137 |
| Total | 4935 | 5435 | 5868 | 5413 | 9745 | 13944 | 14694 | 17133 | 15257 |

† Hargett and Berry (1981).
‡ Mimeographed data, National Limestone Institute, Inc.
§ Assumes 1.8 g of $CaCO_3$ per 1 g of N used (Murphy, 1972).

of N applied was used to estimate the lime equivalent needed for the N. Much of the lime used was nullified by the acidity generated from N applied. In the five eastern states of Illinois, Indiana, Michigan, Ohio, and Wisconsin, which are more humid and thus more subject to leaching, 2.3 times as much lime was applied as the $CaCO_3$ equivalent of the N applied. In contrast, the four most western states of North Dakota, South Dakota, Nebraska, and Kansas had only slightly more than one fourth as much lime applied as the lime equivalent of N. Although many areas of these states have soils with calcareous surface or subsurface horizons, some leaching is occurring in the eastern portions of three of these states (Fig. 2). Also, if this great dominance of acidity generated by N applied over amounts of lime added grows with time, inevitably the time will come when increased liming will be necessary to maintain crop yields in this portion of the region.

The central states in the region, Iowa, Missouri, and Minnesota, provide interesting contrasts. Minnesota applied only one third as much lime as the acidity equivalent of fertilizer N used. This is attributed to Minnesota's large areas of low rainfall, calcareous soils, and hence no need for lime, and to relatively greater areas of corn, small grains, and such miscellaneous crops as sunflower, potato, and sugar beet (Tables 1 and 2) than Missouri has, all of which have high N requirements. In contrast to Minnesota, Missouri applied nearly six times as much lime as the acidity equivalent of fertilizer N, whereas Iowa applied almost nine times as much as the N equivalence. It will also be noted that three times as much N was used in Iowa as in Missouri, yet nearly the same amount of limestone was used in the two states during the 1978 to 1980 period. Iowa has more than seven times as much corn area as Missouri (Table 1), which would explain the N use differential. However, Missouri has a large livestock industry that requires improved pastures and hay. Thus, less N is needed in Missouri and more lime is required to offset the natural acidity of soils so that forage crops can be established and maintained. In contrast, much lime is undoubtedly used in Iowa to offset acidity developed from greater fertilizer N use.

The ratio of 1.8 kg of $CaCO_3$ required $kg^{-1}$ of N (Table 4), which is only half the theoretical amount from nitrification of $NH_4$, seems to be a reasonable average figure to use. Pierre and co-workers (1970, unpublished reports) at Iowa State University have stated that the values are the same for the following forms of N: $NH_4NO_3$, $NH_3$, $NH_4OH$, and urea. The values, of course, would be much higher for $(NH_4)_2SO_4$ and $(NH_4)_2HPO_4$. Pierre et al. (1971) found the acidity values to be generally $< 1.8$ for corn where only grain and cob were removed. They attributed much of the lowering below the theoretical value to denitrification losses of N, but the value increased with rate of N added. Pierre et al. (1970) also noted that the acidity/N ratio increased as cation removal by the crop increasingly exceeded non-N anion removal. For example, the ratio of cation/non-N anion uptake by buckwheat produced acidity equivalent to 6.67 kg of $CaCO_3$ $kg^{-1}$ of N, whereas uptake by oat produced an equivalence of only 2.61. Similarly, removal of corn silage caused 50% more acidity to be produced than harvesting corn for grain. Presumably the acidity from N would be higher for small grains and certainly so for grass hays and pastures, since they would remove relatively more bases per unit of N taken off than corn grain. Unpublished data by Wiese (Woodruff, 1967) had previously shown the value for corn in Wisconsin, based on agricultural limestone, to vary from 3.1 to 5.4 as N was increased from 224 to 1120 kg $ha^{-1}$.

Wolcott et al. (1965) studied changes in LR and exchangeable bases from high rates of N applied to a moderately acid, sandy loam soil in field studies in Michigan. They found soil pH in the upper 25 cm to range from 4.0 to 6.1 depending on the N source, i.e., $(NH_4)_2SO_4$ vs. $NaNO_3$. They observed that the declines in exchangeable bases were one-third to one-half the calculated equivalent of acidifying carriers. In contrast, increases in LR values measured were 1.5 to 6 times that expected from the acidity equivalency of the N materials. Major increases in LR measured occurred below the pH range 5.0 to 5.5, which was ascribed to retention of protons by Al and Fe complexes that are produced at these low pHs. Hence, one should expect a range in LR/N ratio from less than theoretical to much greater than theoretical depending on (i) relative ratio of uptake of $NO_3$/bases as affected by crop and rate of N applied; and (ii) interaction of $H^+$ formed with Al and Fe complexes, which depends on the rate of N and the resulting major changes in pH of a soil. The later effect is thought to be a good reason for liming a soil to maintain its pH above the range where this occurs (Wolcott et al., 1965).

## 3. MEASUREMENT OF LIME REQUIREMENT

The amount of lime required to bring a soil to a given pH or to a selected base saturation, i.e., its LR, can be measured by many methods. It can be done with field liming studies, which require years to complete; with soil-lime incubations of small samples requiring months; with titrations of soils requiring days; with soil-buffer equilibrations and titrations requiring hours; and with quick test, soil buffer equilibrations and ΔpH measurements requiring only minutes. The more time-consuming methods are used to develop, standardize, and correlate quick test methods, which are usually used to determine LR by soil testing laboratories. Hence the amount of lime recommended for a given field is generally determined by such methods. Since pH, OM level, and clay content are factors affecting the LR, it is possible to estimate LR by some function of these factors.

The SMP buffer method for determining soil LR was developed in Ohio because the original Woodruff method (1948a, b) measured only 50% or less of actual LR for most soils. For soils of low LR, the Woodruff method measured more than was needed. It was pointed out by McLean et al. (1966) that the data from the Woodruff method can be regression adjusted against $CaCO_3$ incubation, and others have also recently agreed with this observation (Tran & van Lierop, 1981). Rather large errors can occur in the Woodruff LR by this adjustment simply because of random variations in pH readings (McLean et al., 1966). They found (i) the average ranges in deviations among laboratories for 40 soils was ± 0.39 pH unit, (ii) random variations from the mean of 0.1 and 0.2 pH were common, and (iii) variations of 0.3 and even 0.4 unit occurred occasionally. Such variations where each 0.1 pH represents 2.24 t $ha^{-1}$ or more of LR might not be evident in correlations reported for a given method by one laboratory, but it would be inherent in the methodology used by different laboratories. The recommended methods for estimating the LR of soils in the Midwest is the SMP method as described by the North Central Soil Testing Committee (McLean, 1980). Missouri, however, uses a modification of the original Woodruff buffer method (Brown et al., 1977).

McLean et al. (1978) have proposed a double-buffer adaptation of the SMP buffer. The double-buffer adaptation increases the accuracy of measurement of low LR levels where the error of measurement with the original SMP method is relatively high.

## 4. EFFECTS OF LIME APPLICATION

Even though ground limestone is a salt ($CaCO_3$), it acts as a base in an acid soil. Consequently, as McLean (1971) reported, lime causes the following events to occur in an acid soil, most of which occur simultaneously:

1. Acidity is neutralized.
2. Base (Ca and possibly Mg) saturation of the soil increases.
3. Ratios of basic cations adsorbed and in solution change.
4. Soil pH increases (as soon as the $CO_2$ dissipates away), which in turn affects the solubility of most of the plant nutrients in a soil.
5. Toxic concentrations of $Al^{3+}$, $Mn^{2+}$, and possibly other substances are neutralized (or otherwise inactivated).
6. Acid weathering of primary and secondary minerals is curtailed by the decreased concentration of $H^+$.

7. pH-dependent CEC (i.e., negative charge) increases, adsorbing $Ca^{2+}$ (and $Mg^{2+}$) from which it is hydrolyzed (mobilized) for ready uptake by plants or movement to lower depths in the profile.
8. pH-dependent anion exchange capacity (AEC, i.e., positive charge) decreases, forcing previously adsorbed anions such as $SO_4^{2-}$ into solution.
9. Dinitrogen fixation increases.
10. Nitrogen mineralization from plant residues and soil OM increases.
11. Electrolyte concentration increases with dissolution of lime, and where CEC dominates over AEC, the electrolyte disappears from solution as $CO_2$ volatilizes.
12. Hydroxyl ion concentration increases with dissolution of lime, and where AEC dominates over CEC, the increased $OH^-$ ion concentration neutralizes + charges forcing $SO_4^{2-}$ into solution.

Not all of the above events following liming are beneficial to plants. Under certain conditions, some may be detrimental. Most of them are highly situation dependent. If a crop depends on acid weathering for one or more of the elements it must have, liming may be deterimental to its growth. Also, an adequately fertilized corn crop under continuous culture may not benefit at all from increased $N_2$ fixation or its mineralization from organic residues, whereas a similar crop grown in a rotation under limited fertilization may benefit much from the effect of lime on these microbial-effected events.

## IV. CROP RESPONSES TO LIME

### A. Soil pH and Maximum Yields

The relative yields of corn, soybean, and alfalfa expressed as percentages of the maximum yields obtained by various workers at pH 5.0, 5.5, 6.0, and > pH 6.0 where maximum yields were attained are presented in Table 5. These data are from studies that showed a tendency for yields to increase with one or more levels of liming and where pH data were also given. In cases where salt pHs and not water pHs were given, the former were converted to the approximate water pH by the addition of 0.5. The data are included without regard to level of significance of the yield differences from the treatments imposed in the individual studies. In most cases, crops grown on soils of < pH 5.5 showed some tendency to yield more with additions of lime, but the magnitudes of increase, especially for corn and soybean, were not always very large.

The data were generated from a wide range of field studies reported primarily in extension-type publications. They are generally from studies of continuous culture of comparatively short duration where sufficient fertility was applied for relatively high yields. In most cases, sufficient N was applied to the corn crop to obtain maximum yields without need for (i) N released from decomposition of humic materials or (ii) N from $N_2$ fixed by a legume crop grown earlier in the cropping sequence, both of which are known to be highly pH dependent.

It should be recognized that in these studies the pH where maximum yields occurred was often the highest pH treatment level. Hence, the pH for maximum yield would be slightly higher than the mean of 6.6 or 6.7 obtained for each crop. Brackets around figures (percentages) indicate substantial, but reasonable, extrapolations or interpolations from the actual data reported, whereas omissions (dashes in the table at pH 5) indicate situations where the values would have been suspect due to excessive extrapolations from results obtained at much higher pH levels.

It is evident from the mean percentage yields for each crop at the different pH levels (Table 5) that soybean yields drop off progressively more with decreased pH than do corn yields and that alfalfa yields decline much more than soybean with decrease in pH. We shall not attempt to discuss or to explain all of the causes for these trends. However, if one considers (i) that crops differ in their sensitivities to toxic elements such as Al and Mn at low pH and (ii) that legumes differ in their $N_2$ fixation vs. pH relationship at all pHs, these trends in means with pH for the various crops may well be about what one would expect.

On the basis of the approximate levels for maximum yields of corn grain, soybean grain, and alfalfa hay of 8781, 2688, and 12 320 kg ha$^{-1}$, respectively, the average yield depressions from the maximums for the three crops at the three pH levels are as follows:

| | pH | | |
|---|---|---|---|
| Crop | 6.0 | 5.5 | 5.0 |
| Corn | 237 | 562 | 931 |
| Soybean | 124 | 280 | 481 |
| Alfalfa | 862 | 3487 | 5137 |

These data represent losses in yields from failure to maintain the pH of each soil at a level where maximum yields are attained. The results of these many studies, mostly in the Midwest, seem to collectively support the concept of liming not only to eliminate toxic substances at low pH but to maintain a higher pH where other soil conditions favor maximum yields. This is not to say that there are not conditions in the region where a lower pH should be maintained for maximum yields. For example, maximum yields would not be expected at pH 6.6 in poorly drained soils where major losses of N by denitrification are a threat or in soils providing less than adequate soluble Mn, especially those with calcareous subsoils. In such cases, adjustment of the soil pH to lower levels may be essential for maximizing yields. However, the general tendency for the yields reported to gradually increase with pH to ~6.6 is undeniable. Such experiences probably account for the fact that representatives from most states in the Midwest chose this approximate pH as a minimum soil pH for a considerable number of crops grown in the region (see section IVA3). They evidently make this choice to be on

Table 5—Relative yields (percentage of maximum) for corn, soybean, and alfalfa at various soil pH levels in field experiments in the Midwest.

| State | Soil† | Years | pH 5.0 | pH 5.5 | pH 6.0 | pH >6.0 | Reference |
|---|---|---|---|---|---|---|---|
| | | | Corn | | | | |
| Illinois | Sable (Edinburg), Typic Haplaquolls | 1972–1975 | 96 | (97) | (99) | 100(6.6) | Chamblis & Boone (1976, Table 2) |
| | Ipava (Sable), Aquic Argiudolls | 1972–1975 | (80) | 92 | 100 | -- | Chamblis & Boone (1976, Table 5) |
| | Tama (Ipava and Sable), Typic Argiudolls | 1972–1975 | 88 | (91) | (94) | 100(7.2) | Chamblis & Boone (1976, Table 19) |
| | Cisne (Huey), Mollic Albaqualfs | 1975–1977 | 87 | (90) | 94 | 100(7.1) | Zajicek et al. (1978, Table 3) |
| Indiana | Chalmers, Typic Haplaquolls | 1960–1966 | -- | 97 | 99 | 100(6.2) | Barber (1967, Table 1) |
| Iowa | Floyd, Aquic Hapludolls | 1968–1974 | -- | 86 | 97 | 100(6.4) | Voss & Webb (1977, Table 14) |
| | Floyd | 1967–1974 | -- | 87 | 97 | 100(7.0) | Voss & Webb (1977, Table 12) |
| | Galva, Typic Hapludolls | 1967–1975 | -- | (97) | 98 | 100(6.5) | Voss & Webb (1977, Table 12) |
| | Galva | 1975–1979 | -- | 94 | 96 | 100(7.0) | Webb et al. (1980, Table 15) |
| | Moody, Udic Haplustolls | 1966–1979 | -- | (90) | 94 | 100(6.8) | Webb et al. (1980, Table 14) |
| Missouri | Mexico, Udollic Ochraqualfs | 1962–1967 | 98 | 98 | 99 | 100(7.0) | Fisher (1969, Fig. 8) |
| | Mexico | 1964–1966 | 95 | 97 | 98 | 100(7.2) | Fisher (1969, Fig. 9) |
| | Mexico | 1964–1966 | 95 | 97 | 98 | 100(6.4) | Fisher (1969, Fig. 10) |
| | Edina, Typic Argialbolls | 1959–1967 | 91 | 95 | 98 | 100(6.2) | Fisher (1969, Fig. 11) |
| Nebraska | O'Neill, Typic Haplustolls | 1975–1978 | 94 | 96 | 99 | 100(6.3) | Rehm & Weise (1978, Table 4) |
| Ohio | Loudenville, Ultic Hapludalfs | 1976 | 75 | 88 | 95 | 100(6.5) | McLean et al. (1983, Table 4) |
| Pennsylvania | Murill, Typic Hapludults | 1976–1977 | 94 | 99 | 100 | -- | Fox (1979, Table 2) |
| | Wharton, Aquic Hapludults | 1976–1977 | 85 | 93 | 100 | -- | Fox (1979, Table 2) |
| Wisconsin | Withee, Aeric Glossaqualfs | 1978 | 84 | 94 | 94 | 100(6.7) | Simson et al. (1979a, Table 5) |
| | Mean | | 89.4 | 93.6 | 97.3 | 100(6.6) | |

(continued on next page)

Table 5—Continued.

| State | Soil† | Years | pH 5.0 | pH 5.5 | pH 6.0 | pH >6.0 | Reference |
|---|---|---|---|---|---|---|---|
| | | Soybean | | | | | |
| Illinois | Sable (Edinburg) | 1972–1975 | 91 | 94 | 97 | 100(6.6) | Chamblis & Boone (1976, Table 2) |
| | Ipava (Sable) | 1972–1975 | 71 | 86 | 100 | -- | Chamblis & Boone (1976, Table 5) |
| | Tama (Iapava and Sable) | 1972–1975 | 86 | 89 | 92 | 100(7.2) | Chamblis & Boone (1976, Table 19) |
| | Cisne (Huey) | 1975–1977 | 70 | 87 | 94 | 100(7.1) | Zajicek et al. (1978, Table 3) |
| Indiana | Raub, Aquic Argiudolls | 1977–1978 | -- | 94 | 97 | 100(6.7) | S. A. Barber (unpublished data) |
| Missouri | Mexico | 1963–1967 | 89 | 93 | 97 | 100(6.5) | Fisher (1969, Fig. 1) |
| | Edina | 1959–1967 | (91) | 95 | 99 | 100(6.2) | Fisher (1969, Fig. 2) |
| | Portageville, Vertic Hapquolls | 1964–1966 | 69 | 78 | 86 | 100(7.1) | Fisher (1969, Fig. 3) |
| | Tiptonville, Typic Argiudolls | 1964–1966 | 88 | 91 | 97 | 100(6.3) | Fisher (1969, Fig. 4) |
| Wisconsin | Plano, Typic Argiudolls | 1976–1979 | 84 | 89 | 95 | 100(6.5) | Simson et al. (1979a, Table 4) |
| | Mean | -- | 82.1 | 89.6 | 95.6 | 100(6.7) | |
| | | Alfalfa | | | | | |
| Minnesota | Brainerd, Aquic Fragiochrepts | 1968 | -- | (45) | 90 | 100(6.7) | Overdahl (1972, Fig. 1) |
| | Brainerd | 1969 | -- | 56 | 85 | 100(6.7) | Overdahl (1972, Fig. 1) |
| Missouri | Mexico | 1963–1967 | 71 | 83 | 93 | 100(6.9) | Fisher (1969, Fig. 5) |
| | Mexico | 1964–1967 | 84 | 94 | 98 | 100(6.5) | Fisher (1969, Fig. 6) |
| | Edina | 1959–1967 | 74 | 91 | 99 | 100(6.5) | Fisher (1969, Fig. 7) |
| Ohio | Loudenville, Ultic Hapludalfs | 1979–1980 | 68 | 83 | 94 | 100(6.5) | McLean et al. (1983, Table 4) |
| | Ravenna, Aeric Fragiaqualfs | 1980 | 62 | 81 | 99 | 100(6.2) | J. F. Trierweiler (unpublished data) |
| Wisconsin | Fayette, Typic Hapludalfs | 1979 | 30 | 60 | 87 | 100(6.5) | Schulte et al. (1980, Table 5) |
| | Withee, Aeric Glossaqualfs | 1977 | (19) | 59 | 92 | 100(6.7) | Schulte et al. (1980, Table 7) |
| | Eau Plein, Typic Glossoboralfs | 1977 | -- | 65 | 93 | 100(6.5) | Simson et al. (1979b, Table 4) |
| | Mean | -- | 58.3 | 71.7 | 93.0 | 100(6.6) | |

† Most of the soils are silt loam texture; however, the O'Neil, Brainerd, and Tiptonville are sandy loams, whereas Sable is silty clay and Portageville is clay.

‡ Most of these references are extension-type publications. Hence, they are not included as official references in this paper unless cited elsewhere.

the safe side, even though they know that with adequate fertility supplied, it is difficult to demonstrate significant increases in corn or soybean yields from lime when the soil is already at pH 6. Also, in many cases, the same may be true at pH 5.5.

From the events occurring after liming acid soils, McLean (1971) suggested that a different suite of events may well be important in causing crop yield responses under different weathering conditions. Since most of the soils of the Midwest are the lesser weathered Mollisols and Alfisols with smaller but appreciable areas of Entisols, Ultisols, and Inceptisols, this may be the reason for increased yields to higher pH than is true for more highly weathered soils (Lanyon et al., 1977).

## 1. CORN, SOYBEAN, AND SMALL GRAINS

In general, the previous data does not contradict the statement that there is little evidence for maintaining soil pH above 6.0 for these crops unless they are to be grown in a cropping sequence with forage legumes, in which case the pH should be higher, e.g., 6.5 to 6.8. D. K. Myers (1982, personal communication, Ohio State Univ.) estimated that one half of the wheat (winter) acreage in Ohio is seeded to mixtures of red clover and sweetclover in late winter or early spring as a green manure crop for corn the following spring. Since leaching and crop removal of bases is a continuous process, a soil should be limed to a slightly higher pH than that required for maximum yields.

Both $N_2$ fixation by rhizobia in root nodules and N mineralization from organic residues and soil organic matter are enhanced by liming the soil. Hence, grain crops grown in rotation with forage legumes may, even at pH 6.0 or higher, benefit from liming because of the additionally fixed N and from the enhanced mineralization rate. Even when forage legumes are not in the cropping sequence, the beneficial effects on mineralized N may be obtained by increased soil pH. This, plus differences in the degree of soil weathering, may account for the differences observed between Iowa and Missouri in the soil pH at which maximum yields occur. Fisher (1969) showed little probability of obtaining increases in crop yield from liming soils to a salt pH above 5.5. (A salt pH measured in $0.01M$ $CaCl_2$ of 5.5 will give a soil pH measured in distilled water of 5.9 to 6.1 in most midwestern soils.) In contrast, small increases in corn and soybean yields have appeared in Iowa reports when initial soil pH values were 6.2 or above. Most reports (Table 5) from these and other states in the region have indicated that the soil pH at which maximum corn and soybean yields occur is the same for both crops.

## 2. FORAGE LEGUMES

As indicated above, alfalfa and other forage legumes are expected to reach maximum yield at a somewhat higher pH than that required for corn, soybean, and small grains because their rhizobia react favorably to the higher pH and higher Ca. Hence, it appears that with a few possible exceptions, these legume crops would yield best when the soil is limed and maintained at pH 6.5 to 7.0. Yet in northeast Ohio, maximum yield of alfalfa was obtained recently at pH 6.0 on a soil that had been treated with S as a fungicide for small fruits until it had a pH of 4. In this case, evidently the soil became so highly weathered that a lower pH was adequate for maximum yields. On similar soils nearby, which had never been as low as pH 4, alfalfa responded to lime until pH 7.0 was reached.

## 3. OPTIMUM SOIL pH RANGES

Crops differ in their susceptibilities to soil acidity, and they can be grouped, as Doll (1964) did, according to their acid tolerance (Table 6).

Table 6—Permissible soil pH ranges for various crops growing on mineral soils in Michigan (Doll, 1964).

| Crop | Soil pH range | Crop | Soil pH range |
|---|---|---|---|
| Least acid tolerant | | More acid tolerant | |
| Alfalfa | 6.3–7.8 | Buckwheat | 5.0–7.0 |
| Asparagus | 6.0–8.0 | Oat | 5.0–7.0 |
| Barley | 6.5–7.8 | Potato | 5.2–6.5 |
| Bean | 6.0–7.5 | Raspberry | 5.0–7.0 |
| Pea | 6.0–7.5 | Rye | 5.0–7.0 |
| Red clover | 6.0–7.5 | Strawberry | 5.0–6.5 |
| Soybean | 6.0–7.0 | Vetch | 5.0–7.0 |
| Sugar beet | 6.0–7.5 | | |
| Sweetclover | 6.5–7.9 | | |
| Medium acid tolerant | | Strongly acid soils required | |
| Corn | 5.5–7.5 | Blueberry | 4.0–5.1 |
| Grasses | 5.5–7.5 | Cranberry | 4.2–5.0 |
| Birdsfoot trefoil | 5.5–7.5 | | |
| Wheat | 5.5–7.0 | | |

Table 7—Average minimum pH values considered desirable for various crops as given by persons active in making lime recommendations in various states in the Midwest (Woodruff, 1967).

| Crop or situation | Minimum pH, avg | No. reporting |
|---|---|---|
| Corn, continuous | 6.4 | 10 |
| Soybean | 6.5 | 9 |
| Rotation with legumes | 6.6 | 10 |
| Small grains | 6.5 | 10 |
| Red clover | 6.6 | 10 |
| Alfalfa | 6.6 | 10 |
| Sugar beet | 6.5 | 4 |
| Potato | 5.3 | 3† |
| Cotton | 7.0 | 1‡ |
| Flax | 6.5 | 2 |
| Tobacco | 6.6 | 5§ |
| Canning crops | 6.4 | 3 |

† Does not include Minnesota (reported pH 6.5) where scab is evidently no problem.
‡ One state only, value converted to approximate water pH from salt pH.
§ Does not include Indiana (reported pH 5.3).

Since yields of most crops are also reduced at high pH, the upper pH for near-maximum yields were also given. Except for blueberry and cranberry plants, the upper pH level is generally between 7.0 and 7.5. Of course, one would not intentionally lime to the upper pH. But, on high pH soils, crops should be selected that thrive at these pH levels.

It is evident from the data in Table 7 that the average minimum pH considered desirable is not significantly different for any of the major crops in the region. Only for the relatively minor crops (potato and cotton) is a different minimum pH considered best. Although not evident from these averages, opinions from three states had indicated pH 6.0 was adequate for corn, whereas only one had indicated this pH was adequate for soybean. Examination of the most recent extension publications in the various states shows little tendency to recommend lowering the pH to which soils should be limed, even for corn and soybean. This also seems to be supported by the data in Table 5. In addition, several other factors that relate to maintaining a higher pH than would otherwise be necessary may have been considered. As examples, continuous removal of bases by leaching and cropping, the use of acid-forming N fertilizers, and less than complete dissolution of added lime all cause soil pH to be lower than the goal. From this line of reasoning, a higher target pH is an obvious solution. Furthermore, if the crop depends on an appreciable fraction of its N from the decomposition of crop residues and soil OM, a higher pH may be needed than would otherwise be so.

### B. Basic Cation Saturation Ratio

The basic cation saturation ratio (BCSR) concept of soil test interpretation, as developed in New Jersey (Bear et al., 1945), states that an optimum soil environment for plant growth is created when the cation exchange complex is satisified in the ratio of 65% $Ca^{2+}$, 10% $Mg^{2+}$, 5% $K^{+}$, and 20% $H^{+}$ (Bear et al., 1945). Later, Graham (1959) modified Bear's original concept, stating that crop growth and yield would be little affected by saturations within ranges of 65 to 85% $Ca^{2+}$, 6 to 12% $Mg^{2+}$, and 2 to 5% $K^{+}$, with $H^{+}$ ions occupying remaining sites. The BCSR concept has become a dominant force in shaping lime and fertilizer recommendations in the United States, especially so in the Midwest.

The BCSR concept does possess a certain appeal on theoretical grounds. It has been shown that the percentage base saturation of an individual cation can affect its availability to plants (Allaway, 1945; Chu & Turk, 1949; McLean et al., 1969) and that uptake of one cation may influence uptake of others (Omar & El Kobbia, 1966; Van Itallie, 1938; York et al., 1954). On the practical side, however, research into the effects of varying soil cation ratios on crop yield has been limited and has generally failed to support the concept of an "ideal" ratio. In fact, Bear's original proposals (Bear et al., 1945) were based on lack of response to varying soil Ca/K ratios. Recommendations were based more on the economics of applying each element.

Hunter, who worked with Bear at the inception of the BCSR concept (Hunter et al., 1943; Hunter, 1949), found no best Ca/K or Ca/Mg ratios for alfalfa. Giddens and Toth (1951) found no effects of varying soil cation ratios on the growth of white clover as long as $Ca^{2+}$ was the dominant ion on the exchange complex. Foy and Barber (1958) showed no yield response of corn to varying soil K/Mg ratios despite the appearance of severe Mg deficiency symptoms at wider ratios. McLean and Carbonell (1972) showed no effect of varying soil Ca/Mg ratios on yields of alfalfa or German millet. Simson et al. (1979) concluded from Wisconsin studies that when soil pH was near neutral and sufficient quantities of $K^+$, $Ca^{2+}$, and $Mg^{2+}$ were present, varying soil Ca/Mg ratio from 0.8 to 5.0 had no effect on corn or alfalfa yields. Also, no Ca deficiencies or Mg toxicities could be induced with high rates of Mg (1700 kg $ha^{-1}$).

Fisher (1974) showed earlier that soil test K for maximum yield should increase as a fractional multiple of the CEC, but not as a full multiple of the CEC as Bear et al. (1945) and Graham (1959) had proposed. Fisher showed that the soil test K sufficient for maximum yield could be obtained from the following expression: $K_{SL} = 110 + 2.5$ CEC, where CEC is expressed in cmol (+) $kg^{-1}$ and $K_{SL}$ is in $\mu g$ $g^{-1}$. Since this expression is a distinct deviation for $K^+$ from the BCSR theory and has been substantiated by others (McLean, 1976; Johnson & Wallingford, 1979) as a valid basis for defining soil test $K^+$ for maximum crop yields, the BCSR theory may also be suspect for $Ca^{2+}$ and $Mg^{2+}$.

Although the above-mentioned studies cast doubt on the validity of an "ideal" ratio for maximum crop yield, none was a systematic study of all three cations. Studies in both a growth chamber and the field were recently completed in Ohio (Eckert & McLean, 1981; McLean et al., 1983) in which percentage saturations of $K^+$ and $Mg^{2+}$ were varied across different levels of each other and over different levels of $Ca^{2+}$ saturation (soil pH) in a complete factorial design.

The 6 to 12% $Mg^{2+}$ saturation suggested by Graham (1959) seemed appropriate at least for the growth chamber, since both German millet and alfalfa gave higher yields at the 8% than at the 4% saturation level. However, each crop responded differently to variations in soil $K^+$ saturation within Graham's recommended range of 2 to 5% in that German millet yields increased, whereas those of alfalfa generally decreased with increased $K^+$. The German millet grew best at $Ca^{2+}$ saturations below Graham's minimum level of 65% saturation, whereas maximum alfalfa yields spanned both sides of 65%.

Results from the field studies with six crops indicated no yield response to $Mg^{2+}$ saturation above 4% [~0.3 cmol (½ $Mg^{2+}$) $kg^{-1}$], marked response to $Ca^{2+}$ (or pH) from pH 5 to 6 (no response above pH 6), adverse response of corn to the high $K^+$ level, and favorable response of alfalfa to the high $K^+$ level where crop removal made it difficult to maintain the lower level of soil K. Efforts to statistically segregate linear function of yield vs. Ca/Mg or Mg/K ratio below and above the recommended 6.5 and 2.0, respectively, proved fruitless. Elimination of the only major treatment effect, pH, by omission of pH 5 data generally did not improve the low $r^2$ values but did reveal the decreased corn yields and increased alfalfa yields caused by the high K level. Similarly, when the pH 5 data were omitted, the range in Ca/Mg and Mg/K ratios for the five lowest yielding treatments overlapped those for the five highest yielding treatments of each crop by an average of 83 and 60%, respectively. The only conclusion one can reach from the results of both the growth chamber and field studies and the other studies cited is that there is no ideal basic cation saturation or range of saturations where crop yields are maximum.

### C. Reduced Tillage

Since most agricultural lime used to neutralize acidic soils is ground limestone, it dissolves only as it comes in contact with the acid soil. Hence, where minimum tillage is practiced, it is often recommended that sufficient lime be added to the plow layer to attain the desired pH and that it be incorporated to plow depth prior to initiation of the minimum tillage practice. Others (Murdock & Blevins, 1979) have pointed out that when high rates of N are applied in no-tillage systems, the top 5 to 10 cm of soil tends to become quite acid and reduces yields and herbicide effectiveness. Surface-applied lime raises the pH primarily in the 0- to 5-cm layer but may, under some conditions, affect pH through the upper 20 cm layer. Higher lime rates, based on buffer LR tests, appear to be as effective as lower rates applied more often. Blevins et al. (1979) reported that liming to maintain a favorable soil pH slowed the degradation process and improved the effectiveness of weed control by atrazine and simazine herbicides on no-till corn.

### D. Liming Organic (Muck) Soils

Organic soils should be maintained about 1 to 1.5 pH units lower than that acceptable for mineral soils in the Midwest. Lucas and Davis (1961) in Michigan suggest pH 5.5 to 5.8 for wood-sedge organic soils and pH 5.0 for sphagnum peats. On the other hand, McLean (1980) makes no distinction in the type of organic soils but uses a desired pH of 5.2 for computation of LR by the SMP buffer method. Liming organic soils above pH 5.8 is objectionable because of reduced availabilities of the elements P, Mn, B, and Zn (Lucas & Davis, 1961). In addition, all charges of the cation exchange complex are pH dependent and thus provide readily available $Ca^{2+}$, $Mg^{2+}$, and $K^+$ to the crops even at relatively low pH. The organic system at low pH also provides a chelation mechanism for removal of toxic $Al^{3+}$ and $Mn^{2+}$ from the soil solution.

### E. Lime-induced Nutrient Deficiencies

The availabilities of almost all plant nutrients in soils are affected by pH (Pettinger, 1935; Truog, 1946; Lucas & Davis, 1961), and no two elements act alike in rate and extent of increase or decrease in availability with pH. Hence, it is reasonable such elements as Fe, Mn, B, Cu, and Zn, whose availabilities begin decreasing at relatively low to medium pH values, can become too low to meet a crop's needs if too much lime is added. Such lime-induced nutrient deficiencies generally occur only where marginal levels exist in the unlimed soil. They may also occur more often, and at lower pH, on organic soils where chelation may bind the cations more strongly. D. B. Mengel and C. D. Spies (1981, unpublished data, Purdue Univ.) recommend that dark-colored sands not be limed above pH 6.3 to avoid Mn deficiency in soybean. Lucas and Davis (1961) recommended that organic soils of Michigan not be limed above pH 5.8 lest the elements P, Mn, B, and Zn become deficient.

## V. FACTORS OF ACID SOIL INFERTILITY

Since the development of soil acidity is essentially a removal of basic cations by leaching, an accumulation of $H^+$ ion concentration, and weathering of minerals containing elements toxic to plants, several factors are involved in the acid infertility of a soil.

### A. Toxicities

#### 1. ALUMINUM

Aluminum is a metallic element that exhibits both ionic and covalent bonding. It is the most plentiful of all metallic cations of the earth's crust. It is released from octahedral coordination with oxygen in minerals by weathering, the release of which is quite rapid at pH 4 and below (Jackson, 1960).

The $Al^{3+}$ ion in water assumes octahedral coordination with six $OH_2$ groups, each of which dissociates a $H^+$ ion in sequence as pH increases. The resulting hydroxy-Al ions are adsorbed to the cation exchange complex of the soil. Here, they polymerize on charged surfaces and in the interlayers of the clay minerals, obstructing both the contraction of the clay lattice and the exchange of cations. Since adsorbed or polymerized Al requires base for neutralization, it increases the LR of soils. Exchangeable $Al^{3+}$ has been suggested as an index of LR for leached, mineral soils. This may be adequate for highly weathered soils, but it is unlikely that this is an adequate index for less weathered soils of the Midwest (McLean, 1970; Tran & van Lierop, 1981). Yet, it is probably true that when a soil at pH 5 or less is limed, crop response to lime is due to insolubilization of toxic $Al^{3+}$.

Soluble $Al^{3+}$ appears to be toxic to plants in several ways. It probably has adverse effects on the protoplasm of the cells. It has been observed to cause precipitation of $HPO_4^{2-}$ just inside cell walls. Roots and tops alike are stunted severely in the presence of toxic levels of $Al^{3+}$. The effect on the roots is further characterized by a disorganization of the root cap, the root apex, and the vascular elements (Fleming & Foy, 1968). Different species or varieties of crops, as well as different cultivars of a given species or variety, vary markedly in response to the effects of toxic levels of $Al^{3+}$ (Foy & Brown, 1964). Furthermore, some Al-sensitive varieties tend to lower the soil pH near the roots (Fleming & Foy, 1968). Hence, plant variety differences in Al tolerance may be both because of differences in induced levels of soluble $Al^{3+}$ around their roots and because of differences in specific genetic tolerance to a given concentration of $Al^{3+}$. Since much more work on $Al^{3+}$ toxicity to plants has been done in the southern United States, the reader is invited to examine this section in Chapter 5 of this monograph. The data on toxic $Al^{3+}$ effects on cotton roots reported by Adams and Lund (1966) are especially interesting. They indicate that minimum $Al^{3+}$ level for toxic effects on cotton roots was an activity in solution of 1.5 $\mu M$. Hence, effective concentration in solution is the important parameter, not cmol (+) $kg^{-1}$ or percent saturation of the CEC.

Studies by Fox (1979) on two Ultisols in Pennsylvania indicated that $Al^{3+}$ saturation was a better indicator of corn yields than was soil pH. Aluminum in solution and soluble $Mn^{2+}$ evidently were not measured in this study.

After it was known that different species and cultivars varied in tolerance to $Al^{3+}$, it soon became evident that their performance on acid soils depended frequently on their being selected and developed on sites relatively high in soluble $Al^{3+}$ content. Hence, efforts are underway to select cultivars of high $Al^{3+}$ tolerance as parents for breeding cultivars suitable for soils high in soluble $Al^{3+}$ content. Lafever (1981) recently summarized the situations for genetic differences in plant responses to soil stresses of all types. The practice of breeding or selecting for tolerance of a toxic element, e.g., $Al^{3+}$, or more efficient uptake of a deficient element seems to have its greatest potential in countries where lime and fertilizer applications are inadequate because of their high cost.

## 2. MANGANESE

According to Krauskopf (1972, p. 27),

> The forms that Mn takes in soils can only be guessed: (i) in strongly reduced soils one would expect the element to appear chiefly as absorbed [sic] $Mn^{2+}$, except that in alkaline soils some might form $MnCO_3$; (ii) in soils solutions most of the Mn is probably present as complexes rather than [as] the simple ion; and (iii) in oxidizing soils adsorbed $Mn^{2+}$ may persist in considerable quantity, but a good deal should be present as one or more of the higher oxide minerals. How much Mn may be present as organic compounds is uncertain. Oxidation and reduction reactions in soil, probably aided by bacteria, take place so readily that continual change from oxidized to reduced forms and vice versa should be commonplace.

Excessive quantities of $Mn^{2+}$, which are toxic to plants, commonly occur in strong acid or waterlogged conditions (or both) where the acidic reducing environment favors the solubilization of the $Mn^{2+}$ ion.

Although great differences exist in Mn tolerances among plant species and varieties (Brown et al., 1972), little seems to be known about the plant mechanism involved. However, it is known that the higher tolerance is associated with the ability to tolerate high Mn concentrations in the top of the plant rather than with reduced uptake (Foy et al., 1969; Neenan, 1960) and that at least in alfalfa, it is genetically controlled (Dessureux & Ouellete, 1958).

Adams and Pearson (1967) pointed out that toxic effects from $Al^{3+}$ and $Mn^{2+}$ frequently occur simultaneously on high $Mn^{2+}$ soils. Since plants show essentially no leaf symptoms to $Al^{3+}$ toxicity but show striking leaf symptoms to Mn toxicity, the latter may sometimes be given undue credit for inhibited growth where $Al^{3+}$ may be the main culprit (Foy, 1964). Symptoms of Mn toxicity must be rather severe for plant growth to be inhibited in the absence of $Al^{3+}$ toxicity (Hiatt & Ragland, 1963). The quantity of soluble $Al^{3+}$ should be appreciable in many soils at about the same pH level as that where soluble $Mn^{2+}$ would be, in which case toxic effects on plants would be from both ions. Yet, in most cases, a potential toxic concentration of $Mn^{2+}$ can probably occur at slightly higher soil pH than can a toxic concentration of $Al^{3+}$ (Adams & Pearson, 1967). Karlen et al. (1977) found an acid Marshall silt loam (Typic Hapludolls) in Kansas that had $Mn^{2+}$ level toxic to corn in the field. This was verified in growth chamber experiments where soybean and two crops of corn showed marked decreased concentration of Mn in tissue when grown on limed soil. In contrast, Mitchell and Offner (1981) in Alaska found that when lime, N, and P were added to acid soils, Al extracted with 0.01*M* $CaCl_2$ accounted for 84% of the yield variation of barley; $Al^{3+}$ toxicity symptoms were also evident. Inclusion of extractable $Mn^{2+}$ did not improve the regression analysis, and no evidence of $Mn^{2+}$ toxicity was found.

## B. Deficiencies

### 1. MAGNESIUM

Since the $Mg^{2+}$ ion is more highly hydrated than $Ca^{2+}$, its larger hydrated size makes it weaker bonded, gives it a greater tendency to be adsorbed to pH-dependent sites, and causes a greater percentage to be in solution. Also, more $Mg^{2+}$ than $Ca^{2+}$ can be trapped in nonexchangeable form when highly weathered soils are limed. The phenomena of more $Mg^{2+}$ in solution and more fixed in nonexchangeable form can lead to more $Mg^{2+}$ deficiencies than $Ca^{2+}$. A recent study in Ohio (McLean, 1982) indicated that Mg deficiency occurred in millet and alfalfa at ~0.25 cmol (½ $Mg^{2+}$) $kg^{-1}$, whereas Ca deficiency occurred at ~3.0 cmol (½ $Ca^{2+}$) $kg^{-1}$. Earlier, McLean and Carbonell (1972) reported that 0.58 cmol (½ $Mg^{2+}$) $kg^{-1}$ of soil at pH 5.4 was inadequate for maximum yield of alfalfa, especially at a higher $K^+$ rate.

D. B. Mengel (Agronomy Crops and Soils Notes no. 376, Purdue Univ.) reports that Mg deficiencies in Indiana occur largely on sandy, low CEC soils of northern Indiana and the southern third of the state, usually on acid soils but also on limed soil if low Mg limestone has been used. Once an adequate amount of exchangeable $Mg^{2+}$ is present, it has been shown that the ratio of Ca/Mg can be anywhere in the range of 1:1 to 49:1 without affecting crop yields. Usherwood and Miller (1967) showed consistent increases in corn yields on a Mg-deficient, fine sandy loam soil by liming to higher pH with $CaCO_3$ or by the addition of Mg from several sources. Increased Mg availability at higher pH from $CaCO_3$ is a likely consequence of $Ca^{2+}$ preempting permanent charge sites forcing $Mg^{2+}$ to weaker pH-dependent sites.

Yield responses of corn and soybean were observed in Illinois on a loamy sand at pH 4.1 when dolomitic limestone was applied (Key & Kurtz, 1960). In Indiana, Foy and Barber (1958) reported that Mg deficiency symptoms on corn growing on two sandy loam soils at pH 4.7 and 4.8 were corrected by applications of Mg, although yields were not increased. Magnesium deficiency in varying degrees has occurred in a number of crops in northeast Ohio (Columbiana County) since about 1950. Jones and Haghiri (1963) stated that large acreages of sweet corn were completely lost in this area in 1959 due to Mg deficiency. Symptoms of Mg deficiency in other crops have been noted but only occasionally. Similarly, in Michigan, Tobin and Lawton (1962) found a positive yield response to applied $Mg^{2+}$ in only two out of 126 crop-location field tests with commonly grown field and horticultural crops. In fact, they failed to find $Mg^{2+}$ deficiency at $< 1\%$ saturation of the soil CEC.

If soil pH is low, exchangeable $K^+$ is high, and N is applied as $NH_3$ or $NH_4^+$, more Mg will be required to prevent deficiencies. However, a very low level is adequate if other conditions favor its uptake by the crop. McLean and Carbonell (1972) and Rahmatullah and Baker (1981) found exchangeable $K^+$ was a stronger regulator of Mg uptake by plants than was exchangeable $Mg^{2+}$. Foy and Barber (1958) and Taylor and Smith (1959) also recognized the role of $K^+$ on the availability of $Mg^{2+}$. In view of the clearly demonstrated hazard of Mg deficiency in specific soil areas of the region, the determination of $Mg^{2+}$ is included in the soil testing programs of a number of states. Where it is found to be low, its application as soluble salt or as dolomitic lime is recommended. Also, dolomitic limestone is generally recommended on the acid glacial sands of Minnesota, Wisconsin, and Michigan and on the very acid sands of the older Mississippi River alluvium in the delta area of Missouri. In Indiana, the use of dolomite is suggested for every third limestone application as a general practice.

Although reports elsewhere appear to be inadequate to suggest a specific level where exchangeable $Mg^{2+}$ is sufficient for most crops, R. G. Hanson in Missouri (undated, *Corrective Liming of Missouri Soils,* Sci. and Tech. Guide, no. 9102, Univ. of Missouri, Columbia) suggests that visual Mg deficiencies are not likely to occur when saturation exceeds 5% but that mineral imbalances in some forages may occur on soils with $< 10\%$ $Mg^{2+}$ saturation.

## 2. MOLYBDENUM

Molybdenum deficiency is an ever-present hazard for some legumes growing on acid soils. Increased yields of alfalfa in pot tests were reported for $MoO_4^{2-}$ on 9 of 18 soil types from the major soil areas of Indiana (Foy & Barber, 1959). A response to $MoO_4^{2-}$ was found in only two soils when all were limed according to soil test. Thus, $MoO_4^{2-}$ availability was increased to an adequate level by liming in seven soils but not in two others. In field experiments, Lavy and Barber (1963) observed soybean yield responses ranging up to nearly 538 kg $ha^{-1}$ from seed treatment with $MoO_4^{2-}$. Yield increases occurred where soybean seeds produced on the untreated plots contained 1.6 mg $kg^{-1}$ of Mo or less. In Minnesota, on the other hand, alfalfa grown in pots on 13 different soils did not respond to $MoO_4^{2-}$ additions in yield or protein content (Ahlrichs et al., 1963). However, the lowest pH among these soils was only 5.5. Unpublished field data by J. R. Webb and K. Eik in Iowa (EC-981K, 1975) indicated that $MoO_4^{2-}$ applied to two soils of pH $< 6.0$ increased soybean yields 269 to 336 kg $ha^{-1}$, whereas the same application to two calcareous soils of pH $> 8.0$ decreased the yields by the same amount.

### C. Soil pH and Nutrient Availability

Charts depicting the general effect of soil pH on the relative availability of various plant nutrients have been available more than 45 yr (Pettinger, 1935). Truog (1946) expanded Pettinger's chart from 7 to 11 nutrients but deleted from it the concept of interactions (fixation) of phosphates with $Al^{3+}$ and $Fe^{3+}$ at progressively lower pH and with Ca at progressively higher pH. Most elementary and intermediate soil textbooks have charts adapted from one or the other of these pioneers in this field. Since these charts are considered indicative but not precise and are so widely available, they are not reproduced here. Adams and Pearson (1967) have dealt with this subject based mainly on research from the southeastern United States.

## REFERENCES

Adams, F., and Z. F. Lund. 1966. Effect of chemical activity of soil solution aluminum on cotton root penetration of acid subsoils. Soil Sci. 101:193–198.

Adams, F., and R. W. Pearson. 1967. Crop response to lime in the southeastern United States and Puerto Rico. *In* R. W. Pearson and F. Adams (ed.) Soil acidity and liming. Agronomy 12:161–206. Am. Soc. of Agron., Madison, Wis.

Ahlrichs, L. E., R. G. Hanson, and J. M. MacGregor. 1963. Molybdenum effect on alfalfa grown on thirteen Minnesota soils in the greenhouse. Agron. J. 55:484.

Allaway, W. H. 1945. Availability of replaceable calcium from different types of colloids as affected by degree of calcium saturation. Soil Sci. 59:207–217.

Barber, S. A. 1967. Lime and phosphorus for continuous corn. Res. Progress Rep. 316, Purdue Univ.

Bear, F. E., A. L. Prince, and J. C. Malcolm. 1945. The potassium needs of New Jersey soils. New Jersey Agric. Stn. Bull. 721.

Blevins, R. L., C. E. Rieck, and J. J. Kells. 1979. Influence of soil pH on weed control and corn yields under no-tillage systems. Agron. Abstr., p. 168.

Brown, J. E., J. E. Ambler, R. L. Chaney, and C. D. Foy. 1972. Differential responses of plant genotypes to micronutrients. p. 389–418. *In* J. J. Mortvedt et al. (ed.) Micronutrients in agriculture. Soil Sci. Soc. of Am., Madison, Wis.

Brown, J. R., J. Garrett, and T. R. Fisher. 1977. Soil testing in Missouri. Univ. of Missouri Ext. Div. Circ. 923, Columbia.

Brown, P. E., F. B. Howe, and M. E. Sar. 1914. Soil acidity and the liming of Iowa soils. Iowa Agric. Exp. Stn. Bull. 151.

Call, L. E., and R. I. Throckmorton. 1918. Soil fertility. Kansas Agric. Exp. Stn. Bull. 220.

Chambliss, C. G., and L. V. Boone. 1976. Report of research results. Dep. of Agron., U. of Illinois, Urbana.

Chu, T. S., and L. M. Turk. 1949. Growth and nutrition of plants as affected by degree of base saturation of different types of clay minerals. Michigan Agric. Exp. Stn. Tech. Bull. 214.

Clark, J. S., and W. E. Nichol. 1966. The lime potential, percent base saturation relations of acid surface horizons of minerals and organic soils. Can. J. Soil Sci. 46:281–315.

Coleman, N. T., and G. W. Thomas. 1964. Buffer curves of acid clays as affected by the presence of ferric iron and aluminum. Soil Sci. Soc. Am. Proc. 28:187–190.

Coleman, N. T., and G. W. Thomas. 1967. The basic chemistry of soil acidity. *In* R. W. Pearson and F. Adams (ed.) Soil acidity and liming. Agronomy 12:1–41. Am. Soc. of Agron., Madison, Wis.

Crop Reporting Board. 1982. Crop production—1981 annual summary. CrPr 2-1(82), Statistical Reporting Service, USDA, U.S. Government Printing Office, Washington, D.C.

Dessureaux, L., and G. J. Ouellette. 1958. Tolerance of alfalfa to manganese toxicity in sand culture. Can. J. Soil Sci. 38:8–13.

de Villiers, J. M., and M. L. Jackson. 1967. Cation exchange capacity variations with pH in soil clays. Soil Sci. Soc. Am. Proc. 31:473–476.

Doll, E. C. 1964. Lime for Michigan soils. Michigan Agric. Exp. Stn. Ext. Bull. 471.

Eckert, D. J., and E. O. McLean. 1981. Basic cation saturation ratios as a basis for fertilizing and liming agronomic crops: I. Growth chamber studies. Agron. J. 73:795–799.

Fisher, T. R. 1969. Crop yields in relation to soil pH as modified by liming acid soils. Missouri Agric. Exp. Stn. Res. Bull. 947.

Fisher, T. R. 1974. Some consideration for interpretation of soils tests for phosphorus and potassium. Missouri Agric. Exp. Stn. Res. Bull. 1007.

Fleming, A. L., and C. D. Foy. 1968. Root structure reflects differential aluminum tolerance in wheat varieties. Agron. J. 60:172–176.

Fox, R. H. 1979. Soil pH, aluminum saturation, and corn grain yield. Soil Sci. 127:330–334.

Foy, C. D. 1964. Toxic factors in acid soils of the southeastern United States as related to the response of alfalfa to lime. USDA Production Res. Rep. 80.

Foy, C. D., and S. A. Barber. 1958. Magnesium deficiency and corn yield on two acid Indiana soils. Soil Sci. Soc. Am. Proc. 22:145–148.

Foy, C. D., and S. A. Barber. 1959. Molybdenum response of alfalfa on Indiana soils in the greenhouse. Soil Sci. Soc. Am. Proc. 23:36–39.

Foy, C. D., and J. C. Brown. 1964. Toxic factors in acid soils: II. Differential aluminum tolerance of plant species. Soil Sci. Soc. Am. Proc. 28:27–32.

Foy, C. D., A. L. Fleming, and W. H. Armiger. 1969. Differential tolerance of cotton varieties to excess manganese. Agron. J. 61:690–694.

Giddens, J., and S. J. Toth. 1951. Growth and nutrient uptake of Ladino clover grown on Red and Yellow and Grey-Brown Podzolic soils containing varying ratios of cations. Agron. J. 43:209–214.

Graham, E. R. 1959. An explanation of theory and methods of soil testing. Missouri Agric. Exp. Stn. Bull. 734.

Hargett, N. L., and J. T. Berry. 1981. 1980 Fertilizer summary data. Natl. Fertilizer Development Center, TVA Bull. Y-165.

Harris, J. E. 1914. Soil acidity. Michigan Agric. Exp. Stn. Tech. Bull. 19.

Hiatt, A. J., and J. L. Ragland. 1963. Manganese toxicity of Burley tobacco. Agron. J. 55: 47–49.

Hopkins, C. G., and J. E. Readheimer. 1905. Soil treatment for the lower Illinois glaciation. Illinois Agric. Exp. Stn. Bull. 99.

Hopkins, C. G., and J. E. Readheimer. 1907. Soil improvement for the worn hill lands of Illinois. Illinois Agric. Exp. Stn. Bull. 115.

Hunter, A. S. 1949. Yield and composition of alfalfa as affected by variations in calcium-magnesium ratio in the soil. Soil Sci. 67:53–62.

Hunter, A. S., S. J. Toth, and F. E. Bear. 1943. Calcium-potassium ratios for alfalfa. Soil Sci. 55:61–72.

Jackson, M. L. 1960. Structural role of hydromium in layer silicates during soil genesis. Int. Congr. Soil Sci., Trans 7th (Madison, Wis.) 2:445–455.

Johnson, J. W., and G. W. Wallingford. 1979. Soybeans respond to potash. Better Crops Plant Food 63:3–5.

Jones, J. B., Jr., and F. Haghiri. 1963. Magnesium deficiency on Columbiana County soils. Ohio Agric. Exp. Stn. Res. Cir. 116.

Karlen, D. L., D. A. Whitney, and R. Ellis, Jr. 1977. Nutrient availability, plant growth, and composition as influenced by liming an acid Marshall silt loam. Trans. Kans. Acad. Sci. 80:129–135.

Kenney, D. R., and R. B. Corey. 1963. Factors affecting the lime requirements of Wisconsin soils. Soil Sci. Soc. Am. Proc. 27:277–280.

Key, J. L., and L. T. Kurtz. 1960. Response of corn and soybeans to magnesium fertilizers. Agron. J. 52:300.

Krauskopf, K. B. 1972. Geochemistry of micronutrients. p. 7–40. *In* J. J. Mortvedt et al. (ed.) Micronutrients in agriculture. Soil Sci. Soc. of Am., Inc., Madison, Wis.

Lafever, H. N. 1981. Genetic differences in plant response to soil nutrient stress. J. Plant Nutr. 4:89–109.

Lanyon, L. E., B. Naghshineh-Pour, and E. O. McLean. 1977. Effects of pH level on yields and compositions of pearl millet and alfalfa in soils with differing degrees of weathering. soil Sci. Soc. Am. J. 41:389–394.

Lavy, T. L., and S. A. Barber. 1963. A relationship between the yield response of soybeans to molybdenum applications and molybdenum content of the seed produced. Agron. J. 55: 154.

Lucas, R. E., and J. F. Davis. 1961. Relationships between the pH values of organic soils and availabilities of 12 plant nutrients. Soil Sci. 92:177–182.

Martin, A. E. 1960. Chemical studies of Podzolic illuvial horizons: V. Flocculation of humus by ferric and ferrous iron and nickel. J. Soil Sci. 11:382–393.

Martin, A. E., and R. Reeve. 1960. Chemical studies of Podzolic illuvial horizon: IV. The flocculation of humus by aluminum. J. Soil Sci. 11:369–381.

McCool, M. M., and C. E. Millar. 1918. Some general information on lime and its uses and functions in soils. Michigan Agric. Exp. Stn. Bull. 91.

McLean, E. O. 1970. Lime requirements of soils—inactive toxic substances or favorable pH range? Soil Sci. Soc. Am. proc. 34:363–364.

McLean, E. O. 1971. Potentially beneficial effects from liming: Chemical and physical. Proc. Soil Crop Sci. Soc. Fla. 31:189–196.

McLean, E. O. 1976. Exchangeable K levels for maximum crop yields on soils of different cation exchange capacities. Commun. Soil Sci. Plant Anal. 7:823–838.

McLean, E. O. 1980. Recommended pH and lime requirement tests. p. 5–8. *In* Recommended chemical soil test procedures of the North Central Region. North Central Regional Pub. 221 (rev. ed.), North Dakota Agric. Exp. Stn., Fargo.

McLean, E. O. 1982. Chemical equilibrations with soil buffer systems as bases for future soil testing programs. Commun. Soil Sci. Plant Anal. 13:411–433.

McLean, E. O., and M. D. Carbonell. 1972. Calcium, magnesium, and potassium ratios in two soils and their effects upon yields and nutrient content of German millet and alfalfa. Soil Sci. Soc. Am. Proc. 36:927–930.

McLean, E. O., S. W. Dumford, and F. Coronel. 1966. A comparison of several methods of determining lime requirements of soils. Soil Sci. Am. Proc. 30:26–30.

McLean, E. O., D. J. Eckert, G. Y. Reddy, and J. F. Trierweiler. 1978. An improved SMP soil lime requirement method incorporating double-buffer and quick-test features. Soil Sci. Soc. Am. J. 42:311–316.

McLean, E. O., R. C. Hartwig, D. J. Eckert, and G. B. Triplett. 1983. Basic cation saturation ratios as a basis for fertilizing and liming agronomic crops: II. Field studies. Agron. J. 75:635–639.

McLean, E. O., G. H. Snyder, and R. E. Franklin. 1969. Interaction of pH-dependent and permanent charge of clays: III. Calcium and Rb activities versus soybean root uptake from bentonite and illite suspensions. Soil Sci. Soc. Am. Proc. 33:397–400.

Miller, M. F., C. B. Hutchinson, and R. R. Hedelson. 1915. Soil experiments on the Gray Prairie soils of southwest Missouri. Missouri Agric. Exp. Stn. Bull. 130.

Mitchell, G. A., and J. R. Offner. 1981. Barley response to lime in relation to nutrient availability and potential aluminum toxicity in Susitna Valley soils. *In* Life science in the service of Alaska. Proc. 32nd Alaska Sci. Conf., Fairbanks, Alaska.

Murdock, L. W., and R. L. Blevins. 1979. Lime effects on yield, soil pH, and calcium movement in reduced tillage systems. Agron. Abstr. p. 177.

Murphy, L. S. 1982. Plant-soil-fertilizer relationships. p. 111–135. *In* W. C. White and D. N. collins (ed.) The fertilizer handbook. Fertilizer Inst., Washington, D.C.

Neenan, M. 1960. The effects of soil acidity on the growth of cereals, with particular reference to the differential reactions of varieties thereto. Plant Soil 12:324–328.

Omar, M. A., and T. El Kobbia. 1966. Some observations on the interrelationships of potassium and magnesium. Soil Sci. 101:437–440.

Overdahl, C. J. 1972. Fertilizer experiments with alfalfa on a Brainerd sandy loam soil. Minnesota Agric. Exp. Stn. Misc. Rep. 107.

Pettinger, N. A. 1935. A useful chart for teaching the relation of soil reaction to the availability of plant nutrients to crops. Virginia Agric. Exp. Stn. Bull. 136.

Pierre, W. H., J. Mieisinger, and J. R. Birdhett. 1970. Cation-anion balance in crops as a factor in determining the effect of nitrogen fertilizers on soil acidity. Agron. J. 62:106–112.

Pierre, W. H., J. R. Webb, and W. D. Shrader. 1971. Quantitative effects of nitrogen fertilizer on the development and downward movement of soil acidity in relation to level of fertilization and crop removal in a continuous corn cropping system. Agron. J. 63:291–297.

Rahmatullah, and D. E. Baker. 1981. Magnesium accumulation by corn (*Zea mays* L.) as a function of potassium-magnesium exchange in soils. Soil Sci. Soc. Am. J. 45:899–903.

Rehm, G. W., and R. A. Weise. 1978. Lime and boron for corn. Soil Sci. Res. Rep. 1978, Dep. of Agron., U. of Nebraska, Lincoln.

Schnitzer, M., and J. R. Wright. 1960. Nitric acid oxidation of the organic matter of a Podzol Soil Sci. Soc. Am. Proc. 24:273–277.

Schofield, R. K., and A. W. Taylor. 1955. The measurement of soil pH. Soil Sci. Soc. Am. Proc. 19:164–167.

Schulte, E. E., E. A. Liegel, and J. B. Peters. 1980. Liming Wisconsin soils. Project 1813, Progress Rep., Wisconsin Dep. of Agric., Trade, and Consumer Protection, Madison.

Schwertmann, U., and M. L. Jackson. 1964. Influence of hydroxy-aluminum ions on pH titration curves of hydronium-aluminum clays. Soil Sci. Soc. Am. Proc. 28:179–182.

Shoemaker, H. E., E. O. McLean, and P. F. Pratt. 1961. Buffer methods for determining lime requirement of soils with appreciable amounts of extractable aluminum. Soil Sci. Soc. Am. Proc. 25:274–277.

Simson, C. R., R. B. Corey, and M. E. Sumner. 1979. Effect of varying Ca:Mg ratios on yield and composition of corn (*Zea mays*) and alfalfa (*Medicago sativa*). Commun. Soil Sci. Plant Anal. 10:153–162.

Simson, C. R., K. A. Kelling, and E. A. Liegel. 1979b. Papermill lime-sludge as an alternative liming material. Mimeo. Report, Univ. of Wisconsin, Madison (parts of which were presented in Div. S-5 SSSA meeting, Fort Collins, Colo., August 1979).

Simson, C. R., and E. E. Schulte, and E. A. Liegel. 1979a. The influence of soil pH on crop yields and nutrient availability. Mimeo. Rep. FC 5.1 11/79, Univ. of Wisconsin, Madison.

Soil Survey Staff. 1975. Soil taxonomy: A basic system of soil classification for making and interpreting soil surveys. USDA Agric. Handbook no. 436, U.S. Government Printing Office, Washington, D.C.

Taylor, G. A., and G. B. Smith. 1959. Nutritional studies with sweet corn in Blair County, Pennsylvania. Proc. Am. Soc. Hortic. Sci. 74:454–459.

Thorne, C. E. 1903. The maintenance of fertility. Ohio Agric. Exp. Stn. Bull. 141.

Thornthwaite, C. W. 1948. An approach toward a national classification of climate. Geogr. Rev. 38:55–94.

Tobin, L. W., and K. Lawton. 1962. The magnesium status of Michigan soils and the effect of magnesium sources on the yield and chemical composition of crops. Mich. Agric. Exp. Stn. Q. Bull. 44:443–454.

Tran, T. S., and W. van Lierop. 1981. Evaluation and improvement of buffer-pH lime requirement methods. Soil Sci. 131:178–188.

Truog, E. 1946. Soil reaction influence on availability of plant nutrients. Soil Sci. Soc. Am. Proc. 11:305–308.

Usherwood, N. R., and J. R. Miller. 1967. Effect of soil pH on the availability of magnesium to corn (*Zea mays* L.) from magnesium sulfate and high magnesium liming materials. Soil Sci. Soc. Am. Proc. 31:390–393.

Van Itallie, T. B. 1938. Cation equilibria in plants in relation to the soil. Soil Sci. 46:175–186.

Veitch, F. P. 1902. The estimation of soil acidity and lime requirement of soils. J. Am. Chem. Soc. 24:1120–1128.

Voss, R. D., and J. R. Webb. 1977. Lime for Iowa soils and crops. Coop. Ext. Ser., Iowa State Univ., Ames.

Webb, J. R., A. L. Vogel, and H. E. Myers. 1980. Lime experiments. Annual Progress Rep. 1979, Northwest Res. Center, Iowa State Univ., Ames.

Wheeler, H. J., G. M. Tucker, and B. L. Hartwell. 1896. On the neutralizing action of lime. R.I. Agric. Exp. Stn. Annu. Rep. 9:294–318.

Whitson, A. E., and W. W. Weir. 1913. Soil acidity and liming. Wisconsin Agric. Exp. Stn. Bull. 230.

Wiancko, A. T., S. D. Conner, and S. C. Jones. 1918. The value of lime on Indiana soils. Indiana Agric. Exp. Stn. Bull. 213.

Wolcott, A. R., H. D. Foth, J. F. Davis, and J. C. Shickluna. 1965. Nitrogen carriers: I. Soil effects. Soil Sci. Soc. Am. Proc. 29:405–410.

Woodruff, C. M. 1948a. Determination of the exchangeable hydrogen and lime requirement of the soil by means of the glass electrode and a buffered solution. Soil Sci. Soc. Am. Proc. 12:141–142.

Woodruff, C. M. 1948b. Testing soils for lime requirement by means of a buffered solution and the glass electrode. Soil Sci. 66:53–63.

Woodruff, C. M. 1967. Crop response to lime in the midwestern United States. *In* R. W. Pearson and F. Adams (ed.) Soil acidity and liming. Agronomy 12:207–231. Am. Soc. of Agron., Madison, Wis.

York, E. T., R. Bradfield, and M. Peech. 1954. Influence of lime and potassium on yields and cation composition of plants. Soil Sci. 77:53–63.

Zajicek, F. E., J. W. Biggs, and L. V. Boone. 1978. Report of research results. Dep. of Agron., U. of Illinois, Urbana.

# 7 Crop Response to Lime in the Northeastern United States

**DOUGLAS J. LATHWELL and W. SHAW REID**

*Cornell University*
*Ithaca, New York*

The majority of soils in the northeastern United States (the Northeast) require the regular application of some liming materials for normal plant and crop growth. Exceptions include certain young soils derived from or influenced by limestone parent materials that are still calcareous in their upper horizons. Even in the limestone regions, however, lime applications are usually beneficial because the original limestone has leached from the upper part of the soil profile. On the other hand, some plant species in the region appear to grow best on acid soils. These include potato, under some conditions, blueberry, cranberry, certain ornamentals, and a number of forest species.

Even in early colonial times in America, some farmers recognized that the judicious use of lime or marl on their land was beneficial. In large part, this knowledge came as a result of experiences of early settlers with farming in Europe where the practice of liming had been in use for centuries. Also, it came from observation and reading from the works of the learned European writers who were beginning to seek the principles of plant growth and nutrition. Some of their essays on agriculture were indeed first printed in this country (Anderson, 1799).

Only sporadic use was made of this early knowledge, however, and it remained easier for the farmers to move to areas of more fertile soils than to transport lime the considerable distances often required. Even today with the need for efficient production, the use of lime is less than is generally considered desirable for many farming situations.

## I. SOILS OF THE REGION

The soils of the northeast are extremely variable and generally follow the physiographic and topographic features of the landscape. In the northern portions of the region, the soils are primarily young because they have developed on the most recent glacial materials. Where glaciation did not


*Soil Acidity and Liming*—Agronomy Monograph no. 12 (2nd Edition).

occur, the soils are older and generally are strongly acid except those in some limestone valleys and areas of mixed limestone and shales.

The region has four major topographic divisions that include a number of distinctive physiographic areas. The major divisions are mountains, upland plateau, lowland plain, and ridge and valley (Brady et al., 1957).

The mountainous areas are mostly nonagricultural or, at best, support a limited agriculture. This is especially true of the rugged Adirondacks in New York, the Green Mountains in Vermont, the Berkshire Hills in Massachusetts, and the White Mountains in New Hampshire and Maine. The Allegheny, Catskill, and Taconic mountain ranges, however, have gentler slopes and fairly level valleys that support relatively small farms.

The upland plateaus include the inland nonmountainous sections of New England; northwestern New Jersey; central Maryland; West Virginia; and the southeastern, southwestern, and northern Allegheny plateau areas of Pennsylvania and southern New York.

The lowland plain division may be divided into three major parts: the seacoast and Coastal Plain of New England and some Piedmont sections of Maryland, Delaware, and southern New Jersey; the Mohawk River Valley in New York; and the lake-plain area of western New York and northwestern Pennsylvania.

The ridge and valley division starts in south central Pennsylvania and extends northeastward to include the Hudson River Valley of New York.

The parent materials in the mountains, Coastal Plain, and Piedmont areas are weathered from granites, gneisses, and phyllites. They are hard and break down slowly, leaving a considerable proportion of quartz sand. The plateaus are predominantly of sandstone and shale with limestones or calcareous shales found largely in river valleys and interior lowland areas. There has been considerable mixing of some of these materials along the edges of the different area boundaries due to the mantle of till carried south by the glacier and dropped when the ice melted. It has been further mixed by waters from the retreating glaciers and by wind deposits in certain areas.

The soils of the region are in four orders: Spodosols, Inceptisols, Alfisols, and Ultisols (Fig. 1). These four orders include five major suborders: Orthods, Ochrepts, Udalfs, Udalts, and Aquudults (Soil Survey Staff, 1975). Essentially all of the Orthods of the region fall in the great group of Haplorthods. There are scattered areas of Fragiorthods in northern New York. The Orthods in Maine, New Hampshire, and Vermont and in the higher mountainous sections of Massachusetts and New York are strongly acid. Organic matter accumulates on their surfaces, and because the soils are shallow and the texture is sandy, they tend to be droughty in certain seasons. They respond to good management, lime, and fertilizers. Under these conditions they are good vegetable and potato soils. Those soils lying south of the above-mentioned group in New England are somewhat finer textured, are more fertile, and are strongly to moderately acid.

The Ochrepts of the region fall into two main great groups: the Dystrochrepts and the Fragiochrepts. The Dystrochrepts found in Pennsylvania and West Virginia are low in bases and have no free carbonates in the sub-

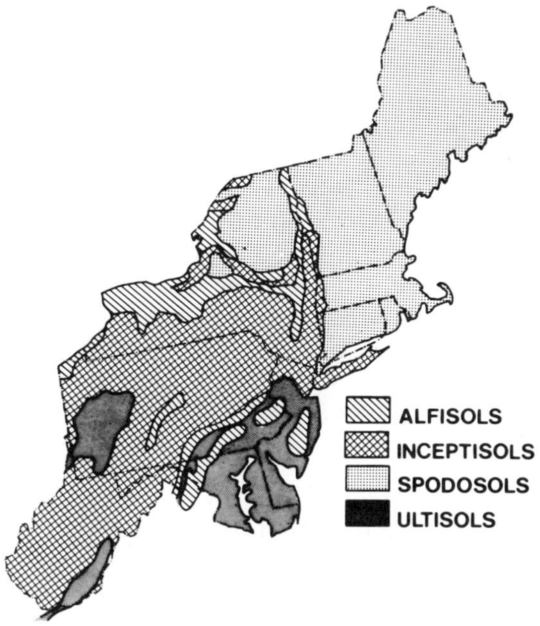


Fig. 1—Soil map of the Northeast.

soil horizons. They are often found on sloping topography and are shallow to bedrock. The Fragiochrepts found mainly in northern Pennsylvania and southern New York are soils low in bases and have a fragipan horizon. Impeded drainage is often a problem on these soils, along with soil acidity. Dairy and general farming is found on the more gentle slopes and in the valleys in the region. The steeper and wetter soils have remained in or have reverted to forest. A few Eutrochrepts are found in northern New York. These soils are high in bases and require little, if any, lime. These are highly productive soils, but the shortness of the growing season is a limitation.

The Udalfs of the region are in the great group Hapludalf, soils with a relatively thin subsurface of clay accumulation. These soils are generally high in bases in the surface horizon as well as in the subsurface horizons, so that their lime requirement for most crops is usually minimal. These soils are adapted to a wide range of crops and are the most productive of the region.

The Udults of the region are Hapludults. These soils have subsurface horizons of clay accumulation that is relatively thin. These soils are low in bases, however, and have a higher lime requirement (LR) than the Hapludalfs. These soils respond well to management, and a wide range of crops are grown on these soils presently.

The Aquults, confined to the coastal areas of New Jersey, Maryland, and Delaware, are in the great group Ochraquults. These low base status soils are seasonally wet and without artificial drainage have limited use. Where drained, a fairly wide range of crops can be grown.

Within these soil groupings, modifications occur that are more or less positional, and certain epipedons develop, depending on conditions. In the North, this frequently takes the form of a histic epipedon. It is most common in the glacial areas where natural drainage channels were obstructed. There is often considerable buildup of peat and muck in these areas, and the resulting soils are strongly acid in reaction.

Texturally, the soils of the region are predominantly sandy loams to silt loams. The sandy loams, loams, and loamy sands are dominant in New England and along the coastal area of the rest of the region. Silt loams prevail throughout the remainder of the region. Many of these soils are high in organic matter in the surface horizons, which contributes significantly to the cation exchange capacity (CEC) of the soils. The sandy soils have CECs that range usually between 5 and 10 cmol ($p^+$) $kg^{-1}$ soil, whereas the CEC range of the silt and clay loams is generally between 10 and 20 cmol ($p^+$) $kg^{-1}$ soil (Prince & Raney, 1961).

## II. CROPS OF THE REGION

Forest products remain major sources of income in the Northeast. High-grade timber, however, is in short supply because the forest lands of the region have been cut two or more times. According to the 1978 agriculture census (U.S. Bureau of the Census, 1978), about 30% or less of the land area is in 200 000 privately owned farms (Table 1), down from 35% and 300 000 farms according to the 1959 census. In 1978, the 5 million ha of harvested cropland constituted 42% of the total land in farms.

Among the states there is wide variation in the proportion of land in farms, ranging from 9% in Maine to 54% in Delaware. The percentage of land in farms from which crops were harvested in 1978 ranged from 16% in West Virginia to 74% in Delaware.

Seven types of farming are generally recognized in the region: (i) dairy-

Table 1—Use of farmland in the Northeast (U.S. Bureau of the Census, 1978).

| State | No. of farms | Area in farms | Cropland harvested | Cropland in pasture | Total pastureland |
|---|---|---|---|---|---|
| | | % | | ha | |
| Connecticut | 4 560 | 16.1 | 72 420 | 22 470 | 51 030 |
| Delaware | 3 632 | 53.5 | 202 670 | 7 560 | 14 200 |
| Maine | 8 158 | 8.1 | 197 620 | 39 720 | 88 785 |
| Maryland | 18 727 | 42.9 | 612 425 | 100 335 | 188 145 |
| Massachusetts | 5 891 | 13.6 | 86 535 | 31 365 | 58 865 |
| New Hampshire | 3 288 | 9.4 | 55 405 | 18 845 | 42 970 |
| New Jersey | 9 895 | 21.8 | 249 205 | 33 590 | 59 475 |
| New York | 49 273 | 32.4 | 1 815 615 | 483 975 | 1 021 725 |
| Pennsylvania | 59 942 | 30.4 | 1 759 340 | 420 145 | 882 075 |
| Rhode Island | 866 | 11.1 | 10 125 | 3 140 | 5 655 |
| Vermont | 7 273 | 29.6 | 235 220 | 102 870 | 217 780 |
| West Virginia | 20 532 | 25.1 | 244 685 | 329 845 | 813 335 |
| Total | 192 137 | | 5 541 260 | 1 593 910 | 3 355 835 |

ing, (ii) dairy and general farming, (iii) dairy and poultry, (iv) dairy and cash crops, (v) special crops and general farming, (vi) fruit and truck crops, and (vii) a nonagricultural area mainly in the mountains of northern and western Maine, northern New Hampshire, southwestern Vermont, and the Adirondack and Catskill highlands of New York.

In the dairy area, the important crops are the forages, with the usually flat valleys and ridge tops typical of the area intensively cropped to corn, alfalfa or clover-grass meadows, and small grains. The side hills are generally devoted to hay and pasture production. The dairy and general farming areas of the southern part of the region and in northern New England have similar cropping practices, but the farming is more diversified with other livestock, poultry, and some fruit and truck crops. Dairy and cash crop areas have smaller farms and devote part of their acreage to truck crops, fruit, and grains that are sold from the farm. Fruit, truck, and dairy farming areas along the Lake Erie and Ontario shores are devoted more intensively to fruit crops, grape, cherry, apple, and peach, and tomato, bean, pea, cucumber, sugar beet, and other high-value crops. Dairy and poultry areas in New England largely specialize in one enterprise or other, but many combine both enterprises along with some tobacco and vegetable crops. Since 1959, however, there has been a marked tendency for farms to specialize only in dairy, if dairying is the main enterprise, or to specialize in fewer crops if the enterprise is developed largely around a cropping enterprise, whether fruit or vegetables or field crops. For example, 85% of the harvested cropland in Delaware was cropped to corn and soybean.

Crop statistics show the general importance of forage crops in all states except Delaware and Maryland, where forage crops occupy only a small percentage of the harvested area (Table 2). In the other states, hay crops and corn silage occupy at least 50 to 80% of the harvested cropland. The total hectarage devoted to hay crops in 1978 has declined in all states in the Northeast compared with 1959. In New York and Pennsylvania, the

Table 2—Hectarage of forage crops in the Northeast (U.S. Bureau of the Census, 1978).

| State | All hay | Alfalfa | Other tame hay | Silage corn | Total corn as silage |
|---|---|---|---|---|---|
| | ha × $10^3$ | | | | % |
| Connecticut | 40 | 9 | 22 | 21 | 93 |
| Delaware | 9 | 3 | 3 | 4 | 6 |
| Maine | 102 | 9 | 68 | 15 | 87 |
| Maryland | 109 | 30 | 59 | 40 | 15 |
| Massachusetts | 54 | 11 | 33 | 16 | 89 |
| New Hampshire | 43 | 8 | 25 | 9 | 92 |
| New Jersey | 56 | 21 | 24 | 14 | 25 |
| New York | 1044 | 389 | 403 | 267 | 52 |
| Pennsylvania | 813 | 323 | 331 | 174 | 26 |
| Rhode Island | 5 | 1 | 2 | 1 | 89 |
| Vermont | 200 | 40 | 104 | 39 | 93 |
| West Virginia | 188 | 32 | 128 | 15 | 36 |
| Total | 2663 | 876 | 1202 | 615 | 55 |

Table 3—Hectarage of selected crops in the Northeast (U.S. Bureau of the Census, 1978).

| State | Corn for grain | Wheat | Oat | Barley | Soybean | Tobacco | Potato |
|---|---|---|---|---|---|---|---|
| | | | | ha × $10^3$ | | | |
| Connecticut | 1 | | | | | 1 | 2 |
| Delaware | 64 | 6 | 1 | 2 | 107 | | 2 |
| Maine | 2 | | 17 | | | | 48 |
| Maryland | 237 | 30 | 8 | 31 | 153 | 11 | 1 |
| Massachusetts | 2 | | | | | | 2 |
| New Hampshire | 1 | | | | | | |
| New Jersey | 41 | 8 | | 5 | 75 | | |
| New York | 243 | 27 | 111 | 2 | 10 | | 20 |
| Pennsylvania | 489 | 79 | 126 | 35 | 31 | 5 | 10 |
| Rhode Island | | | | | | | 2 |
| Vermont | 3 | | 1 | | | | |
| West Virginia | 26 | 2 | 3 | 2 | 1 | 1 | 1 |
| Total | 1109 | 152 | 267 | 77 | 377 | 18 | 88 |

declines were slight, whereas the decline in hay hectarage in other states of the region was more substantial.

Alfalfa is the dominant legume and has tended to increase slightly or to occupy the same percentage of the land in hay since 1959. Alfalfa is confined largely to the well-drained soils of the region to minimize crop failure from excess water and cold damage. Failure of alfalfa seedlings to become inoculated still contributes to crop failure where new seedings are made on freshly limed, strongly acid soils.

Hectarage of corn has increased dramatically from 1959 to 1978, going from approximately 1.1 to 1.7 million ha. The increase has continued beyond 1978. Much of the increase has been in corn grown for grain (Table 3). The dramatic changes are illustrated by the shifts occurring in New York where corn hectarage doubled between 1959 and 1978, and almost all of the increase has been in corn grown for grain. The increase is continuing and has resulted in New York growing almost enough corn grain to meet the needs of its livestock industry. The hectarage of small grains has declined markedly between 1959 and 1978, with the decline in oat being most pronounced. The hectarage of soybean has increased in many states.

Open-field and shade-grown tobaccos are special crops grown in the Connecticut Valley and in Pennsylvania, Maryland, and West Virginia. They require considerable specialization and management. Shade tobacco is grown by the larger companies with hired labor and management on contract acreages.

Potato is important in most of the states and especially in northern Maine, in the southern New England states, in Long Island and upstate New York, and in the southern states of the region. Potato is a relatively high-value, specialized crop that demands high levels of fertilization and special precautions in the use of lime.

Hectarage of vegetable crops has remained relatively constant or declined but little over the 20-yr period, 1959 to 1978 (Table 4). They are

Table 4—Hectarage of vegetable and fruit crops in the Northeast (U.S. Bureau of the Census, 1978).

| State | Vegetables | Fruit |
|---|---|---|
| | ha | |
| Connecticut | 4 680 | 2 554 |
| Delaware | 16 630 | 450 |
| Maine | 4 695 | 3 725 |
| Maryland | 20 945 | 4 660 |
| Massachusetts | 6 225 | 3 470 |
| New Hampshire | 1 715 | 1 935 |
| New Jersey | 30 290 | 9 080 |
| New York | 63 435 | 56 015 |
| Pennsylvania | 20 470 | 28 835 |
| Rhode Island | 910 | 320 |
| Vermont | 650 | 2 630 |
| West Virginia | 625 | 9 330 |
| Total | 171 270 | 123 004 |

grown in every state in the region. Dried beans are an important crop in New York. Fruit crops are grown in all the states but are concentrated along the Great Lakes in New York and Pennsylvania. Local concentrations occur in other states, dependent largely on a favorable climate so that the crop is not reduced by frost.

## III. LIMING PRACTICES IN THE REGION

### A. Past and Present Lime Usage

Early settlers along the eastern seaboard from Maine to southern Maryland and inland to New York, Pennsylvania, and West Virginia found themselves for the most part in acid soil regions with little or no lime within easy hauling distances. Geologists of the time did much to locate deposits of lime and noted the effects of these materials on crops. In his report *Re-examination of the Geology of Massachusetts*, Hitchcock (1838) devoted much of his discussion to the soils of the state with considerable attention to liming and use of lime. Alvord (1888) gives an excellent history of the early work in agriculture at the Maryland Agricultural Experiment Station. One year later, Alvord and Patterson (1889), in describing the work on Maryland soils, refer to state geological reports of Ducatel from 1833 and 1840, two or three of which were primarily descriptions of the marl beds of the eastern seaboard and the use of marl on farms. Marl from deposits in Maryland was compared with burned lime from 1852 to 1858 and was found to be less effective, even at four times the rate of application.

During a portion of this early period, it was possible to obtain a limited amount of seashells and lime byproducts from gas works and tanneries. In the latter part of the 18th century, gypsum was recognized to have a beneficial effect on clovers. Hall (1909) states that Benjamin Franklin applied gypsum in a manner to form the word "plaster" on clover fields by roadsides so travelers would know what caused the improved growth. Kohler (1849) wrote that gypsum was used quite generally on Connecticut Valley meadows with good effect.

By the early years of the 19th century, much literature on agriculture began to appear in the Northeast. An essay by Anderson (1799) gave good coverage of the subject for that time, stating reasons for lime usage and pointing out the differences in burned lime, slaked lime, limestone, chalk, and gypsum. During the short period of its existence, *The Farmers' Register*, published by Edmund Ruffin of Virginia from 1833 to 1842, carried over 163 articles on marl and lime and an additional 24 on gypsum.

Until about the turn of the present century, much of the lime used was burned lime. The burning process served two purposes: (i) it reduced the weight for hauling; and (ii) it was an easy way to bring the product into the required state of subdivision. During this period, rock crushers were run by water power, which often was not convenient either to the rock source or farm areas where the lime would be used. Further, until nearly the present century, few limestone quarries were available to farmers of New England, much of New York, and the coastal areas of the other states. Further inland, farmers used the limestones or marls on their own lands or those of neighbors and frequently burned it for their own use.

Ground limestone is used almost universally today, but because it was costly prior to the development of adequate grinding equipment, burned lime or hydrated lime was more commonly used in earlier times. Both calcitic and dolomitic limestones are used, depending largely on their availability in a given area. Dolomitic stones are preferred and recommended in many areas, especially after some soils were found to be deficient in Mg for potato, tobacco, and certain fruits and vegetables.

The amount of agricultural limestone materials used in the Northeast declined from 1950 to 1970 (Table 5), after a steady and rapid increase in total tonnage of lime used from 1930 to 1950. The period from 1930 to 1950 covers the time when the use of lime was promoted by governmental programs, and a portion of the cost was shared by the federal government.

Table 5—Lime used in the Northeast.

| State | 1950 | 1960 | 1970 | 1978 |
|---|---|---|---|---|
| | $t \times 10^3$ | | | |
| Connecticut | 66 | 41 | 29 | 38 |
| Delaware | 53 | 63 | 95 | 124 |
| Maine | 65 | 64 | 94 | 142 |
| Maryland | 313 | 250 | 227 | 352 |
| Massachusetts | 60 | 57 | 37 | 70 |
| New Hampshire | 35 | 24 | 24 | 21 |
| New Jersey | 165 | 151 | 156 | 206 |
| New York | 656 | 803 | 473 | 580 |
| Pennsylvania | 1197 | 762 | 748 | 1009 |
| Rhode Island | 10 | 9 | 11 | 12 |
| Vermont | 95 | 84 | 81 | 73 |
| West Virginia | 236 | 142 | 154 | 125 |
| Total | 2951 | 2449 | 2129 | 2751 |

Undoubtedly, these programs stimulated the use of lime and account in part for the rapid increase in usage. From a post–World War II low reached in the early 1970s, lime consumption has again risen to levels close to those in 1950. It now approaches 3 million t annually.

At present, the Northeast accounts for about 10% of the total lime used in the United States. This proportion did not change over the period 1950 to 1978, whereas in 1930 it used nearly 20%.

In 1978, between 20 and 40% of the farms in the various states used lime (Table 6). A fewer number of farms used lime in 1978 than in 1959. A higher percentage of farms used lime in 1978, however, since the decline in total farms from 1959 to 1978 was greater than the decline in those using lime. The percentage of harvested cropland that received lime was lower than the percentage of farms using lime and ranged from a low of 6.3% in New York to a high of 27% in Rhode Island. Those states with a heavy emphasis on grain crops (Delaware, Maryland, New Jersey) had a higher percentage of the harvested cropland limed than did the states where forage crops dominated the cropping system (New York, New Hampshire, Pennsylvania). This is probably because of more N being used on grain crops and of the awareness that N fertilizers are acid forming, but it could also be because lower rates are applied more often on annual crops grown on lower CEC soils.

In 1978, farmers in the Northeast applied an average of 3.99 t of liming material for each hectare limed, which is almost identical with the figure of 20 yr earlier. The range was from 2.35 t in Delaware to about 4.50 t in New York and West Virginia. The lower rates applied in Delaware and Maine reflect differences in soils, e.g., coarser texture and lower activity clays (Delaware), and differences in crops, e.g., potato (Maine). In those states dominated by silt loams or finer-textured soils and where forage crops dominate, larger quantities of lime are used. On those soils where smaller quantities of lime are used, the frequency of liming is greater, 25% of the harvested crop-

Table 6—Lime used on farms in 1978 in the Northeast (U.S. Bureau of the Census, 1978).

| State | No. of farms using lime | Area limed | Lime used | Avg rate | Farms using lime | Harvested cropland limed |
|---|---|---|---|---|---|---|
| | | ha × $10^3$ | t × $10^3$ | t $ha^{-1}$ | % | |
| Connecticut | 1 378 | 16.9 | 45.7 | 2.71 | 30.2 | 23.3 |
| Delaware | 1 194 | 44.0 | 103.7 | 2.35 | 32.8 | 21.7 |
| Maine | 2 610 | 37.9 | 95.3 | 2.51 | 32.0 | 19.2 |
| Maryland | 6 880 | 142.0 | 346.7 | 2.44 | 36.7 | 23.2 |
| Massachusetts | 1 878 | 16.9 | 59.6 | 3.54 | 31.9 | 19.5 |
| New Hampshire | 917 | 7.7 | 27.4 | 3.59 | 27.9 | 13.8 |
| New Jersey | 3 203 | 58.7 | 161.2 | 3.76 | 32.4 | 23.6 |
| New York | 10 292 | 115.0 | 515.6 | 4.48 | 20.8 | 6.3 |
| Pennsylvania | 20 214 | 254.2 | 943.9 | 3.72 | 33.7 | 14.4 |
| Rhode Island | 326 | 2.8 | 7.6 | 2.78 | 37.6 | 27.4 |
| Vermont | 2 516 | 33.1 | 121.4 | 3.68 | 34.6 | 14.1 |
| West Virginia | 4 460 | 34.2 | 154.3 | 4.53 | 21.7 | 14.0 |
| Total | 55 868 | 763.3 | 2 582.3 | 3.39 | | |

land limed each year compared with 6 to 15% where higher rates are used. Where forages are part of the cropping system, longer rotations generally are used so that liming frequency likely would be less.

Summaries of soil test in New York made in 1958 and 1963 showed that from 30 to 40% of all samples tested at that time were < pH 6.0. Summaries of soil tests made in 1979 showed that from 30 to 40% of the samples still had soil pH values < 6. Excluded from these summaries were soils collected from potato fields, where low soil pH is maintained for disease control. Information provided with the soil samples support the observations gathered from the 1978 census data. Information supplied by farmers with the soil samples submitted in 1979 showed that in 77% of instances, lime had been applied 4 or more yr previous to soil sampling. In addition, the lime rate applied was between 2.35 and 4.5 t $ha^{-1}$ 68% of the time. Thus, it would appear that little change in soil acidity or in liming practices has occurred since the mid-1950s.

### B. Basis of Lime Recommendations

All states in the Northeast have recommended lime for acid soils throughout all or most of the present century. In the late 19th century, the acid nature of the upland soils in Rhode Island was related to the need for lime (Wheeler, 1895; Wheeler et al., 1895; Wheeler & Tucker, 1896). Prior to that time, the practice of liming had become well established in certain localities based primarily on farmer experience. Field trials with lime were begun in all the states shortly after the establishment of the state agricultural experiment stations. Where other nutrient deficiencies did not occur, yield responses to lime generally were observed. Wheeler and Tucker (1896) in Rhode Island and Miller and Brinkley (1893) in Maryland recorded some of the greatest responses. Noll et al. (1931), in a long-term experiment, found substantial responses to lime in Pennsylvania. Shortly after 1900, crop response to lime had been found in nearly all states of the region, and papers began to appear on the need for liming acid soils and on the toxic effects of Fe and Al. These toxic effects were found to be corrected by the use of lime (Ruprecht, 1915).

Lime recommendations based on measurements of soil acidity were not generally made until the whole question of $H^+$ ion activity and its relation to acid soils came under intensive study between 1915 and 1920. Although litmus was first used as an indicator of acidity in 1865, it had no general application for recommending lime until many years later. The development of other indicators and of electrometric measurement of $H^+$ ion concentration provided a more dependable basis for lime recommendations.

Today, recommendations for liming are usually based on the direct determination of exchangeable acidity in the soil, and much of the recent research has dealt with the problem of obtaining reliable estimates of exchangeable acidity in soils. Exchangeable acidity is related to soil pH, CEC, and percent base saturation at any given soil pH. Thus, if one is to use ex-

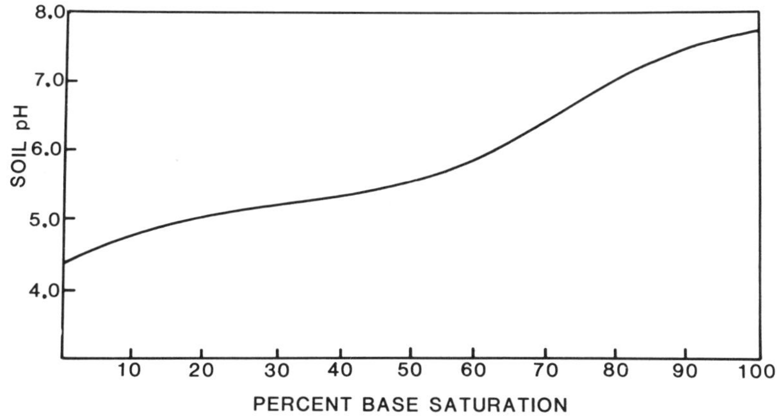


Fig. 2—Relationship between soil pH and percent base saturation of New York soils.

changeable acidity, the relationship between soil pH and percent base saturation must be known.

The relationship between soil pH and percent base saturation for 200 New York soils, including both surface and subsoils, is shown in Figure 2. The pH values, as measured in 1:1 soil/water suspensions, were plotted against the respective base saturations. Maximum deviations from pH values given by the curve for the same base saturation value were about 0.5 pH unit. Such deviations from the graphed soil pH–base saturation relationship may be attributed to differences in the type of clay, the organic matter content, and the concentration of soluble salts in the soil. Recent applications of lime might also contribute to the deviations.

Because the relationship between soil pH and base saturation is known to vary widely with different soils, this curve is valid only for soils similar to those used in obtaining it and obviously should be applied to other soils only after establishing the extent to which the relationship between pH and base saturation deviates from that shown in Fig. 2.

For example the direct determination of the LR by the $BaCl_2$-triethanolamine (TEA) method (Greweling & Peech, 1965) gives the amount of $CaCO_3$ or its equivalent required to bring a soil to 100% base saturation at a pH of 8.2. For optimum crop production, soils are usually limed to increase the pH within the range of 6.0 to 7.0, which corresponds to about 60 to 80% base saturation.

Several buffer methods to estimate LR of acid soils have been developed. In contrast to the TEA method, these methods have been aimed at measuring the quantity of exchange acidity to be neutralized to some given pH (e.g., pH 6.8) rather than to 100% base saturation.

The SMP method (Shoemaker et al., 1961) is a widely used method because it has given reliable estimates of the LR of a wide range of soils when compared with $CaCO_3$ incubation LR of these soils. Fox (1980) compared several buffer methods for LR of agricultural soils in Pennsylvania and concluded that the original SMP method gave the best estimate of LR of these soils. He did suggest that for low LR soils the LR should be adjusted upward slightly to more nearly approximate the LR as determined by $CaCO_3$ incubation. Bartlett (1982) has shown that the determination of reactive Al characterizes the quantity of soil acidity in northern acid soils that must be neutralized to meet the lime needs. He considered that in terms of the LR, reactive Al appears to be a better predictor than the SMP method, although a high degree of correlation existed between the two methods overall.

In Maine, Bourgoin (1980) found that the SMP procedure correlated well with $CaCO_3$ incubation values but that accurately weighed samples were essential for reliable estimates of LR. He did find, however, that the SMP buffer method underestimated the contribution of the organic fraction to the LR of the soil while overestimating the contribution of exchangeable $H^+$.

These studies indicate that either the TEA procedure or the SMP method for determining LR gave satisfactory results. The SMP procedure, because it is more adaptable to routine use in the laboratory, is more widely used to estimate the LR of acid soils in the Northeast.

### C. Methods of Application and Incorporation in Soils

Practically all lime is now spread with truck spreaders. The importance of uniform spreading and thorough mixing of lime with the soil has long been emphasized. Many agronomists contend that a single application of lime is not sufficient to bring about good distribution within a reasonable time. Therefore, they advise applying part of the lime before plowing and the rest just before the land is prepared for seeding. Where moderate quantities of lime are applied (9 t $ha^{-1}$ or less), it should be disked into the seedbed. In most states, a recommended alternative to the split application is to apply the lime in two separate years before seeding perennial legumes and grasses.

In Massachusetts (Weeks & Colby, 1965) and some other states, it is recommended that between 4.5 and 9 t $ha^{-1}$ be applied in a single application for sandy soils. This is done for convenience and economy in spreading and makes it possible to work the land more than once in the liming process to ensure better mixing with the soil.

A study of liming rates (M. E. Weeks, 1960, unpublished data) on Hadley (coarse-silty, mixed, nonacid, mesic Typic Udifluvents), Merrimac (sandy, mixed, mesic Typic Dystrochrepts), and Gloucester sandy loams (sandy-skeletal, mixed Typic Dystrochrepts) in Massachusetts brought out some of the problems of mixing dry, finely ground limestone with moist soil. There was a tendency for both finely ground calcitic and dolomitic stones to absorb enough moisture to resist uniform mixing in the surface layer even with two or three diskings. Areas of lime concentration were found when soil samples were taken from the plots 2 and 3 yr later. In another instance, a field was limed in the spring of 1960 and seeded to forage grasses; samples were taken in November 1961 to study uniformity of mixing (M. Drake & W. G. Colby, 1962, unpublished data). The pH of

Table 7—Effect of different rates of lime on the average soil pH at given period after liming of three fields in eastern Massachusetts.†

| Lime rate, t $ha^{-1}$ | Initial soil pH | Soil pH at intervals after liming | | | |
|---|---|---|---|---|---|
| | | 8 months | 12 months | 24 months | 30 months |
| 4.5 | 5.5 | 6.6 | 6.1 | 6.0 | 6.1 |
| 6.5 | 5.4 | 6.1 | 6.1 | 6.1 | |
| 13.5 | 5.4 | 6.1 | 6.2 | 6.2 | 6.4 |
| 20.0 | 5.5 | 6.1 | 6.2 | 6.2 | 6.3 |
| 24.0 | 5.6 | 6.2 | 6.2 | 6.5 | 6.4 |

† The soils were Merrimac and Narraganset (M. E. Weeks, 1961, unpublished data, Massachusetts Agric. Ext. Serv.).

individual samples taken from various positions in the plow layer varied from 4.5 to 7.9, with the lime concentrated in the 5- to 15-cm zone.

Inadequate mixing of lime with soil results in incomplete neutralization of soil acidity and is probably an important factor in the slow change in soil pH frequently observed after liming. Many soils do not reach maximum pH from a single liming for several years. This is the basis for the recommendations that lime be applied 1 yr or more prior to establishment of alfalfa or other crops sensitive to soil acidity. This was brought out in studies initiated in Massachusetts in 1959 and 1960 (M. E. Weeks, 1961, unpublished data). Liming rates ranged from 4.5 to 24 t $ha^{-1}$ and included both calcitic and dolomitic limestones, about 70% of which passed through a 60-mesh screen. Some difficulty was experienced in mixing the lime adequately with the soil, but in most cases all or part was spread on plowed ground and disked several times. It was found in all the fields that the pH exceeded 6.0 after 8 months, but they seldom reached 6.5 even after 30 months, regardless of rate of application. Table 7 shows average pH values of three experiments located on Merrimac (sandy, mixed, mesic Typic Dystrochrepts) and Narragansett (coarse-loamy, mixed, mesic Typic Dystrochrepts) soils in eastern Massachusetts. The slow reaction rate and the problem of adequate mixing may account, in part, for poor inoculation and attendant growth of alfalfa in some states, notably Maryland (J. R. Miller, 1964, personal communication) and Massachusetts.

Stones, rock outcrops, and other obstructions of topography, as well as no-till crop culture, might prohibit working the lime into the soil. Here, the only alternative is to spread the lime on the surface and wait for it to be carried downward in solution. This system is practiced on no-till crops, on some wet or hill pastures, and often on turf areas. Studies by Blair and Prince (1934), Brown and Munsell (1939), and Beaumont (1941) indicate that there is a gradual downward movement of lime, 1 to 3 cm $y^{-1}$.

In New Hampshire, Percival et al. (1955) made surface applications of lime on sod crops growing in Paxton (coarse-loamy, mixed, mesic Typic Fragiochrepts) and Gloucester soils. The average pH of both soils at the time of liming was 5.6. Downward movement of the lime after 4 and 5 yr is shown in Table 8. Each datum is the mean of samples from 24 plots. There

Table 8—Effect of surface applications of lime on soil pH at various depths several years after treatment with ground limestone (Percival et al., 1955).

| Soil depth† | Paxton | | Gloucester | |
|---|---|---|---|---|
| | Unlimed | 4.4 t of lime | 2.2 t of lime | 6.6 t of lime |
| cm | Soil pH after 4 yr | | Soil pH after 5 yr | |
| 0–5 | 5.6 | 6.9 | 5.8 | 6.6 |
| 5–10 | 5.5 | 6.3 | 5.8 | 6.3 |
| 10–15 | 5.5 | 5.8 | 5.6 | 5.8 |
| 15–20 | 5.6 | 5.7 | 5.5 | 5.6 |

† The initial pH of both soils (0–15 cm) was 5.6. Lime was applied to the Paxton soil in 1947 and to the Gloucester soils in 1948.

was a consistent indication of lime movement down to the 20-cm level, and the pH was 6.0 or above in the surface 10 cm.

In the Northeast, it is generally believed that lime should be applied as far ahead of a new seeding of forage crops as practicable. This usually turns out to be 3 months to 1 yr before establishment, but all too frequently farmers fail to make the application until they prepare the seedbed. In some soils, difficulties have been encountered in obtaining adequate inoculation of alfalfa, resulting in loss of stands after the first year. This is probably a consequence, in some cases at least, of seeding on strongly acid soils that were not limed until just prior to seeding.

For other crops, lime application is often recommended in the fall before a spring crop is planted (Beaumont et al., 1935). No other specifications are usually given for liming except that of convenience. It seems preferable to spread by trucks in late summer or fall when fields are likely to be dry. When soil moisture is high, heavy liming equipment can cause soil compaction, adversely affecting yields for some time.

## IV. CROP RESPONSE TO LIME

Beginning with the very early work reported by Wheeler et al. (1895) and the subsequent reports by Hartwell and Damon (1914, 1919), it was documented that many species of crop plants responded to additions of limestone to acid soils. This is well illustrated by the data in Table 9, where yield increases from liming for crops grown on a Bridgehampton silt loam (coarse-silty, mixed, mesic Typic Dystrochrepts) varied from 46% for barley to nearly 300% for fodder beet.

Table 9—The effect of limestone on 5-yr average yields of different crops on Bridgehampton silt loam (initial pH = 5.2) (Hartwell & Damon, 1919).

| Lime treatment | Alfalfa† | Carrot | Fodder beet | Barley hay |
|---|---|---|---|---|
| | kg ha$^{-1}$ | | | |
| No lime | 2 240 | 9 350 | 6 190 | 2 170 |
| Limestone applied at LR rate | 4 080 | 20 920 | 15 200 | 2 850 |
| Limestone applied at 3 times LR rate | 4 980 | 23 200 | 24 150 | 3 180 |

† One cutting only.

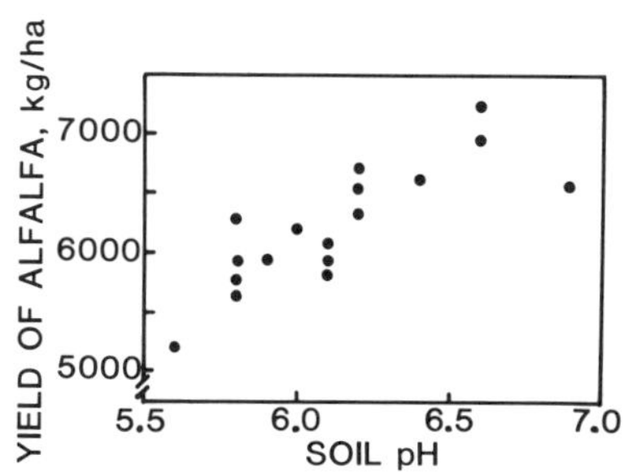


Fig. 3—The relationship between soil pH and yield of alfalfa on a Hublersburg silt loam (A. S. Hunter, 1965, unpublished data, Pennsylvania Agric. Exp. Stn.).

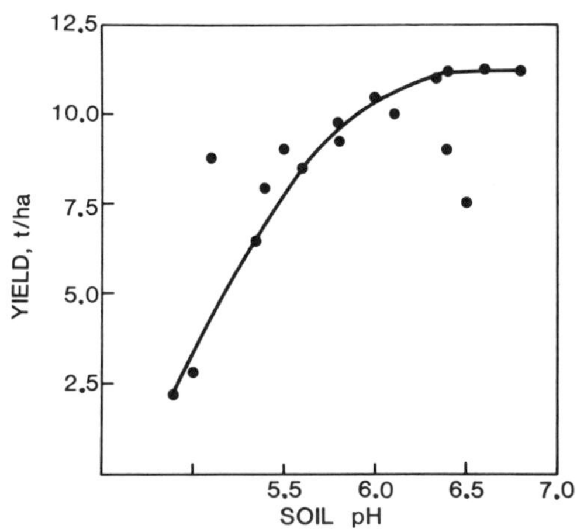


Fig. 4—The relationship between soil pH and yield of alfalfa-birdsfoot trefoil-timothy mixture on Mardin silt loam (W. L. Griffeth, R. Feuer, and R. B. Musgrave, 1967, unpublished data, New York Agric. Exp. Stn.).

In an experiment on a Hublersburg silt loam (clayey, illitic, mesic Typic Hapludults) (A. S. Hunter, 1964, unpublished data) a progressive yield response of alfalfa to increasing soil pH up to at least 6.6 was found (Fig. 3). The same general relationship between soil pH and alfalfa-timothy yields on Mardin silt loam (coarse-loamy, mixed, mesic Typic Fragiochrepts) in New York is shown in Figure 4. Yields increased sharply as soil pH was raised by liming from 4.8 to 6.0, with little additional gain $> 6.5$.

In a study on two acid Hapludults of Pennsylvania, Fox (1979) found that Al saturation was a more accurate indicator than soil pH for predicting reduced corn yields. His data, shown in Table 10, indicate that yield responses to liming were obtained on both soils. Yield of grain dropped sharply when Al saturation exceeded 50% (unlimed soil), and for near optimum yields to be obtained, Al saturation had to be reduced to about 15% or lower.

G. F. Griffen (1981, unpublished data, Connecticut Agric. Exp. Stn.) found a significant yield response of silage corn to liming an acid Fragiochrept, Woodbridge fine sandy loam (coarse-loamy, mixed, mesic Typic

Table 10—Effect of lime rates on soil pH and corn grain yields in Pennsylvania (Fox, 1979).

| Wharton sil (fine-loamy, mixed mesic Aquic Hapludults) | | | | | Murrill sil (fine-loamy, mixed mesic Typic Hapludults) | | | | |
|---|---|---|---|---|---|---|---|---|---|
| | 1976 | | 1977 | | | 1976 | | 1977 | |
| Lime rate | pH | Grain yield† | pH | Grain yield† | Lime rate | pH | Grain yield† | pH | Grain yield† |
| t ha$^{-1}$ | | t ha$^{-1}$ | | t ha$^{-1}$ | t ha$^{-1}$ | | t ha$^{-1}$ | | t ha$^{-1}$ |
| 0 | 4.7 | 2.73a | 4.3 | 2.58a | 0 | 4.9 | 7.53a | 4.3 | 8.63a |
| 2.2 | 4.9 | 4.48b | 4.6 | 5.90b | 2.2 | 5.1 | 7.72a | 4.5 | 9.09a |
| 4.5 | 5.2 | 4.82bc | 4.8 | 6.30b | 6.7 | 5.5 | 8.00a | 5.0 | 9.40ab |
| 9.0 | 5.8 | 5.17cd | 5.3 | 6.99c | 9.0 | 5.8 | 8.00a | 5.0 | 9.40ab |
| 12.3 | 6.2 | 5.51d | 5.6 | 7.61d | 11.2 | 5.8 | 7.74a | 5.5 | 10.25b |
| 15.7 | 6.6 | 5.10bcd | 6.0 | 7.75d | | | | | |

† Values followed by the same letter in vertical columns are not significantly different at the 0.05 probability level, according to Duncan's new multiple range test.

Fragiochrepts), in Connecticut. There were, however, no yield differences ($P < 0.01$) in his experiment due to rates of lime over the 4 yr that yields were measured (Table 11). He found that KCl-extractable Al ranged from 0 to 1.80 cmol (⅓ $Al^{3+}$) kg$^{-1}$ soil in 1978 (2 yr after lime application), and no extractable Al was found at pH 5.5 or above. In general, lower yields occurred when KCl-extractable Al exceeded 0.45 cmol (⅓ $Al^{3+}$) kg$^{-1}$ soil (pH 5.1).

Hunter (1963, unpublished data, Pennsylvania Agric. Exp. Stn.) in Pennsylvania on Gilpin silt loam (fine-loamy, mixed, mesic Typic Hapludults) found a continuing reduction in extractable Al as the rate of liming increased (Table 12). Yield increases of alfalfa-timothy were obtained only to the lowest increment of added limestone, which reduced extractable Al by more than half, and this was likely sufficient to remove Al as a significant contributor to lower yields.

These data indicate that for such crops as corn some measure of potential Al toxicity is the most reasonable criterion to use to estimate the likelihood of lower yield on acid soils. In these soils, extractable Al, regardless of the extractant used, has been reduced to very low values as soon as

Table 11—Corn silage yield on Woodbridge fine sandy loam in Connecticut (G. F. Griffen, 1981, unpublished data, Connecticut Agric. Exp. Stn.).

| Lime treatment | Corn silage yield (30% DM)† | | | | Soil pH in July 1979 |
|---|---|---|---|---|---|
| 1976 | 1976 | 1977 | 1978 | 1979 | |
| | t ha$^{-1}$ | | | | |
| 0 | 29.6 | 31.2 | 30.5 | 26.7 | 4.7 |
| 4.5 | 36.8 | 51.1 | 42.6 | 39.2 | 5.1 |
| 9.0 | 38.6 | 49.3 | 43.9 | 40.8 | 5.5 |
| 13.5 | 41.0 | 48.7 | 44.2 | 40.8 | 5.8 |

† Lime increased yields ($P < 0.01$), but there were no significant differences among lime rates. DM = dry matter.

Table 12—Effect of lime on soil acidity and forage yield of alfalfa-timothy on Gilpin silt loam at Rector, Pa. (A. S. Hunter, 1963, unpublished data, Pennsylvania Agric. Exp. Stn.).

| Lime rate | Final soil pH | Replaceable Al | Avg yield 1959–1962 |
|---|---|---|---|
| t ha$^{-1}$ | | cmol (1/3 $Al^{3+}$) kg$^{-1}$ | kg ha$^{-1}$ |
| 0 | 5.3 | 0.63 | 4770 |
| 3.4 | 5.6 | 0.24 | 5860 |
| 6.7 | 5.9 | 0.08 | 5710 |
| 13.4 | 6.2 | 0.04 | 5860 |

the pH has been raised to 5.3 to 5.5 (Kamprath, 1970; Lathwell & Peech, 1964). On the other hand, yield of alfalfa and other forage legumes has continued to increase at least up to pH 6 and, in some instances, at least to pH 6.5 to 7.0. Obviously, factors other than extractable Al must be limiting yields below pH 6. Extractable Mn may be a possibility. There is evidence that high levels of P in the soil may reduce the LR by reducing extractable Al (Fox, 1979; Bartlett, 1982), and, in fact, Vermont has incorporated a soil test for extractable P into their evaluation of LR.

Various other factors may influence the rate of lime application, the frequency of application, and the method of application to alleviate the deleterious effects of acid soils on plant growth. Lathwell and Peech (1964) showed that the depth to which lime is incorporated into a soil has considerable effect on crop response. The rates of application and depths of incorporation of the limestone are given in Tables 13 and 14, along with the yields obtained in 1948 and 1952 as well as the cumulative yields obtained during the experiment. Incorporation of 2240 kg of limestone ha$^{-1}$ to a depth of 7 cm resulted in higher alfalfa yields than when 2240 kg was incorporated to a depth of 15 cm (Table 13). The same trend was observed when 4480 kg of limestone ha$^{-1}$ were incorporated to a depth of 7, 15, or 30 cm, although the differences were not as great as with the lower rate. Thus, when small quantities of limestone were used, incorporation of the limestone into a smaller volume of soil resulted in a definite yield advantage.

When the rate of limestone was kept constant but the depth of mixing varied, the yield of alfalfa increased markedly with the increase in the depth to which the soil was limed (Table 14). Thus, when 2240 kg of limestone per 15-cm layer of soil were applied to a depth of 60 cm (a total of 8960 kg of limestone ha$^{-1}$), the total yield over the experiment was about six times that of soil that was limed only to a depth of 15 cm. When 4480 kg of limestone per 15-cm layer of soil were applied, no increase in yield was obtained by liming below a depth of 30 cm. When 8960 kg of limestone per 15-cm layer of soil were applied, however, only a slight gain was found by liming below the top 15 cm of soil.

The degree of base saturation and soil pH obviously varied widely as a result of different rates of application and different volumes of soil in which limes were incorporated. However, as larger quantities of limes were applied, the volume of treated soil had less effect on yield.

Table 13—Effect of lime on alfalfa yields on Mardin silt loam as influenced by depth of soil-lime mixing at equal rates of lime per hectare (Lathwell & Peech, 1964).†

| Lime treatment‡ | | Forage yield | | |
|---|---|---|---|---|
| Rate | Depth of mixing | 1948 | 1952 | Cumulative 1948–1952 |
| kg ha$^{-1}$ | cm | | kg ha$^{-1}$ | |
| 2 240 | 0–7 | 2 940 | 860 | 10 620 |
| 2 240 | 0–15 | 2 230 | 640 | 6 240 |
| 4 480 | 0–7 | 5 640 | 2 840 | 22 880 |
| 4 480 | 0–15 | 5 220 | 2 040 | 18 460 |
| 4 480 | 0–30 | 5 040 | 1 500 | 18 080 |
| 8 960 | 0–15 | 7 480 | 5 160 | 35 540 |
| 8 960 | 0–30 | 9 000 | 5 360 | 39 700 |
| 8 960 | 0–60 | 9 150 | 5 850 | 36 720 |
| 17 920 | 0–30 | 9 060 | 6 500 | 40 620 |
| 17 920 | 0–60 | 9 260 | 6 280 | 40 060 |

† Initial pH = 4.8.
‡ Soil and lime were mixed to the depth indicated. Soil in all plots was excavated, mixed to a depth of 60 cm regardless of lime treatment, and replaced prior to seeding.

In field experiments reported by Brown (1961) in Connecticut, yield and stand of alfalfa were more clearly related to rate of liming than to placement. As the rate of liming was decreased, plow-under applications resulted in a poorer stand and lower yield than applications in which the limestone was placed nearer the surface. A surface application that is not worked into the soil is not ordinarily recommended, but in these experiments it proved effective. The yield for surface application was higher than the average of all placements, including plowing under at a 15-cm depth, plowing under at a 7-cm depth, disking in to 10 to 12 cm, and harrowing in

Table 14—Effect of lime on yield of alfalfa on Mardin silt loam as influenced by depth of soil-lime mixing at equal rates of lime per 15-cm layer of soil (Lathwell & Peech, 1964).†

| Lime treatment‡ | | Forage yield | | |
|---|---|---|---|---|
| Lime per 15 cm of soil | Depth of mixing | 1948 | 1952 | Cumulative 1948–1952 |
| kg | cm | | kg ha$^{-1}$ | |
| 2 240 | 0–15 | 2 230 | 640 | 6 240 |
| 2 240 | 0–30 | 5 040 | 1 500 | 18 080 |
| 2 240 | 0–60 | 9 150 | 5 850 | 37 720 |
| 4 480 | 0–7 | 2 940 | 860 | 10 610 |
| 4 480 | 0–15 | 5 220 | 2 040 | 18 460 |
| 4 480 | 0–30 | 9 000 | 5 360 | 39 700 |
| 4 480 | 0–60 | 9 260 | 6 280 | 40 060 |
| 8 960 | 0–7 | 5 640 | 2 840 | 22 880 |
| 8 960 | 0–15 | 7 480 | 5 160 | 35 540 |
| 8 960 | 0–30 | 9 060 | 6 500 | 40 620 |

† Initial pH = 4.8.
‡ Soil and lime were mixed to the depth indicated. Soil in all plots was excavated, mixed to a depth of 60 cm regardless of lime treatment, and replaced prior to seeding.

to 5 to 7 cm. However, it should be noted that this soil had an initial pH of 5.8 or above at all depths < 20 cm. Thus, correction of acidity in the germination zone probably permitted seedlings to become well enough established that primary roots could penetrate the relatively thin, acid soil zone and proliferate in the underlying soil.

Longnecker and Merkle (1952) showed that even when plants grew fairly well in soils with zones of high acidity, the mass of fibrous roots were concentrated in the limed zone. They stated that acid soils should be limed primarily to decrease the solubility of Al and Mn and possibly to increase the availability of P. They also emphasized the importance of mixing lime in the entire soil zone in which root development is expected.

Two choices are available to farmers in developing and carrying out a liming program: (i) they may apply modest but fixed quantities of lime at regular intervals in the cropping system; or (ii) initially they may apply quantities of lime up to the LR and thereafter add modest maintenance applications at infrequent intervals. The influence of four fixed lime rates per 6-yr crop rotation on Mardin soil with an initial pH of 5.1 on the mean yield of crops in the rotation are given in Table 15. All crops, except oat, responded to lime. At these modest rates, wheat and corn responded as well to limestone additions as did the forage legumes.

In a second experiment (Bowen, 1983) on Mardin silt loam, which ran from 1953 to 1975, lime rates up to the LR were applied initially, and selected treatments received reliming at the beginning of the second and third sequence of a 7-yr rotation of corn-oat-alfalfa-timothy (Table 16). In the first sequence, the 5-yr average yield of alfalfa-timothy increased up to the 13.5 t ha$^{-1}$ lime rate, although little was gained in going > 9.0 t ha$^{-1}$. In

Table 15—Influence of lime rate on yields of various crops in a 6-yr rotation on Mardin silt loam.†

| Limestone application‡ | | Avg yields, 1949–1960§ | | | | | | Soil |
|---|---|---|---|---|---|---|---|---|
| Amount per rotation | Time of application | Wheat grain | Mixed hay | Corn silage | Oat grain | Mixed hay | Mixed hay | pH in 1960¶ |
| kg ha$^{-1}$ | | kg ha$^{-1}$ | —— t ha$^{-1}$ —— | | kg ha$^{-1}$ | —— t ha$^{-1}$ —— | | |
| 0 | | 2185 | 6.01 | 31.2 | 2465 | 4.30 | 4.51 | 5.2 |
| 1120 | | 2325 | 6.23 | 31.4 | 2360 | 4.91 | 5.04 | 5.4 |
| 3360 | 2240 kg after plowing for corn | 2495 | 6.64 | 34.3 | 2405 | 5.81 | 5.36 | 5.9 |
| 5600 | 2240 kg before, 2240 kg after plowing for corn | 2610 | 6.82 | 33.6 | 2430 | 6.10 | 5.74 | 6.3 |

† All yield data are 12-yr averages for the second and third rounds of the rotation (R. B. Musgrave, 1961, unpublished data, Cornell Agric. Exp. Stn.).

‡ 560 kg ha$^{-1}$ of limestone were localized with the seed at each hay seeding on all plots except those receiving no lime, thus all of the 1120 kg ha$^{-1}$ of lime applied localized with the seed.

§ Crops were grown in the sequence shown in the table. The following applications per hectare were made on all plots per rotation: 60 t of manure; 65 kg of N; 45 kg of P; and 90 kg of K.

¶ Original soil pH = 5.1.

Table 16—Influence of lime rate on Mardin silt loam on yield of alfalfa-timothy in three separate 7-yr cropping cycles (Bowen, 1983).

| Lime treatment† | | | Forage yield‡ | | |
|---|---|---|---|---|---|
| Initial rate | Relime rate, 2nd cycle | Relime rate, 3rd cycle | 1st cycle | 2nd cycle | 3rd cycle |
| t ha$^{-1}$ | | | kg ha$^{-1}$ | | |
| 0 | 0 | 0 | 3470 | 3900 | 3110 |
| 4.5 | 0 | 0 | 5680 | 4410 | 2985 |
| 9.0 | 0 | 0 | 6550 | 5680 | 3500 |
| 13.5 | 0 | 0 | 6790 | 6360 | 4550 |
| 9.0 | 3.4 | 0 | 6550 | 6555 | 4150 |
| 9.0 | 6.7 | 0 | 6550 | 6530 | 5235 |
| 9.0 | 3.4 | 4.5 | 6550 | 6555 | 6110 |
| 9.0 | 3.4 | 9.0 | 6550 | 6555 | 6110 |
| SE§ | | | 267 | 300 | 298 |

‡ The initial lime treatments were applied as split applications before and after plowing the year prior to alfalfa establishment. The relime treatments were broadcast and disked just prior to the alfalfa establishment for that cycle.

‡ Yearly average of 5 yr of crop in each cycle.

§ Standard error of difference between any two treatment means in a column.

the second cycle, the highest rate of lime resulted in a greater yield advantage, whereas in the third cycle, the yield advantage for the 13.5 t ha$^{-1}$ lime rate was even greater. The effects of lime application at the 13.5 t ha$^{-1}$ rate remained remarkably high even after 20 yr, showing the strong residual effects of liming. Highest yields over the three cycles, however, were obtained where modest amounts of limes were applied at the start of the second and third cycles following the initial 9.0 t ha$^{-1}$ application. Probably because of higher levels of fertility used in this experiment compared with the previous one, lime had little effect on corn silage yield. No effect was noted on oat yields.

Bowen (1983) also found that Ca losses increased with increasing soil pH and increased lime rate. When the annual losses of Ca were regressed on the initial pH values, the equation obtained was

$$\text{Ca lost (kg ha}^{-1}\text{ yr}^{-1}) = -702 + 142\text{ pH}.$$

Thus, both liming programs resulted in significant increases in yield. The choice of a liming program, therefore, may be determined by factors other than yield response to liming. Although the data are not shown here, the responses in these experiments to lime and fertilizer, particularly P, were additive. Since yield response to lime without fertilizer is often as great or greater than to fertilizer without lime, it follows that the lime needs of the soil usually should be satisfied first.

Studies of response of potato to lime on a Plaisted sandy gravelly loam (coarse-loamy, mixed, frigid Typic Fragioorthods) were carried out over an 8-yr period in Maine (Hutchinson et al., 1964). The soil was limed to predetermined pH levels of 5.5, 6.0, and 6.5. In this area, where potato scab was

Table 17—Effect of band and broadcast lime for minimum tillage establishment on yield of alfalfa on Mardin silt loam.†

| | Forage yield§ | | | | | Legume stand | | | | |
|---|---|---|---|---|---|---|---|---|---|---|
| Lime treatment‡ | 1977 | 1978 | 1979 | 1980 | 1981 | 1977 | 1978 | 1979 | 1980 | 1981 |
| | t ha$^{-1}$ | | | | | % | | | | |
| | Trial 1 | | | | | | | | | |
| No lime | 2.7c | 2.2c | 1.6c | 1.8a | 2.5c | 0a | 3a | 7b | 11a | 2a |
| 1700 kg in band | 4.9b | 6.7b | 5.8b | 5.8b | 6.1b | 38b | 53b | 51a | 45b | 35b |
| 9 t ha$^{-1}$ broadcast | 4.9a | 6.7a | 8.1a | 7.8c | 7.8a | 52c | 59c | 59a | 46b | 41b |
| Band + broadcast | 6.5a | 8.5a | 8.1a | 8.3c | 8.1a | 51c | 56c | 55a | 43b | 39b |
| | Trial 2 | | | | | | | | | |
| No lime | | 3.1c | 2.7a | 3.1a | 3.6b | | 10b | 24a | 11a | 6b |
| 1700 kg in band | | 6.7b | 7.0b | 8.1b | 8.1a | | 60a | 50b | 40b | 35a |
| 9 t ha$^{-1}$ broadcast | | 7.8a | 8.3c | 8.3b | 8.7a | | 63a | 61b | 48bc | 43a |
| Band + broadcast | | 8.3a | 8.5c | 9.2b | 8.7a | | 92a | 57b | 54c | 37a |

† R. R. Seaney, W. S. Reid, and R. F. Neugebauer, 1982, unpublished data, Cornell Agric. Exp. Stn.

‡ Band application consists of applying the 1700 kg ha$^{-1}$ of lime with the seed at planting. Band + broadcast received lime at the rates for the separate treatments.

§ Values followed by the same letter in vertical columns are not significantly different at the 0.05 probability level, according to Duncan's new multiple range test.

not a problem, it was found that the higher lime levels caused increased growth and higher yields in the early years, but later, after the fertility level had been built up, differences were negligible. In a rotation for a dairy farm on the same soil type, alfalfa yields were not significantly different except between the lowest and highest rates of lime.

Minimum tillage practices for the establishment of forages complicate incorporation of lime into the soil. About the only choice available to the farmer is to surface apply limestone and seed. Small quantities of limestone may be applied in a band with the seed. Baker (1980), in a study in West Virginia on acid soils unsuited for frequent tillage, found that surface-applied limestone resulted in increased total yield and in the legume component of the mixture. Studies were made in New York on a Mardin silt loam (R. R. Seaney, W. S. Reid, and R. F. Neugebauer, 1982, unpublished data, Cornell Univ. Agric. Exp. Stn.) comparing broadcast with banded limestone under a minimum tillage establishment system. As shown in Table 17, 1700 kg of lime ha$^{-1}$ banded more than doubled yields. Nine tonnes per hectare broadcast resulted in a further yield increase of nearly 2 t ha$^{-1}$ of forage. There was no advantage in using a band application in addition to the broadcast application. The broadcast application of lime gave the highest yields over the 5-yr period, but the band application resulted in surprisingly high and consistent yields. Similarly, little difference in percent legume in the stand was found regardless of lime treatment, although lime was essential for legume to remain in the stand.

In a second experiment on the same site, started 1 yr later, the band application resulted in even higher yields, and in the third and fourth year, yields were almost as high as the broadcast treatments. With minimum tillage systems, it is possible to apply lime in a band at seeding or surface broadcast it and achieve good stands of alfalfa and high yields. Change in soil pH with depth indicates that some $Ca^{2+}$ from the surface broadcasted lime is moving into the soil below the zone of application, accounting, in part at least, for the observed results.

## V. SOIL ACIDITY AND OTHER SOIL FERTILITY FACTORS

### A. Availability

#### 1. MOLYBDENUM

Unlike the other micronutrients, Mo availability increases with increasing soil pH. Evans et al. (1951) found that 18 New Jersey soils ranged from 0.8 to 3.3 mg $kg^{-1}$ of available Mo and that the Mo content of 24 samples of alfalfa from these soils ranged from 0.1 to 1.4 mg $kg^{-1}$. In four of six trials, 1 kg $ha^{-1}$ of $Na_2MoO_4$ produced a marked increase in yield and N content of the crop. They also found response to Mo to be related in some cases to soil pH. On soils limed to pH 6.1 to 6.5, available Mo was 0.7 to 1.1 mg $kg^{-1}$. Further liming to pH 6.9 to 7.6 resulted in 1.6 to 3.0 mg $kg^{-1}$ available Mo.

Although researchers have observed qualitatively that liming acid soils increases root nodulation of legumes and thereby enhances $N_2$ fixation, few quantitative studies on the magnitude of these effects are available. In a greenhouse study using Mardin silt loam soil, Kliewer and Kennedy (1960) determined the response of alfalfa, birdsfoot trefoil, white clover, and red clover to increments of lime as well as to Mo and P. Response to an application of Mo was obtained at low rates of liming but not at the higher rates (Table 18). The addition of either lime or Mo increased the N content of the forage. Factors other than Mo deficiency and a resulting N deficiency were

Table 18—Effect of Mo and lime on the yield, N content, and Mo content of birdsfoot trefoil forage.†

| Lime rate | Soil pH | Yield‡ | | N content | | Mo content§ | |
|---|---|---|---|---|---|---|---|
| t $ha^{-1}$ | | g $pot^{-1}$ | | g $kg^{-1}$ | | mg $kg^{-1}$ | |
| | | − Mo | + Mo | − Mo | + Mo | − Mo | + Mo |
| Lime-pelleted seed | 4.9 | 6.46a | 11.38a | 27.0a | 34.3c | 0.26a | 0.51a |
| 1.12 | 5.1 | 10.85b | 12.63a | 31.8c | 33.3bc | 0.29a | 1.68b |
| 4.5 | 5.6 | 19.46c | 18.50b | 32.6c | 32.0ab | 0.46a | 1.20b |
| 9.0 | 6.0 | 17.37d | 17.14b | 30.1b | 30.9a | 2.80b | 5.24c |
| $LSD_{0.05}$ | | 2.18 | | 1.6 | | 0.52 | |

† Average of two strains of rhizobia and four replicates totaled over five cuttings in a greenhouse experiment (Kliewer & Kennedy, 1960).

‡ Values followed by the same letter in vertical columns are not significantly different at the 0.05 probability level, according to Duncan's new multiple range test. $LSD_{0.05}$ values provided for Mo comparisons.

§ Mo applied at a rate of 0.625 mg of $NaMoO_4 \cdot 2H_2O$ 4 $kg^{-1}$ of soil $pot^{-1}$.

Table 19—Effect of Mo and lime on the yield and N content of the forage of four legumes (average of four replicates totaled over five cuttings) on Mardin silt loam in the greenhouse (Kliewer & Kennedy, 1960).

| Lime rate | Final soil pH | Alfalfa† | | | | Birdsfoot trefoil‡ | | | | White clover | | | | Medium red clover | | | |
|---|---|---|---|---|---|---|---|---|---|---|---|---|---|---|---|---|---|
| | | Yield | | N content | | Yield | | N content | | Yield | | N content | | Yield | | N content | |
| $t\ ha^{-1}$ | | $g\ pot^{-1}$§ | | $g\ kg^{-1}$ | | $g\ pot^{-1}$§ | | $g\ kg^{-1}$ | | $g\ pot^{-1}$§ | | $g\ kg^{-1}$ | | $g\ pot^{-1}$§ | | $g\ kg^{-1}$ | |
| | | −Mo | +Mo | −Mo | +Mo | −Mo | +Mo | −Mo | +Mo | −Mo | +Mo | −Mo | +Mo | −Mo | +Mo | −Mo | +Mo |
| None | 4.9 | 1.85a | 2.14a | 26.6a | 25.3a | 3.85a | 8.99a | 30.2a | 38.9 | 6.80a | 9.60a | 30.1a | 37.4 | 7.11a | 7.78a | 32.0a | 36.8 |
| Lime-pelleted seed | 4.9 | 3.54ab | 5.75ab | 29.5ab | 32.8b | 5.30ab | 8.84a | 29.7a | 38.9 | 8.86a | 12.67b | 30.6a | 38.5 | 9.98b | 12.71b | 35.9b | 39.5 |
| 0.56 kg of lime¶ | 4.9 | 7.73b | 8.81b | 34.0cd | 33.6b | 6.92b | 9.94a | 33.6b | 36.3 | 13.36b | 13.52b | 37.5c | 39.6 | 11.82b | 11.71b | 37.2b | 39.2 |
| 2.2 | 5.3 | 6.13b | 8.98b | 31.3bc | 34.0b | 9.08c | 12.42b | 34.1bc | 36.5 | 11.58b | 17.64c | 34.7b | 38.8 | 16.85c | 18.11c | 34.9b | 38.2 |
| 4.5 | 5.6 | 17.07c | 17.20c | 36.0d | 37.1b | 13.72d | 13.69b | 37.0c | 37.6 | 19.81c | 19.65c | 37.0bc | 37.6 | 21.28d | 19.87c | 36.4b | 37.0 |
| $LSD_{0.05}$ | | NS | | NS | | 2.05 | | 3.0 | | 2.42 | | 2.6 | | NS | | NS | |

† The values within a column not followed by the same letter are significant at the 0.01 level by Duncan's multiple range test. The $LSD_{0.05}$ values are provided for Mo comparisons. Mo × lime significant for birdsfoot trefoil yield and N content and white clover yield and N content. NS = nonsignificant.

‡ Birdsfoot trefoil data are for four cuttings.

§ Mo applied at rate of 1.0 mg as $Na_2MoO_4 \cdot 2H_2O$ 4 $kg^{-1}$ of soil $pot^{-1}$.

¶ Lime banded with seed.

also involved since yields increased from lime application even after the Mo deficiency had been corrected.

The influence of additions of lime and Mo on the yield and N content of several legumes is shown in Table 19. Of the four legumes studies, birdsfoot trefoil gave the greatest response to Mo. Lime applications resulted in an increase in yield for all legumes, whereas N content increased in all legumes except birdsfoot trefoil. These results suggest that on Mardin silt loam, lime increased the uptake of Mo, which in turn increased symbiotic $N_2$ fixation. However, Kliewer and Kennedy (1960) pointed out that 4.5 t $ha^{-1}$ of lime without Mo gave maximum yields for each legume. Therefore, they concluded that liming at this rate had an effect in addition to increasing the availability of Mo of providing a suitable environment for the *Rhizobium*. They also noted that an application of both Mo and lime decreased the number of nodules on birdsfoot trefoil plants, substantiating observations of others that a few large nodules are more efficient in $N_2$ fixation than many small ones.

## 2. OTHER TRACE ELEMENTS

A number of states in the Northeast recommended liming the soil to $<$ pH 6.5 because there is danger of inducing deficiencies of various trace elements at higher pH values. In some cases, however, the deficiency seems to be a temporary condition that lasts only 1 or 2 yr after liming (Midgley, 1932).

A survey of the composition of red clover, oat, and timothy for 10 different elements in samples taken from all parts of Pennsylvania involving hundreds of samples of plant material and soils showed that several elements were markedly affected by the lime level (French et al., 1957).

Purvis and Davidson (1948) found that B deficiency was often lime induced, and Percival et al. (1955) mentioned that concentrations of Cu were reduced in some forages with increased liming. Although the availability of B is reduced by liming, and alfalfa responds to applications of this element on some soils of the region, there is little evidence of B deficiencies in the fine-textured soils. Work in New York has shown that very few soils were deficient in B, and where B deficiency symptoms appear on one crop of alfalfa, they may not appear in subsequent cuttings or years (J. E. Dawson, 1955, personal communication, Cornell Univ. Agric. Exp. Stn.). Midgley and Dunklee (1939) stated that the capacity of a soil to fix B greatly affects the amount of lime that can be safely used in field practices.

Field observations in New York (W. S. Reid, 1981, personal communication, Cornell Univ. Agric. Exp. Stn.) has revealed symptoms of Zn deficiency in corn on heavily limed acid soils, especially where soil P is high. Insufficient information is available to predict how much yields are being affected. Zinc applications may be required, however, where large applications of lime are made to acid soils.

### B. Toxicities

Many acid soils of the Northeast contain large quantities of exchangeable $Al^{3+}$ and $Mn^{2+}$, and the poor growth of plants on acid soils of the region has been attributed primarily to toxicity from these elements. Thus, Longnecker and Merkle (1952) concluded from their experiments on lime placement that the chief function of liming was to decrease the solubility of Al and Mn in the soil. They showed, for instance, that the addition of lime, which changed soil pH from 4.0 to 6.8, reduced the replaceable Al from 23 to 2 mg $kg^{-1}$ and the replaceable Mn from 27 to 2 mg $kg^{-1}$.

Lathwell and Peech (1964) concluded that the beneficial effects of liming acid soils may be attributed primarily to reduction of Al and Mn concentrations in the soil solution rather than to an increase in $Ca^{2+}$ concentration. They showed that the extractable Al and Fe and the available Mn in the soil were markedly reduced as the soil pH was increased.

The data of Fox (1979) showed that toxic levels of Al occurred at Al saturation > 15%. G. F. Griffen (1981, unpublished data, Connecticut Agric. Exp. Stn.) also found that when KCl-extractable Al exceeded 0.45 cmol ($\frac{1}{3}$ $Al^{3+}$) $kg^{-1}$ of soil, Al toxicity was observed in corn.

### C. Lime as Source of Magnesium

Magnesium is present in the majority of the limestones used for liming in the Northeast, although the Mg content of some is very low. Many of the states have favored the use of dolomitic limestones, which contain up to 120 g of Mg $kg^{-1}$. This material has been quite effective in correcting Mg deficiencies in fruits, tobacco, and potato. The preferred practice with potato, and in some cases tobacco, has been to use minimum rates of lime and to apply Mg in the fertilizer as $MgSO_4$ or as Mg–$K_2SO_4$.

McMurtrey (1932) found in some tobacco areas where *sanddrown* (Mg deficiency) was present that the use of $MgSO_4$ did not produce good growth because of the lack of sufficient $Ca^{2+}$. Dolomitic limestones were found to be an efficient way of overcoming this form of Mg-Ca deficiency. Thus, dolomitic limestones have been recommended and preferred in most of the Northeast. Bear and Prince (1948), among others, found that dolomitic limestone should be ground very finely to effectively supply Mg in Mg-deficient soils. At present, most dolomitic stones are ground so that a minimum of 60 to 70% pass through a 60-mesh screen.

There is some evidence from soil testing results in Massachusetts (M. E. Weeks, 1964, unpublished data) and New Hampshire (N. Peterson, 1964, personal communication, New Hampshire Agric. Exp. Stn.) that soil exchangeable $Mg^{2+}$ is built up much more rapidly by a dolomitic limestone than is $Ca^{2+}$. Greenhouse studies by Sasaki (1964) tend to confirm this observation.

A greenhouse study using several acid surface soils from New York (W. S. Reid, 1978, unpublished data, Cornell Univ. Agric. Exp. Stn.) showed greater yield responses of birdsfoot trefoil, alfalfa, and corn to dolomite than to calcitic limestones. In further studies using the Mardin soil, birdsfoot trefoil yields were correlated with the exchangeable $Ca^{2+}/Mg^{2+}$ ratio in the soil. These correlations were not validated in a field trial.

### D. Liming and Plant Disease

Observations of farmers concerning scab (*Streptomyces scabies* Thaxt) prompted some of the early writers in the region to caution against the excessive use of lime on potato and tobacco (Wheeler, 1895). This problem was later studied in more detail (Terman et al., 1948; Hutchinson et al., 1964; Merkle & Hunter, 1962), and in all cases, the development of scab was found to be closely related to the amount of lime applied and the resulting soil pH and Ca content of the soil. Terman et al. (1948) found that scab increased more rapidly and became more severe on the less acid, darker, and poorly drained Easton loam (fine-loamy, mixed, frigid Aquollic Hapludalfs) than on the more acid, lighter-colored, and well-drained Caribau loam (coarse-loamy, mixed, frigid Typic Haplothods). Yields were found to be highest within the soil pH range 4.5 to 6.0. Severe infestations of scab were reduced, and sometimes eliminated, when soil pH was lowered by applications of S or $(NH_4)_2SO_4$.

In a study of lime usage in potato farming systems in Maine, Hutchinson et al. (1964) found a lag of about 3 to 4 yr in the establishment of scab infestation after liming. After 7 yr, they found some scab at the lowest lime level on a Plaisted loam, 50% tuber infection at medium levels, and 95% at high lime levels, although the highest soil pH was only about 6.4. Therefore, the recommendation is that lime be used at rates that limit the soil pH to 5.5 or less to control the scab organism.

## REFERENCES


Alvord, H. E. 1888. First annual report. Maryland Agric. Exp. Stn. Bull. 1.

Alvord, H. E., and H. J. Paterson. 1889. Commercial fertilizers. Maryland Agric. Exp. Stn. Bull. 6.

Anderson, J. 1799. An essay on quick-lime as a cement and as a manure. Printed by Samuel Ethridge for Joseph Nancrede, 49 Marlboro St., Boston, Mass.

Bartlett, R. J. 1982. Reactive aluminum in the Vermont soil test. Commun. Soil Sci. Plant Anal. 13:497–506.

Bear, F. E., and A. B. Prince. 1948. Magnesium needs of New Jersey soils. New Jersey Agric. Exp. Stn. Bull. 739.

Beaumont, A. B. 1941. Downward movement of lime in pasture soils. Massachusetts Agric. Exp. Stn. Annu. Rep. Bull. 378.

Beaumont, A. B., M. E. Snell, W. L. Doran, and A. I. Bourne. 1935. Onions in the Connecticut Valley. Massachusetts Agric. Exp. Stn. Bull. 318.

Blair, A. W., and A. L. Prince. 1934. The influence of lime on the reaction of subsoils. J. Agric. Res. 48:469–473.

Bourgoin, T. L. 1980. Soil acidity and its effect on the fertility and lime requirement of selected Maine soils. M.S. Thesis. Univ. of Maine, Orono.

Bowen, W. T. 1983. Crop yield response to initial and maintenance lime applications over three crop rotation cycles. M.S. Thesis. Cornell Univ., Ithaca, N.Y.

Brady, N. C., R. A. Struchtemeyer, and R. B. Musgrave. 1957. Soils. p. 598–619. *In* Yearbook of Agriculture. USDA, U.S. Government Printing Office, Washington, D.C.

Brown, B. A. 1961. Fertilizer experiments with alfalfa, 1915–1960. Connecticut Agric. Exp. Stn. Bull. 363.

Brown, B. A., and R. I. Munsell. 1939. Soil acidity at various depths as influenced by the time since application, placement and amount of limestone. Soil Sci. Soc. Am. Proc. 3:217–221.

Evans, H. J., E. R. Purvis, and R. E. Bear. 1951. Effect of soil reaction on availability of molybdenum. Soil Sci. 71:117–124.

Fox, R. H. 1979. Soil pH, aluminum saturation, and corn grain yield. Soil Sci. 127:330–334.

Fox, R. H. 1980. Comparison of several lime requirement methods of agricultural soils in Pennsylvania. Commun. Soil Sci. Plant Anal. 11:57–69.

French, C. E., C. B. Smith, H. R. Fortmann, R. P. Pennington, G. A. Taylor, W. W. Hinnish, and R. W. Swift. 1957. Survey of ten nutrient elements in Pennsylvania forage crops. Pennsylvania Agric. Exp. Stn. Bull. 624.

Greweling, T., and M. Peech. 1965. Chemical soil tests. Cornell Univ. Agric. Exp. Stn. Bull. 960 (rev. ed.).

Hall, A. D. 1909. Fertilizers and manures. John Murray, London.

Hartwell, B. L., and S. C. Damon. 1914. The comparative effects on different kinds of plants of liming an acid soil. Rhode Island Agric. Exp. Stn. Bull. 160.

Hartwell, B. L., and S. C. Damon. 1919. A field comparison of hydrated lime with limestone of different degrees of fineness. Rhode Island Agric. Exp. Stn. Bull. 180.

Hitchcock, E. 1838. Report on a reexamination of the economic geology of Massachusetts. Dutton and Wentworth: State Printers, Boston.

Hutchinson, F. E., H. J. Murphy, and D. T. Lewis. 1964. The effect of liming on yield, quality and chemical composition of crops in dairy and potato rotations for seven years on Plaisted soils in Maine. Maine Agric. Exp. Stn. Bull. 617.

Kamprath, E. J. 1970. Exchangeable Al as a criteria for liming leached mineral soils. Soil Sci. Soc. Am. Proc. 34:252–254.

Kliewer, W. H., and W. K. Kennedy. 1960. Studies on response of legumes to molybdenum and lime fertilization on Mardin silt loam soil. Soil Sci. Soc. Am. Proc. 24:377–380.

Kohler, E. 1849. Lime, plaster and other fertilizers. U.S. Patent Office Agric. Rep., U.S. Government Printing Office, Washington, D.C., p. 300–393.

Lathwell, D. J., and M. Peech. 1964. Interpretation of chemical tests. Cornell Agric. Exp. Stn. Bull. 995.

Longnecker, D., and F. G. Merkle. 1952. Influence of placement of lime compounds on root development and soil characteristics. Soil Sci. 73:71–74.

McMurtrey, J. E., Jr. 1932. Relation of calcium and magnesium to the growth and quality of tobacco. J. Am. Soc. Agron. 24:707–716.

Merkle, F. G., and A. S. Hunter. 1962. Potato fertilization in Pennsylvania. Pennsylvania Agric. Exp. Stn. Bull. 698.

Midgley, A. R. 1932. Overliming acid soils. J. Am. Soc. Agron. 24:822–836.

Midgley, A. R., and D. E. Dunklee. 1939. The effect of lime on the fixation of borates in soils. Soil Sci. Soc. Am. Proc. 4:302–307.

Miller, R. H., and E. H. Brinkley. 1893. Corn experiments. Maryland Agric. Exp. Stn. Bull. 25.

Noll, C. F., F. D. Gardner, and C. J. Irwin. 1931. Fiftieth anniversary of the general fertilizer tests. Pennsylvania Agric. Exp. Stn. Bull. 264.

Percival, G. P., D. Josselyn, and K. C. Beeson. 1955. Factors affecting the micronutrient content of some forages in New Hampshire. New Hampshire Agric. Exp. Stn. Tech. Bull. 93.

Prince, A. B., and W. A. Raney. 1961. Some morphological, physical and chemical properties of selected northeastern United States soils. New Hampshire Agric. Exp. Stn. Misc. Pub. 1.

Purvis, E. R., and O. W. Davidson. 1948. Review of the relation of calcium to availability and absorption of certain trace elements by plants. Soil Sci. 65:111–116.

Ruprecht, R. W. 1915. Toxic effects of iron and aluminum salt on clover seedlings. Massachusetts Agric. Exp. Stn. Bull. 161.

Sasaki, S. 1964. Uptake of calcium and magnesium by buckwheat from limestone. Soil Sci. Plant Nutr. (Tokyo) 10:99–106.

Shoemaker, H. E., E. O. McLean, and P. F. Pratt. 1961. Buffer methods for determining lime requirement of soils with appreciable amounts of extractable aluminum. Soil Sci. Soc. Am. Proc. 25:274–277.

Soil Survey Staff. 1975. Soil taxonomy: A basic system of soil classification for making and interpreting soil surveys. USDA Agric. Handbook no. 436, U.S. Government Printing Office, Washington, D.C.

Terman, G. L., F. H. Steinmetz, and A. Hawkins. 1948. Effects of certain soil conditions and treatments upon potato yields and the development and control of potato scab. Maine Agric. Exp. Stn. Bull. 463.

U.S. Bureau of the Census. 1978. 1978 census of agriculture. Superintendent of Documents, U.S. Government Printing Office, Washington, D.C.

Weeks, M. E., and W. G. Colby. 1965. Agronomy guide: A land use and management system for Massachusetts soils. Massachusetts Agric. Exp. Stn. Bull. 384 (rev. ed.).

Wheeler, J. H. 1895. Fertilizers, potatoes, potato scab. Rhode Island Agric. Exp. Stn. Bull. 33.

Wheeler, J. H., B. L. Hartwell, and G. M. Tucker. 1895. A recognition of acidity of upland soils and its bearing upon agricultural practice. Rhode Island Agric. Exp. Stn. Annu. Rep. 8:232–280.

Wheeler, J. H., and G. M. Tucker. 1896. Further observations upon the growth of barley upon an acid upland soil limed and unlimed. Rhode Island Agric. Exp. Stn. Annu. Rep. 9: 242–272.

# 8 Crop Response to Lime in the Western United States

**T. L. JACKSON**

*Oregon State University*
*Corvallis, Oregon*

**H. M. REISENAUER**

*University of California*
*Davis, California*

The western states include a complex combination of geological, topographical, climatic, and vegetative features. The region extends through 1900 km of latitude and contains three major north-south, storm-directing mountain systems. Most of the region is arid (annual precipitation < 50 cm); however, sizeable agriculturally important areas west of the Cascade and Sierra Nevada mountains, and at higher elevations along all the major mountain systems, receive substantial annual precipitation (120–200 cm). Mean annual temperatures vary from 24°C in the desert areas near our border with Mexico to 3°C in the mountain valleys of the Northwest. This wide range of climates superimposed over a variety of parent materials and landforms has favored the development of many widely differing soils. To this variety of soils and climates, farmers have introduced agricultural systems that include a large number of economically important crops grown under management systems that vary from minimal to the most intensive. As a result, acid soils and soil acidification, although not always extensive, are important in every area of the region.

## I. SOILS OF THE REGION

Several factors make the soils and agriculture of the western states different from the soils and agriculture of humid areas east of the Rocky Mountains. The wide variation in soils is shown in Figure 1. First, precipitation is concentrated in the winter, and the summers are dry. Second, west of the Cascade Range, temperature variation through the year is small. July and January average temperatures generally differ < 16.7°C. Third, some areas have very high precipitation, ranging up to > 200 cm $yr^{-1}$. Fourth,


*Soil Acidity and Liming*—Agronomy Monograph no. 12 (2nd Edition).

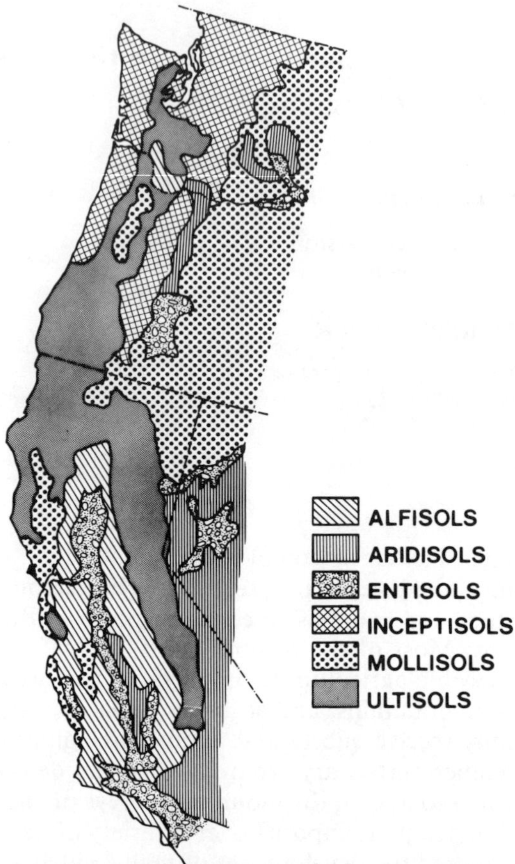


Fig. 1—Soils of the region.

volcanic ash and its weathering products are important components of the soil in many places. These factors result in wide variation in type of clay minerals, percent clay, percent organic matter, and cation exchange capacity (CEC).

### A. Naturally Acid Soils

One major group of acid soils developed from alluvial and lacustrine materials in the Willamette-Puget trough, the low-lying area between the Coast Range and the Cascade Range. In the Willamette Valley, these acid soils are on broad terraces up to 50 km wide. There are floodplains and terraces in narrow strips along many of the smaller streams. Tidal areas on the east side of Puget Sound in Washington include cultivated soils that are relatively level and support a productive and intensive agriculture, including many different crops. Medium and moderately fine textures are dominant, although some soils are sandy and some are clayey. Organic matter content of surface horizons ranges mostly from 20 to 100 g $kg^{-1}$. Basic cation saturation levels range from 40 to 80% in surface horizons and increase with depth. Clay minerals are mostly of the 2:1 lattice type, dominantly montmorillonite and illite. The Haploxerolls, Argiaquolls, Argixerolls, and Albaqualfs are the most common great groups occurring in this area.

The second group consists of upland soils on hills in and at the margins of the Willamette-Puget trough, generally south of the terminal moraine of continental glaciation near Olympia, Washington. This group also extends south into California. Except for some tree fruits and berry crops, these soils are used less intensively than the associated alluvial soils. Organic matter content ranges from 30 to 100 g $kg^{-1}$ in surface horizons. Soil textures are moderately fine in surface horizons and fine in the subsoil. Basic cation saturation levels, when measured with an extracting solution buffered at pH 7.0, range between 10 and 70% in surface horizons and commonly decrease with depth. The pH-dependent charge is significant on most of these soils. This group may be subdivided on the basis of the dominant clay minerals. The soils with 2:1 lattice clay minerals are usually from parent materials derived from sedimentary rock. Moderately fine-textured surface horizons and fine-textured lower horizons are most common. Haploxeralfs and Xeric Haplohumults are the dominant soils, but there are localized areas of Xererts. Alfisols and Ultisols tend to be reddish colored, highly weathered and formed predominantly from sedimentary rock. The soils with dominantly kaolinitic clays are from parent materials that are probably derived from basalt or basic, pyroclastic materials. They contain 100 to 150 g $kg^{-1}$ of hydrous oxides of Fe and Al, are predominantly clayey textures, and are generally low in basic cation saturation.

The third group consists of soils near the Pacific Ocean, primarily in Washington and Oregon and extending into northern California. In this area, the cultivated soils are mostly on terraces, floodplains, and tidal flats, but some are on gently sloping hills. Cropping is limited largely to forages because of cool summer temperatures and excessive fog and precipitation; however, there are small areas of cranberry bogs and other highly specialized crops. Surface horizons commonly have from 100 to 200 g $kg^{-1}$ or more of organic matter. Basic cation saturation levels range from 10 to 40% in surface horizons and generally decrease with depth. The CEC of these soils increases as they are limed from pH 5.0 to 6.0 indicating a significant pH-dependent charge (Janghorbani et al., 1975). Also, the data of Clark (1966), Pratt and Bair (1962), and Blosser and Jenny (1971) show poor relationships between pH and basic cation saturation calculated as percent of CEC at pH 8.2. The variance has been attributed to differences in reactivity of protonated organic and mineral anions. In general, soil organic matter decreases and base saturation increases from north to south. Crystalline clays are mostly of the 2:1 lattice type with some 1:1 kaolinitic clays. Chlorite-like intergrade material or partially interlayered montmorillonite is common. In addition, allophane is present, and hydrous oxide contents of 120 to 150 g $kg^{-1}$ are common. Haplumbrepts and Humaquepts generally occupy the terraces and floodplains, and soils on marine terraces are mainly Haplorthods, Sideraquods, and Haplumbrepts. Soils on tidal flats are mainly Humaquepts and Fluvaquents. Some of the alluvial soils have higher basic cation saturation levels and are much like the alluvial soils east of the Coast Range. Some tidal flat soils contain sulfides that oxidize and lower the pH when the soils are drained. Such soils extend from Puget Sound to San Francisco.

The fourth group consists of soils from glacial till and outwash in the Puget trough around Puget Sound. These soils are commonly sandy loam in texture, or coarser, and many are gravelly. Undisturbed soils have organic matter concentrated in an organic horizon above the mineral profile. This organic matter is vulnerable to loss, and past management practices have resulted in wide variation in the organic matter content of these soils that are cultivated. Soil pH and percent basic cation saturation levels can be changed significantly by moderate rates of lime because CECs are low. Low fertility and low moisture-holding capacity levels limit the use of these soils, especially in comparison with the associated soils from alluvial and lacustrine parent materials. These soils are dominantly Durochrepts, Xerochrepts, Fluvaquents, and Xerumbrepts.

## B. Acid Soils Caused by Cultural Practices

In addition to the naturally acid or "geologically" acid soils of the region, wide areas of both dryland and irrigated soils have been made acid by agricultural practices—notably fertilization, irrigation, and basic cation removal in harvested crops. Intense acidification from cultural practices was first noted in Washington orchards on coarse-textured soils to which $(NH_4)_2SO_4$ had been applied in the drip line of the trees (Benson & Vandecaveye, 1951; Benson & Woodbridge, 1961). In some cases, soil acidity was limiting nitrification and uptake of N by the trees. Similar results have been reported from California (Aldrich et al., 1945) and Oregon (Westwood et al., 1964) and are still observed throughout the area. Corrective practices include applications of lime and shift to nitrate sources of N, principally $Ca(NO_3)_2$, and foliar applications of urea.

Within the past few years, rapidly increasing acidity of the plow layer of the soils of the dryland, wheat-growing areas of eastern Washington and northern Idaho has been documented (Mahler, 1982). Data of the soil testing laboratories of these states indicate that the pH of the plow layer of the majority of the soils of that area has dropped from their original near-neutral reaction to values $< 6$. For the most part, this dramatic shift in soil pH has taken place since 1960. It is attributed to greatly increased use of $NH_4^+$ fertilizers and removal of basic cations by the higher yielding crops. The use of N fertilizers has expanded dramatically since 1950, with 18 000 t of N, 4% of the U.S. total, used by the three Pacific Coast states in 1950 compared with 939 000 t of N, 9% of the U.S. total, in 1980.

The agriculturally acidified soils of the region differ from the geologically (naturally occurring) acid soils in that a lower fraction of the exchange acidity (Yuan, 1959) is exchangeable Al. This is illustrated in Figure 2 as the ratios of exchangeable Al/exchange acidity for the different groups of soils. At pH 4.5, $< 50\%$ of the exchange acidity was exchangeable Al in the

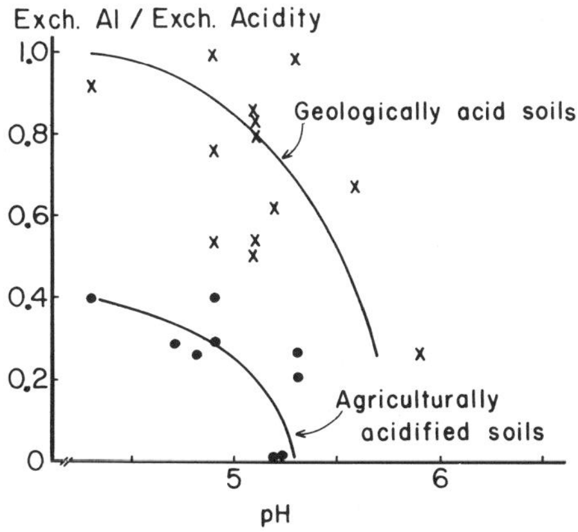


Fig. 2—Exchangeable Al as a fraction of "exchangeable acidity" for geologically acid and agriculturally acidified soils (Janghorbani et al., 1975; Brown, 1975).

agriculturally acidified soils, whereas in the geologically acid soils this fraction was > 90%. The difference is assumed to result from the low rates of accumulation of Al in the decomposition of acid clays or from the effects of the phosphate level (Lindsay, 1979) or other ligands on $Al^{3+}$ activity. Although measurements were not made, high levels of soluble phosphate would be expected in these intensively managed, acidified soils. The relationships of Figure 2 also illustrate a source of the difficulties encountered in assessing lime requirements of the acid soils of the region.

Acidification of the intensively cropped, irrigated soils of the region is an equally serious problem and one less subject to generalization. Management practices altering soil cation levels include the acidifying effects of acid-forming fertilizers, removal of basic cations in harvested crops, and transport, with subsequent precipitation or loss, of basic cations out of the surface soil. These acidifying effects are countered by additions of alkaline carbonates in the irrigation waters or as liming materials, release of basic cations in the weathering reactions forming soil, and the use of basic fertilizers. The rate of pH change is determined by the relative contributions of these acidifying and alkalizing reactions and by the buffering characteristics of the soil.

Western farmers have a long history of preferred use of S-supplying, acidifying fertilizer materials. This choice was not without cause because western soils are inherently low in available S and responses to soil acidification are well documented (Lorenz & Johnson, 1953; Jackson & Carter, 1976). The data of Table 1 illustrate the preferred use of ammoniacal fertilizers in the Pacific states. From 1950 to 1980, fertilizer use in the region accounted for < 10% of the total U.S. consumption of N fertilizers but for about 50% of the nation's use of $(NH_4)_2SO_4$ and $(NH_4)H_2PO_4$. This usage of strongly acidifying nutrient sources has disproportionately amplified fertilizer-caused soil acidification in the region.

Other significant acidifying actions influencing the soils of the region include crop removal of basic cations and the transfer of basic cations from

Table 1—Fertilizer use in Pacific states (Scholl & Wallace, 1950; Hargett & Berry, 1981).

| Year | Fertilizer | Pacific states | |
|---|---|---|---|
| | | Usage | Fraction of U.S. total |
| | | $t \times 10^3$ | % |
| 1949–1950 | Ammonium sulfate | 111 | 62 |
| | Ammonium phosphate | 82 | 58 |
| | All N | 20 | 4 |
| 1959–1960 | Ammonium sulfate | 245 | 45 |
| | Ammonium phosphate | 157 | 26 |
| | All N | 382 | 14 |
| 1969–1970 | Ammonium sulfate | 440 | 56 |
| | Ammonium phosphate | 237 | 37 |
| | All N | 646 | 9 |
| 1979–1980 | Ammonium sulfate | 346 | 41 |
| | Ammonium phosphate | 273 | 46 |
| | All N | 1036 | 9 |

surface to subsoil layers with percolating waters. Crop removal of basic cations is significant principally with crops for which a large fraction of the plant is removed at harvest, e.g., hay, silage, and pasture. An 18 t $ha^{-1}$ crop of alfalfa hay removes an average of 23.2 mmol of basic cations (+) $dm^{-2}$ (Pierre & Banwart, 1973). This is an important loss of cations from western soils, because animal feed crops occupy up to 15% of the cultivated acreage. Acidification from transfer of basic cations (largely as $NO_3^-$ salts) to subsurface layers is difficult to evaluate. Efficient use of irrigation water involves an extended dry-down period between applications during which plant uptake of water occurs mostly from the upper portions of the rooting depth. Under the high evaporative demand of desert regions, as many as 10 to 12 irrigations $yr^{-1}$ are required for long-season crops. These regular, efficient transfer actions with subsequent plant uptake of the translocated $NO_3^-$ leave the surface depleted of bases while the subsoils are enriched with lime and other less soluble salts. These acidifying actions are further enhanced by the low buffering capacity of many of the soils common to the region's intensive agriculture.

Basic cations are added to western soils as alkaline carbonates in irrigation waters and liming materials, by weathering of rocks, and by applications of basic fertilizers. Irrigation waters, though generally recognized as a source of accumulating salts and nutrients such as N and S, are not commonly considered as a source of basic salts. The titratable alkalinity ($CO_3^{2-}$ plus $HCO_3^-$) of western irrigation waters varies from trace amounts to levels exceeding 20 mmol (−) $L^{-1}$. Most surface-derived waters contain 1 to 3 mmol ($HCO_3^-$) $L^{-1}$, whereas well waters may contain as much as 5 to 10 mmol $L^{-1}$. An average seasonal irrigation (Stewart, 1975) of 1 m of a water containing 1 mmol ($HCO_3^-$) $L^{-1}$ supplies 10 mmol of basic cations (+) $dm^{-2}$ (equivalent to 500 kg $ha^{-1}$ of $CaCO_3$). Thus, water characteristics may be a dominant factor in assessing potential pH changes in irrigated soils. The rate of basic cation inputs via soil weathering should also be included in evaluating the effects of management systems on soil properties. This contribution can be expected to vary widely over different parent materials and climates. As estimated by the mean of the values reported by Barshad (1964), Cole et al. (1967), and Silkworth and Grigal (1982), weathering of minerals releases 5.3 ± 1.3 mmol of basic cations (+) $dm^{-2}$ $yr^{-1}$.

For an established alfalfa field producing 18 t $ha^{-1}$ of hay, the approximate cation balance sheet would be:

| Acid or base source | Amount produced [mmol (+) or (−) $dm^{-2}$] |
|---|---|
| Excess bases in harvest crop | 23.2 of acid |
| $NO_3^-$, 100 kg of N $ha^{-1}$, from organic sources | 7.2 of acid |
| Rock weathering | 5.3 of base |
| Net effect | 25.1 of acid |

The bases needed to avoid soil acidification could be supplied by irrigation water containing at least 2.5 mmol (+) $dm^{-2}$ as $HCO_3^-$ salts or by liming materials. Estimates for other crops can be made similarly. In most cases, some transport of $NO_3^-$ salts out of the surface soil should be anticipated.

The factors contributing to agricultural acidification of arid soils have also intensified soil acidity problems on intensively managed, naturally acid soils. Grass seed fields of the Willamette Valley are frequently found with soil pH values of 4.5 and 2 to 3 cmol of [½ (Ca + Mg)] $kg^{-1}$ of soil where 100 to 150 kg of N $ha^{-1}$ $yr^{-1}$ has been used for 30 to 40 yr.

## II. CROPS OF THE REGION

The wide range of soils and climates of the region, along with extensive irrigation, allows production of over 200 commercial crops. Pasture, grain, hay, and seed crops dominate the dry-farmed areas. In contrast, the irrigated areas produce a wide variety of crops, including fruit, nuts, vegetables, grapes, alfalfa, cotton, sugar beet, grain crops, and irrigated pasture. Production of irrigated crop is especially favored by the mediterranean-type climate of the area, which facilitates cultural and harvest operations unhampered by inclement weather. The nonfarmed areas provide forest products, range, and recreation and include desert and brush lands.

The production of processing crops is concentrated on the recent alluvial and the older, valley floor soils that are well drained. Almost all of this acreage is irrigated. Berries, tree fruits, and nuts have been important crops for this area for many years. Although the total area occupied by these crops is relatively small, they are of major economic importance to the area due to the high per-hectare return. Tree fruits, potato, hops, sweet corn, and onion are intensively grown, irrigated crops of major importance east of the Cascades. Vineyards have long been a major California crop and are emerging as an important crop in central Washington.

The agriculture of California is particularly diverse, producing over 200 different crops on its varied soils and climates. Over two thirds of its more than 3 million ha of irrigated farmlands are in the central valley, producing a wide variety of crops, many in double-crop rotations. The Imperial Valley in southern California, with its year-long frost-free growing season, produces many off-season, high-value crops. Production of cool-season vegetable crops, quality wine grapes, pasture, and timber products is allowed by the sea-moderated climates of the coastal valleys. Individual crops, listed in order of decreasing cash farm income for 1980, were cotton, grapes, alfalfa, nursery products, almond, tomato, rice, flowers and foliage, lettuce, wheat, and orange (California Dep. of Food and Agric., 1981).

Crop rotations are seldom followed in the Pacific Northwest or in California. Grain farmers in the area between the Cascades and the Rocky Mountains typically have grain as the single major cash crop. Winter barley and wheat are grown in sequence with tomato and other row crops in many central California areas. Pea is used as an alternate cash crop for a portion of the acreage in the foothills of the Blue Mountains and in the Palouse area of eastern Washington and northern Idaho. Potato production is concentrated in the semiarid soils that have been brought under irrigation in the Columbia Basin in central Washington, the Klamath Basin in south-central Oregon and northern California, the irrigated areas in central Oregon, and the Snake River Valley in southern Idaho. Early-harvest potato is concentrated in central and southern California.

Forage crops are produced for livestock throughout the West. Dairy enterprises are concentrated in western Oregon and Washington and in the irrigated areas of central Washington, southern Idaho, and California. The northeastern parts of Washington and northern Idaho support mixed livestock enterprises, cereal production, and grass seed production. Poor nodulation and low production of legumes have been problems on the moderately acid soils throughout this latter area for many years (Harder et al., 1962; Reisenauer, 1963; Pumphrey & Moore, 1973).

In the areas west of the Cascades, crimson and red clover seed crops are generally produced in a cropping sequence with winter wheat. Oat and grass seed crops followed by winter wheat minimizes wheat diseases and has developed as a good crop sequence in western Oregon. The farmers that produce small fruits and vegetable crops frequently specialize in intensive crops and do not follow rotations. Winter annual cover crops are often grown in association with the more intensive, annual processing crops for erosion control and soil organic matter maintenance. Vegetable crops followed by winter wheat has developed as a very desirable crop sequence for those vegetable growers with enough hectares of cultivated soils to allow efficient wheat production. Root disease problems are minimized on both crops, weed problems are reduced, and the fertilizer carryover from the vegetable crop results in maximum wheat yields with relatively low production costs.

A major portion of the lime that is applied in the Pacific Northwest is for the production of vegetable crops and forage legumes. Alfalfa, red clover, crimson clover, white clover, and subterranean clover are the main legumes grown in this area.

## III. LIMING PRACTICES

### A. Lime Use and Method of Application

Records from the state departments of agriculture in Oregon, Washington, and California show that annual consumption of lime has averaged about 225 000 t $yr^{-1}$ in Oregon and about 135 000 t $yr^{-1}$ in Washington and 150 000 t $yr^{-1}$ in California. Minimal usage of lime in the region has been attributed to its comparatively high cost and the use of phosphate fertilizer to moderate the acid soil problems of intensively cropped areas. A major portion of the limestone marketed in the Pacific Northwest is calcitic limestone. A small tonnage of dolomitic limestone is shipped to Washington and Oregon from Oakland, California, for use on soils that are low in Mg.

Lime is most frequently broadcast before or during the latter stages of seedbed preparation and worked into the seedbed with a disk. Mixing of lime throughout the upper 15 to 30 cm of soil is achieved during seedbed preparation and cultivation operations for the production of succeeding crops.

Recent research in western Oregon (Kauffman & Gardner, 1978) has shown that limited mixing of low lime rates in the surface 5 to 10 cm of soil results in pockets of soil with a desirable pH and maximizes yields with limited expense for the immediate crop. However, subsequent tillage operations provide more thorough mixing and reduce responses from low rates, e.g., 1 t $ha^{-1}$ applications. In some cases, plowing to a depth of 25 to 30 cm for the next crop, following a surface lime application and light disking, will place most of the lime at the bottom of the plow furrow, leaving the surface 20 cm at the original low pH and with the original soil acidity problems for seedling establishment.

Low rates of lime combined with minimum soil mixing may be an alternative production practice where land is rented and only one crop is produced every 4 to 5 yr that has a low tolerance for acid soils (e.g., lettuce, spinach, table beet, and cabbage family crops).

More than one plowing plus disking is necessary to provide complete mixing of lime with the soil. If complete mixing is important, it may be preferable to apply lime 6 months to 1 yr before planting to allow time for additional tillage operations during normal crop production practices to mix lime throughout the soil.

The frequency with which lime should be applied has not been established for western conditions. The lack of definite rotations that have a fairly specific lime requirement and the wide variation in the amount of acid-forming fertilizers applied make it difficult to establish a recommendation for frequency of application that would cover the range of conditions encountered. It is presently recommended that farmers submit soil samples from each field to a soil testing laboratory every 2 to 6 yr for an estimation of the lime that should be applied for the specific crops or crop rotation to be used. It is important to recognize that where intensive agricultural production is being followed with high N rates (100–200 kg of N $ha^{-1}$ $yr^{-1}$) as $NH_4^+$-N, animal manure, or sewage sludge, decreases in soil pH of 0.5 to 1.0 unit in 6 to 7 yr with a corresponding reduction in basic cations throughout the surface 25 to 30 cm of soil should be expected where pH values are already $< 6.0$.

### B. Soil Acidity and Lime Recommendations

Soil pH, exchangeable $Ca^{2+}$, and percent basic cation saturation have been used as a basis for recommending application of lime in this area. However, no single measurement has proved to be a reliable criterion for predicting response from liming under a wide range of soil conditions. This would be anticipated when the range of soil acidity conditions, soils, and the range in the tolerance of different crops to soil acidity factors are considered. Responses to lime in the region have been associated with reduction of excessive quantities of Mn, Fe, and Al in the soil (Jackson et al., 1966; Janghorbani et al., 1975; Lingle et al., 1961; Stephenson, 1929), increased availability of soil P, and correction of Mo deficiency (James et al., 1967; Reisenauer, 1955; 1963).

Extension workers, consultants, and most growers recognize that lime recommendations are not precise and that soil acidification is a continuing process. Whenever a visual response from lime is observed, crops with comparable tolerance to soil acidity have been suffering reduced yields on that field for a number of years. Recommendations normally are made to apply enough lime to achieve a specific soil pH or percent base saturation, depending on the crops that will be grown. The most common rates recommended for one application are 3 to 5 t of lime $ha^{-1}$. Rates of 6 to 10 t $ha^{-1}$ are sometimes required for alfalfa on hill soils that have not been limed previously and where growers want to raise soil pH to 6.0 or higher to help control clubroot on cabbage, broccoli, and cauliflower and to produce vegetable crops with limited tolerance to soil acidity.

Application rates of 1.0 to 2 t $ha^{-1}$ of lime are recommended for strawberry plants where soils are low in Ca and are acid (Kirsch & Jackson, 1959). Dolomitic lime is recommended where soils are low in Mg or have questionable Mg levels (Mortensen et al., 1961).

## IV. LIME AND PLANT NUTRIENT INTERACTIONS

Lime-phosphorus interactions have been observed over a wide range of soil and cropping conditions. Increasing the pH of acid soils for legume production increases both P and Mo availability. Increased $N_2$ fixation by rhizobia is associated with higher levels of P and Mo; also, the soil pH range for optimum efficiency of some $N_2$-fixing rhizobia is generally recognized as 6.0 to 8.0. All of these factors complicate the problem of identifying specific cause and effect relationships associated with crop response to liming.

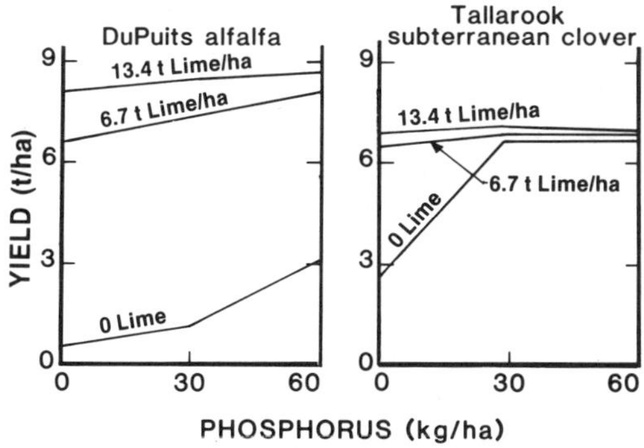


Fig. 3—Lime and P effects on yield of alfalfa and subterranean clover, Red Soils Exp. Stn., 1956–1961.

Responses of alfalfa and subterranean clover to liming illustrate (Fig. 3) the marked differences in tolerance to soil acidity of different legume species (Jackson et al., 1964). Response to liming and lime-P interactions for white and red clover are generally similar to that for subterranean clover.

A number of vegetable crops, with limited tolerance to both Al and Mn, frequently show marked increases in P uptake following lime application in western Oregon. Table beet, lettuce, and spinach are some of the less acid-tolerant crops that exhibit this response pattern (Jackson et al., 1974; Hemphill & Jackson, 1982). Rutabaga represents the response of more acid-tolerant root crops (Table 2).

The marked increases in yields of table beet and spinach to liming are associated with increases in P concentration of leaf samples and marked decreases in Mn concentrations. Rutabaga response from lime was much smaller and the leaves showed smaller changes in P and Mn concentrations. The decrease in Ca concentration in table beet and spinach leaves is associated with marked increases in yield and is a dilution of the Ca content of low-yielding plants. This decrease in leaf-Ca concentration with the first

Table 2—Chemical analyses of leaf samples and yield with different rates of lime, North Willamette Exp. Stn. (Jackson et al., 1974).

| Crop | Treatment† | pH | Ca | Mg | P | K | Zn | Mn | Yield |
|---|---|---|---|---|---|---|---|---|---|
| | | | g kg$^{-1}$ | | | | mg kg$^{-1}$ | | kg m$^{-1}$ |
| Table beet | $L_0$ | 5.8 | 20.4 | 12.4 | 5.1 | 69.1 | 71 | 328 | 1.34 |
| | $L_9$ | 6.3 | 17.7 | 9.3 | 6.3 | 70.8 | 57 | 78 | 1.88 |
| | $L_{18}$ | 6.6 | 21.9 | 9.5 | 6.5 | 67.5 | 45 | 71 | 1.77 |
| Spinach | $L_0$ | 5.8 | 19.5 | 11.7 | 4.6 | 63.0 | 164 | 204 | 0.06 |
| | $L_9$ | 6.3 | 18.2 | 10.8 | 6.7 | 78.0 | 137 | 70 | 0.25 |
| | $L_{18}$ | 6.6 | 20.8 | 9.4 | 6.8 | 76.2 | 113 | 63 | 0.25 |
| Rutabaga | $L_0$ | 5.8 | 21.8 | 2.8 | 5.3 | 51.1 | 34 | 70 | 1.11 |
| | $L_9$ | 6.3 | 24.5 | 2.5 | 5.2 | 57.8 | 31 | 42 | 1.22 |
| | $L_{18}$ | 6.6 | 22.8 | 2.2 | 5.8 | 47.0 | 32 | 42 | 1.29 |

† kg × $10^3$ ha$^{-1}$ of lime applied.

increment of lime applied emphasizes that very few lime responses are associated with deficiencies of Ca as a nutrient.

A range of yield responses was reported for snap bean, carrots, and lettuce (Hemphill & Jackson, 1982) on this same soil series. Lettuce response was comparable to that of spinach, whereas response of carrot tended to follow the rutabaga response discussed earlier. Snap bean was intermediate in response. There was a 25 to 30% yield increase when the soil pH was increased from 5.1 to 5.6 or 5.7, but yield increase was limited with additional application of lime. Lettuce yields were increased 30 to 50% when the soil pH was increased from 5.6 or 5.7 to > 6.0 with applications of lime. These increases in yield were associated with significant increases in leaf P and decreases in leaf Mn. Calcium concentrations in lettuce leaf samples were consistently decreased when the first increment of lime increased yields two to three times. In contrast, moderate increases in snap bean and carrot yields from liming resulted in small, but consistent, increases in leaf-Ca concentrations.

The important relationship between soil pH and response to P applications was illustrated in greenhouse experiments with 448 different California soils of pH 5.0 to 8.0 (Jenny et al., 1950). Growth response of 'Romaine' lettuce to applications of N and P was evaluated. Lettuce responded to P applications in about 80% of the soils of pH < 5.0. Response occurred in 40% of the soils of pH about 6.5 and in 30% when pH was 7.0 to 7.3. Obviously, residual P also influenced response from P, but the soil pH-P response relationship was striking.

The increase in soil P availability associated with liming acid soils requires a moderate to high P soil test value. Guerrero et al. (1967) found that actual increases in yield of 10 different grass species from P application were greater on two acid grassland soils after these soils were limed. There was a response from both P and lime; the lime plus P treatments yielded five to eight times more than the lime-zero, P-zero treatment in these experiments.

## V. MANGANESE TOXICITY

Recent research to evaluate crop production practices on poorly drained soils in the Willamette Valley has shown that Mn toxicity is one of the major factors limiting production of certain crops on these soils. This work (Jackson et al., 1966) indicated Mn toxicity occurred in bush snap bean when the most recently matured trifoliate leaf contained 600 to 800 mg $kg^{-1}$ or more of Mn. Liming poorly drained soils from pH 4.9 to 6.3 reduced the Mn content of snap bean leaves from 1010 to 620 mg $kg^{-1}$. Improved drainage and aeration also reduced the levels of exchangeable $Mn^{2+}$ and the amount of Mn taken up by bean plants grown on these soils. Thus, Mn toxicity on acid soils is accentuated by restricted drainage.

When alfalfa is grown on different western Oregon soils, marked differences have been observed in the relationship between soil pH, or percent basic cation saturation, and response of alfalfa to lime. This range in response from lime has been related, in part, to differences in levels of exchangeable and readily reducible $Mn^{2+}$ in these soils (Dawson, 1958). About 40 mg $kg^{-1}$ of $NH_4OAc$–exchangeable $Mn^{2+}$ were present in two soils where there was a marked reduction in Mn content of alfalfa associated with the increases in yield from liming in greenhouse experiments. The Mn content of the alfalfa foliage was 100 to 130 mg $kg^{-1}$ with 60% base saturation and 49 to 55 mg $kg^{-1}$ with 100% base saturation. Other research has also identified that excessive Al levels limit alfalfa production on a number of these soils (Janghorbani et al., 1975; Vlamis & Williams, 1962).

Benson and Vandecaveye (1951) reported Mn toxicity on 'Delicious' apple trees where surface soils were below pH 5.0 and Mn content of leaves collected in midsummer or later was 500 mg $kg^{-1}$ or more. The critical level for Mn toxicity in Delicious apple leaves collected in midsummer or later was later revised downward to 300 mg $kg^{-1}$ by Benson and Woodbridge (1961). These authors pointed out that all orchards with surface soils of pH 5.0 did not result in Mn toxicity and that waterlogging also contributed to increased availability of soil Mn. They recommended adding enough lime to raise the soil pH to 6.0 on the area within a 2-m radius of young trees showing Mn toxicity. A comprehensive evaluation would be required to identify the relative contribution of low soil pH and reduced drainage to Mn toxicity, especially where low soil pH is associated with drip line applications of acid-forming fertilizers. Relatively high levels of hydrous oxides of Fe and Al have undoubtedly contributed to the response observed from application of lime in some of these situations.

## VI. SUMMARY

We tend to think of the calcareous soils in the western United States with micronutrient deficiency problems associated with high soil pH levels. However, there are extensive areas of naturally occurring acid soils west of the Cascade mountains, in or associated with other mountain ranges, along the California coast, and where the use of acidifying materials on intensively managed soils have decreased the soil pH and induced soil acidity problems. There is a definite need for increasing the application of liming materials in many parts of the region.

## REFERENCES

Aldrich, D. G., E. R. Parker, and H. D. Chapman. 1945. Effects of several nitrogenous fertilizers and soil amendments on the physical and chemical properties of an irrigated soil. Soil Sci. 59:299–312.

Barshad, I. 1964. Chemistry of soil development. p. 1–70. *In* F. E. Bear (ed.) Chemistry of the soil. Van Nostrand Reinhold Co., New York.

Benson, N. R., and S. C. Vandecaveye. 1951. Soil fertility conditions in the apple orchards of north central Washington. Washington Agric. Exp. Stn. Bull. 527.

Benson, N. R., and C. G. Woodbridge. 1961. Apple measles. p. 153–155. *In* Proc. 57th Annual Meeting Washington State Horticulture Assoc., Wenatchee, Wash. 3–5 December. Washington State Horticulture Assoc., Wenatchee.

Blosser, D. L., and H. Jenny. 1971. Correlations of soil pH and percent base saturation as influenced by soil forming factors. Soil Sci. Soc. Am. Proc. 35:1017–1018.

Brown, B. D. 1975. Dissimilarities among naturally occurring and agriculturally acidified soils. M.S. Thesis. Univ. of California, Davis.

California Department of Food and Agriculture. 1981. Fertilizing materials tonnage report 1980. California Dep. of Food and Agric., Sacramento.

Clark, J. S. 1966. The relation between pH and soluble and exchangeable Al in some acid soils. Can. J. Soil Sci. 46:94–96.

Cole, D. W., S. P. Gessell, and S. F. Dice. 1967. Distribution and cycling of nitrogen, phosphorus, potassium and calcium in a second growth Douglas fir ecosystem. p. 198–232. *In* H. E. Young (ed.) Symposium on primary production and mineral cycling in natural ecosystems. Univ. of Maine Press, Orono.

Dawson, M. D. 1958. Influence of base saturation and calcium levels on yield and mineral content of alfalfa. Soil Sci. Soc. Am. Proc. 22:328–333.

Guerrero, F. P., W. A. Williams, and W. E. Martin. 1967. Yield and mineral composition of grass species grown on acid grassland soils. J. Range Manage. 20:107–108.

Harder, R. W., G. R. Anderson, and C. T. Brackney. 1962. The influence of lime on the growth of alfalfa in northern Idaho. Idaho Agric. Res. Progress Rep. 60.

Hargett, N. L., and J. T. Berry. 1981. 1980 fertilizer summary data. Tennessee Valley Authority Bull. Y-165.

Hemphill, D. D., Jr., and T. L. Jackson. 1982. Effect of soil acidity and nitrogen on yield and elemental concentration in bush bean, carrot, and lettuce. J. Am. Soc. Hortic. 107:740–744.

Jackson, T. L., and G. E. Carter. 1976. Nutrient uptake by Russet Burbank potatoes as influenced by fertilization. Agron. J. 68:9–12.

Jackson, T. L., H. H. Rampton, and J. McDermid. 1964. The effect of lime and phosphorus on the yield and phosphorus content of legumes in western Oregon. Oregon Agric. Exp. Stn. Tech. Bull. 83.

Jackson, T. L., W. A. Sheets, N. S. Mansour, and H. J. Mack. 1974. Lime: Response in spinach and other vegetables. Oreg. Veg. Dig. 23(2):1–2.

Jackson, T. L., D. T. Westermann, and D. P. Moore. 1966. The effect of chloride and lime on the manganese uptake by bush beans and sweet corn. Soil Sci. Soc. Am. Proc. 30:70–73.

James, D. W., T. L. Jackson, and M. E. Harward. 1967. Effect of molybdenum and lime on the growth and molybdenum content of alfalfa grown on acid soils. Soil Sci. 105:397–402.

Janghorbani, M., S. Roberts, and T. L. Jackson. 1975. Relationship of exchangeable acidity to yield and chemical composition of alfalfa. Agron. J. 67:350–354.

Jenny, H., J. Vlamis, and W. E. Martin. 1950. Greenhouse assay of fertility of California soils. Hilgardia 20:1–8.

Kauffman, M. D., and E. H. Gardner. 1978. Segmental liming of soil and its effect on the growth of wheat. Agron. J. 70:331–336.

Kirsch, R. K., and T. L. Jackson. 1959. Fertilizing strawberries in the Willamette Valley. Oregon Agric. Exp. Stn. Circ. of Information 594.

Lindsay, W. L. 1979. Chemical equilibria in soils. John Wiley & Sons, Inc., New York.

Lingle, J. C., R. H. Sciaroni, B. Lear, and J. R. Wight. 1961. The effect of soil liming and fumigation on the manganese content of Brussels sprouts leaves. Proc. Am. Soc. Hortic. Sci. 78:310–318.

Lorenz, O. A., and C. J. Johnson. 1953. Nitrogen fertilization as related to the availability of phosphorus in certain California soils. Soil Sci. 75:119–129.

Mahler, R. L. 1982. Increasing soil acidity could reduce yields in the inland Pacific Northwest. Better Crops Plant Food 66:37–40.

Mortensen, W. P., A. S. Baker, J. C. Dodge, P. C. Crandall, and M. W. Carstens. 1961. Western Washington strawberry fertilizer recommendations. Washington State Univ. FR-21.

Pierre, W. H., and W. L. Banwart. 1973. Excess-base and excess-base/nitrogen ratio of various crop species and parts of plants. Agron. J. 65:91–96.

Pratt, P. F., and F. L. Bair. 1962. Cation-exchange properties of some acid soils of California. Hilgardia 33:689–706.

Pumphrey, F. V., and D. P. Moore. 1973. Sick alfalfa. Oregon Agric. Exp. Stn. Circ. of Information 644.

Reisenauer, H. M. 1955. Molybdenum content of alfalfa in relation to deficiency symptoms and response to molybdenum fertilization. Soil Sci. 81:237–242.

Reisenauer, H. M. 1963. The effect of sulfur on the absorption and utilization of molybdenum by peas. Soil Sci. Soc. Am. Proc. 27:553–555.

Scholl, W., and H. M. Wallace. 1950. Commercial fertilizers consumption in the United States, 1949–50. USDA-ARS, Bureau of Plant Industry, Soils and Agric. Engineering, Beltsville, Md.

Silkworth, D. R., and D. F. Grigal. 1982. Determining and evaluating nutrient losses following whole-tree harvesting of aspen. Soil Sci. Soc. Am. J. 46:626–631.

Stephenson, R. E. 1929. Colloidal properties of Willamette Valley soils. Soil Sci. 28:235–247.

Stewart, J. I. 1975. Irrigation in California. Rep. to California State Water Resources Control Board, Agreement no. 2-2-65.

Yuan, T. L. 1959. Determination of exchangeable hydrogen in soils by a titration method. Soil Sci. 88:164–167.

Vlamis, J., and D. E. Williams. 1962. Liming reduces aluminum and manganese toxicity in acid soils. Calif. Agric. 16(9):6–7.

Westwood, M. N., E. S. Degman, and J. H. Grim. 1964. Effects of long-term fertilizer and management practices on growth and yield of pears grown on a clay adobe soil. Oregon Agric. Exp. Stn. Tech. Bull. 82.

# 9 Crop Response to Lime on Soils in the Tropics

**EUGENE J. KAMPRATH**
*Department of Soil Science*
*North Carolina State University*

Soils in the humid tropics in their natural state are generally acid with reactions of < pH 5 (Pearson, 1975). In many early studies to increase food production, these soils were often limed to near neutrality because this was the practice followed on the highly productive Mollisols and Alfisols in the temperate regions. Often poor results were obtained, and with some soils, large, uneconomical amounts of lime were required to bring the soil to pH 7.

The concepts about the liming of highly weathered soils underwent considerable change after the findings in the 1950s that acid soils contained more exchangeable $Al^{3+}$ than $H^{+}$. Primary consideration was given to removal of the soil acidity factors limiting growth rather than liming to pH 7.

## I. SOILS

### A. Climatic Regimes

Climate is an important factor in soil formation through its effects on weathering of parent material, leaching of basic cations, and biological activity. High rainfall is generally associated with the tropics. Only about one fourth of the tropics has a humid climate (Table 1). The soil moisture regimes in the humid regions are udic, which indicates that moisture stress is generally not a problem most of the year. The predominant vegetation is rain forest. The principal land areas in the humid regime are the upper Amazon basin, Congo basin, Indonesia, and Malaysia (Sanchez, 1976). Over three fourths of the soils in the humid regime are highly weathered and leached. These soils are expected to be acid and low in bases.

The subhumid climatic regime encompasses about one half of the land area in the tropics (Table 1). The soil moisture regime is ustic, which indicates a marked dry season of 3 to 6 months duration. The predominant vegetation in the subhumid regions are savanna and deciduous forests. The main geographical areas that have a subhumid climate are the central


*Soil Acidity and Liming*—Agronomy Monograph no. 12 (2nd Edition).

Table 1—The proportion of the land area in the tropics in the humid and subhumid regions (Sanchez, 1976).

| Climate | Soil moisture regime | Predominant vegetation | Proportion of land area, % |
|---|---|---|---|
| Humid | Udic | Rain forest and forest | 24 |
| Subhumid | Ustic | Savanna and deciduous forest | 49 |

plateau of Brazil, Llanos of Colombia and Venezuela, the area of Africa south of the Sahara Desert and north of the Kalahari Desert except for most of the Congo basin, and most of India (Sanchez, 1976). Approximately two thirds of the soils in the subhumid region are highly weathered and leached and are expected to have acidity problems (Table 2).

### B. Soil Taxonomy

Base saturation has been used as a criterion for classification at the great group level for Inceptisols and Oxisols and as a criterion of the Ultisol order. Except for the great group Euthrorthox, only the base saturation of a subsurface horizon is used for classification. Generally when the subsurface is acid, it is a good indication that the surface soil is acid in the natural state, and if the soil has been limed, it will again become acid with time.

Some general conclusions can be made about the use of soil taxonomy for identifying soils with acidity problems (Table 3). Ultisols in their natural state will be acid and will need to be limed for intensive agriculture. The great groups with the prefixes "dystr" and "hapl" can be acid, ranging from highly to slightly, because the subsurface base saturation can range from 0 to 35 or 50 or 60%. The Ultic subgroups of the Alfisol order also have a low base saturation in the subsurface horizon and therefore can have acidity problems.

The suborder "acr" identifies soils that are highly weathered and have a very low cation exchange capacity (CEC) and are low in exchangeable bases. Three suborders of the Oxisols are in this criterion (Table 3). These soils are very low in exchangeable $Ca^{2+}$ and $Mg^{2+}$ and also can have a relatively low pH. Because of their very low CECs, these soils may be very low in exchangeable acidity, where crop response to liming would be primarily to additions of Ca and Mg rather than pH changes.

Table 2—Proportion of soils in the humid and subhumid regions of the tropics that are highly weathered and leached (Sanchez, 1976).

| Climatic regime | Proportion of highly weathered, leached soils | |
|---|---|---|
| | Within a climatic regime | All tropical soils |
| | % | |
| Humid | 78 | 19 |
| Subhumid | 64 | 31 |
| Total | | 50 |

Table 3—Soils in tropical regions that are inherently acid or low in bases.

| Acid soils | Low CEC and low bases |
|---|---|
| Ultisols | Acrustox |
| Haplorthox | Acrorthox |
| Haplustox | Acrohumox |
| Dystropepts | |
| Dystrandepts | |
| Ultic Paleustalfs | |
| Ultic Haplustalfs | |
| Ultic Hapludalfs | |

## C. Cation Exchange Properties

The colloids of highly weathered soils have a constant surface potential, which is characteristic of kaolinite, Al-interlayered chlorites, hydrous oxides of Fe and Al, and Al–organic matter complexes (Bell & Gillman, 1978). The surface potential is determined by the activity of the potential-determining $H^+$ and $OH^-$ ions in the bulk solution (Van Raij & Peech, 1972). As a consequence, the CEC of highly weathered soils is pH dependent. Therefore, measurement of CEC of highly weathered soils using neutral $NH_4OAc$ gives values that are valid only at pH 7. More realistic values of the CEC are obtained by summing the exchangeable $Ca^{2+}$, $Mg^{2+}$, $K^+$, and the KCl-extractable acidity. This has been called the effective cation exchange capacity (ECEC) and more accurately reflects the active CEC of the soil at the particular soil pH.

The ECECs of highly weathered soils in the tropics are quite low (Table 4). The CEC of B horizons of Oxisols was lower than that of the A horizon. Adjusting the surface soils to pH 7 increased their ECECs by as much as fourfold. The large pH-dependent charge of the highly weathered soils has important significance as to the type of liming practices to follow. Considerably more lime will be required to adjust the pH of these acid soils to neutrality than is required to neutralize exchangeable $Al^{3+}$.

Table 4—The effective cation exchange capacity (ECEC) of Oxisols and Ultisols as influenced by pH.

| Soil | Horizon | Clay | C | pH | ECEC | Reference |
|---|---|---|---|---|---|---|
| | | — g kg$^{-1}$ — | | | cmol(p$^+$) kg$^{-1}$ | |
| Typic Haplorthox | A | 123 | 17 | 4.2 | 2.2 | Morais et al. (1976) |
| | | | | 7.0 | 5.6 | |
| | B | 201 | 6 | 4.9 | 0.7 | |
| | | | | 7.0 | 2.6 | |
| Haplic Acrorthox | A | 500 | 40 | 3.8 | 3.7 | Morais et al. (1976) |
| | | | | 7.0 | 14.1 | |
| | B | 666 | 6 | 4.8 | 1.0 | |
| Ustoxic Paleustults | A | -- | -- | 4.7 | 1.72 | Juo & Uzu (1977) |
| | | | | 7.1 | 5.95 | |
| Oxic Paleudults | A | 170 | 104 | 5.0 | 2.72 | Juo & Ballaux (1977) |
| | | | | 6.0 | 2.83 | |

Table 5—The amount of Al complexed by organic matter in acid soils of the tropics (Juo & Kamprath, 1979).

| Soil | Location | Organic matter | pH | Complexed Al |
|---|---|---|---|---|
| | | $g\ kg^{-1}$ | | $cmol(p^+)\ kg^{-1}$ |
| Haplorthox | Zaire | 31 | 3.7 | 1.54 |
| Haplorthox | Brazil | 99 | 3.5 | 4.19 |
| Haplorthox | Brazil | 32 | 4.6 | 3.65 |
| Haplustox | Colombia | 41 | 4.7 | 9.0 |
| Paleudults | Brazil | 27 | 4.7 | 3.08 |
| Paleudults | Nigeria | 18 | 4.2 | 2.39 |
| Paleustults | Nigeria | 21 | 4.6 | 2.46 |

One reason for the low CEC of acid soils in the tropics is that a large part of the negative charge of organic matter is countered by $Al^{3+}$ (Table 5). The Al complexed by organic matter is not exchangeable with cations of unbuffered salt solutions. However, with an increase in soil pH on the addition of lime, the complexed Al undergoes hydrolysis and results in an increase in the effective CEC. The Al complexed by organic matter is an important buffer in surface soils, particularly after the KCl-exchangeable $Al^{3+}$ has been neutralized.

## II. FACTORS OF ACID SOIL INFERTILITY

### A. Hydrogen Toxicity

Poor plant growth on acid soils is generally associated with a low soil pH. Islam et al. (1980), however, point out that the effects of soil pH on growth are complex and that it is difficult to separate the direct effects of $H^+$ from the indirect effects associated with changes in the solubility and availability of various elements affecting plant growth. Solution-culture experiments provide a means for studying the direct effects of $H^+$. Islam et al. (1980) concluded from a review of the literature and from their own experiments that there were two main effects of $H^+$ on growth: (i) injury to roots and (ii) decreased uptake of cations.

Root injury was observed at pH 4 and lower. At these low levels, lateral root development was suppressed and in some cases root tips were killed (Islam et al., 1980). The roots were a discolored brown or a dull gray. At low pH and low Ca concentrations, damage to root membranes that results in leakage of ions from the roots can occur.

Cation uptake is suppressed in the presence of high $H^+$ concentrations. Numerous studies have shown that as $H^+$ concentration increases, the uptake of Ca, Mg, and K is decreased (Islam et al., 1980). At pH 5.5, however, they found no suppression of cation uptake.

Relatively high $H^+$ ion concentration could be an important factor limiting plant growth in certain soils of the tropics. Some soils in their natural condition are extremely acid, having pH values of 4 or less (Table 6). The pH of acid soils can also be lowered considerably when fertilizers

Table 6—The exchangeable $Al^{3+}$, Al saturation, and pH of representative acid soils in the tropics.

| Region | No. of samples | Soil | pH range | Exchangeable $Al^{3+}$ Range | Exchangeable $Al^{3+}$ Avg | Al saturation Range | Al saturation Avg |
|---|---|---|---|---|---|---|---|
| | | | | cmol ($p^+$) $kg^{-1}$ | | % | |
| Brazil | | | | | | | |
| Sao Paulo† | 5 | Typic Haplustox | 4.0–4.6 | 1.8–3.2 | 1.7 | 38–90 | 74 |
| | 3 | Oxisols | 4.1–4.2 | 0.7–1.9 | 1.2 | 83–90 | 88 |
| Amazon‡ | 3 | Haplorthox | 3.6–3.8 | 1.9–3.8 | 2.7 | 73–85 | 79 |
| Colombia§ | 26 | Oxisols | 4.2–5.1 | 0.6–5.8 | 3.1 | 29–86 | 68 |
| Panama¶ | 6 | Oxisols | 4.8–5.2 | 0.3–5.8 | 2.5 | 44–68 | 57 |
| Puerto Rico# | 5 | Ultisols | 3.9–4.6 | 4.5–9.9 | 6.6 | 36–70 | 57 |
| | 3 | Oxisols | 4.2–4.8 | 0.4–1.1 | 0.8 | 18–40 | 33 |
| West Africa†† | 21 | Ultisols | 5.1–5.4 | 0 –2.0 | 0.5 | 0–43 | 13 |
| | 26 | Ultisols | 4.5–5.0 | 0.05–3.7 | 1.3 | 2–69 | 41 |
| | 27 | Ultisols | 4.0–4.4 | 0.05–2.8 | 1.2 | 1–67 | 35 |
| Malaysia‡‡ | 8 | -- | 3.6–4.2 | 0.7–12.4 | 3.4 | 34–86 | 61 |

† Pratt and Alvahydro (1966); ‡ Ramos and Velloso (1977); § Vargas (1964); ¶ Mendez and Kamprath (1978); # Abruna et al. (1975); †† Juo (1977); ‡‡ Singh and Talibudeen (1969).

are applied because of the replacement of Al from exchange sites and subsequent hydrolysis of the Al in solution. In these situations, root injury due to $H^+$ could be expected. Also, at pH 5 and below, $H^+$ concentration could suppress uptake of both Ca and Mg and result in deficiencies of these elements, particularly in soils with very low CECs.

## B. Aluminum Toxicity

The exchangeable cation population in very acid soils of the tropics consists primarily of KCl-exchangeable $Al^{3+}$. A large portion of the effective CEC of these acid soils is countered by $Al^{3+}$ (Table 6). When the exchangeable $Al^{3+}$ saturation was > 60%, there was a large increase in soil solution Al (Nye et al., 1961; Evans & Kamprath, 1970). Concentration of soil solution Al in an Oxisol from Brazil increased very markedly when the Al saturation was > 45% (Fig. 1).

The poor growth of plants in acid soils has been associated with the concentration of Al in the soil solution. Growth of corn was reduced when the concentration of Al in the soil solution was > 0.4 mmol ($p^+$) $L^{-1}$ (4 ppm) (Evans & Kamprath, 1970; Friesen et al., 1980b). Soybean growth was decreased at Al concentrations > 0.2 mmol ($p^+$) $L^{-1}$ (2 ppm) (Evans & Kamprath, 1970). The growth of six *Stylosanthes* species was severely restricted at a soil solution Al concentration of 0.17 mmol ($p^+$) $L^{-1}$ (1.5 ppm) (de Carvalho et al., 1980).

The soil solution Al concentrations of very acid soils in the tropics are at toxic levels (Table 7). These solution levels of Al were associated with ex-

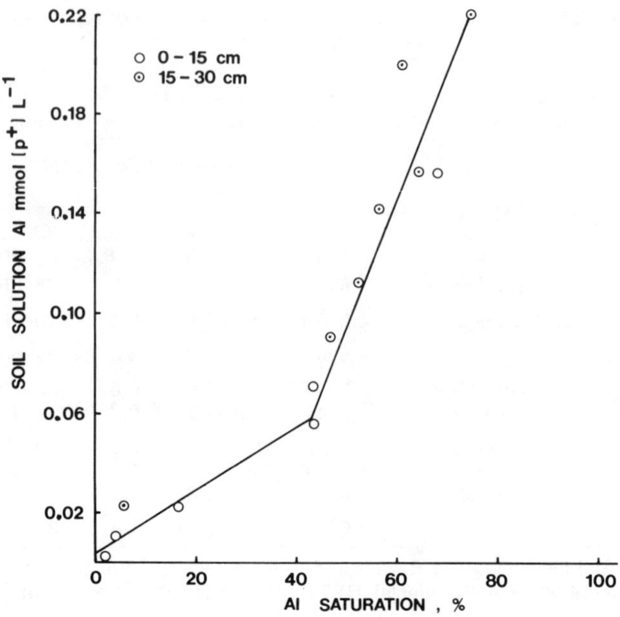


Fig. 1—Relationship between Al saturation and concentration of Al in the soil solution of an Oxisol (Gonzalez-Erico et al., 1979).

changeable $Al^{3+}$ saturation of 60% or greater. One important point that needs to be considered in the management of these soils is that when recommended amounts of N and K were applied for high yields, there was a large increase in the soil solution Al concentration (Table 7). Thus, at relatively low rates of fertilization, certain genotypes may appear to be tolerant to Al, but at high rates of fertilization, growth may be very much depressed relative to limed treatments because of the large increase in solution Al.

Table 7—Exchangeable $Al^{3+}$ and $Ca^{2+}$ and soil-solution Al and Ca of acid, highly weathered, surface soils.

| Area | Soil | pH | Exchangeable $Ca^{2+}$ | Exchangeable $Al^{3+}$ | Saturation Ca | Saturation Al | Soil solution Ca | Soil solution Al |
|---|---|---|---|---|---|---|---|---|
| | | | cmol $(p^+)$ $kg^{-1}$ | | — % — | | mmol $(p^+)$ $L^{-1}$ | |
| Brazil† | Haplustox | 4.5 | 0.45 | 1.15 | 26 | 70 | 1.14 | 0.16 |
| Puerto Rico‡ | | 4.1 | 0.44 | 5.35 | 6 | 77 | 3.24 | 0.31 |
| Nigeria§ | Paleudults | 4.2 | 0.26 | 1.83 | 9 | 64 | -- | 0.30 |
| Nigeria¶ | Ustoxic | | | | | | | |
| | Paleustults | | | | | | | |
| | Unfertilized | 4.7 | 0.21 | 1.88 | 7 | 64 | -- | 0.07 |
| | Fertilized | 4.2 | -- | -- | -- | -- | -- | 1.23 |
| | Oxic Paleudults | | | | | | | |
| | Unfertilized | 4.3 | 0.21 | 1.79 | 6 | 60 | -- | 0.30 |
| | Fertilized | 4.1 | -- | -- | -- | -- | -- | 1.13 |

† Gonzalez-Erico et al. (1979); ‡ Brenes and Pearson (1973); § Juo (1977); ¶ Friesen et al. (1980a).

Table 8—Growth of plants in acid soils with a high saturation of exchangeable $Al^{3+}$ compared with limed treatments.

| Area | Soil | Al saturation | Crop | Relative growth | References |
|---|---|---|---|---|---|
| | | % | | % | |
| Brazil | Oxisols | 72 | Corn | 48 | Gonzalez-Erico et al. (1979) |
| Puerto Rico | Ultisols | 65 | Corn | 20 | Abruna et al. (1975) |
| Nigeria | Ultisols | 45 | Corn | 30 | Friesen et al. (1980b) |
| Brazil | Oxisols | 64 | Wheat | 56 | Martini et al. (1977) |
| Brazil | Oxisols | 53 | Soybean | 44 | Martini et al. (1974) |
| Puerto Rico | Ultisols | 65 | Soybean | 3 | Abruna et al. (1978) |
| Puerto Rico | Ultisols | 60 | Bean | 30 | Abruna et al. (1975) |
| Puerto Rico | Ultisols | 66 | Sweet potato | 22 | Abruna et al. (1979) |
| Colombia | Oxisols | 95 | *Brachiaria decumbens* Stapf | 35 | CIAT (1980) |

Soils with 60% Al saturation or greater have soil solution Al concentrations that are detrimental to plant growth. The growth of various crops on acid soils with a high Al saturation compared with the limed soils is given in Table 8. In most instances, growth on the acid soils was < 50% of that on limed soils when the Al saturation was > 60%. It must be kept in mind, however, that the tolerance to Al can vary widely among varieties of the same species. The data from the studies cited in Table 8 strongly support the idea that Al toxicity is a major cause of poor growth on acid soils.

## C. Manganese Toxicity

Manganese toxicity is one of the factors that may limit plant growth on certain acid soils. Soils that have a large sesquioxide content often contain high amounts of Mn. Oxisols in Puerto Rico contain large amounts of easily reducible Mn compared with Ultisols (Table 9).

Exchangeable $Mn^{2+}$ of acid Oxisols and an Oxic Paleudult decreased very markedly as the soil pH was increased by liming (Table 10). Plants growing on soils with a high content of exchangeable $Mn^{2+}$ accumulate large amounts of Mn in their tissues and growth is reduced (Table 11). Liming these soils to pH 5.5 and above decrease the solubility of Mn and uptake of Mn sufficiently to eliminate the toxicity and increase growth (Table 11).

Table 9—Easily reducible Mn of Ultisols and Oxisols in Puerto Rico (Abruna et al., 1974).

| Soil | Texture | Easily reducible Mn |
|---|---|---|
| | | mg $kg^{-1}$ |
| Tropohumults | clay | 120 |
| Tropodults | clay | 210 |
| Tropohumults | clay | 40 |
| Haplorthox | clay | 700 |
| Haplorthox | clay | 638 |
| Haplorthox | sand | 144 |

Table 10—Exchangeable $Mn^{2+}$ content of acid soils as related to soil pH.

| Region | Soil | pH | Exchangeable $Mn^{2+}$ | Reference |
|---|---|---|---|---|
| | | | mg kg$^{-1}$ | |
| Brazil | Oxisols | 4.7 | 205 | Martini et al. (1974) |
| | | 5.7 | 59 | |
| Hawaii | Oxisols | 5.1 | 108 | Jones & Fox (1978) |
| | | 5.7 | 49 | |
| Puerto Rico | Oxisols | 4.7 | 137 | Abruna et al. (1975) |
| Nigeria | Oxic Paleudults | 5.3 | 43 | Juo & Uzu (1977) |
| | | 5.7 | 21 | |

Table 11—Response to liming where Mn toxicity was a factor on soils in the tropics.

| Crop and location | pH | Al sat. | Exch. $Mn^{2+}$ | Growth† | Plant Mn | Reference |
|---|---|---|---|---|---|---|
| | | % | mg kg$^{-1}$ | | mg kg$^{-1}$ | |
| | | | | Greenhouse | | |
| Soybean (Brazil) | 4.7 | 10 | 205 | 2.61 g (75) | 1068 | Martini et al. (1974) |
| | 5.7 | 0 | 59 | 3.50 g (100) | 193 | |
| | | | | Field | | |
| | 4.9 | 10 | -- | 2 178 kg ha$^{-1}$ (59) | -- | |
| | 5.5 | 1 | -- | 3 676 kg ha$^{-1}$ (100) | -- | |
| | | | | Greenhouse | | |
| Corn (Nigeria) | 5.3 | -- | 43 | 3.35 g (83) | 325 | Juo & Uzu (1977) |
| | 5.7 | -- | 21 | 4.05 g (100) | 105 | |
| | | | | Field | | |
| Napier grass (Puerto Rico) | 4.0 | -- | 96 | 25 309 kg ha$^{-1}$ (76) | 347 | Abruna et al. (1964) |
| | 5.0 | -- | 28 | 33 428 kg ha$^{-1}$ (10) | 141 | |

† Values in parentheses are percent growth.

In many instances, when plants on acid soils with low Al saturations respond to liming, it is due to alleviation of Mn toxicities. Soybean yields on an Oxisol in Brazil at pH 4.9 and 10% Al saturation were increased considerably by raising the soil pH to 5.5 (Table 11). Subsequent greenhouse studies showed that the soil contained 205 mg kg$^{-1}$ (205 ppm) of exchangeable $Mn^{2+}$ at pH 4.7, and soybean plants contained extremely high levels of Mn. Liming the soil to pH 5.7 reduced Mn to nontoxic levels.

### D. Calcium Deficiency

Many acid soils in the tropics have very low amounts of exchangeable $Ca^{2+}$ and a low saturation of the ECEC (Table 7). At these low levels of available Ca, the uptake of Ca may also be inhibited by Al.

The Ca concentration of the ear leaf of corn was very low at 9% Ca saturation of the ECEC (Table 12). The growth of corn on this Ultisol was severely depressed at this low level of Ca saturation. The low leaf Ca was probably a result of both low Ca supply and the inhibitory effect on Ca

Table 12--Concentration of Ca in ear leaf of corn at silking as related to Ca saturation of soil and relative yield.

| Region | pH | Al saturation | Ca saturation | Exch. $Ca^{2+}$ | Leaf Ca | Relative yield |
|---|---|---|---|---|---|---|
| | | % | | cmol $(p^+)$ $kg^{-1}$ | g $kg^{-1}$ | % |
| Nigeria† | 4.3 | 64 | 9 | 0.26 | 2.1 | 35 |
| (Ultisols) | 5.0 | 31 | 63 | 1.71 | 3.4 | 88 |
| | 5.4 | 7 | 83 | 2.31 | 3.8 | 100 |
| Brazil‡ | 4.5 | 68 | 27 | 0.45 | 3.5 | 45 |
| (Oxisols) | 5.0 | 43 | 49 | 0.85 | 3.5 | 84 |
| | 5.5 | 4 | 89 | 2.25 | 3.7 | 90 |
| | 5.9 | 0 | 95 | 4.35 | 4.2 | 100 |

† Juo and Ballaux (1977).
‡ Gonzalez-Erico et al. (1979).

uptake due to the high Al saturation (Table 12). An application of 1.3 t of $CaCO_3$ $ha^{-1}$ increased leaf Ca to adequate levels (Juo & Ballaux, 1977).

Leaf-Ca concentration of corn grown on an Oxisol was adequate at 27% Ca saturation, although growth was depressed because of the high Al saturation (Table 12). Raising the Ca saturation to 49% brought about a large increase in grain yield without a change in leaf Ca. This further indicates that Ca supply itself was adequate at a Ca saturation of 27%, which resulted in a soil solution concentration of 1.14 mmol $(p^+)$ $L^{-1}$ (Gonzalez-Erico et al., 1979).

The growth of tropical and temperate legume species, as influenced by Ca supply, was studied on a very sandy soil of pH 5.5 and extremely low in exchangeable $Ca^{2+}$ (Andrew & Norris, 1961). Maximum growth of most species was obtained at an exchangeable $Ca^{2+}$ content of 0.87 cmol $(p^+)$ $kg^{-1}$, which was 26% saturation of the CEC at pH 7 (Table 13). A number of very

Table 13—Growth and tropical and temperate legumes as influenced by Ca saturation of soil (Andrew & Norris, 1961).

| Legume species | Percent Ca saturation 3 (0.1)† | 14 (0.47) | 26 (0.87) | 37 (1.23) | Plant-Ca content at maximum growth |
|---|---|---|---|---|---|
| | relative growth, % | | | | g $kg^{-1}$ |
| | Tropical | | | | |
| *Desmodium uncinatum* (Jacq.) DC. | 27 | 43 | 78 | 100 | 12.0 |
| *Indigofera spicata* Forsk. | 43 | 73 | 100 | -- | 14.5 |
| *Centrosema pubescens* Benth. | 58 | 70 | 100 | -- | 13.5 |
| *Stylosanthes bojeri* | 66 | 87 | 100 | -- | 15.8 |
| *Phaseolus lathyroides* L. | 73 | 80 | 100 | -- | 9.1 |
| | Temperate | | | | |
| *Medicago tribuloides* Desr. ex Lam. | 1 | 60 | 100 | -- | 12.9 |
| *Medicago sativa* L. | 2 | 63 | 100 | -- | 13.0 |
| *Trifolium fragiferum* L. | 3 | 56 | 86 | 100 | 14.0 |
| *Trifolium repens* L. | 1 | 72 | 100 | -- | 10.2 |

† Values in parentheses are exchangeable $Ca^{2+}$, cmol $(p^+)$ $kg^{-1}$.

Table 14—Critical Ca and Al saturations for tropical grasses (CIAT, 1980).

| Grass species | Critical levels for 75% relative growth† | | |
|---|---|---|---|
| | Al saturation | Ca saturation | Plant Ca |
| | % | % | g $kg^{-1}$ |
| *Pennisetum purpureum* Schumach. | 69 | 29 | 4.0 |
| *Panicum maximum* Jacq. | 72 | 24 | 6.0 |
| *Brachiaria decumbens* Stapf | 74 | 20 | 4.2 |
| *Melinis minutiflora* Beauv. | 86 | 12 | 3.2 |

† Critical levels for 75% relative growth of that on limed soil.

pertinent conclusions were drawn from this study. First, the growth of the tropical legumes was relatively good at the lowest Ca level, whereas the temperate legumes made essentially no growth. Second, the same level of available Ca was required for maximum growth by both the tropical and temperate species.

Studies with tropical grasses indicate that good growth is obtained at relatively low Ca saturations (Table 14). These grasses are Al tolerant and have adequate Ca concentrations even though Al saturations are very high.

Some soils in tropical regions have very low CEC, and their capacity to retain Ca is very limited. An example of this is a Latosol from Panama that had an ECEC at pH 4.8 of 0.48 cmol ($p^+$) $kg^{-1}$ (Mendez & Kamprath, 1978). No response to liming was obtained on the soil when adequate Ca was supplied by the addition of superphosphate. With soils such as this of very low CEC, the supply of Ca is probably the main factor affecting growth.

Although it is often difficult to separate the direct effects of H and Al from those of Ca on plant growth, a number of points can be made about Ca in highly weathered soils. A Ca saturation of approximately 25 to 30% appears to be adequate for supplying the Ca requirements of most plants. Growth may not be maximum at this Ca saturation because some other factor is limiting yield. A Ca level of 1.0 cmol ($p^+$) $kg^{-1}$ appears to be a minimum value for good growth of many plants.

## III. CROP RESPONSES TO LIMING

Response of crops to liming is mainly a result of neutralization of Al, Mn, and H toxicities and the supplying of Ca, Mg, or both. With acid soils of the tropics, any one of these factors may be related to increased growth. In many of the early studies on liming, only pH was measured so that the causative factor responsible for growth could not be identified. Pearson (1975) compiled an excellent review of crop response to liming in the tropics. A summary of his conclusions is presented in this section along with an examination of selected studies on crop responses to liming in various tropical areas where acid infertility factors were identified.

### A. Previous Reviews

An excellent in-depth review of the results of liming studies in humid tropical regions was made by Pearson (1975). The reader is referred to that review for information about specific studies. The following conclusions were drawn by Pearson (1975) regarding the nature of the liming responses of important crops in the tropics:

1. With regard to the cereals corn, sorghum, and rice, corn responded to liming when the soil pH was < 5 and exchangeable $Al^{3+}$ saturation was > 15%. Sorghum responses to lime can be expected under about the same soil conditions as corn, although a higher pH appears to be desirable for sorghum. Rice grown under flooded conditions seldom responds to liming. Upland rice responds similarly to corn.
2. Soybean responses to lime have been obtained on moderately to strongly acid soils. Responses occurred when Al saturations were about 10% or more. Yield increases to liming > pH 5.5 were probably due to increased Mo availability.
3. Sugarcane yield responses to lime are expected on highly weathered soils when the pH is < 5. Responses to liming were obtained on Andepts in Hawaii when the exchangeable $Ca^{2+}$ was < 400 kg $ha^{-1}$ in the top 30 cm (Baver & Ayers, 1962).
4. Cotton responded to liming on acid Oxisols in Brazil that had a range in Al saturation from 45 to 85% (McClung et al., 1961). No response of cotton to liming was obtained in Uganda above pH 5.2 (soil was 100% base saturated) (Foster, 1970).
5. Peanut plants growing on the highly leached, poorly buffered soils of the tropics will respond to liming because of the unique Ca requirements of the developing fruit and the increased availability of Mo as the pH is increased.
6. Forage grasses adapted to the tropics are generally very tolerant of soil acidity. However, with high rates of ammoniacal forms of N, soil pH can be decreased to very low levels. Under these conditions, responses to lime can be obtained.
7. Regarding the tropical forage legumes, many are rather tolerant of low soil pH except where high levels of Mn are present. Responses of forage legumes to very low applications of lime on very acid Oxisols indicate Ca and Mg deficiencies.
8. Food legumes are responsive to liming. This appears to be related to elimination of toxicities and a favorable pH for rhizobia.
9. Coffee is very tolerant of acid soils when Mn toxicity is not a factor.

The results of field experiments conducted in Puerto Rico, Brazil, Peru, and Ghana to measure crop response to liming of acid Ultisols and Oxisols were summarized by Lathwell (1979). The major conclusions were (i) soils on which lime responses were obtained generally had a high proportion of the ECEC saturated with Al; (ii) lime rates that neutralized Al resulted in maximum yields of most crops; and (iii) to eliminate Al toxicity, one must apply 2 t of $CaCO_3$ $ha^{-1}$ for each 1 cmol (⅓ $Al^{3+}$) $kg^{-1}$.

## B. Crop Response to Liming and Acid Infertility Factors

The expanded research to increase food production in the tropics has resulted in new research on liming to identify those acid soil factors limiting growth. These studies have provided the necessary data for developing sound liming practices for soils in the tropics.

### 1. WHEAT

Wheat is an important crop grown on Oxisols in Brazil, and varieties relatively tolerant to Al have been developed. Response to lime was related to Al saturation of the ECEC and soil pH (Table 15). The largest responses in yield were obtained at a location where the Al saturation was 65% (Table 15). When acid soils containing appreciable amounts of exchangeable $Al^{3+}$ were limed to raise the pH to 4.7 or above, maximum yields were obtained (Table 15). This generally reduced the exchangeable $Al^{3+}$ to $<$ 1.5 cmol ($p^+$) $kg^{-1}$. When the Al saturation was $<$ 10%, essentially no response was obtained with liming even though the soil pH was 4.9. Liming of these soils to pH 5.7 and above resulted in an increase in the occurrence of *Ophiobolus graminis* (take-all) and a soilborne mosaic disease (Martini et al., 1977).

The critical level of available P for optimum yield of wheat on an Oxisol was considerably less when the Al saturation was reduced to zero, which occurred at pH 5.2 (Fig. 2). Yields of wheat at pH 4.4 approached that at pH 5.2 and 5.8 when considerably more fertilizer P had been applied to give a much higher level of available P. Apparently, one of the benefits of liming was more efficient use of available P. This was probably due to greater root exploitation of the limed soil (Friesen et al., 1980b).

Lime rates of 3.5 t $ha^{-1}$ were adequate on these clayey Oxisols for maximum wheat yields (Table 15). With this rate of lime, Al saturations were re-

Table 15—Response of wheat to liming of Oxisols as related to pH and Al saturation (Martini et al., 1977).

| Location | Lime | Grain | Relative yield | pH | Al | Al saturation |
|---|---|---|---|---|---|---|
| | t $ha^{-1}$ | kg $ha^{-1}$ | % | | cmol ($p^+$) $kg^{-1}$ | % |
| 1 | 0 | 1323 | 56 | 4.6 | 2.8 | 64 |
| | 3.5 | 2370 | 100 | 4.8 | 1.0 | 31 |
| 2 | 0 | 783 | 62 | 4.4 | 2.2 | 31 |
| | 3.5 | 1089 | 100 | 4.7 | 0.6 | 8 |
| 3 | 0 | 1535 | 100 | 4.9 | 0.3 | 9 |
| | 3.5 | 1498 | 100 | 5.5 | 0.1 | 3 |
| 4 | 0 | 1757 | 94 | 4.9 | 0.4 | 4 |
| | 3.5 | 1872 | 100 | 5.4 | 0 | 0 |
| 5 | 0 | 1943 | 77 | 4.6 | 3.5 | 46 |
| | 3.5 | 2514 | 100 | 4.9 | 1.5 | 23 |

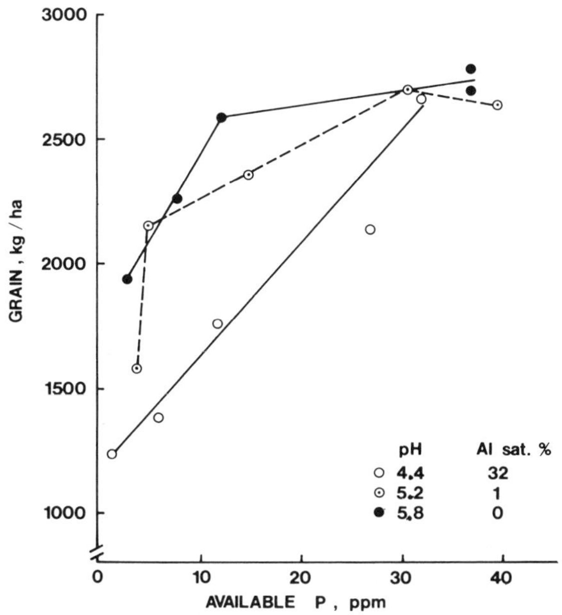


Fig. 2—Effect of soil pH on the response of wheat to 37 kg of P ha$^{-1}$ at different levels of available soil P of an Oxisol (de Magalhoes et al., 1980).

duced to levels that were not detrimental to those wheat varieties relatively tolerant to Al.

## 2. SOYBEAN

Soybean is an important crop in the tropics. A large yield response to liming was obtained on an Oxisol in Brazil when the Al saturation was > 50% (Table 16). Similar results were obtained in Puerto Rico (Table 17). Maximum yields were generally obtained when Al saturation was < 10% and soil pH levels were 5.1 or above (Table 16).

Table 16—Response of soybean to liming of Oxisols as related to pH and Al saturation (Martini et al., 1974).

| Location | Lime | Grain | Relative yield | pH | Al | Al saturation |
|---|---|---|---|---|---|---|
| | t ha$^{-1}$ | kg ha$^{-1}$ | % | | cmol (p$^+$) kg$^{-1}$ | % |
| 1 | 0 | 1324 | 44 | 4.7 | 2.9 | 53 |
| | 3.5 | 2363 | 78 | 5.0 | 1.7 | 27 |
| | 7.0 | 3031 | 100 | 5.2 | 1.0 | 15 |
| 2 | 0 | 2275 | 72 | 4.9 | 0.7 | 13 |
| | 3.5 | 3140 | 100 | 5.5 | 0.1 | 2 |
| 3 | 0 | 2505 | 100 | 5.4 | 0.1 | 1 |
| | 3.5 | 2410 | 100 | 5.8 | 0 | 0 |
| 4 | 0 | 1653 | 82 | 4.4 | 2.2 | 31 |
| | 7.0 | 2011 | 100 | 5.1 | 0.4 | 5 |

Table 17—Response of several crops in Puerto Rico to liming as related to pH and percent Al saturation.

| Soil | Crop | pH | Al saturation | Yield | Relative yield | Reference |
|---|---|---|---|---|---|---|
| | | | % | kg ha$^{-1}$ | % | |
| Aquic Tropudults | Soybean | 4.5 | 65 | 62 | 3 | Abruna et al. (1978) |
| | | 4.8 | 26 | 585 | 28 | |
| | | 5.0 | 12 | 1 311 | 63 | |
| | | 5.6 | 3 | 2 081 | 100 | |
| Tropeptic Haplorthoxs | Soybean | 4.3 | 36 | 2 540 | 71 | Abruna et al. (1978) |
| | | 4.7 | 16 | 2 997 | 84 | |
| | | 5.3 | 5 | 3 555 | 100 | |
| Typic Tropohumults | Corn | 3.9 | 66 | 1 150 | 22 | Abruna et al. (1975) |
| | | 4.5 | 36 | 4 090 | 77 | |
| | | 4.7 | 17 | 4 420 | 83 | |
| | | 5.3 | 3 | 5 340 | 100 | |
| Typic Tropohumults | Bean | 3.9 | 66 | 1 810 | 26 | Abruna et al. (1975) |
| | | 4.5 | 36 | 4 870 | 70 | |
| | | 5.3 | 3 | 6 930 | 100 | |
| Aquic Tropudults | Bean | 4.3 | 68 | 4 310 | 39 | Abruna et al. (1975) |
| | | 4.6 | 39 | 8 130 | 73 | |
| | | 4.7 | 18 | 8 890 | 80 | |
| | | 5.4 | 2 | 11 170 | 100 | |
| Aquic Tropudults | Sweet potato | 4.2 | 66 | 3 100 | 22 | Abruna et al. (1979) |
| | | 4.4 | 50 | 6 800 | 49 | |
| | | 4.6 | 33 | 9 100 | 66 | |
| | | 4.8 | 26 | 12 300 | 89 | |
| | | 5.0 | 12 | 13 100 | 95 | |
| | | 5.6 | 3 | 13 800 | 100 | |
| Ultisols | Napier grass | 4.0 | 79 | 16 242 | 55 | Abruna et al. (1964) |
| | | 4.7 | 18 | 29 551 | 100 | |
| Ultisols | Sugarcane | 4.0 | 73 | 2 500 | 25 | Abruna et al. (1967) |
| | | 4.8 | 20 | 10 000 | 100 | |

In certain instances, responses to lime have been obtained on Oxisols where the Al saturation was very low (Table 11). Responses to liming in these situations have been associated with an elimination of Mn toxicity. Another factor affecting the response of soybean to lime is the increased availability of soil Mo.

## 3. CORN

Responses of corn to liming have been obtained throughout the humid tropics. Numerous field studies in Puerto Rico have shown that large yield increases in corn yields were obtained by liming when the Al saturation was $> 35\%$ (Table 17) (Abruna et al., 1974). Similar results were obtained in Brazil where grain yields were substantially increased with an application of 2 t of $CaCO_3$ ha$^{-1}$ to an Oxisol with 1.1 cmol (⅓ $Al^{3+}$) kg$^{-1}$ and 71% Al saturation (Gonzalez-Erico et al., 1979).

Incorporation of lime to a depth of 30 cm resulted in higher grain yields than when lime was incorporated only in the top 15 cm (Gonzalez-Erico et al., 1979). Root length in the unlimed 15- to 30-cm depth averaged 1.5 cm $cm^{-3}$ of soil compared with 3.5 cm $cm^{-3}$ of soil in the limed 15-to 30-cm depth. The increased root length with liming resulted in a greater amount of soil moisture being used from the limed soil than from the nonlimed.

Growth responses of corn to lime were obtained in a greenhouse study with a Nigerian Ultisol containing 1.83 cmol (⅓ $Al^{3+}$) $kg^{-1}$ and 64% Al saturation (Juo & Ballaux, 1977). Maximum growth was obtained with a lime rate equivalent to 3 t of $CaCO_3$ $ha^{-1}$. Another study in Nigeria indicated that maximum growth of corn was obtained in two well-fertilized Ultisols when the Al saturation was $<$ 28% (Friesen et al., 1980b).

Yield responses of corn to lime on acid soils in the Philippines were small when Al saturations were $<$ 20%, but large yield increases were obtained on soil with 46% Al saturation (Samonte & Ocampo, 1977).

Yield responses to liming were obtained in Ghana on an Ultisol even though soil pH was 5.4 and Al saturation was only 4% (Lathwell, 1979). However, when P was applied to the nonlimed treatments, yields were similar to the limed treatments receiving no P. Lathwell (1979) concluded that increased mineralization of organic P at higher pH was the reason for the response to lime.

## 4. PEANUT

Peanut plants are grown throughout the humid tropics, and responses to lime are almost always reported on these acid, low Ca-saturated soils. Maximum yields of peanuts were obtained on acid, low CEC Ultisols in Venezuela with an application of 0.5 t of $CaCO_3$ $ha^{-1}$, which reduced Al saturation to 30% (Sanchez, 1977). The Ca saturation of these soils was $<$ 15%, and the response to liming was due not only to a reduction of Al saturation but also to an increase in Ca supply.

Data from Senegal suggested that nodulation of peanut was inhibited when the Al saturation was $>$ 30% (Pieri, 1974).

Results from Indonesia substantiate that peanut plants are relatively tolerant of Al. Maximum yields of peanut plants were obtained with lime rates equivalent to 50% of the exchangeable $Al^{3+}$, whereas corn required a lime rate equivalent to 100% of the exchangeable $Al^{3+}$ (Soepardi & Idris, 1977).

## 5. BEAN

Yields of green bean in Puerto Rico were markedly increased by liming acid soils that had a high Al saturation (Table 17). Maximum yields were obtained when Al saturation was reduced to zero (Abruna et al., 1975).

The increased Ca availability with liming also was beneficial in ameliorating the detrimental effect of high Mn concentrations in bean leaves (Abruna et al., 1975). As the Ca/Mn ratio of bean leaves increased, bean yields increased significantly.

Yields of dried beans in Uganda and Venezuela were maximum with lime rates that neutralized all exchangeable $Al^{3+}$ (Foster, 1970; Sanchez, 1977).

### 6. COWPEA

Cowpea, an important grain legume in the tropics, is very responsive to low amounts of lime on acid soils. In a field trial in Colombia on an Oxisol, pH 4.2 and 78% Al saturation, yields were essentially doubled by an application of 0.5 t of $CaCO_3$ $ha^{-1}$, which reduced Al saturation to 67% and gave a Ca saturation of 22% (Spain et al., 1975).

Responses to liming by a number of cowpea cultivars were measured in a greenhouse experiment with an acid Ultisol in Nigeria (Edwards et al., 1981). Aluminum-tolerant cultivars produced 74% of their maximum growth at pH 4.2, 63% Al saturation, and 12% Ca saturation, whereas Al-sensitive cultivars produced only 48% of their maximum growth. Application of lime equivalent to 0.5 t of $CaCO_3$ $ha^{-1}$ resulted in an average of 85% of the maximum growth for both Al-tolerant and Al-sensitive cultivars. The 0.5 t of $CaCO_3$ $ha^{-1}$ rate reduced Al saturation to 40%. It was concluded that Al toxicity was the major growth-limiting factor corrected by liming.

### 7. OTHER CROPS

Responses to liming by sweet potato, sugarcane, and napier grass were related to a decrease of Al saturation (Table 17). Maximum yields of sweet potato were obtained on four acid soils in Puerto Rico when the Al saturation was $< 20\%$ (Abruna et al., 1979). Sweet potato plants did not respond to liming on soils in Uganda at pH 5.2 (Foster, 1970). Exchangeable $Al^{3+}$ was not present in these soils at pH 5.2. The results from Puerto Rico and Uganda indicate that responses of sweet potato plants to liming will be primarily on soils with an Al toxicity problem.

Sorghum responded to lime rates that were sufficient to reduce Al saturation of an Ultisol in Venezuela to zero (Sanchez, 1977). Sorghum roots were much more sensitive than corn to a given activity of soil solution Al in acid Ultisols and Oxisols from Puerto Rico (Brenes & Pearson, 1973). This, again, indicates that response to lime is due to neutralization of toxic levels of Al.

## IV. LIME REQUIREMENT

### A. Criteria for Liming

For many years, soil pH was the sole criterion used to determine whether a soil should be limed or not. However, the pH of a given soil can vary, depending on such factors as soil/water ratio and electrolyte concentration. Furthermore, the relationship between soil pH and base saturation (based on CEC determined with $NH_4OAc$) varies among soils, depending on the kind of clay minerals present.

A more realistic measure of the CEC of soils, i.e., the exchange sites controlling the cation composition of soil solution, is the sum of exchangeable $Ca^{2+}$, $Mg^{2+}$, $K^{+}$, $Na^{+}$, and unbuffered salt-extractable acidity (Coleman et al., 1959). In mineral soils, the unbuffered salt-extractable acidity is primarily $Al^{3+}$. This measure of CEC, the ECEC, is a more realistic measure of the active exchange sites in a soil at a given pH and is the appropriate CEC for calculating percent base saturation.

The percentage of the ECEC saturated with Al was proposed as the criterion for determining whether mineral soils needed to be limed or not (Kamprath, 1967, 1970). Soils with low Al saturation have very low concentrations of Al in the soil solution and do not require liming. The results of liming studies reported in section III support this approach of using percent Al saturation as the criterion for determining whether lime is needed or not. When the Al saturation is zero, it is not often that a response to liming will be obtained.

The use of soil pH in diagnosis of acidity problems should be based on the relationship between pH and Al saturation of the ECEC. This relationship needs to be determined for soils with various chemical properties so that better interpretations can be made of soil pH for a given soil.

## B. Predicting Lime Requirement

Coleman et al. (1958) stated that the most significant experiment one can make on an acid soil is to leach with a neutral salt, such as 1*M* KCl. Unbuffered, neutral salt solutions, e.g., KCl, replace cations bound electrostatically to soil surfaces. In strongly acid mineral soils, $Al^{3+}$ is the major cation countering active exchange sites. Exchangeable acidity can be readily measured by extracting a soil sample with 1*M* KCl and titrating the extract with a base to the phenolphthalein end point. The acidity measured by this procedure in acid mineral soils is in many instances mostly $Al^{3+}$.

Exchangeable $Al^{3+}$ has been suggested as the criterion for determining the amount of lime to apply to acid soils (Mohr, 1960; Kamprath, 1967, 1970; Reeve & Sumner, 1970). The relationship between the amount of $CaCO_3$ required to neutralize a given amount of exchangeable $Al^{3+}$ is given by the equation $CaCO_3$ equivalent (t ha$^{-1}$) = Factor $X$ cmol (⅓ $Al^{3+}$) kg$^{-1}$.

The factor in the above equation has been found to range from 1.5 to 3.3, with most values in the range of 1.5 to 2 (Table 18). The reason that the

Table 18—Tons of $CaCO_3$ per hectare required to neutralize each cmol (⅓ $Al^{3+}$) kg$^{-1}$.

| Area | Soil | $CaCO_3$, t ha$^{-1}$ for each cmol (⅓ $Al^{3+}$) kg$^{-1}$ | Reference |
|---|---|---|---|
| Brazil | Ultisols | 2 | Bloise et al. (1979) |
| Brazil | Oxisols | 3 | Sfredo et al. (1978) |
| Brazil | Oxisols | 2 | Muzilli and Godoy (1979) |
| Brazil | Oxisols | 2 | Gonzalez-Erico et al. (1979) |
| Colombia | Oxisols | 2 | Spain et al. (1975) |
| Panama | Oxisols | 1.5 | Mendez & Kamprath (1978) |
| India | -- | 1.5–2.0 | Pradhan & Khera (1976) |
| Nigeria | Ultisols | 1.5 | Juo & Ballaux (1977) |
| Natal | Oxisols | 3.3 | Reeve & Sumner (1970) |

factor in the equation is > 1 is because of the pH-dependent acidity that consumes $OH^-$ ions as the soil pH increases (Kamprath, 1970). A large portion of this acidity, which is not extractable with an unbuffered neutral salt, is Al bound by organic matter and is an important component of the buffer capacity of soils (Juo & Kamprath, 1979). Studies with a large number of acid soils showed that the exchangeable $Al^{3+}$ and organic matter content were the two main factors determining the amount of lime required to bring the soils to pH 6 (De Souza et al., 1980).

The $CaCO_3$ equivalent, as calculated by the expression 1.5 to 2 times the cmol (⅓ $Al^{3+}$) $kg^{-1}$, will reduce the Al saturation to approximately zero and will supply adequate amounts of Ca or Ca plus Mg. Liming rates that are based on exchangeable $Al^{3+}$ will give maximum yields.

Cochrane et al. (1980) developed a lime requirement equation that takes into account the Al tolerance of the cultivar. The amount of $CaCO_3$, t $ha^{-1}$, is given by the equation [1.8 cmol (⅓ $Al^{3+}$) $kg^{-1}$] = [RAS (cmol [$p^+$] $kg^{-1}$) of Al + Ca + Mg]. The term *RAS* is defined as the required Al saturation. This approach would be applicable where lime is difficult to obtain and rather costly and Al-tolerant cultivars are available.

## REFERENCES

Abruna, F., R. W. Pearson, and R. Perez-Escolar. 1975. Lime response of corn and soybeans grown on typical Oxisols and Ultisols of Puerto Rico. Soil Manage. Trop. An., Proc. Semin. 261–282.

Abruna, F., R. Perez-Escolar, J. Vicente-Chandler, R. W. Pearson, and S. Silva. 1974. Response of corn to acidity factors in eight tropical soils. J. Agric. Univ. P.R. 58:59–77.

Abruna, F., J. Rodriguez, J. Badillo-Feliciano, S. Silva, and J. Vicente-Chandler. 1978. Crop response to soil acidity factors in Ultisols and Oxisols in Puerto Rico: Soybeans. J. Agric. Univ. P.R. 62:90–11.

Abruna, F., and J. Vicente-Chandler. 1967. Sugarcane yields as related to acidity of a humid tropic Ultisol. Agron. J. 59:330–331.

Abruna, F., J. Vicente-Chandler, and R. W. Pearson. 1964. Effects of liming on yields and composition of heavily fertilized grasses and on soil properties under humid tropical conditions. Soil Sci. Soc. Am. Proc. 28:657–661.

Abruna, F., J. Vicente-Chandler, J. Rodriguez, J. Badillo, and S. Silva. 1979. Crop response to soil acidity factors in Ultisols and Oxisols in Puerto Rico: V. Sweet potato. J. Agric. Univ. P.R. 63:250–267.

Andrew, C. S., and D. O. Norris. 1961. Comparative responses to calcium of five tropical and four temperate pasture legume species. Aust. J. Agric. Res. 12:40–55.

Baver, L. D., and A. S. Ayers. 1962. Soil analyses as bases for fertilizer recommendations in sugarcane. p. 835–841. *In* Int. Soc. Soil Sci., Trans. Comm. IV and V (Palmerston, N.Z.).

Bell, L. C., and G. P. Gillman. 1978. Surface charge characteristics and soil solution composition of highly weathered soils. p. 37–57. *In* C. A. Andrew and E. J. Kamprath (ed.) Mineral nutrition of legumes in tropical and subtropical soils. CSIRO, Melbourne, Australia.

Bloise, R. M., J. F. Dynia, and G. N. C. Moriera. 1977. Fertility of soils in the region of the trans-Amazonic highway: I. Neutralization of the exchangeable aluminum in a Red-Yellow Podzolic soil. Pesqui. Agropecu. Bras. 12:71–74.

Brenes, E., and R. W. Pearson. 1973. Root responses of three Gramineae species to soil acidity in an Oxisol and an Ultisol. Soil Sci. 116:295–302.

CIAT. 1980. Tropical pasture program report. CIAT, Cali, Colombia.

Cochrane, T. T., J. G. Salinas, and P. A. Sanchez. 1980. An equation for liming acid mineral soils to compensate crop aluminum tolerance. Trop. Agric. (Trinidad) 57:133–140.

Coleman, N. T., E. J. Kamprath, and S. B. Weed. 1958. Liming. Adv. Agron. 10:475–522.

Coleman, N. T., S. B. Weed, and R. J. McCracken. 1959. Cation exchange capacities and exchangeable cations in Piedmont soils of North Carolina. Soil Sci. Soc. Am. Proc. 23:146–149.

de Carvalho, M. M., C. S. Andrew, D. G. Edwards, and C. J. Asher. 1980. Comparative performance of six *Stylosanthes* species in three acid soils. Aust. J. Agric. Res. 31:61–76.

de Magalhaes, J. C. A., E. Lobato, and L. H. Rodrigues. 1980. Calagem e adubacão fosfatada para dois cultivares de trigo em solo de Cerrado. Rev. Bras. Cienc. Solo 4:160–164.

De Souza, D. M. G., L. N. De Miranda, E. Lobato, and H. J. Klieman. 1980. Evaluation of methods for estimating lime requirements of Cerrado soils of Goias and the Federal District. Rev. Bras. Cienc. Solo 4:144–148.

Edwards, D. G., B. T. Kang, and S. K. A. Danso. 1981. Differential response of six cowpea (*Vigna unguiculata* L. Walp.) cultivars to liming in an Ultisol. Plant Soil 59:61–73.

Evans, C. E., and E. J. Kamprath. 1970. Lime response as related to percent Al saturation, solution Al, and organic matter content. Soil Sci. Soc. Am. Proc. 34:893–896.

Foster, H. L. 1970. Liming continuously cultivated soils in Uganda. East Afr. Agric. For. J. 36:58–69.

Friesen, D. K., A. S. R. Juo, and M. H. Miller. 1980a. Liming and lime-phosphorus-zinc interactions in two Nigerian Ultisols: I. Interactions in the soil. Soil Sci. Soc. Am. J. 44: 1221–1226.

Friesen, D. K., M. H. Miller, and A. S. R. Juo. 1980b. Liming and lime-phosphorus-zinc interactions in two Nigerian Ultisols: II. Effects on maize root and shoot growth. Soil Sci. Soc. Am. J. 44:1127–1232.

Gonzalez-Erico, E., E. J. Kamprath, G. C. Naderman, and W. V. Soares. 1979. Effect of depth of lime incorporation on the growth of corn on an Oxisol of central Brasil. Soil Sci. Soc. Am. J. 43:1155–1158.

Islam, A. K. M., D. G. Edwards, and C. J. Asher. 1980. pH optima for crop growth: Results of a flowing solution culture experiment with six species. Plant Soil 54:339–357.

Jones, J. P., and R. L. Fox. 1978. Phosphorus nutrition of plants influenced by manganese and aluminum uptake from an Oxisol. Soil Sci. 126:230–236.

Juo, A. S. R. 1977. Soluble and exchangeable aluminum in Ultisols and Alfisols in West Africa. Commun. Soil Sci. Plant Anal. 8:17–35.

Juo, A. S. R., and J. C. Ballaux. 1977. Retention and leaching of nutrients in a limed Ultisol under cropping. Soil Sci. Soc. Am. J. 41:757–761.

Juo, A. S. R., and E. J. Kamprath. 1979. Copper chloride as an extractant for estimating the potentially reactive aluminum pool in acid soils. Soil Sci. Soc. Am. J. 43:35–38.

Juo, A. S. R., and F. O. Uzu. 1977. Liming and nutrient interactions in two Ultisols from southern Nigeria. Plant Soil 47:419–430.

Kamprath, E. J. 1967. Soil acidity and response to liming. Int. Soil Testing Tech. Bull. 4, North Carolina State Univ., Raleigh.

Kamprath, E. J. 1970. Exchangeable aluminum as a criterion for liming leached mineral soils. Soil Sci. Soc. Am. Proc. 24:252–254.

Lathwell, D. J. 1979. Crop response to liming of Ultisols and Oxisols. Cornell Int. Agric. Bull. 35, Cornell Univ., Ithaca, N.Y.

Martini, J. A., R. A. Kochhann, E. P. Gomez, and F. Langer. 1977. Response of wheat cultivars to liming in some high Al Oxisols of Rio Grande do Sul, Brasil. Agron. J. 69:612–616.

Martini, J. A., R. A. Kochhann, O. J. Siqueira, and C. M. Borkert. 1974. Response of soybeans to liming as related to soil acidity, aluminum and manganese toxicities, and phosphorus in some Oxisols of Brazil. Soil Sci. Soc. Am. Proc. 38:616–620.

McClung, A. C., L. M. M. de Freitas, D. S. Mikkelsen, and W. L. Lott. 1961. Cotton fertilization on Campo Cerrado soils, State of Sao Paulo, Brazil. IBEC Res. Inst. Bull. 27.

Mendez, J., and E. J. Kamprath. 1978. Liming of Latosols and the effect on phosphorus response. Soil Sci. Soc. Am. J. 42:86–88.

Mohr, W. 1960. A influencia da acidez sobre a fertilidade dos solos. p. 61–76. *In* Anais I. Congreso Nacional de Conservacão do Solo, Campinas, Brazil.

Morais, F. I., A. L. Page, and L. T. Leend. 1976. The effect of pH, concentration and nature of electrolytes on the charge characteristics of Brazilian tropical soils. Soil Sci. Soc. Am. J. 40:521–527.

Muzilli, O., and O. P. Godoy. 1979. Soil acidity correction on a red-yellow distrophic Latosol from the west of Parana State, Brasil. Pesqui, Agropecu, Bras. 14:159–163.

Nye, P., D. Craig, N. T. Coleman, and J. L. Ragland. 1961. Ion exchange equilibrium involving aluminum. Soil Sci. Soc. Am. Proc. 25:14–17.

Pearson, R. W. 1975. Soil acidity and liming in the humid tropics. Cornell Int. Agric. Bull. 30, Cornell Univ., Ithaca, N.Y.

Pieri, C. 1974. First experimental results on the susceptibility of ground nut to aluminum toxicity. Agron. Trop. (Paris) 29:685.

Pradhan, H. R., and M. S. Khera. 1976. Lime needs of soils of Sikkim based upon exchangeable aluminum test. Bull. Indian Soc. Soil Sci. 11:291–299.

Pratt, P. F., and R. Alvahydo. 1966. Cation exchange characteristics of soils of Sao Paulo, Brasil. IRI Res. Inst. Bull. 31.

Ramos, D. P., and A. C. X. Velloso. 1977. Caracteristicas de alguns solo da região do médio-Amazonas. Pesqui. Agropecu. Bras. 12:97–104.

Reeve, N. G., and M. E. Sumner. 1970. Lime requirement of Natal Oxisols based on exchangeable aluminum. Soil Sci. Soc. Am. Proc. 34:595–598.

Samonte, H. P., and A. M. Ocampo. 1977. Liming and its residual effect on corn-planted soils. Philipp. Agric. 60:420–430.

Sanchez, C. 1977. Encolamiento de Ultisoles de Sabana. Univ. de Oriente, Jusepin Monagas, Venezuela.

Sanchez, P. A. 1976. Properties and management of soils in the tropics. John Wiley & Sons, Inc., New York.

Sfredo, G. J., B. V. Defelipo, V. H. Alvarez V., and J. M. Braga. 1978. Calcium/magnesium equilibrium for correcting the acidity and production of dry matter in sorghum on an Oxisol (Latossolo Bruno Distrofico). Rev. Ceres 25:491–498.

Singh, M., and O. Talibudeen. 1969. Thermodynamic assessment of the nutrient status of rubber growing soils. J. Rubber Res. Inst. Malays. 21:240–249.

Soepardi, G., and K. Idris. 1977. Liming experiments on acid latosol in West Java, Indonesia. p. 480–486. *In* Proc. Int. Semin. Soil Environ. Fertil. Manage. Intensive Agric. Soc. Sci. Soil and Manure, Japan. Tokyo, Japan.

Spain, J. M., C. A. Francis, F. H. Howeler, and F. Calvo. 1975. Differential species and varietal tolerance to soil acidity in tropical crops and pastures. p. 308–329. *In* E. Bornemisza and A. Alvarado (ed.) Soil management in tropical America. North Carolina State Univ., Raleigh.

Van Raij, B., and M. Peech. 1972. Electrochemical properties of some Oxisols and Alfisols of the tropics. Soil Sci. Soc. Am. Proc. 36:587–593.

Vargas, E. 1964. El alumino cambio en suelos de los Llanos Orientales. EE-9 Inst. Geografico "Augustin Codazzi," Bogota, Colombia.

# GLOSSARY

## Common and Scientific Names of Plants Referred to in this Monograph

| | |
|---|---|
| alfalfa | *Medicago sativa* L. |
| almond | *Prunus amygdalus* Batsch |
| apple | *Malus sylvestris* Mill. |
| asparagus | *Asparagus officinalis* L. |
|   garden | *Asparagus officinalis* L. |
| azalea | *Rhododendron calendulaceum* (Michx.) Torr. |
| | |
| bahiagrass | *Paspalum notatum* Fluegge |
| barley | *Hordeum vulgare* L. |
| bean | *Phaseolus* sp. |
|   bush | *Phaseolus vulgaris* L. |
|   dried | *Phaseolus* sp. |
|   french | *Phaseolus vulgaris* L. |
|   green | *Phaseolus vulgaris* L. |
|   mung | *Vigna radiata* (L.) Wilczek var. *radiata* |
|   phasey | *Macroptilium lathyroides* (L.) Urb. |
|   red kidney | *Phaseolus vulgaris* L. |
|   snap | *Phaseolus vulgaris* L. |
| beet | *Beta vulgaris* L. |
|   red | *Beta vulgaris* L. |
|   sugar | *Beta vulgaris* L. |
|   table | *Beta vulgaris* L. |
| bermudagrass | *Cynodon dactylon* (L.) Pers. |
| bilberry | *Vaccinium myrtillis* L. |
| birch | |
|   European white | *Betula pendula* Roth. |
|   paper | *Betula papyrifera* Marsh. |
|   river | *Betula nigra* L. |
| blueberry | *Vaccinium* sp. |
| bluegrass | *Poa pratensis* L. |
| broadbean | *Vicia faba* L. |
| broccoli | *Brassica oleracea* var. *botrytis* L. |
| brussels sprout | *Brassica oleracea* var. *gemmifera* DC. |
| buckwheat | *Fagopyrum esculentum* Gaertn. |
| buffelgrass | *Cenchrus ciliaris* L. |
| | |
| cabbage | *Brassica oleracea* var. *capitata* L. |
| canarygrass, reed | *Phalaris arundinacea* L. |
| cantaloupe | *Cucumis melo* L. |
| carnation | *Dianthus caryophyllus* L. |
| carrot | *Daucus carota* L. ssp. *sativus* (Hoffm.) Arcang. |
|   wild | *Daucus carota* L. |
| cassava | *Manihot esculenta* Crantz |
| castorbean | *Ricinus* sp. |
| cauliflower | *Brassica oleracea* var. *botrytis* L. |
| celery | *Apium graveolens* L. var. *dulce* (Mill.) Pers. |
| cherry | *Prunus cerasus* L. and other species |

| | |
|---|---|
| citrus | *Citrus* sp. |
| clover | *Trifolium* sp. |
| arrowleaf | *Trifolium vesiculosum* Savi |
| crimson | *Trifolium incarnatum* L. |
| Kenya white | *Trifolium semipilosum* Fresen. |
| red | *Trifolium pratense* L. |
| subterranean | *Trifolium subterraneum* L. |
| white | *Trifolium repens* L. |
| cocklebur | *Xanthium* sp. |
| cocoa | *Theobroma cacao* L. ssp. *cacao* |
| coffee | *Coffea arabica* L. |
| collards | *Brassica oleracea* var. *acephala* DC. |
| corn | *Zea mays* L. |
| sweet | *Zea mays* L. ssp. *mays* or *Z. mays* L. ssp. *saccharata* |
| cotton | *Gossypium hirsutum* L. |
| upland | *Gossypium hirsutum* L. |
| cowpea | *Vigna sinensis* Endl. |
| cranberry | *Vaccinium oxycoccus* L. |
| cress, mouse-ear | *Arabidopsis thaliana* L. |
| cucumber | *Cucumis sativus* L. |
| dallisgrass | *Paspalum dilatatum* Poir. |
| dogwood, flowering | *Cornus florida* L. |
| eggplant | *Solanum melongena* L. |
| eucalyptus | *Eucalyptus gummifera* (Gaertn.) Hochr. |
| fern, sensitive | *Onoclea* sp. |
| fescue | |
| sheep | *Festuca ovina* L. |
| tall | *Festuca arundinacea* Schreb. |
| flax | *Linum usitatissimum* L. |
| flaxseed | *Linum usitatissimum* L. |
| foxtail | *Setaria sphacelata* (Schum.) Stapf & C.E. Hubb ex M.B. Moss 'Narok' |
| ginger | *Zingiber officinale* Roscoe |
| glycine | *Glycine wightii* (Grah. ex Wight & Arn.) Verdc. |
| grape | *Vitis vinifera* L. |
| grass | |
| napier | *Pennisetum purpureum* Schumach |
| scurvy | *Cochlearia anglica* L. |
| hops | *Humulus lupulus* L. |
| kale | *Brassica oleracea* L. |
| kikuyugrass | *Pennisetum clandestinum* Hochst. |
| lespedeza | *Lespedeza striata* Thunb. and *L. bicolor* Turcz. |
| sericea | *Lespedeza cuneata* (Dum.) G. Don |
| lettuce | *Lactuca sativa* L. |
| lily, Easter | *Lilium longiflorum* Thunb. |
| lotononis, Miles | *Lotononis bainesii* |

| | |
|---|---|
| lovegrass, weeping | *Eragrostis curvula* (Schrad.) Nees |
| lucerne | *Medicago sativa* L. |
| lupine | *Lupinus* sp. |
| blue | *Lupinus angustifolia* L. |
| white | *Lupinus albus* L. |
| | |
| maize | *Zea mays* L. |
| mangel | *Beta vulgaris* L. |
| mangrove | *Rhizophora mangle* L. |
| maple | |
| pagoda | *Acer micranthum* |
| redvein | *Acer rufinerve* |
| siebold | *Acer sieboldianum* |
| sugar | *Acer saccharum* Marsh. |
| marrow, garden | *Cucurbita pepo* L. |
| melon | *Cucumis melo* L. |
| millet | *Panicum miliaceum* L. |
| browntop | *Panicum ramosum* L. |
| German | *Setaria italica* (L.) Beauv. |
| pearl | *Pennisetum americanum* (L.) Leeke |
| morning glory | *Pharbitis nil* (L.) Choisy |
| Japanese | *Pharbitis nil* (L.) Choisy |
| muskmelon | *Cucumis melo* L. |
| | |
| oat | *Avena sativa* L. |
| onion | *Allium cepa* L. |
| Welsh | *Allium fistulosum* L. |
| orange | *Citrus sinensis* (L.) Osb. |
| orchardgrass | *Dactylis glomerata* L. |
| | |
| palm, betel | *Areca triandria* |
| pangolagrass | *Digitaria decumbens* Stent |
| pea | *Pisum sativum* L. |
| dried | *Pisum* sp. |
| garden | *Pisum sativum* L. |
| green | *Pisum sativum* L. |
| peach | *Prunus persica* L. Batsch |
| peanut | *Arachis hypogaea* L. |
| pecan | *Carya illinoensis* (Wangenh.) K. Koch |
| pepper | *Capsicum* sp. |
| green | *Capsicum annuun* L. var. *annuum* |
| peppermint | *Mentha* × *piperita* L. |
| pine | *Pinus* sp. |
| poisonsumac, peelberry | *Rhus trichocarpa* |
| poplar, yellow | *Liriodendron tulipifera* L. |
| poppy | *Papaver somniferum* L. |
| potato | *Solanum tuberosum* L. |
| sweet | *Ipomoea batatas* (L.) Lam. |
| | |
| rape | *Brassica napus* L. |
| raspberry | *Rubus occidentalis* L. |
| rhodesgrass | *Chloris gayana* Kunth |
| rice | *Oryza sativa* L. |
| upland | *Oryza sativa* L. |

| | |
|---|---|
| rose | *Rosa fortumana, R. odorata,* and other species |
| rutabaga | *Brassica campestris* var. *napobrassica* DC. |
| rye | *Secale cereale* L. |
| ryegrass | *Lolium* sp. |
| perennial | *Lolium perenne* L. |
| satsuma | *Citrus reticulata* Blanco |
| sensitive plant | *Mimosa* sp. |
| sorghum | *Sorghum bicolor* (L.) Moench |
| forage | *Sorghum bicolor* (L.) Moench |
| soybean | *Glycine max* (L.) Merr. |
| spiderwort | *Tradescantia virides* |
| spinach | *Spinacia oleracea* L. |
| spruce | |
| Norway | *Picea abies* (L.) Karst. |
| Sitka | *Picea sitchensis* (Bong.) Carr. |
| squash | *Cucurbita mixta* Pang. |
| strawberry | *Fragaria* × *ananassa* Duch. |
| garden | *Fragaria* × *ananassa* Duch. |
| stylo, Townsville | *Stylosanthes humilis* H.B.K. |
| sudangrass | *Sorghum sudanense* (Piper) Stapf |
| sugarcane | *Saccharum officinarum* L. |
| sunflower | *Helianthus annuus* L. |
| sweetclover | *Melilotus officinalis* Lam. |
| tea | *Camellia sinensis* (L.) Ktze. |
| tickclover, Spanish | *Desmodium uncinatum* (Jacq.) DC. |
| timothy | *Phleum pratense* L. |
| tobacco | *Nicotiana tabacum* L. |
| burley | *Nicotiana tabacum* L. |
| tomato | *Lycopersicon esculentum* Mill. |
| trefoil | *Lotus corniculatus* L. |
| birdsfoot | *Lotus corniculatus* L. |
| triticale | X *Tritosecale* |
| turnip | *Brassica rapa* L. |
| uncinatum, silverleaf | *Desmodium uncinatum* (Jacq.) DC. |
| vetch | *Vicia sativa* L. |
| watermelon | *Citrullus lanatus* (Thunb.) Matsum. & Nakai |
| wheat | *Triticum aestivum* L. and other species |
| willow | *Salix viminalis* L. |
| yucca | *Yucca flaccida* Haw. |

# SUBJECT INDEX